Lecture Notes in Computer Science   3296

Commenced Publication in 1973
Founding and Former Series Editors:
Gerhard Goos, Juris Hartmanis, and Jan van Leeuwen

Editorial Board

David Hutchison
  Lancaster University, UK
Takeo Kanade
  Carnegie Mellon University, Pittsburgh, PA, USA
Josef Kittler
  University of Surrey, Guildford, UK
Jon M. Kleinberg
  Cornell University, Ithaca, NY, USA
Friedemann Mattern
  ETH Zurich, Switzerland
John C. Mitchell
  Stanford University, CA, USA
Moni Naor
  Weizmann Institute of Science, Rehovot, Israel
Oscar Nierstrasz
  University of Bern, Switzerland
C. Pandu Rangan
  Indian Institute of Technology, Madras, India
Bernhard Steffen
  University of Dortmund, Germany
Madhu Sudan
  Massachusetts Institute of Technology, MA, USA
Demetri Terzopoulos
  New York University, NY, USA
Doug Tygar
  University of California, Berkeley, CA, USA
Moshe Y. Vardi
  Rice University, Houston, TX, USA
Gerhard Weikum
  Max-Planck Institute of Computer Science, Saarbruecken, Germany

Luc Bougé   Viktor K. Prasanna (Eds.)

High Performance Computing – HiPC 2004

11th International Conference
Bangalore, India, December 19-22, 2004
Proceedings

 Springer

Volume Editors

Luc Bougé
IRISA/ENS Cachan, Campus Ker Lann
35170 Bruz Rennes, France
E-mail: luc.bouge@bretagne.ens-cachan.fr

Viktor K. Prasanna
University of Southern California, Department of Electrical Engineering
Los Angeles, CA, 90089-2562, USA
E-mail: prasanna@usc.edu

Library of Congress Control Number: 2004116658

CR Subject Classification (1998): D.1-4, C.1-4, F.1-2, G.1-2

ISSN 0302-9743
ISBN 3-540-24129-9 Springer Berlin Heidelberg New York

This work is subject to copyright. All rights are reserved, whether the whole or part of the material is concerned, specifically the rights of translation, reprinting, re-use of illustrations, recitation, broadcasting, reproduction on microfilms or in any other way, and storage in data banks. Duplication of this publication or parts thereof is permitted only under the provisions of the German Copyright Law of September 9, 1965, in its current version, and permission for use must always be obtained from Springer. Violations are liable to prosecution under the German Copyright Law.

Springer is a part of Springer Science+Business Media

springeronline.com

© Springer-Verlag Berlin Heidelberg 2004
Printed in Germany

Typesetting: Camera-ready by author, data conversion by Scientific Publishing Services, Chennai, India
Printed on acid-free paper   SPIN: 11369158   06/3142   5 4 3 2 1 0

Message from the Program Chair

Welcome to the proceedings of the 11th International Conference on High Performance Computing, HiPC 2004!

This year, we were delighted to receive 253 submissions to this conference from more than 25 different countries, including (besides India!) countries in North and South America, Europe, Asia, and Australia. This is a major increase on last year (169 submissions from 20 countries). Eventually, 48 submissions (the same number as last year) from 13 different countries were selected for presentation at the conference and publication in the conference proceedings.

This sharp increase in the number of submissions required adapting the regular selection process used in the previous years. First, all submitted papers were carefully considered by the Program Chair and Vice-Chairs to check their consistency with the minimal syntactic requirements for acceptance. At the end of this first stage, we were left with 214 submissions, which were further considered by the Program Committee. Each of these papers was reviewed by three Program Committee members. As many as 632 reviews were collected (2.95 per paper on average) and each paper was discussed at the online Program Committee meeting. Finally, 48 out of 214 (22%) were accepted for presentation and publication in the proceedings. Among them, two outstanding papers were selected as "Best Papers"; one in the algorithms and applications area and the other in the systems area. They will be presented in a separate plenary session and each paper will be awarded a prize sponsored by InfoSys. Here is a general summary of the results with respect to the origins of the submissions:

Submission origin	Reviewed	Accepted	Acceptance rate
Overall	214	48	22%
India	43%	31%	16%
Asia except India	17%	15%	19%
North America (mainly USA)	20%	42%	46%
Elsewhere (mainly Europe)	20%	12%	13%
Total	100%	100%	

These figures show that the selection process was highly competitive. We were pleased to accommodate eight (parallel) technical sessions of high-quality contributed papers, plus the special plenary "Best Papers" session. In addition, this year's conference also featured a Poster Session, an Industrial Track Session, six Keynote Addresses, six Tutorials and six Workshops.

It was a pleasure putting together this program with the help of five excellent Program Vice-Chairs and their 73 Program Committee members. The hard work of all the Program Committee members is deeply appreciated, and I especially wish to acknowledge the dedicated effort put in by the Vice-Chairs: Frédéric Desprez (Algorithms), Ramesh Govindan (Communication Networks), Thilo Kielmann (System Software), Frank Mueller (Applications), and Per Stenström (Architecture). Without their help and timely work, the quality of this program would not be as high nor would the process have run so smoothly.

I also wish to thank the other supporting cast members who helped in putting together this program, including those who organized the keynotes, tutorials, workshops, awards, poster session, industrial track session, and those who performed the administrative functions that are essential to the success of this conference. The work of Sushil K. Prasad in putting together the conference proceedings is also acknowledged, as well as the support provided by Mathieu Jan and Sébastien Monnet, PhD students at IRISA, in maintaining the CyberChair online paper submission and evaluation software. Last, but certainly not least, I express heartfelt thanks to our General Co-chairs, Viktor Prasanna and Uday Shukla, and to the Vice-General Chair, David A. Bader, for all their useful advice.

The preparation of this conference was unfortunately marked by a very sad and unexpected event: the sudden demise of Dr. Uday Shukla, who was a strong supporter of HiPC over the past ten years. He passed away on July 20, 2004 after a very brief illness. Dr. Shukla had been involved in organizing HiPC since the beginning. In addition to his encouragement in organizing HiPC, Dr. Shukla was a strong supporter of research activities in computer science and information technology in India. We will miss a friend of HiPC.

I would like to end this message by thanking the Conference General Co-chairs for giving me the opportunity to serve as the Program Chair of this conference. This truly was a very rewarding experience for me. I trust the attendees found this year's program to be as informative and stimulating as we endeavored to make it. I hope they enjoyed their HiPC 2004 experience, and I hope they also found time to enjoy the rich cultural experience provided by the fascinating city of Bangalore, India!

December 2004   Luc Bougé

Message from the Steering Chair

It is my pleasure to welcome you to the proceedings of the 11th International Conference on High Performance Computing, held in Bangalore, the IT capital of India.

I would like to single out the contributions of Luc Bougé, Program Chair, for organizing an excellent technical program. We received a record number of submissions this year, surpassing our previous high set last year. Also, the submissions were from a record number of countries. I am grateful to him for his efforts and thoughtful inputs in putting together the meeting program.

Many volunteers continued their efforts in organizing the meeting. While I thank them for their invaluable efforts, I would like to welcome R. Badrinath, India Publicity Chair, Susamma Barua, Registration Chair, and Sally Jelinek, Local Arrangements Co-chair, to the "HiPC family." Bertil Schmidt acted as our Cyber Chair. Rajeev Muralidhar of Intel India, though not listed in our announcements, was a great asset to us in handling meeting arrangements and interfacing with local institutions. I would like to thank all our volunteers for their tireless efforts. The meeting would not have been possible without the enthusiastic commitment of these individuals.

Major financial support for the meeting was provided by several leading IT companies in India. I would like to acknowledge the following individuals for their support: N.R. Narayana Murthy, Infosys; Venkat Ramana, Hinditron Infosystems; Uday Shukla, IBM India; Dinakar Sitaram, HP India; and V. Sridhar, Satyam.

Finally, I would like to thank Animesh Pathak at USC for his continued assistance and enthusiasm in organizing the meeting. He, along with the volunteers listed earlier, pulled together as a team to meet the several challenges presented this year.

This message would not be complete without the posthumous acknowledgement of our debt to Uday Shukla whose contributions remain pivotal to this event as a "home-grown" undertaking to showcase India's IT accomplishments. We honor his spirit by carrying forward the organization and presentation of this program.

December 2004   Viktor K. Prasanna

MESSAGE FROM THE VICE-GENERAL CHAIR

Welcome to the proceedings of the 11th International Conference on High Performance Computing, held in Bangalore. It was an honor and a pleasure to be able to serve the international community by bringing together researchers, scientists, and students, from academia and industry, to this meeting in the technology capital of India.

First let me recognize Manish Parashar for his help publicizing this conference, and Sushil K. Prasad for serving as the publications chair. Srinivas Aluru did an excellent job organizing tutorials presented by leading experts. HiPC 2004 included six tutorials in areas likely to be at the forefront of high-performance computing in the next decade, such as storage and file systems with InfiniBand, pervasive computing, grid computing for e-science and e-business, new networking technologies, network security, and embedded system design.

I wish to thank all of the conference organizers and volunteers for their contributions in making HiPC 2004 a great success. I would especially like to thank the general co-chairs, Viktor K. Prasanna and Uday Shukla, for their enormous contributions steering and organizing this meeting. It is to their credit that this meeting has become the premier international conference for high-performance computing. With deep sorrow, we will miss Dr. Uday Shukla, whose leadership and strong support of research activities in computer science and information technology in India was remarkable. Special thanks are also due to the program chair, Luc Bougé, for his hard work assembling a high-quality technical program that included contributed and invited papers, an industrial track, keynote addresses, tutorials, and several workshops.

December 2004   David A. Bader

A Tribute to Dr. Uday Shukla

Dr. Uday Shukla
1951–2004
In tribute to his pioneering
leadership and contributions to India's
advanced computing technology

Uday Shukla passed away on July 20, 2004 at the age of 53. He leaves behind his wife Rekha, son Nitish and daughter Vinita, and we extend to them our deepest sympathy and condolences.

In his passing, the information technology industry has lost not only a visionary but also a rare individual with a capacity to inspire a technology-savvy generation to innovate. For HiPC, his loss has been profound, with our organization having come to rely on his leadership in India and on the world-wide recognition and alliances that he brought to the event. Here we pay tribute to his many accomplishments, as highlighted in a brief resume of his professional associations and undertakings.

Dr. Shukla joined IBM India (then Tata Information Systems Ltd., a joint venture between IBM and Tata) in 1994 as the head of the Systems Group. He was the director of IBM Software Labs, India and IBM Engineering & Technology Services, India when he passed away. Prior to joining IBM, he was the head of R&D at Tata Elxsi following his tenure as the location head of the Centre for the Development of Advanced Computing, Bangalore. He received his PhD in aerospace engineering from the Indian Institute of Science, Bangalore. He was a fellow of the Institution of Engineers (India), and a senior member of IEEE.

In the years following, Dr. Shukla took on the task of moving IBM's Indian operations to the cutting edge of technology. His focus was on creating a climate that would nurture young talent and set them on the path of pursuing and implementing innovative ideas. His efforts in this direction included the creation of an R&D group, a Technology Incubation Centre, the Centre for Advanced Studies, a University Relationships program, an affiliate of the IBM Academy of Technology, and an in-house lecture series on science and technology. The in-house lecture series covers topics like quantum computing, molecular biology, formal mathematical systems, and algorithms.

Collectively these activities helped generate science-based intellectual property and a research environment that is rather unique. A large number of patent applications filed under his leadership have been in areas such as compiler optimization, molecular biology, and operating systems. These initiatives were based on his faith in the unusual problem-solving capabilities of the people he was leading in India. While the journey to achieving technological excellence is long and tedious, Shukla was able to firmly establish basic elements that motivate talented researchers to stay the course, through an awareness of the importance of basic science in the development of technology and the importance of an ethical environment. He had the pleasure of seeing at least a dozen of his colleagues well on the path to becoming prolific inventors and dozens more filing their first patent application. In addition, his encouragement emboldened a few very young engineers to make a mark as researchers in molecular biology.

His association with HiPC was an important part of his dream, and he was involved in organizing HiPC from its beginning. As the co-chair of the Workshop on Cutting Edge Computing since 2001, he sought to enhance the image of HiPC by inviting papers from experts in new and emerging technologies. He was also active in bringing together colleagues in India to form a National Advisory Committee for the event, often working in the background with organizers to develop local participation and support. He gave generously of his professional wisdom and and organizational energy, serving as the general co-chair of HiPC in 2002 and then again signing on to co-chair HiPC 2004. In addition to his encouragment for organizing HiPC, Dr. Shukla remained a strong supporter of research activities in computer science and information technology in India. We will miss this friend of HiPC.

October 2004   Viktor K. Prasanna

CONFERENCE ORGANIZATION

General Co-chairs
Viktor K. Prasanna, University of Southern California, USA
Uday Shukla, IBM India, India

Vice-General Chair
David A. Bader, University of New Mexico, USA

Program Chair
Luc Bougé, IRISA/ENS Cachan, France

Program Vice-Chairs
Algorithms
Frédéric Desprez, INRIA Rhône-Alpes, France

Applications
Frank Mueller, North Carolina State University, USA

Architecture
Per Stenström, Chalmers University of Technology, Sweden

Communication Networks
Ramesh Govindan, University of Southern California, USA

Systems Software
Thilo Kielmann, Vrije Universiteit, The Netherlands

Steering Chair
Viktor K. Prasanna, University of Southern California, USA

Workshops Chair
C.P. Ravikumar, Texas Instruments, India

Poster/Presentation Chair
Rajkumar Buyya, University of Melbourne, Australia

Scholarships Chair
Atul Negi, University of Hyderabad, India

Finance Co-chairs
Ajay Gupta, Western Michigan University, USA
B.V. Ramachandran, Software Technology Park, Bangalore, India

Tutorials Chair
Srinivas Aluru, Iowa State University, USA

Awards Chair
Arvind, MIT, USA

Keynote Chair
Rajesh Gupta, University of California, San Diego, USA

Industry Liaison Chair
Sudheendra Hangal, Sun Microsystems, India

Publicity Chair
Manish Parashar, Rutgers, State University of New Jersey, USA

Publications Chair
Sushil K. Prasad, Georgia State University, USA

Cyber Chair
Bertil Schmidt, Nanyang Technological University, Singapore

Local Arrangements Chair
Sally Jelinek, Electronic Design Associates, Inc., USA

Local Arrangements Co-chair
Rajeev D. Muralidhar, Intel, India

Registration Chair
Susamma Barua, California State University, Fullerton, USA

Steering Committee

R. Badrinath, HP, India
José Duato, Universidad Politecnica de Valencia, Spain
N.S. Nagaraj, Infosys, India
Viktor K. Prasanna, University of Southern California (Chair), USA
N. Radhakrishnan, US Army Research Lab, USA
Venkat Ramana, Cray-Hinditron, India
Shubhra Roy, Intel, India
Sartaj Sahni, University of Florida, USA
Dheeraj Sanghi, IIT, Kanpur, India
Assaf Schuster, Technion, Israel Institute of Technology, Israel
Uday Shukla, IBM, India
V. Sridhar, Satyam Computer Services Ltd., India

PROGRAM COMMITTEE

Algorithms

Mikhail Atallah, Purdue University, USA
Michael A. Bender, State University of New York at Stony Brook, USA
Andrea Clematis, IMATI, CNR, Genoa, Italy
Jose Fortes, University of Florida, USA
Isabelle Guerin-Lassous, INRIA, INSA Lyon, France
Mahmut Kandemir, Pennsylvania State University, USA
George Karypis, University of Minnesota, Minneapolis, USA
Ran Libeskind-Hadas, Harvey Mudd College, USA
Muthucumaru Maheswaran, McGill University, Canada
Sato Mitushisa, University of Tsukuba, Japan
Sushil K. Prasad, Georgia State University, USA
Arnold A. Rosenberg, University of Massachussetts at Amherst, USA
Christian Scheideler, Johns Hopkins University, USA
Ramin Yahyapour, University of Dortmund, Germany
Albert Y. Zomaya, University of Sydney, Australia

Applications

Rupak Biswas, NASA Ames Research Center, USA
Franck Capello, INRIA, Orsay, France
Siddhartha Chatterjee, IBM T.J. Watson Research Center, USA
Chen Ding, University of Rochester, USA
Nikil Dutt, University of California, Irvine, USA
Rudi Eigenmanm, Purdue University, USA
Jesus Labarta, Technical University of Catalonia, Spain
Dave Lowenthal, University of Georgia, USA
Xiaosong Ma, North Carolina State University, USA
Manish Parashar, Rutgers, State University of New Jersey, USA
Keshav Pingali, Cornell University, USA
Jeff Vetter, Oak Ridge National Laboratory, USA
Xiaodong Zhang, William and Mary College, Williamsburg, USA

Architecture

Ricardo Bianchini, Rutgers, State University of New Jersey, USA
Mats Brorsson, KTH, Stockholm, Sweden
José Duato, University of Valencia, Spain
Michel Dubois, University of Southern California, USA
Rama Govindarajan, Indian Institute of Science, Bangalore, India
Wolfgang Karl, University of Karlsruhe, Germany
Josep Llosa, UPC Barcelona, Spain
Sang-Lyul Min, Seoul National University, Korea
Li Shuan Peh, Princeton University, USA
Partha Ranganathan, HP Western Research Laboratory, USA
Martin Schulz, Cornell University, USA

Olivier Temam, University of Paris Sud, France
Stamatis Vassiliadis, Delft University, The Netherlands

Communication Networks

Bengt Ahlgren, SICS, Kista, Sweden
Suman Banerjee, University of Maryland, USA
Erdal Cayirci, Istanbul Technical University, Turkey
Sonia Fahmy, Purdue University, USA
Paul Havinga, University of Twente, The Netherlands
Ahmed Helmy, University of Southern California, USA
Abhay Karandikar, IIT Bombay, India
Amit Kumar, IIT Delhi, India
Krishna Sivalingam, University of Maryland, Baltimore County, USA
C. Siva Ram Murthy, IIT Madras, India
Yoshito Tobe, Tokyo Denki University, Japan
Yu-Chee Tseng, National Taiwan University, Taiwan
Daniel Zappala, University of Oregon, USA

Systems Software

Olivier Aumage, INRIA, Bordeaux, France
Thomas Fahringer, University of Innsbruck, Austria
Phil Hatcher, University of New Hampshire, USA
Shantenu Jha, University College London, UK
Laxmikant V. Kale, University of Illinois at Urbana Champaign, USA
Anne-Marie Kermarrec, Microsoft Research, Cambridge, UK
Koen Langendoen, Technical University of Delft, The Netherlands
Shikharesh Majumdar, Carleton University, Ottawa, Canada
Ludek Matyska, Masaryk University, Brno, Czech Republic
Raju Pandey, University of California, Davis, USA
CongDuc Pham, LIP, ENS Lyon, France
Ana Ripoll, Universitat Autônoma de Barcelona, Spain
Martin Swany, University of Delaware, USA
Osamu Tatebe, AIST Tsukuba, Japan
Ramin Yahyapour, University of Dortmund, Germany

Workshop Organizers

Workshop on Cutting Edge Computing

Chair
  Rajendra K. Bera, IBM Software Lab, India

Workshop on Dynamic Provisioning and Resource Management

Co-chairs
  Sharad Garg, Intel Corp., USA
  Jens Mache, Lewis & Clark College, USA

Trusted Internet Workshop

Co-chairs
  G. Manimaran, Iowa State University, USA
  Krishna Sivalingam, Univ. of Maryland Baltimore County, USA

Workshop on Performance Issues in Mobile Devices

Co-chairs
  Rajat Moona, IIT Kanpur, India
  Gopal Raghavan, Nokia, USA
  Alexander Ran, Nokia, USA

Workshop on Software Architectures for Wireless

Co-chairs
  S.H. Srinivasan, Satyam Computer Services Ltd., India
  Srividya Gopalan, Satyam Computer Services Ltd., India

Workshop on New Horizons in Compiler Analysis and Optimizations

Co-chairs
  R. Govindarajan, IISc, Bangalore, India
  Uday Khedker, IIT, Bombay, India

LIST OF REVIEWERS

Ahlgren, Bengt
Al-Ars, Zaid
Albertsson, Lars
Alessio, Bertone
Antoniu, Gabriel
Armstrong, Brian
Atallah, Mikhail
Aumage, Olivier
Bader, David A.
Bahn, Hyokyung
Bai, Liping
Banerjee, Suman
Basumallik, Ayon
Bavetta, Bayard
Baydal, Elvira
Bender, Michael A.
Bianchini, Ricardo
Bian, Fang
Biswas, Rupak
Bougé, Luc
Bourgeois, Anu
Bouteiller, Aurénn
Brezany, Peter
Brorsson, Mats
Capello, Franck
Cayirci, Erdal
Chadha, Vineet
Chakravorty, Sayantan
Chatterjee, Siddhartha
Chélius, Guillaume
Chen, Jianwei
Chen, Yu
Chintalapudi, Krishna
Choi, Woojin
Chung, Sung Woo
Clematis, Andrea
Corana, Angelo
Cores, Fernando
Crisu, Dan
Cuenca, Pedro
D'Agostino, Daniele

Dandamudi, Sivarama
Datta, Jayant
Davison, Brian
de Langen, Pepijn
Desprez, Frédéric
Dhoutaut, Dominique
Ding, Chen
Djilali, Samir
Drach, Nathalie
Duato, José
Dubois, Michel
Duranton, Marc
Dutt, Nikil
Eigenmann, Rudi
Ernemann, Carsten
Ersoz, Deniz
Fahmy, Sonia
Fahringer, Thomas
Fedak, Gilles
Ferreira, Renato
Flouris, Michail
Fortes, José
Fouad, Mohamed Raouf
Francis, Paul
Galizia, Antonella
Galuzzi, Carlo
Gelenbe, Erol
Gioachin, Filippo
Glossner, John
Gluck, Olivier
Gnawali, Omprakash
Goglin, Brice
Gore, Ashutosh
Govindan, Ramesh
Govindarajan, Rama
Graham, Peter
Guérin-Lassous, Isabelle
Gummadi, Ramakrishna
Guo, Minyi
Gustedt, Jens
Hainzer, Stefan

Harting, Jens
Harvey, Matt
Hatcher, Phil
Havinga, Paul
Helmy, Ahmed
Hérault, Thomas
Hernandez, Porfidio
Hoffman, Forrest
Hurfin, Michel
Jain, Mayank
Jajodia, Sushil
Jégou, Yvon
Jha, Shantenu
Jiao, Xiangmin
Johnson, Troy A.
Jorba, Josep
Joseph, Russell
Jugravu, Alexandru
Kale, Laxmikant V.
Kandemir, Mahmut
Karandikar, Abhay
Karl, Wolfgang
Karypis, George
Kavaldjiev, Nikolay
Kaxiras, Stefanos
Kermarrec, Anne-Marie
Keryell, Ronan
Kielmann, Thilo
Kim, Young Jin
Kommareddy, Christopher
Kothari, Nupur
Kumar, Amit
Kwon, Minseok
Labarta, Jesús
Lacour, Sébastien
Lai, An-Chow
Langendoen, Koen
Law, Y.W.
Lee, Sang-Ik
Lee, Seungjoon
Lee, Sheayun
Legrand, Arnaud
Lemarinier, Pierre
Lhuillier, Yves
Libeskind-Hadas, Ran

Lijding, Maria
Lim, Sung-Soo
Lin, Heshan
Li, Xiaolin
Li, Xin
Llosa, Josep
López, Pedro
Lowenthal, Dave
Madavan, Nateri
Maheswaran, Muthucumaru
Majumdar, Shikharesh
Maniymaran, Balasubramaneyam
Marchal, Loris
Margalef, Tomas
Matsuda, Motohiko
Matyska, Ludek
Ma, Xiaosong
Meinke, Jan
Min, Sang-Lyul
Mishra, Arunesh
Mishra, Minal
Mitton, Nathalie
Mitushisa, Sato
Mouchard, Gilles
Moure, Juan Carlos
Mueller, Frank
Naik, Piyush
Nam, Gi-Joon
Nandy, Biswajit
Nieberg, Tim
Oliker, Leonid
Panda, Preeti Ranjan
Pandey, Raju
Parashar, Manish
Peh, Li Shuan
Pérez, Christian
Pham, CongDuc
Pingali, Keshav
Porter, Andrew
Prakash, Rajat
Prasad, Sushil K.
Prasanna, Viktor K.
Preis, Robert
Pu, Calton
Ramamritham, Krithi

Ranganathan, Partha
Ren, Xiaojuan
Riccardo, Albertoni
Ripoll, Ana
Robert, Yves
Rosenberg, Arnold A.
Sainrat, Pascal
Salodkar, Nitin
Sanyal, Soumya
Scheideler, Christian
Schulz, Martin
Senar, Miguel Angel
Shenai, Rama
Sherwood, Rob
Shi, Zhijie
Sips, H.J.
Sivakumar, Manoj
Sivalingam, Krishna
Siva Ram Murthy, C.
Stenström, Per
Subramani, Sundar
Sundaresan, Karthikeyan
Suppi, Remo
Swany, Martin
Tammineedi, Nandan
Tatebe, Osamu
Temam, Olivier
Thierry, Eric
Tobe, Yoshito
Trigoni, Niki
Trystram, Denis
Tseng, Yu-Chee

Utard, Gil
Vachharajani, Manish
Vadhiyar, Sathish S.
Vallée, Geoffroy
Van der Wijngaart, Rob F.
van Dijk, H.W.
van Gemund, A.J.C.
Vassiliadis, Stamatis
Vetter, Jeff
Villazon, Alex
Wang, Hangsheng
Welzl, Michael
Wieczorek, Marek
Wilmarth, Terry
Wu, Jian
Wu, Yan
Wu, Yunfei
Xu, Jing
Yahyapour, Ramin
Younis, Ossama
Yu, Ting
Yu, Yinlei
Zappala, Daniel
Zelikovsky, Alex
Zhang, Guangsen
Zhang, Hui
Zhang, Qingfu
Zhang, Xiangyu
Zhang, Xiaodong
Zhao, Ming
Zomaya, Albert Y.

Table of Contents

Keynote Addresses

Plenary Session - Best Papers

Session I - Wireless Network Management

Session II - Compilers and Runtime Systems

Session III - High-Performance Scientific Applications

Session IV - Peer-to-Peer and Storage Systems

Session V - High-Performance Processors and Routers

Session VI - Grids and Storage Systems

Session VII - Energy-Aware and High-Performance Networking

Session VIII - Distributed Algorithms

Rethinking Computer Architecture Research

Arvind

Computer Science and Artificial Intelligence Laboratory (CSAIL),
Massachusetts Institute of Technology
arvind@mit.edu

Abstract. The prevailing methodology in architecture research is to propose a mechanism, incorporate it in some existing execution-driven software simulator and collect statistics related to some standard benchmarks to determine the merits of the architectural proposal. In recent architecture conferences as many as ninety five percent of the papers have followed this methodology.

Some of the pitfalls of this approach are well known: frequent omissions (either deliberately or inadvertently) of important aspects of the system, total disregard for implementation complexity, inability to model time accurately in situations where interactions are nondeterministic and thus, potentially deeply affected by timing assumptions, and inability to run realistic applications with large enough data sets, especially in parallel systems. Architects of real systems have generally ignored such simulation studies because of aforementioned weaknesses. They are driven primarily by constraints such as power and clock speeds, and by compatibility with their own older systems. The relative importance of these constraints changes from time to time and sometimes that can affect the microarchitecture under consideration. For example, the next generation technology will offer so much variability in power and speed that it may require us to reconsider the whole notion of "timing closure".

There are two new developments that together enables a more "implementation aware" study of microarchitectures. Availability of FPGA's with as many as 6-million gates, multiple RISC cores and 256K bytes of memory make it possible to implement very complex devices (e.g., a complex 64-bit processor) on a single FPGA. Furthermore, high-level synthesis tools, such as Bluespec, make it possible for a small team to generate RTL for complex devices within a matter of weeks, if not days. A proper setup with FPGA's and high-level synthesis tools can enable universities to architectural studies that are much more rewarding and useful than pure software simulations. The same high-level synthesis techniques can also be directed toward ASICs and custom circuits where physical layout and other backend issues can be considered.

We will illustrate this new style of "implementation aware" research through several ongoing projects at MIT and CMU.

Biography: Arvind is the Johnson Professor of Computer Science and Engineering at the Massachusetts Institute of Technology. As the Founder and President of Sandburst, a fabless semiconductor company, Arvind led the Company from its inception in June 2000 until his return to

L. Bougé and V.K. Prasanna (Eds.): HiPC 2004, LNCS 3296, pp. 1–2, 2004.
© Springer-Verlag Berlin Heidelberg 2004

MIT in August 2002. His work at MIT on high-level specification and description of architectures and protocols using Term Rewriting Systems (TRSs), encompassing hardware synthesis as well as verification, laid the foundations for Sandburst and more recently Bluespec Inc. Previously, he contributed to the development of dynamic dataflow architectures, and together with Dr R.S.Nikhil published the book "Implicit Parallel Programming in pH".

Event Servers for Crisis Management

K. Mani Chandy

California Institute of Technology
mani@cs.caltech.edu

Abstract. Organizations need to respond rapidly to threats and opportunities in the extended environment. Examples of threats are terrorism, chemical spills, intrusion into networks, and delayed arrivals of components. Opportunities include arbitrage and control of distributed systems such as the electrical power grid and logistics networks. The increasing prevalence of sensors such as RFID (radio frequency ID) and Web sites that offer measured information, coupled with information in databases and applications allows organizations to sense critical conditions in the extended environment. A response to a critical condition is to send an alert to appropriate people or send a message that initiates an application. A critical condition is a predicate on the history of global states of a system. An event server is a software component with sensors, responders, and computational elements. Computations are incremental in the sense that computations incorporate each new event in an online algorithm as opposed to repeatedly computing predicates over the entire history. This talk discusses the problem space, proposes an abstract mathematical model of the problem, suggests approaches to solutions, describes an implementation, and relates this work to research in data stream management systems and rules engines.

Biography: Mani Chandy received his B.Tech from IIT Madras in 1965 and his PhD from MIT in 1969 in Electrical Engineering. He worked in industry and taught at the University of Texas at Austin and is now at the California Institute of Technology. He has published papers in computer performance modeling, distributed systems, program correctness and event systems. He received the IEEE Koji Kobayashi Award and the CMG Michelson Award for contributions to computer performance modeling, and he is a member of the US National Academy of Engineering.

L. Bougé and V.K. Prasanna (Eds.): HiPC 2004, LNCS 3296, p. 3, 2004.
© Springer-Verlag Berlin Heidelberg 2004

DIET: Building Problem Solving Environments for the Grid

Frédéric Desprez

LIP Laboratory / GRAAL Project,
CNRS, ENS Lyon, INRIA, Univ. Claude Bernard Lyon, France
Frederic.Desprez@ens-lyon.fr

Abstract. DIET (Distributed Interactive Engineering Toolbox) is a set of hierarchical components to design Network Enabled Server systems. These systems are built upon servers managed through distributed scheduling agents for a better scalability. Clients ask to these scheduling components to find servers available (using some performance metrics and information about the location of data already on the network). Our target architecture is the grid which is highly heterogeneous and dynamic. Clients, servers, and schedulers are better connected in a dynamic (or peer-to-peer) fashion.

In this keynote talk, we will discuss the different issues to be solved for the efficient deployment of Network Enabled Servers systems on the grid. These issues include the automatic deployment of components, performance evaluation, resource localization, scheduling of requests, and data management. See http://graal.ens-lyon.fr/DIET/ for further information.

Biography: Frédéric Desprez is a director of research at INRIA and holds a position at LIP laboratory (ENS Lyon, France). He received is PhD in C.S. from the Institut National Polytechnique de Grenoble in 1994, his MS in C.S. from the ENS Lyon in 1990, and his habilitation in 2001. His research interests include parallel libraries for scientific computing on parallel distributed memory machines, problem solving environments, and grid computing. See http://graal.ens-lyon.fr/~desprez for further information.

L. Bougé and V.K. Prasanna (Eds.): HiPC 2004, LNCS 3296, p. 4, 2004.
© Springer-Verlag Berlin Heidelberg 2004

The Future Evolution of High-Performance Microprocessors

Norman P. Jouppi

Hewlett Packard,
Palo Alto, USA
norm.jouppi@hp.com

Abstract. The evolution of high-performance microprocessors is fast approaching several significant inflection points. First, the marginal utility of additional single-core complexity is now rapidly diminishing due to a number of factors. The increase in instructions per cycle from increases in sizes and numbers of functional units has plateaued. Meanwhile the increasing sizes of functional units and cores are beginning to have significant negative impacts on pipeline depths and the scalability of processor clock cycle times.

Second, the power of high performance microprocessors has increased rapidly over the last two decades, even as device switching energies have been significantly reduced by supply voltage scaling. However future voltage scaling will be limited by minimum practical threshold voltages. Current high-performance microprocessors are already near market limits of acceptable power dissipation. Thus scaling microprocessor performance while maintaining or even reducing overall power dissipation benefit of appreciable further voltage scaling will prove especially challenging.

In this keynote talk we will discuss these issues and propose likely scenarios for the future evolution of high-performance microprocessors.

Biography: Norman P. Jouppi is a Fellow at HP Labs in Palo Alto, California. From 1984 through 1996 he was also a consulting assistant/associate professor in the department of Electrical Engineering at Stanford University. He received his PhD in Electrical Engineering from Stanford University in 1984.

He started his contributions to high-performance microprocessors as one of the principal architects and the lead designer of the Stanford MIPS microprocessor. While at Digital Equipment Corporation's Western Research Lab he was the principal architect and lead designer of the MultiTitan and BIPS microprocessors. He has also contributed to the architecture and implementation of graphics accelerators, and has conducted extensive research in telepresence. He holds more than 25 U.S. patents and has published over 100 technical papers. He currently serves as ACM SIGARCH Chair and is a Fellow of the IEEE.

L. Bougé and V.K. Prasanna (Eds.): HiPC 2004, LNCS 3296, p. 5, 2004.
© Springer-Verlag Berlin Heidelberg 2004

Low Power Robust Computing

Trevor Mudge

University of Michigan, Ann Arbor
tnm@eecs.umich.edu

Abstract. In a recent speech Intel founder Andrew Grove argued that Moore's law will not slow down for at least a decade. By that time, integrated circuits will have feature sizes of 30 nanometers, allowing for integration of billions of devices on a single die and enabling unforeseen computational capabilities. However, with growing levels of integration, power densities will also skyrocket to hundreds of Watts. In fact, Grove cites power consumption as a major show stopper with off-state current leakage "a limiter of integration."

In addition to the power consumption crisis, aggressively scaled feature sizes also result in increased process variability and poor reliability. Hence, Grove mentions that at 30nm design will enter an era of "probablisitic computing," with the behavior of logic gates no longer being deterministic. To take advantage of scaling, it will be necessary to compute in the presence of various types of errors.

Our talk will present recent results in robust low power computing. The perspective will be microarchitectural: what can the microarchitect do to reduce the dependency on power and improve robustness. We will discuss recent academic and commercial proposals to limit power consumption. Finally, we will review some techniques to improve robustness based on recent ideas in timing speculation exemplified by our Razor research.

Biography: Trevor Mudge is the Bredt Family Professor of Electrical Engineering and Computer Science at the University of Michigan. He received a Ph.D. in computer science from the University of Illinois, Urbana-Champaign. His research interests include computer architecture, CAD, and compilers. He has chaired over 30 theses and authored over 250 articles in these research areas. In addition, he is the founder of Idiot Savants, a chip-design consultancy. He is a Fellow of the IEEE and a member of the ACM, the IEE, and the British Computer Society.

L. Bougé and V.K. Prasanna (Eds.): HiPC 2004, LNCS 3296, p. 6, 2004.
© Springer-Verlag Berlin Heidelberg 2004

Networks and Games

Christos Papadimitriou

University of California, Berkeley
christos@cs.berkeley.edu

Abstract. The Internet is the first computational artifact that was not designed by a single entity, but emerged from the complex interaction of many. As a result, it must be approached as a mysterious object, akin to the universe, the brain, the market, and the cell, to be understood by observation and falsifiable theories. The theory of games promises to play an important role in this endeavor, since the entities involved in the Internet are interacting selfish agents in various and varying degrees of collaboration and competition.

We survey recent work by the speaker and collaborators considering networks and protocols as equilibria in appropriate games, and trying to explain phenomena such as the power law distributions of the degrees of the Internet topology in terms of the complex optimization problems faced by each node.

Biography: Christos H. Papadimitriou is C. Lester Hogan Professor of Computer Science at UC Berkeley. Before Berkeley he taught at Harvard, MIT, Athens Polytechnic, Stanford, and UCSD. He has written four textbooks and many articles on algorithms, complexity, and their applications to optimization, databases, AI, economics, and the Internet. He holds a PhD from Princeton, and honorary doctorates from ETH (Zurich) and the University of Macedonia (Thessaloniki). He is a member of the American Academy of Arts and Sciences and of the National Academy of Engineering, and a fellow of the ACM. His novel Turing (a novel about computation), was published by MIT Press in 2003.

L. Bougé and V.K. Prasanna (Eds.): HiPC 2004, LNCS 3296, p. 7, 2004.
© Springer-Verlag Berlin Heidelberg 2004

An Incentive Driven Lookup Protocol for Chord-Based Peer-to-Peer (P2P) Networks*

Rohit Gupta and Arun K. Somani

Department of Electrical and Computer Engineering,
Iowa State University, Ames, IA 50011, USA
{rohit, arun}@iastate.edu

Abstract. In this paper we describe a novel strategy for carrying out lookups in Chord-based peer-to-peer (P2P) networks, wherein nodes are assumed to behave selfishly. This is in contrast to the traditional lookup schemes, which assume that nodes cooperate with each other and truthfully follow a given protocol in carrying out resource lookups. The proposed scheme also provides efficient and natural means for preventing free-riding problem in Chord without requiring prior trust relationships among nodes. In addition, we evaluate the performance of Chord for providing routing service in a network of selfish nodes and prove that it has good structural properties to be used in uncooperative P2P networks.

1   Introduction

Almost all the current research in P2P systems is based on a cooperative network model. It is generally assumed that although there can be rogue nodes in a system, most of the nodes are trustworthy and follow some specific protocol as suggested by the network designer. We believe that such assumptions do not always hold good in large-scale open systems and have to be done away with in order to make P2P systems reliable, robust, and realize their true commercial potential. Moreover, it has been pointed out that free-riding is one of the most significant problems being faced by today's P2P networks [2].

We consider a Chord [1] based P2P network and describe a novel strategy for carrying out lookups in such networks. Our proposed scheme provides an efficient and natural means for preventing free-riding problem in Chord without requiring prior trust relationships among nodes. Therefore, it incurs low overhead and is highly robust, and unlike other schemes it does not rely on any centralized entity or require specialized trusted hardware at each node. The protocol proposed here is essentially an *incentive driven lookup* protocol that ensures that reward received by intermediate nodes and resource provider is maximized by following the protocol steps. It is in contrast to other lookup schemes, which assume that nodes cooperate with each other in finding data and faithfully follow a given protocol for carrying out resource lookups irrespective of whether they are currently overloaded or not, for example.

---

* The research reported in this paper is funded in part by Jerry R. Junkins Chair position at Iowa State University.

L. Bougé and V.K. Prasanna (Eds.): HiPC 2004, LNCS 3296, pp. 8–18, 2004.
© Springer-Verlag Berlin Heidelberg 2004

We evaluate the performance of Chord for providing routing service in a network of selfish nodes. We show that in a large network, unless nodes have privilege information about the location of network resources, following Chord is a good strategy provided that everyone else also follow the Chord protocol.

The paper is structured as follows. Section 2 is on related work, Section 3 describes the network model. Section 4 gives a detailed description of the proposed lookup protocol. Section 5 presents a resource index replication strategy that is useful in dealing with selfish nodes in Chord. Section 6 explains why Chord is a good protocol to be used in a network with selfish nods. We conclude the paper in Section 7.

2 Related Work

The need for developing protocols for selfish agents (nodes) in P2P systems has often been stressed before (see [3, 4]). The research in [5, 6] provides solution to avoid free-riding problem in P2P networks. The basic approach in all of these is to make sure that nodes indeed share their resources before they themselves can obtain services from a network. Also, most of these solutions rely on self-less participation of groups of trusted nodes to monitor/police the activities of individual nodes, and ensure that everyone contributes to the system.

To the best of our knowledge, none of the existing solutions that deal with the problem of free-riding in P2P networks also address the more basic question of why nodes would route messages for others. Since these nodes belong to end users, they may in order to conserve their bandwidth and other resources, such as buffer space, memory etc., may drop messages received for forwarding.

The problem of selfish routing has been encountered and addressed in the context of mobile ad-hoc networks (see [7, 8]). Some of these proposals can also find application in P2P networks.

3 Network Model

The model of network assumed here is a Chord [1] based P2P network. The nodes are assumed to be selfish. By *selfish* we mean that nodes try to maximize their profits given any possible opportunity. The profit from a transaction (or an activity) is equal to the difference between the reward that a node earns and the cost that it incurs by participating in the transaction. The reward can be anything that is deemed to have value, the possession of which adds to a node's utility.

An example of a transaction is a lookup process, i.e. the process of searching for and downloading a desired resource object. The cost in the form of bandwidth, memory etc. that a node x incurs by participating in a transaction is referred to as its marginal cost (MC_x). The cost incurred by server S (and also the intermediate nodes) increases in proportion to the amount of traffic it is handling and any request offering less than its current MC_S value is not fulfilled.

We assume that for each resource there is a single server, i.e. caching and replication of data does not take place. Nodes that store the index of a resource are called the *terminal* nodes for that resource. For resource R these nodes are denoted by $T_{R_i} \ \forall i \in \{1, \ldots, k\}$, where k is the index replication factor. The terminal nodes maintain a mapping (i.e. an index) from the resource name, represented by R, to the IP address of the server that provides the resource. The terminal nodes are the Chord successors of the mappings of a resource onto the Chord network. The method by which these mappings are determined is explained in Section 5.

Unless otherwise specified, all message communication is assumed to provide message non-repudiation. Our protocol relies on message non-repudiation to ensure that nodes do not go back on their *commitment* as suggested by the content of the messages sent by them. We assume that there is a mechanism in place to punish nodes if it can be proven that they did not fulfill their commitments.[1]

4   Incentive Driven Lookup Protocol

We now describe how routing of lookup messages is performed when nodes behave selfishly and how prices for resources are set with minimum additional overhead on the system. To simplify our discussion, we take an example of a lookup process and see how it is carried out under the given protocol.

4.1   Parallel Lookup Towards the Terminal Nodes

The client (C) before initiating the lookup process estimates its utility (U_R^C) of the resource (R) to calculate the maximum price that it can offer for the resource. Since the locations of the resource indices can be calculated by using the same mechanism as used by the server (S) to store them, C sends a separate lookup message towards each of them. Together these parallel lookup messages can be considered as constituting a single lookup process and the routing of an individual lookup message is done using the Chord routing protocol.

Each lookup message Msg_{lookup} contains the following information, as included by the client - address of one of the k terminal nodes (T_{R_i}), the resource ID (R), the maximum price offered (P_C), the marginal cost (MC_{total}), the request IDs $(Reqid_{private}$ and $Reqid_{public})$.

$Reqid_{public}$ identifies the lookup process such that S (and intermediate nodes) on receiving multiple lookup messages knows that the messages pertain to the same lookup process. Thus, the same value of $Reqid_{public}$ is included in all the lookup messages. On the other hand, a unique value of $Reqid_{private}$ is included in each of the lookup message. In Section 4.3, we illustrate the significance of $Reqid_{private}$. MC_{total} value is the sum of C's marginal cost MC_C and the marginal cost of the next hop neighbor (also called the successor) to which the message is forwarded.[2] C before sending the lookup

---

[1] In an enterprise computing environment there might be a central authority one can report to in order to identify and punish the cheating node. For large-scale open systems one can use reputation mechanisms to ensure that cheating nodes are accurately identified and isolated from receiving services.

[2] The term "successor" as used here is the same as used in the description of the Chord protocol, where it referred to the node which immediately succeeds an ID value in a Chord ring or network. Also, the term "predecessor" refers here to a previous hop node along the lookup path.

message inquires its successor about its marginal cost. The received value is added by C to its own marginal cost and stored in MC_{total}. Likewise, each intermediate node on receiving the lookup message updates the MC_{total} value by adding to it the marginal cost of its successor.

Intermediate nodes for all the lookup requests route the received lookup messages to the next hop neighbors and this process continues till the messages reach the desired terminal nodes. Since the terminal nodes store the IP address of S, they contact S in order to obtain the resource. S receive k such requests and from the $Reqid_{public}$ values knows that all the requests pertain to the same lookup process. S then holds a second price sealed-bid auction (also called Vickrey auction [9, 3]) with all the terminal nodes as the bidders. S provides the resource to the terminal node that offers it the highest price.

4.2   Bidding for the Resource by the Terminal Nodes

In Vickrey auction, the highest bidder wins the auction, but the price that it has to pay is equal to the second highest bid. Vickrey auction has several desirable properties, such as existence of truth revelation as a dominant strategy, efficiency, low cost etc. Vickrey auction in its most basic form is designed to be used by altruistic auctioneers, which are concerned with overall system efficiency or social good as opposed to self-gains. Self-interested auctioneer is one of the main reasons why Vickrey auction did not find widespread popularity in human societies [10].

Since S behaves selfishly and tries to maximize its own profit, the auction process needs to ensure the following.

– Selecting the highest bidder is the best strategy for S.
– The price paid by the highest bidder is indeed equal to the second highest bid, i.e. S should reveal true second highest bid to the highest bidder.
– Collusion among S and the bidders should not be possible.

In view of the above requirements, we provide a two-phase secure Vickrey auction protocol.

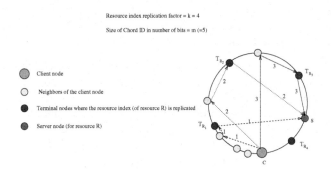

Fig. 1. Lookup message propagation in the *incentive driven lookup* protocol

In summary, the incentive driven lookup strategy involves sending lookup messages to all the terminal nodes of a resource, such that at most one message is sent out for all the terminal nodes that go through the same next hop neighbor (the terminal nodes selected is one which is closest to that neighbor). For example, in Fig. 1, C sends a single lookup request message towards T_{R_3} instead of sending towards both T_{R_3} and T_{R_4}. Due to the nature of the Chord routing protocol, with high probability, the number of hops required to go from C to T_{R_3} is less than or equal to that required for going to T_{R_4}. Therefore, the number of terminal nodes that are contacted during a lookup process may be less than the total number of terminal nodes for a resource. However, for simplicity, we assume that all the terminal nodes for a resource are contacted and participate in the lookup process.

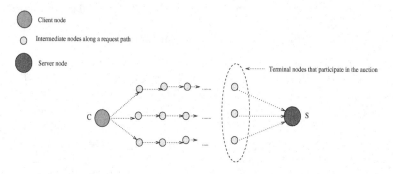

Fig. 2. Formation of request chains due to the propagation of lookup requests

Fig. 2 depicts an equivalent representation of Fig. 1 and shows different request chains that are formed due to the parallel lookup process. The request chain containing the highest bidder, i.e. the winning terminal node, is called the winning request chain (*WRC*). In subsequent discussion, we denote the highest and second highest bids by M_1 and M_2, respectively. The price offered by a terminal node to S is equal to $P_C - MC_{total}$. The amount of profit made by the *WRC* is equal to $M_1 - M_2$. This profit is shared among the nodes of the *WRC* in proportion to their marginal costs, i.e. nodes with higher marginal costs get a higher proportion of the total profit, and vice versa.

4.3   Secure Vickrey Auction to Determine Resource Prices

S employs a two-phase Vickrey auction to select the highest bidder and determine the price at which the resource is provided. In the first phase, the bidders send encrypted copies ($E(randKey_i; b_i)$) of their bids in message Msg_{bid} to S. Here $E(randKey_i; b_i)$ is the encryption of bid value b_i of terminal node T_{R_i} using a randomly chosen secret key $randKey_i$. Each message Msg_{bid} also includes $Reqid_{public}$ value received by a terminal node, so that S can determine that the bids pertain to the same lookup process. The received encrypted bids are sent by S back to all the bidders in message $Msg_{bid-reply}$. Since after receiving $Msg_{bid-reply}$, the bidders have encrypted copies of all the bids (total k such bids), S is unable to (undetectedly) alter existing or add fake bids.

In the next and last phase of the auction, each bidder after receiving the message $Msg_{bid-reply}$, sends its secret key in message Msg_{key} to S. The received key values are now sent by S back to all the bidders in message $Msg_{key-reply}$. At the end of this phase, S and all the bidders are able to open the encrypted bids and find out about the highest and second highest bids.

S then sends a message Msg_{cert} to the winning terminal node, denoted by T_{RWRC}, certifying that it has won the auction. The received certificate is forwarded along the reverse path, i.e. opposite to that followed by the lookup request, till it reaches C. C then finds out that the resource has been looked up and is available at a price within its initial offer of P_C. Msg_{cert} contains the following information - the highest bid M_1, the second highest bid M_2, the total marginal cost MC_{total} (received by S in Msg_{bid}), and the IP addresses and Chord IDs (each Chord ID is represented by m number of bits, which is also the size of the routing table in Chord) of all the terminal nodes that participated in the auction. The terminal nodes are contacted by C to verify the auction results, and the $Reqid_{private}$ values initially sent in the lookup messages are used to authenticate the terminal nodes. (Note that $Reqid_{private}$ values are not sent by the terminal nodes to S in Msg_{bid}). In [11] we provide a detailed analysis of the robustness of the proposed lookup protocol, including the various threat models it is designed to withstand.

The information in messages Msg_{cert} and Msg_{lookup} allow the intermediate nodes, including T_{RWRC}, to calculate their reward for being part of the WRC.[3] The knowledge of the auction result also enables C to determine the price it finally has to pay for R. The calculation of exact payoff values are discussed below.

4.4   Rewarding Nodes of the WRC

Msg_{cert} includes the total marginal cost value MC_{total} of all the nodes in the WRC. This information along with the highest and second highest bids determines each WRC node's payoff. For example, node x's payoff Pay_x is calculated as follows.

$$Pay_x = MC_x + (\frac{MC_x}{MC_{total}} * (M_1 - M_2)) \tag{1}$$

The amount received by S is equal to M_2 ($> MC_S$). The profit share of C, i.e. the portion of its initial offer that it saves or gets to keep, is similarly calculated as given below.

$$Profit_C = (\frac{MC_C}{MC_{total}} * (M_1 - M_2)) \tag{2}$$

C after keeping its profit share gives the remaining, i.e. $U_R^C - Profit_C$, to its successor. The successor node in turn after keeping its payoff gives the remainder to its successor, and so on.

5   Resource Index Replication

The proposed resource pricing scheme utilizing Vickrey auction is based on competition among different chains of nodes attempting to forward the lookup request and delivering the resource back to the client. Higher the competition among the nodes is (i.e. more

---

[3] The possession of messages Msg_{cert} and Msg_{lookup} serves as a contract between a node and its predecessor regarding the reward that it is entitled to receive (from the predecessor).

disjoint the request chains are), higher is the robustness of the pricing scheme. If normal Chord index replication is used, i.e. storing the resource index values at the k Chord successors of the ID where the resource hashes to, then with high probability lookups to all these replicas pass through a single (or a small group) node. Such a node can easily control the lookup process and charge arbitrarily high payoffs for forwarding the requests. To avoid such a monopolistic situation and ensure fair competition in setting prices, we propose that resource indices be replicated uniformly around the Chord network at equal distances from each other. In other words, resource Chord ID mappings should span the entire Chord ID space; this ensures that the lookup paths to different index replicas are maximally disjoint, and are not controlled by any single or a group of small nodes. Below we give a mechanism for determining the location for storing index replicas in the network.

If resource R hashes to Chord ID RI (i.e. the output of the hash function, whose input ID is R, is RI),[4] then the k resource index replicas map to the following Chord IDs.

$$RI_{R_i} = (RI + \frac{2^m}{k} * (i - 1))mod(2^m), \forall i \in \{1, \ldots, k\} \qquad (3)$$

The index values are then stored at the Chord successors (the terminal nodes) of the IDs represented by $RI_{R_i} \forall i \in \{1, \ldots, k\}$. The intent of replication in Chord is to simply obtain fault-tolerance, while in our protocol the intent is to obtain both fault-tolerance as well as fair pricing. Uniformly spacing the resource index ensures that the lookup paths for different index copies are as node disjoint as possible. This is evident from the results of Fig. 3, where we find that the replication strategy described above decreases the probability that the same nodes are included in multiple paths to reach the index replicas. Similar results would be obtained for a network of any size and any replication factor.

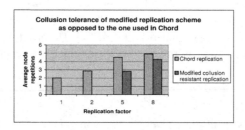

Fig. 3. Average number of repetitions of intermediate nodes that appear in multiple lookup paths to the resource index replica copies. Size of the network, $N = 500$

6   Selfish Network Topology

So far we have assumed that nodes form a Chord network and the lookup messages are forwarded in accordance with the Chord routing protocol. Now we investigate how

---

[4] The hash function used for computing resource Chord ID mapping is the same as that used for determining Chord IDs of the nodes.

correct is the assumption that nodes truthfully follow the Chord protocol. Since nodes are selfish and join a network in order to obtain resources and maximize their profits, the manner in which they select neighbors has a bearing on how successful they are in achieving these goals. This argument definitely holds true for our protocol, since intermediate nodes take their *cut* (equal to their marginal costs) before forwarding a lookup request. Thus, fewer intermediate nodes generally translate to higher profits for the client.

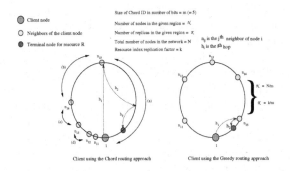

Fig. 4. Comparison of Chord and the greedy routing approach

A node can make higher profits by being close to terminal nodes of as many different resources that it requires as possible. However, if the location of those terminal nodes is not known beforehand, then it might be advantageous for a node to greedily choose neighbors around the Chord network distributed at equal distances from each other. This strategy seems appropriate, especially since the resource indices are also uniformly distributed around the network. Consider the network shown in Fig. 4 in which node *1* fill the m (= 5) entries in its routing table as per the greedy routing approach instead. So if node *1* needs to send a message to the terminal node for resource R, it can do it in two hops as opposed to three hops required using Chord. In general, one can see that in at least half of the cases, when the resource to be looked up has Chord ID mapping in region (*a*), the greedy routing scheme guarantees that the number of hops required to reach the corresponding terminal node is less than (or at most equal to) what is required by Chord. (This assumes that the network size, N, is large and the nodes are uniformly distributed around the Chord network.) Even for the other regions, the greedy approach appears to perform comparably to Chord. This is because node *1* always first sends a lookup message to its neighbor that is closest (and with lower Chord ID) to the resource Chord ID mapping, and from there on the message is routed using the Chord protocol.

From the above discussion it appears that nodes do not have motivation to follow the Chord protocol, and can make higher profits by utilizing the fact that other nodes follow Chord. In such a scenario the whole routing service would break down, i.e. instead of $O(logN)$ routing provided by Chord, $O(N/k)$ hops would be required due to the resulting sequential search for the resource indices. However, minimization of the number of routing hops by the greedy approach when everyone else follow Chord is not always correct. We prove below that in a large network, on average the performance of greedy routing approach is no better than that provided by Chord.

Theorem 1. *In a large network, on average the greedy routing approach for any node (say node 1 in Fig. 4) requires the same number of hops as that required by the Chord routing approach.*

Proof. For the ease of discussion, we assume that the routing table size is fixed for all the nodes in both the approaches and is equal to m (same as it is in Chord).

In Chord, the average number of hops required to reach any node from a given node is $1/2 * (log N)$. Therefore, the average number of hops required to reach any one of the given set of n nodes (with Chord IDs in the range $0 - 2^{m-1}$) is $\frac{1}{2} * log(\frac{N/2}{n/2}) = \frac{1}{2} * log(\frac{N}{n})$. These n nodes are assumed to be located at equal distances from each other. Using these results we obtain the following values.

Average number of hops required by the greedy routing approach to reach one of the k terminal nodes: The neighbors (i.e. the entries in the routing table) of node *1* in this case completely span the entire Chord ID space, i.e. they are uniformly located at equal distances from each other around the Chord network. Therefore, node *1* requires the same number of average hops to reach any of the terminal nodes. (The client first sends the request to its neighbor, which then follows the Chord routing protocol to further route the request.)

Thus, the average number of hops taken by the greedy routing approach to reach any resource index replica are given as follows.

$$(1 + \frac{1}{2} * log(\frac{N}{k})) \tag{4}$$

Average number of hops required by the Chord routing protocol to reach one of the k terminal nodes: Now we calculate the average number of hops required to reach a resource index replica when node *1* (and everyone else) follows the Chord protocol. It will be $(1 + \frac{1}{2} * log(\frac{N}{k}))$ when the terminal node is located in region (a), $(1 + \frac{1}{2} * log(\frac{N}{k}))$ when in region (b)[5], and so on (up to m such terms).

Therefore, the total average hops needed to reach any of the resource index replicas are given as follows:

$$(1 + \frac{1}{2} * log(\frac{N}{k})), \tag{5}$$

which is same as the number of hops given by Equation 4.

The results in Fig. 5 from our simulations also confirms the fact that a node cannot benefit by selecting the greedy routing strategy. Fig. 5 gives the difference in the observed number of hops when greedy routing is used as opposed to normal Chord routing. We did simulations for varying number of total nodes and averaged the number of hops for several lookups performed for each network size. As can be seen, there is a difference of at most one hop in the two routing strategies and this is true for both small as well as large network sizes. Thus, a node does not gain an advantage by not following Chord.

The following lemma follows directly from this result.

Lemma 1. *If others in a large network follow the Chord protocol, it is a good strategy to do the same in order to maximize one's payoff.*

---

[5] $(1 + \frac{.}{.} * log(\frac{N/.}{k/.}))$.

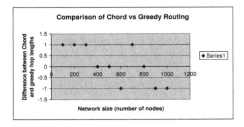

Fig. 5. Comparison of Chord and the greedy routing approaches with regards to the hop-length

7   Conclusion

In this paper we have presented an *incentive driven lookup* protocol for searching and trading resources in Chord-based P2P networks. Our proposed protocol takes selfish behavior of network nodes into account. We used Vickrey auction for setting resource prices in P2P networks and described how it can implemented in a completely distributed and untrusted environment.

We also investigated the applicability of Chord network topology in forming connectivity among selfish nodes. We proved that in the absence of privilege network information, the best strategy for a node is to follow Chord, provided that everyone else also follow the Chord protocol.

References

1. I. Stoica, R. Morris, D. Karger, M. F. Kaashoek, and H. Balakrishnan. Chord: A Scalable Peer-to-peer Lookup Protocol for Internet Applications. *In Proceedings of the 2001 ACM SIGCOMM Conference*, 2001.
2. A. Eytan, and A. H. Bernardo. Free Riding on Gnutella. First Monday, vol. 5, No. 10, Oct. 2000.
3. N. Nisan. Algorithms for Selfish Agents: Mechanism Design for Distributed Computation. *In Proceedings of the 16th Symposium on Theoretical Aspects of Computer Science, Lecture Notes in Computer Science, volume 1563, Springer, Berlin, pages 1-17*, 1999.
4. D. Geels, and J. Kubiatowicz. Replica Management Should Be A Game. *In Proceedings of the SIGOPS European Workshop*, 2002.
5. V. Vishumurthy, S. Chandrakumar, and E. G. Sirer. KARMA: A Secure Economic Framework for Peer-to-Peer Resource Sharing. *In Proceedings of the 2003 Workshop on Economics of Peer-to-Peer Systems, Berkeley CA*, 2003.
6. S. M. Lui, K. R. Lang, and S. H. Kwok. Participation Incentive Mechanism in Peer-to-Peer Subscription Systems. *In Proceedings of the 35th Annual Hawaii International Conference on System Sciences (HICSS'02), vol. 9*, January 2002.
7. S. Zhong, J. Chen, and Y. R. Yang. Sprite: A simple, cheat-proof, credit-based system for mobile ad-hoc networks. *In Proc. of IEEE INFOCOM, vol. 3, pages 1987-1997*, March 2003.
8. L. Buttyan and J. P. Hubaux. Stimulating cooperation in self-organizing mobile ad-hoc networks. *In Proc. of ACM Journal for Mobile Networks (MONET), special issue on Mobile Ad Hoc Networks*, summer 2002.

9. W. Vickrey. Counterspeculation, auctions and competitive sealed tenders. *Journal of Finance, pages 8-37*, 1961.
10. T. Sandholm. Limitations of the Vickrey Auction in Computational Multiagent Systems. *In Proceedings of the 2nd International conference on Multi-Agent Systems, pages 299-306. Kyoto, Japan*, December 1996.
11. Technical   Report.   http://ecpe.ee.iastate.edu/dcnl/DCNLWEB/Publications/pub-research_tech-rep.htm

A Novel Battery Aware MAC Protocol for Ad Hoc Wireless Networks

S. Jayashree, B. S. Manoj, and C. Siva Ram Murthy*

Department of Computer Science and Engineering,
Indian Institute of Technology, Madras, India
{sjaya, bsmanoj}@cs.iitm.ernet.in, murthy@iitm.ernet.in

Abstract. A major issue in the energy-constrained ad hoc wireless networks is to find ways that increase their lifetime. The communication protocols need to be designed such that they are aware of the battery state. They should also consider the presence of varying sizes and models of the nodes' batteries. Traditional MAC protocols of these networks are designed considering neither the state nor the characteristics of the batteries of the nodes. Major contributions of this paper are: (a) a novel distributed heterogeneous Battery Aware MAC (HBAMAC) protocol, that takes benefit of the chemical properties of the batteries and their characteristics, to provide fair node scheduling and increased network and node lifetime through uniform discharge of batteries, (b) theoretical analysis, using discrete time Markov chains, for batteries of the nodes, and (c) a thorough comparative study of our protocol with IEEE 802.11 and DWOP (Distributed Wireless Ordering Protocol) MAC protocols.

1 Introduction

The nodes of an ad hoc wireless network, a group of uncoordinated heterogeneous nodes (Heterogeneous nodes, in this paper, refer to a set of nodes with varying battery models) that self organize themselves to form a network, have constrained battery resources. In addition, in the case of heterogeneous networks, traditional protocols perform poorly as they do not consider the underlying behavior of the batteries. In such scenarios, there exists a need for battery (energy) aware protocols at all the layers of the protocol stack. On the other hand, ad hoc wireless networks, with characteristics such as the lack of a central coordinator and mobility of the nodes (as in the case of battle-field networks), require nodes with a very high energy reserve. However, advances in the battery technologies are negligible when compared to the recent advances that have taken place in the field of mobile computing and communication. The increasing gap between power consumption requirements and energy density (energy stored per unit weight of a battery) tends to increase the size of the batteries and hence increases the need for energy management (energy-aware design of ad hoc networks and their protocols, which efficiently utilize the battery charge of the nodes and minimize the energy consumption). The lifetime

---

* Author for correspondence. This work was supported by the Department of Science and Technology, New Delhi, India.

L. Bougé and V.K. Prasanna (Eds.): HiPC 2004, LNCS 3296, pp. 19–29, 2004.
© Springer-Verlag Berlin Heidelberg 2004

of ad hoc networks can be defined as the time between the start of the network (when the network becomes operational) to the death of the first node. We use this definition because in ad hoc networks, the death of even a single node may lead to partitioning of the network and hence may terminate many of the ongoing transmissions.

The rest of the paper is organized as follows. Section 2 provides an overview of battery characteristics and existing work in this area. In Section 3, the description of the proposed HBAMAC protocol is provided and Section 4 presents a theoretical analysis for the batteries. In Section 5, we present the simulation results and a comparative study of theoretical and simulation results. Finally, Section 6 summarizes the paper.

2   Overview of Battery Characteristics and Related Work

A battery consists of an array of one or more electro-chemical cells. It can be characterized either by its voltages (open circuit, operating, and cut-off voltages) or by its initial and remaining capacities. The behavior of the batteries is governed by the following two major chemical effects.

Rate and Recovery Capacity Effects: As the intensity of the discharge current increases, an insoluble component develops between the inner and outer surfaces of the cathode of the batteries. The inner surface becomes inaccessible as a result of this phenomenon, rendering the cell unusable even while a sizable amount of active materials still exists. This effect termed as the rate capacity effect depends on the actual capacity of the cell and the discharge current. The recovery capacity effect is concerned with the recovery of charges under idle conditions. Due to this effect, on increasing the idle time of the batteries, one may be able to completely utilize the theoretical capacity of the batteries.

Battery Capacities: The amount of active materials (the materials that react chemically to produce electrical energy when battery is discharged and restored when battery is charged) contained in the battery refers to its theoretical capacity (T) and hence total number of such discharges cannot exceed the battery's theoretical capacity. Whenever the battery discharges, the theoretical capacity of the battery decreases. Nominal (standard) capacity (N) corresponds to the capacity actually available when the battery is discharged at a specific constant current. Whenever the battery discharges, nominal capacity decreases, and increases probabilistically as the battery remains idle (also called as recovery state of the battery). This is due to the recovery capacity effect. The energy delivered under a given load is said to be the actual capacity of the battery. A battery may exceed the actual capacity but not the theoretical capacity. This is due to the rate capacity effect. By increasing the idle time one may be able to utilize the maximum capacity of the battery. The lifetime of a battery is the same as its actual capacity.

Battery Models: Battery models depict the characteristics of the batteries used in real life. The following battery models are discussed in [1]. The authors of [1] summarize the pros and cons of each of them: Analytical models, Stochastic models, Electric circuit models, and Electro-chemical models.

Recent work in [2] and [3] suggest that a proper selection of power levels for nodes in an ad hoc wireless network leads to power saving. The authors of [4] have shown that the pulsed discharge current applied for bursty stochastic transmissions improves the battery lifetime than the constant discharge. The authors of [5] have assumed each node to contain a battery pack with L cells and have proposed three battery scheduling policies for scheduling them. In [6], Kanodia *et al.* proposed DWOP, a MAC scheduling protocol which tries to provide, for the nodes of the network, a fair share access to the channel. Since this protocol schedules the nodes in a *round-robin* fashion, it introduces indirectly a uniform discharge of the batteries for their nodes. However, the authors do not consider the state of the battery into the scheduling process. In our earlier work [7], we have proposed a battery aware MAC protocol, BAMAC(k), for homogeneous ad hoc wireless networks, which uses a round-robin scheduling of the nodes to attain a uniform discharge of their batteries and hence an increased lifetime.

3   Our Work

In this paper, we propose a novel heterogeneous MAC protocol, HBAMAC, that tries to utilize the battery in an efficient manner. We also show how battery awareness influences throughput, fairness and other factors that describe the performance of the network. Existing MAC protocols do not consider the nodes' battery state in their design. While our earlier work in [7] provides a battery aware MAC protocol for homogeneous ad hoc networks, to the best of our knowledge, there has been no such reported work till date for the heterogeneous ad hoc wireless networks. In this paper, we propose a novel distributed battery aware MAC scheduling scheme, where we consider nodes of the network, contending for the common channel, as a set of heterogeneous batteries and schedule the nodes in order to provide a uniform discharge of their batteries. The key idea behind our protocol lies in calculating the back-off period for the contending nodes which can be stated as follows: "higher the remaining battery capacity, lower the back-off period". That is, a node with higher remaining battery charge backs-off for a longer duration than the one with lower battery charge. This ensures a uniform discharge of batteries. The basic principle behind HBAMAC protocol is to provide proportionally more number of recovery slots (the minimum amount of time required for an idle battery to recover one charge) for the nodes with lesser battery recovery capacity effect than those with higher recovery capacity. We now discuss this protocol in detail.

3.1   HBAMAC

In real life, end users (mobile nodes) of the ad hoc wireless network may have heterogeneous nodes, which may vary in voltages, recovery capacity effect, current rating, lifetime, operational environment, and weight. We model the heterogeneity based on voltage and recovery capacities. Traditional MAC protocols which assume the nodes to behave in a homogeneous manner may perform poorly in the presence of heterogeneous nodes. We, in this paper, propose HBAMAC protocol which extends the basic BAMAC(k) protocol [7] to work in the presence of heterogeneous nodes. As explained earlier, on increasing the idle time of the battery, the whole of its theoretical capacity can be completely utilized. Equation 1 (explained later in this section) clearly shows that this effect will be higher when the battery has higher remaining capacity and decreases with decrease in the remaining battery capacity. Thus, HBAMAC protocol tries to provide enough idle time for the nodes of ad hoc wireless networks by scheduling the nodes in an appropriate manner. It tries to provide uniform discharge of the batteries of the nodes that contend for the common channel. This can be effected by using a round-robin scheduling (or fair-share scheduling) of these nodes if the nodes are homogeneous. However, in the case of heterogeneous battery scheduling, a different strategy has to be adopted to attain a uniform discharge of batteries. We now give a brief description of the BAMAC(k) protocol and the variations introduced in HBAMAC.

In BAMAC(k) protocol, to attain a round-robin scheduling of the nodes in a distributed manner, each node maintains a battery table that contains information about the remaining battery charge of each of its two-hop neighbor nodes. The entries in the table are arranged in the non-increasing order of the remaining battery charges. The RTS, CTS, Data, and ACK packets carry the following information: remaining theoretical (in terms of remaining battery voltage) and nominal capacities of the battery and the time of last usage of the battery (the time at which the battery underwent its last discharge) of the node that originated the packet. The neighbor nodes, on listening to these packets, make a corresponding entry in their battery table. The objective of the back-off mechanism used in BAMAC protocol is to provide a near round-robin scheduling of the nodes. The back-off period is given by

$$back-off = Uniform[0, (2^x \times CW_{min}) - 1] \times rank \times (T_t + T_{SIFS} + T_{DIFS}).$$

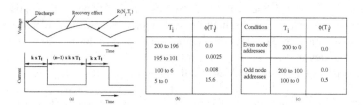

Fig. 1. Illustration of battery discharge of nodes using BAMAC

where CW_{min} is the minimum size of the contention window and $rank$ is the position of that entry in the battery table of the nodes which are arranged based on the following rule: "The battery table is arranged in descending order of its theoretical capacity of the nodes. Any tie, that arises, is broken by choosing the one with higher nominal capacity and then by choosing the one with least value for the time of last usage. Further ties are broken randomly". T_{SIFS} and T_{DIFS} represent the short inter-frame spacing and DCF inter-frame spacing durations, respectively. Their values are the same as those used in IEEE 802.11. T_t is the longest possible time required to transmit a packet successfully, including the RTS-CTS-Data-ACK handshake. The nodes follow the back-off even for the retransmission of the packets. When this back-off scheme is followed, nodes with lesser $rank$ values back-off for smaller time durations compared to those with higher $ranks$. $Uniform[0, (2^x \times CW_{min}) - 1]$ returns a random number distributed uniformly in the range 0 and $(2^x \times CW_{min} - 1)$, where x is the number of transmission attempts made so far for a packet. Thus the nodes are scheduled based on their remaining battery capacities. The higher the remaining battery capacity, the lower the back-off period and vice versa. Hence, a uniform rate of battery discharge is guaranteed across all the nodes. This provides a discharge time of $k \times T_t$ and an average recovery time of $(n-1) \times k \times T_t$ for the nodes as shown in Figure 1(a), where n is the number of nodes and k is the number of packets transmitted continuously by a node on gaining access to the channel. In each idle/recovery slot, the battery recovers one charge unit with a probability $R_{i,j}$ (explained in Section 4). This improves the lifetime of the battery as it gains more idle time to recover charge because of the recovery capacity effect. In the BAMAC(k) protocol, whenever the node attempts to gain access to the channel, it waits for DIFS duration before transmitting the first packet. If no other neighbor transmits in this duration, the active node (the node that gains access to the channel) initiates its transmission. For transmitting each of the next $k-1$ packets, it waits only for an SIFS duration; since the channel remains idle during this SIFS duration, the active node proceeds with the transmission of the packet. This ensures a continuous packet transmission for a node.

As mentioned earlier, in BAMAC(k) protocol, maximum lifetime of a network can be achieved when a round-robin scheduling of nodes is carried out. This actually results in a uniform discharge of batteries. However, the same round-robin scheduling may not be directly applicable in the case of ad hoc networks with heterogeneous nodes. For example, let us consider a network with n nodes and node A amongst the n nodes be of a different battery model. Let us assume its battery does not recover the charges as quickly as the rest of the $n-1$ nodes in the network. If an ideal round robin scheduling of nodes is used, all the nodes of the network irrespective of their battery characteristics would be given same number of recovery slots. When $n-1$ nodes of the n nodes recover more charges in the recovery slot time they gain, node A would recover only fewer charges due to its lesser battery recovery capacity effect and thus dies sooner. For a uniform discharge of batteries, node A should gain access to the channel less frequently compared to others.

The basic principle behind HBAMAC algorithm is to give proportionally higher number of recovery slots for the nodes with lesser battery recovery capacity effect than those with higher recovery capacity. HBAMAC uses the same structure for the battery table and back-off function as that of BAMAC(k) protocol. In addition, it uses the same fields for control and data packets as that of BAMAC(k). In HBAMAC, the battery table is always arranged in the decreasing order of the remaining battery capacities of the nodes as in BAMAC(k). The difference between these protocols lies in the values of $\phi(T_i)$ in Equation 1. The values we have assumed in our protocol are shown in Figure 1(c). We have assumed $k=1$ in the case of HBAMAC protocol for simplicity. This assumption was taken from the research work of Adamou and Sarkar in [8]. However, the probability of recovery also depends on the remaining nominal capacity of the batteries. As the ϕ value decreases, the probability of recovery of a charge increases. Unlike BAMAC(1) protocol [7], wherein all the nodes uniformly increase in their position in the table by one for every recovery slot, in HBAMAC, the arrangement of the nodes in the battery table occurs such that, the nodes with higher recovery probabilities will have lesser $rank$ value than those with lower recovery probabilities. Hence, the nodes with higher recovery probabilities have more transmissions than recoveries. Whereas the nodes with lesser probability for recovery have more recovery slots than transmission slots. This ensures a uniform discharge of batteries and hence maximum lifetime for the network. We have provided the theoretical analysis for HBAMAC protocol for a network with two kinds of battery models.

4   Modeling the Batteries Using Discrete-Time Markovian Chains

The behavior of the batteries of the nodes, which uses HBAMAC protocol for transmission, is represented using a Markov model as shown in Figure 2. The state of the battery in the Markov model represents the remaining nominal capacity of the battery. Hence, the battery can be in any of the states from 0 to N, where N is the maximum nominal capacity of the battery. The battery model assumes that, in any Δt time unit, the battery can remain in any one of the two main states – transmission state (Tx) or the reception state (Rx), where Δt is the sum of the average back-off value and the time taken for one packet transmission. In each time unit Δt, if a node remains in the Tx state, it transmits a packet and the battery discharges two units of its charge; if it remains in the Rx state, the neighbor nodes transmit and if it does not receive any packets, the battery recovers one unit of the charge with probability R_{N_i,T_i}, where R_{N_i,T_i} is given by

$$R_{N_i,T_i} = \begin{cases} e^{-g \times (N-N_i) - \phi(T_i)} & \text{if } 1 \leq N_i \leq N, 1 \leq T_i \leq T, \\ 0 & : \text{otherwise} . \end{cases} \qquad (1)$$

where g is a constant value and $\phi(T_i)$ is a piecewise constant function of number of charge units delivered which are specific to the battery's chemical properties. An example of the values of the piecewise constant function $\phi(T_i)$ is shown in Figure 1(b). This value affects the battery recovery drastically. If the battery receives a packet in the Rx-state, it discharges one unit of its charge. In the model shown above, Rx_{ij} (Tx_{ij}) represents the battery in the Rx (Tx) state at time unit i and j represents the remaining nominal capacity of the battery. Rx_{I0} and Tx_{I0} represent the battery in its dead (absorbing) state with nominal capacity 0 at any time unit I. Here, we assume Δt as one basic time unit and one cycle time as the time between two successive entries in to Tx state or Rx state. State of a battery is denoted by the tuple $< N_i, T_i >$ and the initial state is given by the tuple $< N, T >$. In Δt, a battery which is in state $< N_i, T_i >$, goes to state $< N_{i-1}, T_{i-1} >$ if it is in Rx state and receives a packet. If the battery remains idle in Rx state, it reaches $< N_{i+1}, T_i >$ or $< N_i, T_i >$ with probabilities R_{N_i,T_i} and I_{N_i,T_i} respectively, where the probability to remain in the state on being idle is given by $I_{N_i,T_i} = 1 - R_{N_i,T_i}$. Hence, the battery can be modeled differently in each of these two states and the battery flip-flops between these two states. Let us consider an ad hoc network with $n + 1$ nodes. Each node alternates between periods of recovery and discharge. Let us assume that x out of n nodes be of battery model A and the remaining $n - x$ be of battery model B. Let the batteries that belong to model A be y times more efficient than those of model B. In other words, for recovery of a charge, batteries of model A require at least one recovery slot, whereas those of model B require at least y slots. This can be explained using an example. Let, node A with battery model A and another node B with battery model B have T_i, remaining theoretical capacity value, 50. Let N and N_i of these batteries be 25 and 10, respectively. If each of these nodes gets one recovery slot, the probability of

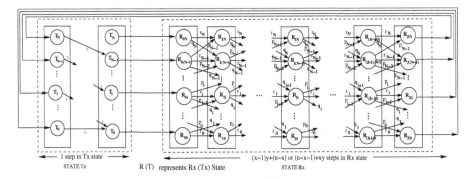

Fig. 2. Discrete-time Markov model representing battery states for heterogeneous nodes

recovery for these two nodes, which can be calculated from Equation 1, is found to be 0.4724 and 0.2865, respectively. This implies that nodes with battery model B require twice ($y = 2$) the number of recovery slots as that of model A. Hence, in each cycle, nodes using batteries of model A should be made to gain access to the channel y times more often than those of model B. This is ensured by the battery table arrangement and the equation for the recovery probability. Now, for each node with battery model A, the number of recovery and discharge slots can be calculated as follows. Any node containing battery model A will have $x - 1$ nodes of its own type and the remaining $n - x$ of type B. When $k = 1$, the total recovery time for nodes with battery model A is $(x - 1)y + (n - x)$ and that of battery model B is $(n - x - 1) + xy$ recovery slots. The discharge time for all the nodes remains one slot. Thus, the probability matrix for nodes with battery model A is given by, $P_A = Tx \times Rx = Trans \times Rec^{(x-1)y+(n-x)}$ and for battery model B is $P_B = Trans \times Rec^{(n-x-1)+xy}$ [7], where $Trans$ and Rec represent the probability matrices for one basic time unit of the battery in Tx and Rx states, respectively. We assume the matrix index to start from 0 for ease of denoting the 0th state or the dead state.

$$P = \begin{bmatrix} 1 & 0 & 0 & \ldots & 0 & 0 \\ 1 & 0 & 0 & \ldots & 0 & 0 \\ 1 & 0 & 0 & \ldots & 0 & 0 \\ 0 & 1 & 0 & \ldots & 0 & 0 \\ \vdots & \vdots & \vdots & & \vdots & \vdots \\ \cdot & \cdot & \cdot & \ldots & \cdot & \cdot \\ \cdot & \cdot & \cdot & & \cdot & \cdot \\ 0 & \ldots & 1 & 0 & 0 & 0 \\ 0 & \ldots & 0 & 1 & 0 & 0 \end{bmatrix} \times \begin{bmatrix} 1 & 0 & 0 & 0 & \ldots & 0 & 0 \\ q_1 & r_1 & p_1 & 0 & \ldots & 0 & 0 \\ 0 & q_2 & r_2 & p_2 & 0 & \ldots & 0 \\ \vdots & \vdots & \vdots & \vdots & & \vdots & \vdots \\ \cdot & \cdot & \cdot & \cdot & \ldots & \cdot & \cdot \\ 0 & 0 & \ldots & q_{N-2} & r_{N-2} & p_{N-2} & 0 \\ 0 & 0 & \ldots & q_{N-1} & r_{N-1} & p_{N-1} & \\ 0 & 0 & 0 & \ldots & 0 & q_N & r_N \end{bmatrix}^{\mathbf{Z}}$$

$$Z = \begin{cases} (x - 1)y + (n - x) & for\ model\ A \\ (n - x - 1) + xy & for\ model\ B \end{cases}$$

The probability value $P_{i,j}$ refers to the probability that the battery goes to state j from i. We can now calculate the maximum number of packets transmitted by the nodes with heterogeneous batteries, by substituting the corresponding values of probability matrices (P_A and P_B) in the place of P in the following. (a) Given any probability matrix P, matrix $Q=[Q_{(i,j)}]$, where i and j represent only the transient states, is calculated. Hence, in our protocol, $Q_{(i,j)} = P_{(i+1,j+1)}$. (b) Matrix $M = (I - Q)^{-1}$, where I is the identity matrix, is calculated. (c) Now $M_{(i,j)} \times \Delta t$ represents the total number of times the battery enters state j if the starting state is i and Δt is the time duration the Markov model spends in state j once it enters it. That is matrix $Q=[Q_{(i,j)}]$ is calculated from P_A (P_B) for nodes with battery model A (B), where i and j represent only the transient states. Hence, in our protocol, $Q_{(i,j)} = P_{(i+1,j+1)}$. Now, $M = (I - Q)^{-1}$, where $M_{(i,j)} \times \Delta t$ represents the total number of times the battery enters state j if the starting state is i. Let, T_{active} of a battery model give the total active time of batteries. Now the time for which battery remains active or the lifetime of the battery can be given by, $T_{active} = \sum_{i=1}^{i=N} M_{N,i}$. This is nothing but the total number of transitions (left, right, and stationary) in the model from the starting state N till the battery reaches state 0. The left, right and stationary transitions denote the battery state as being – discharge, recovery and idle (remains in the same state after recovery) respectively. T_{active} is the sum of the elements present in the Nth row of the matrix M, which is equal to the number of times battery enters into states 1 to N if the starting state is N. The number of left transitions in the model represents the actual number of discharges. Hence, the total number of packets transmitted is the total number of left transitions by a battery starting from state N is given by, $T_{left} = \sum_{i=1}^{N} \times \sum_{j=1}^{i-1} P_{i,j}$, here $P_{i,j}$ corresponds to the entry at the ith row and jth column of matrix P. Thus, we can calculate the total number of packets transmitted by a node given its battery model and the probability matrix using the above equation.

5   Performance Analysis

The proposed HBAMAC protocol was implemented using GloMoSim simulator. All the nodes send packets with same transmission power. The routing protocol used was Dynamic Source Routing (DSR) protocol and the channel bandwidth was assumed to be 2 Mbps. Results were derived by simulation using the following values for the parameters: $T=2000$, $N=250$, $g=0.05$. We have compared our protocol with the IEEE 802.11 and DWOP [6] (see Section 2) MAC protocols. In the following discussion, capacity refers to nominal battery capacity unless otherwise specified. We have assumed a data packet size of 512 bytes. We make an assumption that receiving takes half the amount of power spent as that of transmission. We, at this point of time, neglect the power spent by the nodes for control packets (RTS, CTS, Data, and ACK) transmission and reception. We assume that listening to the channel consumes negligible amount of power and if a battery idles for one unit of time (recovery slot time), that is, if the node neither transmits nor receives a packet in one time slot, it recovers one unit of charge with probability $R_{i,j}$. It is assumed to be equal to the sum of the transmission time slot and the average back-off value for the node. We assume the availability of a small alternate battery to power up the electronic components of the node while the node resides in the idle mode. Since, the power required by these electronic components is very small compared to the power spent in transmission and reception, we do not consider the effect of this battery on the nodes' lifetime calculation. We make use of the same basic assumptions as in BAMAC(k) protocol. Here, we assume the value of $\phi(T_i)$ to be varying as a piecewise constant function as shown in Figure 1(c). That is, half of the nodes are assumed to have battery model A and the rest of model B.

Figure 3 shows the total number of packets transmitted by the nodes using HBA-MAC, IEEE 802.11, and DWOP protocols. Here, HBAMAC transmits more packets than IEEE 802.11 and DWOP protocols. Unlike the homogeneous case, here we find that DWOP transmits lesser number of packets than IEEE 802.11 MAC protocol. This is because DWOP tries to establish a round-robin scheduling of the flows, which ultimately fails in the presence of heterogeneous nodes. Figure 4 shows the total remaining battery capacity of all the nodes in the network (either transmitter or receiver). HBAMAC shows

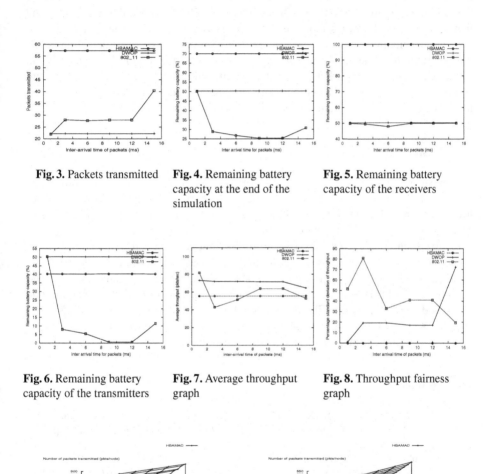

Fig. 3. Packets transmitted

Fig. 4. Remaining battery capacity at the end of the simulation

Fig. 5. Remaining battery capacity of the receivers

Fig. 6. Remaining battery capacity of the transmitters

Fig. 7. Average throughput graph

Fig. 8. Throughput fairness graph

Fig. 9. Theoretical results with varying N and n

Fig. 10. Simulated results with varying N and n

better performance even in terms of the remaining battery capacity. Though nodes of the network using DWOP protocol have higher remaining battery capacity, the reason behind transmitting lesser number of packets is due to the non-uniform discharge of batteries, which can be observed from Figures 5 and 6 which show the remaining battery capacity of the receivers and transmitters, respectively. Unlike IEEE 802.11, where nodes try to transmit in a greedy manner, in DWOP, the nodes use a fair scheduling for transmission of their packets. Due to the presence of heterogeneity, nodes gain a non-uniform discharge of their batteries. Figure 7 shows the throughput achieved by the nodes. As shown in the figure, throughput achieved remains higher for DWOP compared to HBAMAC and that of IEEE 802.11 varies randomly with inter-arrival time of packets. Though, the throughput remains less in the case of HBAMAC protocol, standard deviation of the throughput across the nodes also remains less as shown in Figure 8. Figures 9 and 10 show the number of packets transmitted calculated through theoretical analysis and simulations, respectively. The graphs show that there exists some discrepancy between them, which is due to the random nature of back-off values.

6 Summary

In this paper, we proposed a novel energy efficient heterogeneous battery aware MAC protocol (HBAMAC), the main aims of which were minimal power consumption, longer life, and fairness for the nodes of ad hoc wireless network. Traditional ad hoc wireless MAC protocols are designed without considering the battery state and their character-istics. We found that even those MAC protocols, that perform better in the presence of homogeneous nodes, performs worser for heterogeneous networks. We, thus, provided a MAC protocol to work in the presence of nodes with heterogeneous batteries. We found that our protocol performs better, in terms of power consumption, fairness, and lifetime of the nodes, compared to IEEE 802.11 and DWOP MAC protocols. A discrete Markov chain was used to theoretically analyze the batteries of the nodes, the correctness of which was verified through extensive simulation studies.

References

1. K. Lahiri, A. Raghunathan, S. Dey, and D. Panigrahi, "Battery-Driven System Design: A New Frontier in Low Power Design", *Proceedings of IEEE ASP-DAC/VLSI Design 2002*, pp. 261-267, January 2002.
2. R. Wattenhofer, L. Li, P. Bahl, and Y. M. Wang, "Distributed Topology Control for Power-Efficient Operation in Multi-Hop Wireless Ad Hoc Networks," *Proceedings of IEEE INFOCOM 2002*, vol. 3, pp. 1388-1397, April 2001.
3. R. Zheng, J. C. Hou, and L. Sha, "Asynchronous Wakeup for Ad Hoc Networks," *Proceedings of ACM MOBIHOC 2003*, pp. 35-45, June 2003.
4. C. F. Chiasserini and R. R. Rao, "Pulsed Battery Discharge in Communication Devices", *Proceedings of ACM MOBICOM 1999*, pp. 88-95, November 1999.
5. C. F. Chiasserini and R. R. Rao, "Energy Efficient Battery Management", *Proceedings of IEEE INFOCOM 2000*, vol. 2, pp. 396-403, March 2000.

6. V. Kanodia, C. Li, A. Sabharwal, B. Sadeghi, and E. Knightly, "Ordered Packet Scheduling in Wireless Ad Hoc Networks: Mechanisms and Performance Analysis", *Proceedings of ACM MOBIHOC 2002*, pp. 58-70, July 2002.
7. S. Jayashree, B.S. Manoj, C. S. R. Murthy, "On Using Battery State for Ad hoc Wireless Networks," *Proceedings of ACM MOBICOM 2004*, September 2004.
8. M. Adamou and S. Sarkar, "A Framework for Optimal Battery Management for Wireless Nodes", *Proceedings of IEEE INFOCOM 2002*, vol. 3, pp. 1783-1792, June 2002.

Dynamic Topology Construction
in Bluetooth Scatternets

Rajarshi Roy[1], Mukesh Kumar[4], Navin K. Sharma[2], and Shamik Sural[3]

[1] Dept. of Electronics & Electrical Communication Engineering,
[2] Dept. of Civil Engineering,
[3] School of IT, Indian Institute of Technology, Kharagpur, 721302, India
{royr@ece, shamik@cse}.iitkgp.ernet.in
[4] Dept of CSE, Institute of Technology, Banaras Hindu University, Varanasi, India

Abstract. We propose a topology construction algorithm for Bluetooth devices in which a piconet or a scatternet is assumed to be a dynamic network structure. A new node can join the scatternet or an existing node can shut down at any point of time, without affecting the rest of the topology. The algorithm works bottom up by first combining individual nodes into small piconets, the small piconets then joining to form larger piconets. Finally, the large piconets join to form a scatternet. All these operations are conducted in a way such that none of the nodes need to act as a coordinator. No a priori information about the total number of nodes in the network is required for successful topology construction.

1 Introduction

After power up, a new Bluetooth node discovers other nodes by synchronizing the frequency hopping patterns before it is able to exchange data or voice packets [1]. An important step in the successful operation of a Bluetooth network (either a single piconet or a scatternet) is the formation of a network topology – a configuration in which each node in a piconet is synchronized with its master and multiple piconets are connected by bridge nodes. Bluetooth nodes are symmetric so that any node can act as a master or as a slave [2,3].

One of the early attempts at connected topology construction in Bluetooth was made by Salonidis et al [4]. They introduce a scatternet formation algorithm—Bluetooth Topology Construction Protocol (BTCP). Miklos et al suggest the use of a number of heuristics to generate scatternets [5]. Tan [6] proposes a distributed Tree Scatternet Formation (TSF) protocol. Law et al [7] present a Bluetooth scatternet formation protocol in which they try to optimize some of the quality measures, namely, the number of piconets, maximum degree of the nodes and the network diameter. Basagni and Petrioli propose a three-phase scatternet formation protocol [8] in which a "Bluestar" piconet is initially formed. Multiple Bluestar piconets then join

L. Bougé and V.K. Prasanna (Eds.): HiPC 2004, LNCS 3296, pp. 30–39, 2004.
© Springer-Verlag Berlin Heidelberg 2004

together to form BlueConstellations. Yun et al [9] also propose a three-phase protocol for the construction of their version of Bluestar. Basagni et al [10] compare the scatternet formation protocols for networks of Bluetooth nodes – Bluetree, Bluenet and the Yao protocol. Guerin et al [11] investigate the problem from an algorithmic perspective.

A careful analysis of the above-mentioned research works reveals a few commonalities among the various approaches. These include (i) Assumption of a leader election process [4,6] (ii) Topology optimization starting with a fixed set of Bluetooth nodes [4,9,12] (iii) Deferring the problem of topology "reconstruction" as a future extension [4,8,11] and most importantly (iv) Approach the topology construction problem as a stand-alone problem and not as an outcome of specified properties of Bluetooth nodes [5,7-9,11,12]. We approach the topology construction problem from a different perspective. Instead of considering it as a one-time, stand-alone requirement, we integrate topology formation specific properties with the normal operations of Bluetooth nodes. It should be kept in mind that the Bluetooth nodes are power-constrained devices due to their portability. Hence, we cannot possibly run elaborate optimization algorithms like that proposed by Marsan et al [12] on such nodes in practical applications. We feel that it is more practical to develop an engineering solution for dynamic scatternet formation. To achieve this, we propose a Bluetooth topology construction protocol, which is simple, implementable and does not alter the baseband specifications.

In the next section, we present the dynamic topology construction scheme. This is followed by simulation results in section 3 and we conclude in section 4 of the paper.

2 Dynamic Topology Construction Algorithm

The proposed algorithm attempts to form a fully connected and balanced network. By fully connected we mean a scatternet configuration in which each piconet is connected to each other piconet by a bridge. It is evident that if the number of nodes is high, we cannot obtain a fully connected scatternet. Such a situation is quite rare in practice and if it happens, we try to minimize the number of hops to connect one piconet to another. A balanced network is one in which each piconet has similar number of nodes. In the rest of the paper, we use the term optimum to denote a scatternet configuration in which piconets are either fully connected or are connected through minimum number of hops and the scatternet is balanced. However, with every perturbation of an existing network, we do not try to minimize the number of piconets immediately since this leads to high message complexity and device power wastage. Instead, the topology keeps getting modified leading towards an optimum scatternet.

In our approach to topology construction, we consider the situation in which any node can go to the ON state at any instant of time. We use the term ON state in a broader sense to mean that a node is within radio proximity of a piconet and it tries to communicate with other nodes in order to become a member of the piconet. An ON node can take up any one of the four roles, namely, Isolated (I), Master (M), Slave (S) or Bridge (B). Similarly, any node may go to the OFF state at any instant of time. OFF state means that the node is no longer able to communicate either because it is powered off or it has moved out of the radio range and thus should not be an active part of a piconet. A node that has become OFF with respect to one piconet may become ON with respect to another piconet due to mobility.

The complete algorithm for dynamic scatternet formation consists of five routines: Start-up Routine (SUR), Next State Routine (NSR), Piconet Formation and Modification Routine (PFMR), Scatternet Formation and Modification Routine (SFMR) and Normal Communication Routine (NCR). Each routine resides within every Bluetooth node and gets called depending on the current role of the node and discovery of other nodes.

When a node is powered ON, it first calls the Start-up Routine shown in Fig. 1. Inside this routine, the current role of the node, denoted by the ROLE variable, is assigned a value of I – an Isolated node. In this role, the node performs either Inquiry or Inquiry-scan. In order to determine whether the node should enter Inquiry or Inquiry-scan state, the Next State Routine shown in Fig. 2 is called. The NSR routine selects Inquiry as the next state of an Isolated node with probability PI and selects Inquiry-scan with probability 1-PI. During Inquiry or Inquiry-scan, when the isolated node latches with another node, they first exchange configuration information. If the second node is also isolated, they together form a piconet by calling the Piconet Formation and Modification Routine of Fig. 3. However, if the second node is the master or the slave of an existing piconet, there are two possibilities. The piconet may be able to accommodate the new node, in which case, the isolated node is added to it by calling the Piconet Formation and Modification Routine. Otherwise, the existing scatternet is modified to include the new node by calling the Scatternet Formation and Modification Routine of Fig. 4. The new node is assigned the role of Master or Slave depending on its memory and available power and that of the master of the piconet in which the node joins.

In the role of Master or Slave, a node calls the Normal Communication Routine of Fig. 5 and takes part in the usual data/voice exchange. In the Master Communication Protocol of NCR, after some regular intervals, a master signals its most idle non-bridge slave to participate in node/piconet discovery by going to the Inquiry-scan state. If a slave receives this signal while executing the Slave Communication Protocol of NCR, it goes to Inquiry-scan state. The master may also participate in node/piconet discovery by going to the Inquiry or the Inquiry-scan state at certain intervals. When a master or a slave detects an isolated node or another piconet, either the two piconets are merged together or they start sharing a bridge to extend the scatternet using the PFMR or the SFMR routine. In the SFMR routine, some of the slaves are assigned as bridge nodes. The bridge nodes execute the Bridge Communication Protocol of the Normal Communication Routine. It should be noted that a slave cannot discover another slave as both of them perform node discovery in the Inquiry-scan state.

In the NCR, if a master detects that one of its slaves is not responding to polling packets for a long time, the master assumes that the slave has become OFF. The master then removes this slave from its piconet and notifies all the remaining slaves. Similarly, if a slave does not receive polling packets from its master for a number of cycles, the master is assumed to be OFF. The slave considers itself to be isolated and tries to join a new piconet by invoking the Start-up Routine. If a bridge node does not receive a polling packet from one of its masters while executing the Bridge Communication Protocol of NCR, it assumes that particular master to be OFF. The bridge then becomes a non-bridge slave of its other master. Finally, a master may swap its role with a slave if its power falls below a certain threshold. However, it should be noted from the PFMR and SFMR routine descriptions that, usually the more powerful (in terms of memory and power availability) nodes have higher chance of becoming master in the first place.

```
function SUR ()
ROLE (i) = I              // ROLE of node i set to Isolated
While (ROLE(i)==I)        // Stays in role I till it gets connected
        j = NSR (I)       // Call Next State Routine
        If (j==1)
            While elapsed time < TI    // Stay in Inquiry for TI to latch with another node
                If node i receives message from node j
                    If ROLE(j)== I     // j is also an Isolated node
                        Call PFMR(I, I)
                    Else
                        Let n(j) be the number of nodes in the piconet to which j belongs
                        If ROLE(j) = M and n(j) < 7 Call PFMR(I, M)
                        If ROLE(j) = M and n(j) = 7 Call SFMR(I, M)
                        If ROLE(j) = S and n(j) < 7 Call PFMR(I, S)
                        If ROLE(j) = S and n(j) = 7 Call SFMR(I, S)
        Else
            While elapsed time < TIS   // Stay in Inquiry-scan for TIS to latch with another node
                If node i receives message from node j
                    Call PFMR or SFMR based on role of node j as shown above
```

Fig. 1. Start-up Routine (SUR)

```
function NSR (node_role)
Generate a random number r between 0 and 1
If ( node_role == I)
        If (r <=PI)               // An Isolated Node goes to Inquiry State with probability PI
                return 1
        Else                      // An Isolated Node goes to Inquiry-scan State with probability 1-PI
                return 0
If ( node_role ==M)
        If (r <=PM)               // A Master goes to Inquiry State with probability PM
                return 1
        Else                      // A Master goes to Inquiry-scan State with probability 1-PM
                return 0
```

Fig. 2. Next – State Routine (NSR)

function (node_role_i, node_role_j)

Case 1 – An isolated node detects another isolated node
If (node_role_i == I AND node_role_j == I)
    Let F(m,p) denote a function of the memory 'm' and power 'p' of a node
    If (F(m_i,p_i) >= F(m_j,p_j))
                // a piconet is formed with node i as master and node j as slave
        ROLE(i) = M; i calls NCR(M)
        i sends message to j to do the following
            ROLE(j) =S; j calls NCR(S)
    Else
        Same as above with roles of i and j reversed.

Case 2 – An isolated node detects master of another piconet with less than 7 slaves
If (node_role_i == I AND node_role_j == M)
    If (F(m_i,p_i) >= F(m_j,p_j))
        ROLE(i) = M   // Node i becomes the new master
        i calls NCR(M)
        Let P(j) denote the piconet to which node j belongs
        Broadcast new master information to all the slaves of P(j)
        i sends message to j to do the following
            ROLE (j) =S; j calls NCR(S)
    Else
        Same as above with roles of i and j reversed.

Case 3 – An isolated node detects a slave of another piconet with less than 7 slaves
If (node_role_i == I AND node_role_j == S)
    Let M(j) be the master of a piconet to which node j belongs
    Notify M(j) about new node discovery
    Let k = M(j)
    i sends message to j to do the following
        j calls NCR(S)   // Goes back to normal slave communication state
    i joins P(j) either as a slave or as the new master similar to Case 2 above

Case 4 – A Master detects master of another piconet with a total of less than 7 slaves
If (node_role_i == M AND node_role_j == M)
    If (F(m_i,p_i) >= F(m_j,p_j))
        Merge P(i) and P(j)   // The master of the combined piconet is node i
        i sends message to j to do the following
            ROLE (j) =S; j calls NCR(S)
        i sends message to B(i, j) to do the following
            ROLE (B(i,j)) =S   //Bridge between P(i) and P(j) no longer required
            B(i,j) calls NCR(S)
        ROLE (i) =M; i calls NCR(M)
    Else
        Same as above with roles of i and j reversed.

Case 5 – A Master detects a slave of another piconet with a total of less than 7 slaves
If (node_role_i == M AND node_role_j == S)
    Notify M(j) about new piconet discovery
    i sends message to j to do the following
        j calls NCR(S)   // Goes back to normal slave communication state
    P(i) and P(j) are merged similar to Case 4 above

Fig. 3. Piconet Formation and Modification Routine (PFMR)

function SFMR (node_role_i, node_role_j)

Case 1 - An isolated node detects master of another piconet with 7 slaves
If (node_role_i == I AND node_role_j == M)
    Let NBS(j) denote the total number of non-bridge slaves of the piconet whose Master is j
        If (NBS (j)!=0)   //P(j) has some non-bridge slaves that can be shared
            Form a new piconet with i as the master
            Transfer upto 4 non-bridge slaves from P(j) to P(i) //To keep the piconets balanced
            Designate any node b as B(i,j)
            i calls NCR(M)
            i sends message to b to do the following
                b calls NCR(B)
       else
            i calls SUR() / / i starts Inquiry Inquiry-scan again

Case 2 - An isolated node detects a slave of another piconet with 7 slaves
If (node_role_i == I AND node_role_j == S)
    Notify master M(j) about new isolated node discovery
    Let k = M(j)
    i sends message to j to do the following
        j calls NCR(S) // Goes back to normal slave communication state
    i and k form a scatternet of two piconets similar to Case 1 above

Case 3 – A Master detects master of another piconet with a total of 7 to 13 slaves
If (node_role_i == M AND node_role_j == M)
    If ((n(i)==7 AND NBS(i)==0) OR (n(j)==7 AND NBS(j)==0))
        i calls NCR(M)
        i sends message to j to do the following
            j calls NCR(M)
   Else
      If B(i,j) exists
        If (n(i)+n(j)==7)
            Call PFMR(M,M)   // If the piconets are merged, Bridge node is not required
        If (n(i)>=n(j) AND NBS(i)>=0)
            Let n= (n(i)-n(j))/2
            if (NBS(i)>=n)
                n non-bridge slaves from P(i) join P(j)
            else
                NBS(i) non-bridge slaves from P(i) join P(j)
        If (n(j)>=n(i) AND NBS(j)>=0)
            Transfer non-bridge slaves from P(j) to P(i) as shown above
      Else
        Assign slave k with maximum F(m_i,p_l) from P(i) and P(j) as B(i,j)
        Non-bridge slaves join P(j) from P(i) or join P(i) from P(j) as shown above
        Both i and j broadcast new piconet structure to their slaves
        i calls NCR(M)
        i sends message to j to do the following
            j calls NCR(M)

Case 4 - A Master detects a slave of another piconet with a total of 7 to 13 slaves
If (node_role_i == M AND node_role_j == S)
    Notify master M(j) about new piconet discovery
    Let k = M(j)
    i sends message to j to do the following
        j calls NCR(S) // Goes back to normal slave communication state
    i and k form a scatternet of two piconets similar to Case 3 above

Fig. 4. Scatternet Formation and Modification Routine (SFMR)

function NCR (node_role)

// Case 1 – Master Communication Protocol
If (node_role == M)
    Let m be the master node
    m polls slaves for data/voice exchange
    Let j be the non-bridge slave of P(m) with the lowest packet arrival rate
    M signals j to go in IS state for a random period of time
    Call NSR(M) and go to I or IS state for new node/piconet discovery
    If m receives message from any node k in IS or I state
        If ROLE(k) = I and $n(m) < 7$ Call PFMR(I, M)
        If ROLE(k) = I and $n(m) = 7$ Call SFMR(I, M)
        If ROLE(k) = M and $n(m)+n(j) < 7$ Call PFMR(M, M)
        If ROLE(k) = M and $7 <= n(m)+n(j) < 14$ Call SFMR(M, M)
        If ROLE(k) = S and $n(m)+n(j) < 7$ Call PFMR(M, S)
        If ROLE(k) = S and $7 <= n(m)+n(j) < 14$ Call SFMR(M, S)
    If a slave l of P(m) does not respond to a large number of polling packets
        Remove l from the piconet   //Assumed l is OFF
        Notify all the other slaves of P(m)
    Receive $F(m_i,p_i)$ for each of n(m) slaves of P(m)
    If $F(m_m,p_m)$ falls below Min $(F(m_i,p_i))$ for all $i \in P(m)$
        Swap master role with the non-bridge slave k with highest $F(m,p)$
        i sends message to k to do the following
            k calls NCR(M)
        i calls NCR(S)
// Case 2 – Slave Communication Protocol
If (node_role == S)
    Let s be the slave node
    s sends data/voice packets to the master
    If packet arrival rate is low
        Go to IS state for node discovery by notifying the master
        If s receives message from any node j in I state
            If ROLE(j) = I and $n(s) < 7$ Call PFMR(I, S)
            If ROLE(j) = I and $n(s) = 7$ Call SFMR(I, S)
            If ROLE(j) = M and $n(s)+n(j) < 7$ Call PFMR(M, S)
            If ROLE(j) = M and $7 <= n(s)+n(j) < 14$ Call SFMR(M, S)
    If polling packet not received from master for a number of polling cycles
        ROLE(s) = I   //s assumes master is OFF. Goes to Isolated Node role
        s calls SUR()
// Case 3 – Bridge Communication Protocol
If (node_role == B)
    Let b be the bridge node
    b sends and routes data/voice packets between two masters
    If polling packet not received from one master for a number of polling cycles
        ROLE(b) = S   //b becomes a non-bridge slave of the other piconet
        b calls NCR(S)

Fig. 5. Normal Communication Routine (NCR)

3   Simulation Results

We have developed a Bluetooth topology formation and scheduling environment using 'C' on the Linux platform [13].

After setting the parameter values through initial simulation, we measured the performance of the algorithm through extensive simulation. In Fig. 6(a), we plot the average number of slaves per piconet vs. number of nodes. We notice that the average number of slaves per piconet is close to the ideal average number of slaves per piconet. It may be observed that as the number of nodes in the scatternet increases, our algorithm closely tracks the ideal number.

In Fig. 6(b), we plot number of piconets vs. number of nodes. We also plot the ideal number of piconets vs. number of nodes. We note that the number of piconets produced by the proposed algorithm is close to the ideal number. The number of piconets is either equal to the ideal number or is one greater than the ideal. Fig. 7(a)

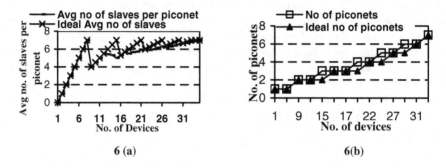

6 (a)   6(b)

Fig. 6. (a) Average number of slaves per piconet Versus number of nodes (b) Number of piconets Versus number of nodes

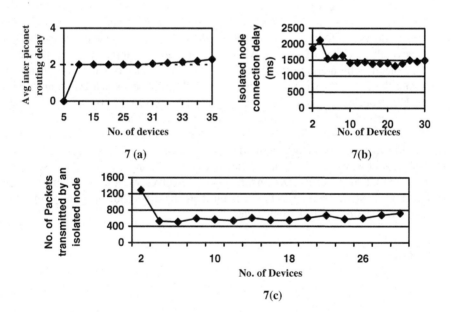

7 (a)   7(b)

7(c)

Fig. 7. (a) Average inter-piconet routing delay Versus number of nodes (b) Isolated node connection delay Versus number of nodes (c) Average number of packets transmitted by an isolated node during connection establishment Versus number of existing nodes in the scatternet

shows that inter-piconet routing delay (in number of hops) increases *slowly* as the number of nodes increases. This is due to the fact that our algorithm tries to maintain full one-hop connectivity between all the piconets. Here, by number of hops between two piconets we mean hop counts between the masters of those piconets.

The set up delay and the number of packets used for connecting an isolated node to an existing scatternet are shown in Fig. 7(b) and (c), respectively. We notice that the set up delay is initially high when the number of existing nodes in the scatternet is low. This is because, only one or two masters alternate in Inquiry and Inquiry-scan states. The slaves that could perform node discovery through Inquiry-scan may not be able to go in the IS state due to less number of slaves in the piconet. However, as the number of nodes goes up, there are a larger number of masters and slaves trying to perform node discovery. As a result, the time and the number of packets for link set up goes down and then remain almost constant as we increase the number of nodes. Since we try to achieve full connectivity between piconets, with higher number of nodes, a higher percentage of slaves take up the role of bridge nodes and do not join the node discovery process. Thus, the set up time or the number of set up packet exchanges attains a steady state.

4 Conclusions

In this paper, we have described a Bluetooth scatternet formation algorithm. Our algorithm addresses a dynamic scenario where nodes may join and leave the scatternet at any point of time. We try to maintain full minimum hop connectivity between piconets and form a scatternet where the piconets are as balanced as possible in terms of the number of slaves. We re-organize the scatternet and try to create balanced piconets by merging small piconets with larger ones if necessary when new nodes join the network. Simulation results show that the nodes achieve connectivity and take reasonable amount of time and message exchanges to attain the same. Average inter-piconet routing delay also increases *slowly* as the number of nodes increases. Thus, the proposed algorithm achieves a scatternet with balanced number and size of piconets while maintaining desirable set-up time, message complexity and inter-piconet communication delay. The algorithm operates in a fully distributed manner without any leader election process. It involves evaluation of only a few branch conditions and simple arithmetic. Further, the algorithm operates within the existing specifications of Bluetooth nodes.

References

1. The Bluetooth Special Interest Group. http://www.bluetooth.com. Specification of the Bluetooth system, Volume 1, Core.
2. B. A. Miller, C. Bisdikian, Bluetooth Revealed: The Insider's Guide to an Open Specification for Global Wireless Communications, Prentice Hall, USA, 2000.
3. J. Haartsen, Bluetooth - the universal radio interface for ad-hoc, wireless connectivity, *Ericsson Review*, 3 (1998) 110–117.

4. T. Salonidis, P. Bhagwat, L. Tassiulas, R. LaMaire, Distributed topology construction of Bluetooth personal area networks, Proc. Infocom (2001).
5. G. Miklos, A. Racz, Z. Turanyi, A. Valko, P. Johansson, Performance aspects of Bluetooth scatternet formation, Proc. The First Annual Workshop on Mobile Ad-hoc Networking and Computing, (2000) 147–148.
6. G. Tan, Self-organizing Bluetooth scatternets, Master's thesis, Massachusetts Institute of Technology, January 2002.
7. C.Law, A.K.Mehta, K-Y Siu, A New Bluetooth Scatternet Formation Protocol, ACM/Kluwer Journal on Mobile Networks and Applications (MONET), Special Issue on Mobile Ad Hoc Networks, 8 (2003).
8. S. Basagni, C. Petrioli, Multihop Scatternet Formation for Bluetooth Networks, Proc. VTC (2002) 424–428.
9. J. Yun, J. Kim, Y-S Kim, J.S. Ma, A three-phase ad-hoc network formation protocol for Bluetooth Systems, The 5th International Symposium on Wireless Personal Multimedia Communications, (2002) 213–217.
10. S.Basagni, R.Bruno, A Petrioli, A Performance Comparison of Scatternet Formation Protocols for Networks of Bluetooth Devices, IEEE International Conference on Pervasive Computing and Communications (PerCom'03) (2003) 341–350.
11. R. Guerin, J. Rank, S. Sarkar, E. Vergetis, Forming Connected Topologies in Bluetooth Adhoc Networks, International Teletraffic Congress (ITC18), Berlin, Germany (2003).
12. M. A. Marsan, C. F. Chiasserini, A. Nucci, G. Carello, L. De Giovanni, Optimizing the Topology of Bluetooth Wireless Personal Area Networks, Proc. Infocom (2002).
13. R. Roy, M. Kumar, N.K. Sharma, S. Sural, P^3-A Power-aware Polling Scheme with Priority for Bluetooth. Proc. International Conf. on Parallel Processing (ICPP) Workshops, 2004, Montreal, Canada (to appear).

Efficient Secure Aggregation in Sensor Networks

Pawan Jadia and Anish Mathuria

Dhirubhai Ambani Institute of Information and Communication Technology,
Gandhinagar, Gujarat, India
{pawan_jadia, anish_mathuria}@da-iict.org

Abstract. In many applications of sensor networks, readings from sensor nodes are aggregated at intermediate nodes to reduce the communication cost. The messages that are relayed in the data aggregation hierarchy may need confidentiality. We present a secure data aggregation protocol for sensor networks that uses encryption for confidentiality, but without requiring decryption at intermediate nodes. A salient feature of the protocol is the use of two-hop pairwise keys to provide integrity while minimizing the communication required between the base station and sensor nodes. We analyze the performance of our protocol and compare its efficiency with a protocol proposed by Hu and Evans.

1 Introduction

In many sensor network applications, the base station needs to find the summarized statistics of the whole network or part of the network. To support this requirement, each node can send its reading to the base station and then the base station can compute the aggregate of the data. This approach is not very energy efficient since each reading is communicated to the base station. Instead if we combine several sensor readings at intermediate sensor nodes as the readings are routed towards the base station, we can save on communication at the cost of some extra computation. This process is called aggregation or in-network aggregation [1]. Since aggregation reduces the amount of data to be transmitted through the network, it increases the lifetime of the energy constrained sensor nodes [2, 3].

Security is a critical requirement in data aggregation. A dishonest intermediate node can read, modify, or drop the data as well as sending false values to mislead the base station [4]. Therefore authentication and integrity are important requirements [5]. Data confidentiality is an equally important concern as authentication and integrity, especially in battlefield management where most of the information is sensitive. If data is not kept confidential then an adversary or opponent in the battlefield can get sensitive data.

To provide confidentiality in data aggregation, one can use a key sharing approach in which messages are encrypted using keys that are shared pairwise between nodes. Under this approach it is essential that a node that aggregates a secure message be able to decrypt the message. Therefore, an attacker who compromises a neighboring node by obtaining its keying material can obtain

L. Bougé and V.K. Prasanna (Eds.): HiPC 2004, LNCS 3296, pp. 40–49, 2004.
© Springer-Verlag Berlin Heidelberg 2004

the sensor readings. From a security point of view, the readings should not be visible at intermediate nodes. Another disadvantage of the above key sharing approach is that it requires decryption at intermediate nodes to process the data, followed by re-encryption. This three steps process (decryption, aggregation and re-encryption) is not very computationally efficient [6]. In this paper, we propose a protocol for secure data aggregation that does not require decryption at intermediate nodes. A salient feature of our protocol is the use of two-hop pairwise keys to provide integrity while minimizing the communication required between the base station and sensor nodes.

The LEAP protocol [6] provides confidentiality in aggregation process. However, it requires each node to possess a unique cluster key to be established to secure its message, which the neighboring nodes use to decrypt and authenticate the message. Our protocol does not require cluster or group keys to be established. Unlike LEAP, our protocol has low computation cost at intermediate nodes since it does not require decryption for aggregation.

The rest of the paper is organized as follows. In Section 2 we discuss the security requirements for sensor networks. In Section 3 we review an existing protocol of Hu and Evans [7]. In Section 4 we discuss the security mechanisms used in our protocol. In Section 5 we describe our protocol. In Section 6 we give a security analysis of the protocol as well as the cost analysis.

2   Requirements for Sensor Network Security

The following basic requirements need to be considered while designing security protocols for sensor networks [5, 8, 9].

- *Data confidentiality:* The neighboring nodes or the attacker should not be able to read the data. We should ensure that the attacker is not able to deduce the plaintext, even if he sees multiple encryption of the same plaintext. This property can be achieved by changing the encryption key each time a message is encrypted. In order to generate a different encryption key, we can derive the encryption key as a function of some nonpersistent quantity like a counter value which changes on every query. To lower the communication overhead, we can maintain a counter at both transmitting and receiving ends which is incremented each time the base station injects a query. The encryption key will be the function of counter value at that instant and the master secret shared between the transmitter and receiver.
- *Data authentication and integrity:* Authentication ensures the receiving node that the data is coming from the claimed source only. It guarantees that an adversary is not injecting the messages instead of a trusted sensor node. To achieve data authentication (as well as integrity) we can use a hash based MAC.
- *Data Freshness:* Since the sensor nodes send readings over time, we must also guarantee that data is fresh in order to avoid the replay of messages by an adversary. One common method of ensuring freshness is challenge-

response exchange using nonces. However, this method is not suitable because it causes more communication overhead. Changing the encryption keys provides data freshness without the extra communication overhead.

In a sensor network, nodes are generally deployed in remote areas and the environment is generally unattended and hostile. So nodes can be compromised with non-zero probability. One very important design goal is to confine the security impact of a node compromise to immediate network neighborhood of the compromised node [6]. A single node should not be able to misrepresent the readings of a large part of the sensor network. If a sensor node forwards the wrong aggregation result of its descendants, then the base station should be able to detect and exclude spoofed data without ignoring genuine readings originating from the same subtree.

3   Hu and Evans Protocol

This is a lightweight security protocol that provides data integrity, authentication and data freshness. The protocol makes the following assumptions:

- The base station can directly send messages to all nodes whereas the sensor nodes can only communicate with neighboring nodes.
- Each node shares a secret key with the base station. That is, every node trusts the base station.

The protocol consists of two rounds. In the first round, shown in Figure 1, the sensor readings are transmitted and aggregated at intermediate nodes. In the second round, the base station reveals the keys needed to authenticate the messages. These two rounds are described below.

1. Data Transmission Round:
    - Each leaf node α sends its reading P_α along with the MAC of P_α under a one-time key K_{α_i} shared between α and the base station.
    - The parent node forwards the data and the MACs it receives to the grandparent node along with the MAC over the aggregation. It does not send the aggregated data.
    - The grandparent node saves the MACs computed by the grandson nodes. It aggregates the data received from each child node and sends the aggregate value along with the MAC computed by the child node. It sends a MAC of the aggregated data of all child nodes under the one-time key shared with the base station.
    - After the grandparent node, each node in the routing hierarchy performs aggregation.
2. Data Validation Round:
    - In this round the base station reveals the keys using the $\mu - Tesla$ broadcast authentication protocol [5]. These keys are received by all nodes.

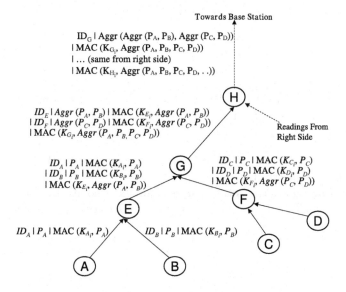

Fig. 1. Hu and Evans Protocol

The above protocol ensures that the parent node of a data originating node does not tamper with the data. The leaf nodes send their readings to the grandparent node through the parent node (unchanged). As noted earlier, the parent node calculates the aggregate value but sends only a MAC of the aggregated value. Since the grandparent node can independently calculate the aggregate value, the parent node cannot change the actual readings without being noticed. If any parent node tries to change the readings, it will get detected by its immediate parent during the data validation round.

The Hu-Evans protocol has the following deficiencies:

- It does not provide confidentiality.
- Because of delayed authentication, sensor nodes need to buffer the data to authenticate it later. Thus this protocol has extra memory overhead.
- Since the keys revealed by the base station are received by all nodes, the nodes waste energy in receiving keys that are not intended for them [10].
- It requires the base station to be involved in the data validation step.

4   Security Mechanisms

This section describes the security mechanisms that we will use to provide confidentiality, and to minimize the involvement of the base station.

4.1   Processing over Encrypted Data

There exist several lightweight methods for processing encrypted data without decryption [11]. We adopt the method where encryption is performed by adding the data to a sufficiently long random encryption key. A word size big enough to prevent data overflow is assumed. Consider as example a 4-node topology with two leaf nodes A and B, one intermediate node C, and a base station S. All data sent by the leaf nodes to S is routed via C. We will use the following notations.

- K_A, K_B, K_C are the encryption keys shared by nodes A, B, C with the base station, respectively.
- P_A, P_B, P_C are the actual sensor readings of nodes A, B, C, respectively.
- X_A, X_B, X_C are the encrypted data of nodes A, B, C, respectively.

Consider the scenario where A and B send their readings in encrypted form to C. Encryption of actual readings at node A, B and C is done simply as

$$X_A = (P_A + K_A)$$
$$X_B = (P_B + K_B)$$
$$X_C = (P_C + K_C)$$

Then node C can perform aggregation over encrypted data by calculating the value $Z = X_A + X_B + X_C$. Upon receiving this value, S can obtain the aggregate sum by calculating $Z - (K_A + K_B + K_C)$.

4.2   One-Hop and Two-Hop Pairwise Keys

Our motivation for using pairwise keys is to replace the data validation step of the Hu-Evans protocol with some other mechanism that does not require unnecessary key reception by all nodes. Instead of relying on keys shared between the base station and sensor nodes for authentication, we use pairwise keys shared between nodes. Use of pairwise keys reduces the memory overhead since it does not require messages to be stored for later verification. In our protocol we make use of one-hop as well as two-hop pairwise keys. There are various mechanisms to establish one-hop and two-hop pairwise keys [6], [12], [13], [10], [14].

5   Protocol for Aggregating Encrypted Data

We will use the following notations to describe our protocol.

- A, B, C, D are the sensor nodes that transmit their readings.
- ID_α is the unique identity of sensor node α.
- K_{α_i} is the ith encryption key of node α synchronized with the base station. We assume there is some master secret K_α that is shared between α and

the base station. If C_i is the ith counter value synchronized with the base station, then $K_{\alpha_i} = \mathrm{PRF}(K_\alpha, C_i)$, where PRF is a suitable psuedo-random function.

- P_{α_i} is the reading of node α in response to the ith query. X_{α_i} is the encryption of P_{α_i} using key K_{α_i}, that is, $X_{\alpha_i} = P_{\alpha_i} + K_{\alpha_i}$.
- $K_{\alpha\beta}^i$ is the ith two-hop (or one-hop) pairwise key shared between nodes α and β, where β is the grand-parent node (or parent node) and α is the grandson node (or child node). This ith key is the result of $\mathrm{PRF}(K_{\alpha\beta}^0, C_i)$, where $K_{\alpha\beta}^0$ is a pairwise key deployed initially. On each query injected by the base station, the counter value i is incremented by the nodes α and β to ensure that they both use the same key.
- $\mathrm{MAC}(K_{\alpha\beta}^i, M)$ denotes the message authentication code of M under key $K_{\alpha\beta}^i$, where $K_{\alpha\beta}^i$ is one-hop or two-hop pairwise key and M is either some encrypted message or aggregated encrypted data.

5.1 Protocol Steps

Figure 2 shows the steps of our protocol.

1. Base station injects a query in the network. Whenever a sensor node receives the query, it increments its counter value by one. We assume that the base station does not send the next query until it receives all responses to the last query. This is just to ensure self-synchronization of sensor nodes amongst themselves as well as with the base station.

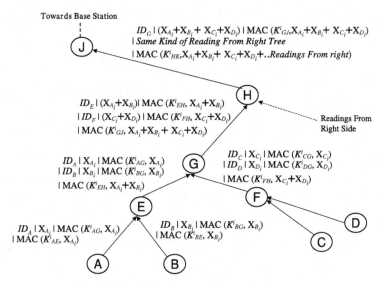

Fig. 2. Illustration of Protocol Steps

2. The data originating nodes encrypt their readings with the encryption keys that they share with the base station. The encryption method is same as the one described in Section 4.1. For example, node A will perform following operation: $X_{A_i} = (P_{A_i} + K_{A_i})$. For further steps this encrypted reading (X_{A_i}) will act as the data.

3. The data originating nodes send their encrypted readings to their respective parents along with their IDs and two MAC values, a MAC of the encrypted data under the two-hop key shared with the grandparent node and a MAC of encrypted data under the one-hop key shared with the parent node. For example:

$$A \Rightarrow E: ID_A \mid X_{A_i} \mid MAC(K_{AG}^i, X_{A_i})$$
$$\mid MAC(K_{AE}^i, X_{A_i})$$

4. The parent node receives the message and verifies the origin of the data using the one-hop pairwise shared key. It performs the aggregation over encrypted data but does not transmit this aggregated value. It calculates a MAC of this aggregated value using the two-hop pairwise key shared with its grandparent node and transmits it along with the encrypted sensor readings and the MACs that it received from its children (minus the MAC intended for itself). The parent node does not attach its own node id with the last MAC, since we are assuming that node G will already know this. For example:

$$E \Rightarrow G: ID_A \mid X_{A_i} \mid MAC(K_{AG}^i, X_{A_i})$$
$$\mid ID_B \mid X_{B_i} \mid MAC(K_{BG}^i, X_{B_i})$$
$$\mid MAC(K_{EH}^i, X_{A_i} + X_{B_i}).$$

5. The node G receives the message and checks the authenticity and integrity of X_{A_i} and X_{B_i} in the received message using the keys K_{AG}^i and K_{BG}^i. If X_{A_i} and X_{B_i} are modified in transit, node G will detect and notify the base station. Otherwise it aggregates the encrypted data of each branch separately and sends the aggregated values along with the corresponding MACs that it received from nodes E and F. It also sends a MAC of the encrypted data aggregated over all branches. For example:

$$G \Rightarrow H: ID_E \mid (X_{A_i} + X_{B_i}) \mid MAC(K_{EH}^i, X_{A_i} + X_{B_i})$$
$$\mid ID_F \mid (X_{C_i} + X_{D_i}) \mid MAC(K_{FH}^i, X_{C_i} + X_{D_i})$$
$$\mid MAC(K_{GJ}^i, X_{A_i} + X_{B_i} + X_{C_i} + X_{D_i})$$

6. Similarly node H verifies that node G is sending the correct aggregation (X_{A_i} + X_{B_i}) and ($X_{C_i} + X_{D_i}$) with the help of $MAC(K_{EH}^i, X_{A_i} + X_{B_i})$ and $MAC(K_{FH}^i, X_{C_i} + X_{D_i})$ respectively. Again if any misbehavior is detected, node H will notify the base station. Then the same process is repeated at node H. In this way aggregated readings (encrypted) are forwarded to base station. In the whole process, no intermediate node can modify the data that it receives from its child nodes without being detected.

7. The base station receives the final encrypted aggregated value. It checks the authenticity of the received messages and then subtracts the encryption keys shared with the data originating nodes to get the aggregated data in plaintext.

6   Protocol Analysis

6.1   Security Analysis

We assume that the data originating nodes (in our case leaf nodes) are not compromised, that is, that attacker does not possess the keying material of leaf nodes. Otherwise, there is no way to differentiate between a compromised node and a legitimate node. We only consider attacks involving intermediate nodes.

- *Protection Against Data Theft.* Our protocol protects against aggregated data theft from unattended sensor nodes. It achieves security similar to end-to-end data encryption. The data is only encrypted at the originating end and finally decrypted at the base station. No decryption is required at intermediate nodes. Thus the protocol prevents theft of data at intermediate nodes.
- *Detection of Compromised Node.* If a node is compromised, its key are revealed. The adversary is then able to spoof the MAC intended for the grandparent node. Assuming two consecutive nodes are never compromised, the parent node that receives the message will be able to detect the compromised node. After a compromised node is found, the parent node can notify the base station.
- *Replay Protection.* Since encryption and authentication keys are changed with every message, replay attacks are not possible.
- *False Node Attack.* A node not having valid keys termed a false node. A false node may try to transmit the sensor readings (encrypted) of a legitimate child node, but it will not be able to generate a valid MAC. Thus, it will be detected by the legitimate grandparent node.
- *Damage Associated with a Compromised Node.* A node in the upper hierarchy may attempt to misrepresent the readings of its descendants. In the event of this, the legitimate grandparent node knows that either any or some of child or grandchild nodes are misbehaving, thus it can exclude readings from only those descendants.
- *Collusion Resistance.* Our protocol is weak against this type of attack. If the parent and child nodes collude, they can misrepresent the readings of whole subtree without being detected. To protect against this attack, node snooping can be employed [6].

6.2   Cost Analysis and Comparisons

We consider the case that only leaf nodes transmit their sensor readings. A similar analysis can be made for the more general case where all or some of the intermediate nodes also transmit their readings. Following Hu and Evans [7], we assume a general tree hierarchy in which every node has b children and the depth of the tree is d, that is, leaf nodes are d nodes away from the base station. Suppose the length of data field (encrypted or aggregated reading) is x bits, node id is y bits, and the length of MAC is z bits. The leaf nodes will send their encrypted readings (x) along with the node id (y) and two MACs ($2z$).

The total number of leaf nodes is b^d. Thus the total number of message bits sent by all nodes at depth d will be $b^d(x+y+2z)$. A node at depth $d-1$ will send $[b(x+y+z)+z]$ message bits and there are b^{d-1} such nodes. So the total number of message bits sent by nodes at this level will be $b^{d-1}[b(x+y+z)+z]$. Similarly nodes at $d-2$ level will send $b^{d-2}[b(x+y+z)+z]$ message bits and so on. The total number of bits transmitted in our protocol will be

$$b^d(x+y+2z)+b^{d-1}[b(x+y+z)+z] +b^{d-2}[b(x+y+z)+z]+...+ b[b(x+y+z)+z]$$
$$= b^d(x+y+2z) +(b^d\text{-}b)[b(x+y+z)+z]/(b\text{-}1)$$

If we compare this result with Hu and Evans [7], we need one extra MAC per leaf node, that is, $b^d(z)$ extra bits to be transmitted. Besides this our protocol also requires one extra bit per leaf node for overflow protection. So our protocol requires $b^d(z) + b^d$ extra bits over the Hu-Evans protocol. We require extra bits at intermediate nodes for overflow protection. The Hu-Evans protocol too requires these extra bits because of arithmetic sum over unencrypted data. One example scenario given in [7] ($b = 4$, $d = 5$, 22-byte messages, 2-byte node id and 6-byte MAC) requires approximately 75 KB data to be transmitted for secure aggregation. In our case extra transmission will be $b^d(z)+b^d = 1024(6)+1024 = 7$ KB, that is total transmission will be 82 KB as compared to 75 KB. Our protocol saves energy that the nodes would waste in the Hu-Evans protocol in receiving keys from the base station. For the case $b = 4$ and $d = 5$, the total number of nodes (excluding the base station) will be $\dfrac{b^{d+1} - b}{(b - 1)}$, which in our case equals 1364. So a total of 1364 keys need to be disclosed (broadcast) by the base station and all the 1364 nodes will listen to these messages (although later they discard most of the packets), so total $1364^2 = 1860496$ key receptions will be there and a large amount of energy will be wasted. As compared to this our protocol does not require keys to be disclosed. It does not require any extra data to be flooded in the network for synchronization. Every node is automatically synchronized whenever it receives some query from the base station.

7   Conclusions

We proposed a secure aggregation protocol for sensor networks that provides confidentiality as well as integrity guarantees. Our protocol aggregates encrypted data directly, without requiring decryption at intermediate nodes. The protocol does not require intermediate nodes to buffer data for later authentication; this results in reduced delays. The protocol is not suitable for certain kinds of queries like Min, Max. This is part of our future work.

Acknowledgements

We thank the three anonymous reviewers for their helpful criticisms on a draft of this paper.

References

1. Karl, H., Loebbers, M., Nieberg, T.: A data aggregation framework for wireless sensor networks. In: Proc. Dutch Technology Foundation ProRISC Workshop on Circuits, Systems and Signal Processing, Veldhoven, Netherlands (2003)
2. Deng, J., Han, R., Mishra, S.: Security support for in-network processing in wireless sensor networks. In: Proc. 1st ACM Workshop on the Security of Ad Hoc and Sensor Networks (SASN). (2003) 83–93
3. Zhao, J., Govindan, R., Estrin, D.: Computing aggregates for monitoring wireless sensor networks. In: Proc. IEEE International Workshop on Sensor Net Protocols and Applications. (2003) 139–148
4. Przydatek, B., Song, D., Perrig, A.: SIA: secure information aggregation in sensor networks. In: Proc. 1st International conference on Embedded networked sensor systems. (2003) 255–265
5. Perrig, A., Szewczyk, R., Wen, V., Culler, D., Tygar, D.: SPINS: security protocols for sensor networks. Wireless Network journal (WINE) **8** (2002) 521–534
6. Zhu, S., Setia, S., Jajodia, S.: LEAP: efficient security mechanisms for large-scale distributed sensor networks. In: Proc. 10th ACM Conference on Computer and Communications Security (CCS). (2003) 62–72
7. Hu, L., Evans, D.: Secure aggregation for wireless network. In: Proc. IEEE Symposium on Applications and the Internet Workshops (SAINT'03). (2003) 384–394
8. Zhou, L., Haas, Z.J.: Securing ad hoc networks. IEEE network Magazine **13** (1999) 24–30
9. Cam, H., Ozdemir, S., Muthuavinashiappan, D., Nair, P.: Energy-efficient security protocol for wireless sensor networks. In: Proc. 58th IEEE Vehicular Technology Conference. (2003) 2981–2984
10. Liu, D., Ning, P.: Establishing pairwise keys in distributed sensor networks. In: Proc. 10th ACM Conference on Computer and Communications Security (CCS'03). (2003) 52–61
11. Ahituv, N., Lapid, Y., Neumann, S.: Processing encrypted data. Communications of the ACM **30** (1987) 777–780
12. Chan, H., Perrig, A., Song, D.: Random key predistribution schemes for sensor networks. In: Proc. IEEE Symposium on Security and Privacy. (2003) 197–213
13. Du, W., Deng, J., Han, Y.S., Varshney, P.: A pairwise key pre-distribution scheme for wireless sensor networks. In: Proc. 10th ACM Conference on Computer and Communications Security (CCS). (2003) 42–51
14. Zhu, S., Xu, S., Setia, S., Jajodia, S.: Establishing pairwise keys for secure communication in ad hoc networks: A probabilistic approach. In: Proc. 11th IEEE International Conference on Network Protocols (ICNP'03). (2003) 326–335

Optimal Access Control for an Integrated Voice/Data CDMA System

Shruti Mahajan, Manish Singh, and Abhay Karandikar

Information Networks Laboratory, Department of Electrical Engineering,
Indian Institute of Technology Bombay, Powai, Mumbai-76, India
`karandi@ee.iitb.ac.in`

Abstract. Access control (call admission and data transmission policies) becomes essential for optimum resource utilization with traffic having varying Quality of Service (QoS) requirements. In this paper, we determine optimal algorithms for joint admission control of voice calls and transmission of data packets in an integrated voice and data dual-class CDMA system. The objective is to minimize the cost due to blocking of voice calls and the average delay suffered by the data packets. The problem is formulated as an optimization problem over an infinite horizon, and the solution turns out to be a stationary policy for a given set of bit rate, Signal to Interference Ratio (SIR) and outage probability requirements.

1   Introduction

Next generation wireless networks need to be suitably engineered to support a heterogeneous mix of traffic with varying Quality of Service (QoS) requirements. CDMA has been acclaimed as the predominant medium access control (MAC) protocol for next generation systems for its effectiveness in high capacity multi-class traffic systems. The capacity of a CDMA air interface is determined by the capacity of the reverse link, which is typically interference-limited. In this work, we focus on access control of traffic on such a reverse link.

Quite a few attempts have been made to characterize a dual-class CDMA system, with a delay-intolerant voice class and a delay-tolerant data class with higher SIR and bit-rate requirements. Probability of blocking the arriving calls is a direct measure of customer satisfaction for voice subscribers, while the data traffic can be characterized by the queuing delay seen by the packets.

Since the reverse link of a CDMA system is interference limited, the power transmitted by a Mobile Station (MS) is controlled by the Base Station (BS) using sophisticated power control algorithms [1]. In [2], a necessary condition for the feasibility of transmit power assignment for given bit-rate and SIR requirements has been derived. In a perfectly power controlled system, if this condition is not satisfied, SIR demands of all users will not be met, an event termed as *outage*. This condition has been employed by the authors of [3] to schedule data packets in periods of low voice activity. Voice Activity Detection (VAD) capability is assumed to differentiate between voice activity intervals. Though we model

L. Bougé and V.K. Prasanna (Eds.): HiPC 2004, LNCS 3296, pp. 50–59, 2004.
© Springer-Verlag Berlin Heidelberg 2004

our traffic types differently and consider an open system with arrivals and departures, we largely retain their notations for the similarity with their system model and parameters.

The system model in [4] considers two classes (similar to voice and data) of user traffic, with complementary delay tolerance and bit-rate requirements. Their objective is to maximize the throughput of delay tolerant users while minimizing the sum of transmit powers of all mobiles. This work has been extended to include minimization of inter-cell interference in [5]. Both these problems are formulated from a "power allocation" point of view, with a given number of users and do not consider call admission control. For other related work, the readers are referred to [6], [7], [8], [9], [10], [11].

Despite several interesting proposals described above, none of them attempt 'optimal' management of the intrinsic trade-off between admitting more voice calls and delayed scheduling of data packets that are queued up. The primary contribution of our work is to determine a joint admission control and scheduling algorithm which is optimal with respect to the long term average cost due to blocking of voice arrivals and queuing delay of data packets.

2   System Model

We consider a single cell slotted CDMA system. The slot duration is t_0, and is equal to the transmission time of a packet. All arrivals into or departures from the system take place at slot boundaries, and so do any changes in voice activity state of the admitted users.

Voice Traffic. There are N voice subscribers, out of which a maximum of M can be supported by the system at any time so as to satisfy the outage probability constraint, which is the probability that target SIR values of all users are not achieved. A voice subscriber who initiates a call and is subsequently accepted into the system is termed as "active". Subscribers not making calls are called "inactive" or "idle". Users which are present in the system just before the start of slot n (where n is the slot index) are denoted by $K_V(n)$. Assuming Voice Activity Detection (VAD), only $V(n)$ of these (active) users might be ON. ON users transmit fixed length packets with a rate of R_V bits/s, OFF users don't transmit (in a practical system, they might transmit at a minimal rate though). We employ geometric probability distributions for the period for which a user remains in the system and for voice activity intervals to model the discrete nature of the system. We define the following probabilities, for events at the start of a slot:

p = probability with which a voice user initiates a call

q = probability with which a voice user departs from the system

s = probability with which a voice user continues to be ON in this slot

u = probability with which a voice user continues to be OFF in this slot

Arrival and departure processes are discrete time batch processes. Thus, at the start of slot n, the number of voice call arrivals, $A_V(n)$, and departures, $G(n)$, are observed. Assume $a_c(n)$ is the number of voice call arrivals accepted, with the decision being instantaneous. These admitted users are considered to be ON users in the current slot n. In addition, users that depart are also considered to be ON in the previous slot. This assumption simplifies our analysis while it introduces a negligible error in the computations.

Users in the system are assumed to change their voice activity state at the beginning of a slot. We denote by $V_{ON}(n)$ the number of voice users that become ON, and by $V_{OFF}(n)$ the number of voice users that become OFF at the start of slot n.

Data Traffic. A simple data model with a single queue of maximum length of L packets is considered. Packets of fixed length arrive according to a Poisson process with an average arrival rate of λ_D. A new packet is placed at the end of the queue. It is assumed that the rate is low enough so that the probability of queue length exceeding the maximum length is negligible. Each packet takes exactly one slot to transmit at a rate of R_D bits/s. Multiple packets may be scheduled for transmission on the reverse link in a slot. We assume that r CDMA codes are alloted to the data queue, so that a maximum of r simultaneous transmissions of data packets are possible. This would be possible in a system implementing multi-code CDMA. In case of outage, the transmitted packets are lost. However, the analysis can be easily modified to incorporate the case when the errored packets are retransmitted. Just before the start of slot n, the queue length is denoted by $Q(n)$. At the beginning of the slot, $A_D(n)$ packets arrive, out of which $D(n)$ are scheduled to be transmitted in slot n. Again, the feedback time from BS to the mobile is assumed to be negligible.

An Outage Event. Consider the nth slot, i.e., the interval $[n, n+1)$. Just before start of this slot, we have $V(n)$ ON voice users in the system. At the beginning of this slot, $a_c(n)$ new voice users are admitted and $D(n)$ data packets are scheduled for transmission. Considering the departures and the users that change their ON/OFF state (as given in the evolution equation below), we'll now have $V(n+1)$ users that'll be ON during the interval $[n, n+1)$. Suppose the target SIR values for voice and data packets are γ_V and γ_D respectively. With $V(n+1)$ ON voice users and $D(n)$ data packets in transmission in slot n, a feasible power assignment will exist [2][1] if

$$P\left\{\frac{V(n+1)}{a_V} + \frac{D(n)}{a_D} < 1\right\} \text{ where } a_V = W/(R_V\gamma_V) + 1$$

$$a_D = W/(R_D\gamma_D) + 1 \text{ and } W = \text{spread-bandwidth}$$

The event when this condition is violated is termed as outage. The probability of outage is constrained by a given number δ, which is typically a specified system parameter. The constraint

---

[1] Note that our definition of variables is different from that of [2].

$$P_{out} = P\left\{\frac{V(n+1)}{a_V} + \frac{D(n)}{a_D} > 1\right\} \le \delta \tag{1}$$

has to be satisfied on a per slot basis. We note here the assumption of perfect power control. For obtaining optimal design constraints, we satisy Equation 1 with equality in our analysis.

The maximum number of voice users that can be admitted into the system, M, can be calculated using the outage condition (1) and the stationary probability distributions of different states of a voice user.

3   Formulation as a Markov Decision Process

The evolution equations for the system variables can be written as

$$K_V(n+1) = K_V(n) + a_c(n) - G(n), \quad Q(n+1) = Q(n) + A_D(n) - D(n)$$
$$V(n+1) = V(n) + a_c(n) - G(n) + V_{ON}(n) - V_{OFF}(n)$$

Define the system state by the 4-tuple $\mathbf{x_n} = \{K_V(n), V(n), A_V(n), Q(n)\}$. The number of states is finite, their number denoted by S. At every state, we take an action, denoted by the 2-tuple $\mu(\mathbf{x_n}) = \{a_c(n), D(n)\}$. The space of all possible actions, $U(\mathbf{x})$, is state-dependent. Given the system is in state $\mathbf{x_n}$ at the observation epoch n (the beginning of nth slot), if action $\mu(\mathbf{x_n}) = \mathbf{u_n}$ is taken, some immediate cost $g_n(\mathbf{x_n}, \mathbf{u_n})$ is incurred (which we quantify later), and the system moves to the next state $\mathbf{x_{n+1}}$ with a probability $p_{\mathbf{x_n x_{n+1}}}(\mathbf{u_n})$. Such a controlled dynamic system is a Markov Decision Process [12]. The sequence of actions $\Gamma = \{\mu_1, \mu_2, ...\}$ is termed as a *policy*.

Our system satisfies the Markovian assumption, that the next state to be visited depends only on the present state of the system. The transition probability is controlled by taking actions. The transition probabilities are stationary in our problem, which also implies the stationarity of the policy so computed [13].

3.1   State Transitions

Since the ordering among the states does not matter, without loss of generality, we index the states for convenience of notation as follows:

Let $i = \{K_V, V, A_V, Q\}$, $j = \{K'_V, V', A'_V, Q'\}$ and $\mathbf{u} = \{a_c, D\}$. We also drop the time index n. Then, we have,

$$K'_V = K_V + a_c - G, \quad Q' = Q - D + A_D$$
$$V' = V + a_c - G + V_{ON} - V_{OFF}$$

where, as mentioned earlier, we have assumed that arrivals and departures are all from the pool of ON users. The state transition probabilities can be derived as follows.

$$p_{ij}(\mathbf{u}) = P\{\mathbf{x_{n+1}} = j | \mathbf{x_n} = i, \mathbf{u_n} = \mathbf{u}\}$$
$$= P\{K_V', V', A_V', Q' | (K_V, V, A_V, Q), (a_c, D)\}$$
$$= P\{K_V' | K_V, a_c\} P\{V' | K_V', K_V, V, a_c\} P\{A_V' | K_V'\} P\{Q' | Q, D\} \quad (2)$$

Now

$$P\{K_V' | K_V, a_c\} = P\{G = K_V - K_V' + a_c\}$$
$$= \binom{V}{G} q^G (1-q)^{V-G} \quad (3)$$

Letting $z = V' - V - a_c + G = V' - V + K_V - K_V'$ then

$$P\{V' | K_V', K_V, V, a_c\} = \sum_{x=0}^{V-G} \binom{V-G}{x} (1-q-s)^x s^{V-G-x}$$
$$\binom{K_V - V}{z+x} (1-u)^{z+x} u^{K_V - V - z - x} \quad (4)$$

Voice arrivals have a geometric distribution while the data packets arrive according to a Poisson process. Hence we have,

$$P\{A_V' | K_V'\} = \binom{N - K_V'}{A_V'} p^{A_V'} (1-p)^{N - K_V' - A_V'} \quad (5)$$

$$P\{Q' | Q, D\} = P\{A_D = Q' - Q + D\} = \frac{(\lambda t_0)^{Q' - Q + D} \exp(-\lambda t_0)}{(Q' - Q + D)!} \quad (6)$$

Substituting the numerical values from (3), (4), (5) and (6) in (2), we can compute the transition probability $p_{ij}(\mathbf{u})$.

3.2   Cost Formulation

At every decision epoch, the decision that is taken incurs a cost for the system. The immediate cost is defined as

$$g_n(\mathbf{x_n}, \mathbf{u_n}) = \frac{A_V(n) - a_c(n)}{pN} + \beta \frac{Q(n) - D(n)}{\lambda_D} \quad (7)$$

The first term is the cost incurred due to rejection of voice arrivals, while the latter takes into account the cost due to queuing of data packets. β is a scalar and has been employed to assign desired weights to the two cost terms. This enables us to give priority to voice or data according to the system requirements; e.g. the lower the value of β, the lower will be the fraction of voice calls blocked on arrival. This we verify later through numerical results.

For a finite observation interval, of length say K slots, the total cost incurred will be

$$J_K = g_K(\mathbf{x_K}, \mathbf{u_K}) + \sum_{n=0}^{K-1} g_n(\mathbf{x_n}, \mathbf{u_n})$$

However, the system under consideration can be assumed to evolve over a very long period of time, effectively over an infinite horizon. We seek to minimize the average cost per unit time, defined as [14]

$$
\lim_{K \to \infty} \frac{1}{K} \left\{ \sum_{n=0}^{K} g(\mathbf{x_n}, \mu_{\mathbf{n}}(\mathbf{x_n})) \right\}
$$

$$
= \lim_{K \to \infty} \frac{\sum_{n=0}^{K} (A_V(n) - a_c(n))}{KpN} + \lim_{K \to \infty} \beta \frac{\sum_{n=0}^{K} (Q(n) - D(n))}{K\lambda_D}
$$

$$
= \text{blocking probability} + \beta(\text{average delay of data packets})
$$

where the Little's Theorem has been used for computing the average delay. Thus, we attempt to minimize a function of the long-term blocking probability for voice arrivals and the average delay seen by the data packets.

4   Computing the Optimal Policy

The above optimization problem can be solved using Dynamic Programming techniques. Our system has a large state space, and *Value Iteration Algorithm* is well suited for solving such a system over an infinite horizon. The recursion equation for Value Iteration can be written as

$$
J_{k+1}(i) = \min_{\mathbf{u} \in U(i)} \left\{ g(i, \mathbf{u}) + \sum_{j=1}^{S} p_{ij}(\mathbf{u}) J_k(j) \right\}
$$

Here i is the current state and $\mathbf{u}$ is one of the feasible actions. After convergence, the equation can be interpreted as follows. The cost of $k + 1$ stages is broken into immediate cost and the cost incurred over the next k stages. We note that the time indexing has been reversed here, and $J_k(i)$ can be interpreted as the expected "cost-to-go" with k periods left over the time horizon, when the current state is i and a terminal cost of $J_0(j)$ is incurred when the system ends up in state j. This suggests that for large k, the one-step difference $J_k(i) - J_{k-1}(i)$ will come very close to the minimal average cost per unit time [12].

Let the set of admissible (feasible) policies for a state be denoted by $\Pi(\mathbf{x_n})$. Natural constraints that arise in our system are

$$
D(n) \leq \min\{Q(n), r\} \quad \text{and} \quad a_c(n) \leq A_V(n)
$$

The number of ON voice users along with the number of data packets in transmission should satisfy the outage probability constraint in every slot. For slot n, taking the limiting value of outage probabilty

$$
P_{out} = P \left\{ \frac{V(n+1)}{a_V} + \frac{D(n)}{a_D} > 1 \right\} = \delta \quad \Rightarrow \quad P\{V(n+1) > C_n\} = \delta \tag{8}
$$

$$
\text{where } C_n = \lfloor a_V (1 - D(n)/a_D) \rfloor
$$

At the start of slot n, before deciding on the action, we know $K_V(n)$, the number of active voice users, and $V(n)$, the number of ON voice users, that were there in the previous slot. Now, for a given value of $a_c(n)$, we can determine the maximum number of data packets that can be scheduled so that P_{out} does not exceed δ.

$$V(n+1) = V(n) + a_c(n) - G(n) + V_{ON}(n) - V_{OFF}(n)$$
$$\Rightarrow P_{out} = P\{V_{ON}(n) - V_{OFF}(n) - G(n) > C_n - V(n) - a_c(n)\}$$

Denoting $K_V(n)$ by K_v and $V(n)$ by V,

$$P_{out} = \sum_{j=C_n}^{K_v+a_c} \sum_{z=0}^{V} \sum_{y=0}^{K_v-V} \sum_{x=0}^{V-z} \binom{V}{z} q^z (1-q)^{V-z}$$
$$\binom{K_v - V}{y}(1-u)^y u^{K_v-V-y} \binom{V-z}{x}(1-q-s)^x$$
$$s^{V-z-x} \delta_f(j - (V + a_c - z + y - x))$$

where j, z, y, x count $V(n+1), G(n), V_{ON}(n), V_{OFF}(n)$ respectively and $\delta_f(x)$ denotes the impulse function. We can thus determine C_n and the corresponding $D(n)$. Let us call this number D_{max}. Accordingly, we modify the admissible values of $D(n)$ to

$$D(n) \leq \min\{D_{max}, Q(n), r\}$$

Iterating on the cost J_k, the algorithm searches for the minimum cost policy among the admissible set $\Pi(\mathbf{x})$ for every state $\mathbf{x}$. Let $c_k(\mathbf{x}) = J_{k+1}(\mathbf{x}) - J_k(\mathbf{x})$. The average cost, in the long run, is given by

$$\lim_{k \to \infty} c_k(\mathbf{x}) = \lim_{k \to \infty} (J_{k+1}(\mathbf{x}) - J_k(\mathbf{x}))$$

The convergence criteria used, for a desired level of accuracy ϵ is,

$$\max_{\forall \mathbf{x}} (c_{k+1}(\mathbf{x}) - c_k(\mathbf{x})) < \epsilon \tag{9}$$

The policy so computed is a stationary policy- given that the system is in a particular state, the optimal action is fixed in time. This has a lot of practical utility, since the mapping from states to actions can be computed and stored in the case where traffic characteristics vary slowly with time.

5   Results and Observations

Table 1 gives the values of parameters used in the computations, unless specified otherwise. We consider a representative system with 15 voice subscribers and a

Table 1. Values of parameters used for computations

t.	W	γ_V	γ_D	a_V	a_D	p	q	s	u	δ	λ_D	N	L, r	M	β
20ms	1.25MHz	7dB	10dB	10	7	0.04	0.04	0.67	0.5	0.01	20	15	10	10	0.5

data queue with a buffer capacity of 10 packets. The system can support 10 voice users if no data packets were scheduled for a slot. The values of s and u used correspond to mean ON time of 3 slots (60 ms) and mean OFF time of 2 slots (40 ms) respectively. Similarly, $q = 0.04$ corresponds to a mean call holding time of 25 slots, which is 500 ms. The small value of λ_D has been taken so that the average queue length is not more than 5 for a reasonable delay. e.g. for a delay of 5 slots, the average queue length is 2 packets.

In Figure 1(a) and Figure 1(b)we plot the number of accepted calls (a_c) and the number of data packets scheduled (D), respectively, as we vary the number of voice call arrivals (A_V) and the queue length (Q), while the number of users in the system (K_V) and the number of ON voice users (V) are kept constant. The fairness of the optimal policy is evident when, for the same number of active users, ON users and arrivals in the system, lesser number of voice calls are accepted and more data packets are scheduled as the queue length increases. To take an example, for 5 users in the system out of which 3 are ON, and 5 arrivals, $a_c = 5, D = 0$ when $Q = 3$, while we get $a_c = 2, D = 1$ when $Q = 4$.

On convergence of the algorithm, the differential cost so obtained ($J_{k+1} - J_k$) is the minimum cost per stage of running the system. We plot the optimal cost per stage versus the parameter λ_D of the Poisson arrival process of data packets. As shown in Figure 2, increasing the probability of voice call initiation (p) results in higher operating costs. This is expected because with increased load on the system, the average queue length and the fraction of rejected voice calls increases and hence the average cost will increase.

The value of β affects the relative priority of voice arrivals or queued data packets. Figure 3 illustrates how the number of accepted voice calls decrease while the number of scheduled data packets increase, respectively, as we increase β, the scaling factor of the data cost term. Values of other parameters are same as given in the table.

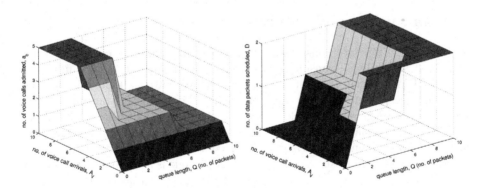

Fig. 1. (a) No. of accepted calls as a function of no. of arrivals and queue length (for $K_V = 5, V = 3$) (b) No. of data packets scheduled as a function of no. of arrivals and queue length (for $K_V = 5, V = 3$)

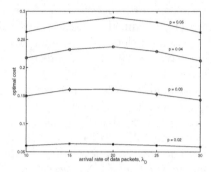

Fig. 2. Optimal cost vs data arrival rate for different values of call initiation probability p

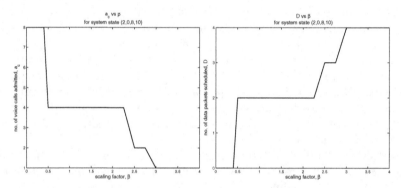

Fig. 3. (a) No. of voice calls accepted vs scaling factor (b) No. of data packets scheduled vs scaling factor

6   Conclusions and Future Directions

In this paper, we have considered the problem of admission control of voice users and scheduling (transmission) of data packets on the reverse link of a slotted CDMA system for a single cell scenario. We have computed an optimal policy for access control (voice call admission and data packet transmission) by minimizing a weighted sum of long term blocking probability for voice arrivals and the average delay seen by data packets. The optimization problem has been formulated over an infinite horizon, and solved using Value Iteration. The results illustrate the fairness of such a policy - the data queue is not starved of resources even if the voice load on the system increases. We observe that by increasing the weight of the cost term attributed to the delay of data packets, the number of data packets transmitted (keeping values of other variables same) increases, and vice versa.

In spite of the practical utility of such an algorithm, we note that the solution does not give us actual values of call blocking probability (p_B) and average delay

(τ) that are achieved by the system. We can address this shortcoming by solving a constrained optimization problem, that of minimizing p_B while τ is upper bounded by a specified value Υ. The solution thus obtained not only gives us the required policy, but the optimal cost is a lower bound for p_B. Due to lack of space, this formulation has not been discussed here.

References

1. Andrew J. Viterbi *CDMA: Principles of Spread Spectrum Communication* Addison Wesley Longman Inc., 1995
2. Ashwin Sampath, P. Sarath Kumar, Jack M. Holtzman, "Power Control and Resource Management for a Multimedia CDMA Wireless System" *Proceedings of PIMRC'95* Toronto, Canada, September 1995
3. Ashwin Sampath, Jack Holtzman, "Access Control of Data in Integrated Voice/Data CDMA Systems: Benefits and Tradeoffs" *IEEE Journal on Selected Areas in Communications* Vol.15, No.8, October 1997
4. Sudhir Ramakrishna, Jack M. Holtzman, "A Scheme for Throughput Maximization in a Dual-Class CDMA System " *IEEE Journal on Selected Areas in Communications* Vol.16, No.6, August 1998
5. Sudhir Ramakrishna, Jack M. Holtzman, "Throughput Maximization in a DS-CDMA System via Time Scheduling with Inter-Cell Interference Constraints" *http://winwww.rutgers.edu/pub/about/people/student/rsudhir/jsac2.ps*
6. Dongxu Shen, Chuanyi Ji, "Admission Control of Multimedia Traffic for Third Generation CDMA Network" *Proceedings of IEEE INFOCOM 2000* Volume: 3, 26-30 Mar 2000
7. C. Comaniciu, N. B. Mandayam, D. Famolari, P. Agrawal, "QoS Guarantees for Third Generation (3G) CDMA Systems via Admission and Flow Control" *Vehicular Technology Conference (VTC), 2000* Volume: 1 , 2000
8. Yue Ma, James J. Han, Kishor S. Trivedi, "Call Admission Control for Reducing Dropped Calls in CDMA Cellular Systems" *IEEE Infocom*, pp.1481-1490, 2000.
9. Mohsen Soroushnejad, Evaggelos Geraniotis, "Multi-Access Strategies for an Integrated Voice/Data CDMA Packet Radio Networks" *IEEE Transactions on Communications* Vol.43, No.2/3/4, February 1995
10. Taekyoung Kwon, Sooyeon Kim, Yanghee Choi, M. Naghshineh "Threshold-type Call Admission Control in Wireless/Mobile Multimedia Networks using Prioritised Adaptive Framework" *Electronics Letters* Volume: 36 Issue: 9 , 27 Apr 2000
11. Jihyuk Choi, Taekyoung Kwon, Yanghee Choi, Mahmoud Naghshineh "Call Admission Control for Multimedia Services in Mobile Cellular Networks: a Markovian Decision Approach" *Proceedings of Fifth IEEE Symposium on Computers and Communications (ISCC), 2000.*
12. Henk C. Tijms *Stochastic Modelling and Analysis- A Computational Approach* John Wiley & Sons
13. Cyrus Derman *Finite State Markovian Decision Processes* Academic Press 1970
14. Dmitri P. Bertsekas *Dynamic Programming and Optimal Control* Vol.I, Athena Scientific, Massachusetts, 1995

Adaptive Load Balancing of Cellular CDMA Systems Considering Non-uniform Traffic Distributions

Kuo-Chung Chu[1, 2] and Frank Yeong-Sung Lin[1]

[1] Department of Information Management, National Taiwan University,
[2] Department of Information Management,
Jin-Wen Institute of Technology, Taipei, Taiwan
d5725003@im.ntu.edu.tw

Abstract. In this paper, we investigate the load balancing problem by jointly considering sectorization and hybrid F/CDMA scheme in the scenario of non-uniform traffic distributions. The problem is formulated as a mathematical optimization model, and is solved by Lagrangean relaxation approach. The model objective is to minimize weighted blocking probability in terms of distribution diversity. To evaluate the model, performance analysis of adaptive load balancing is conducted by proposed bandwidth segmentation scheme; it is denoted adaptive scheme (AS). Non-adaptive (NA) approach by common power control scheme is compared. Combining sectorization and bandwidth segmentation scheme provides novel adaptive load balancing. Performance improvement that proposed adaptive scheme outperforms common power control scheme is about 50%.

1   Introduction

Code-division multiple access (CDMA) technique has been a promising technique for next generation wireless communication systems since the same bandwidth is shared by all users in the system, i.e. reuse of unity, soft channel capacity, and so forth. Users transmitting in the same frequency band are identified by user-specific code. For the sake of imperfect code orthogonality, interferences are incurred. In a multi-cellular environment, whenever a particular CDMA cell becomes increasingly loaded and the user increases, it will unavoidably affect all users in the system, not only the users in the home cell but also those in neighboring cells, especially in the scenario of hot spot and non-uniform traffic distributions. Hence, the system performance decreases as the number of active users increases.

Generally, a solution to the interference problem is power control mechanism for uniform traffic distribution environment, which attempts to achieve constant received mean power from each mobile within a cell [1]. However, with hot-spot cell, powering up all users in the cell to accommodate more users will result in excessive interference to users in neighboring cells to maintain sufficient signal-to- interference ratio (SIR) levels at their cell sites. Considering linear distribution, as in highway, power control may not be appropriate approach in the scenario. For non-uniform traffic distribution, sectorization is an effective way to maximize the network capacity

L. Bougé and V.K. Prasanna (Eds.): HiPC 2004, LNCS 3296, pp. 60–70, 2004.
© Springer-Verlag Berlin Heidelberg 2004

[2][3]. The goal of dynamic sectorization is similar to load balancing in previous studies [4][5]. Adaptive load-shedding scheme combines the power control and soft handoff functionality to enforce some users farthest away from cell/base station (two terms are used in turn hereafter) enter forced soft handoffs, and transfer to neighboring cells that are lightly loaded. In such a way, heavily loaded cells dynamically down size their coverage area in order to serve traffics, while adjacent cells that are less heavily loaded increase their coverage to accommodate the extra traffics.

A hybrid F/CDMA scheme has been proposed to moderately mitigate interferences [6]. In consideration of F/CDMA, capacity analysis in multi-band overlaid CDMA is proposed, and maximum bandwidth utilization is obtained [7][8]. Especially, the multi-band spectrum is to provisioning heterogeneous services requirements with sub-bands [8]. In this paper, we investigate the load balancing to maximize system capacity by jointly considering sectorization and allocating appropriate sub-spectrum in a cell. CDMA background is given in section 2. Section 3 presents the model of adaptive load balancing as well as solution approach. Section 4 illustrates the computational experiments. Finally, Section 5 concludes this paper.

2   CDMA Background

2.1   Sectorization and Hybrid F/CDMA Scheme

The common way to reducing the interference between users is sectorization using directional antennas. Sectorization utilizes the spatial domain to introduce orthogonalization [2][3] to the system. Since only a subset of the users is received at each antenna, the interference that each user incurs is less compared to a single antenna system. Without loss of generality, the interferences between users can be treated as the interferences between sectors. If the interference indicator function between sectors is pre-calculated, the interference from other users/cells is easily analyzed.

To do so, sector candidates probably configured in base station (BS) must be defined. Denote K the set of sector configurations. Denote S the set of sector candidates, each sector candidate $s_{k,i}$ is defined by both sector configuration (k) and sector identity (i). Denote B the set of BSs, in this paper, two configurations are given at base station ($|K|$=2), they includes one sector ($360°$ with omni-directional antenna) and three sectors ($120°$ per sector), and assign k =1, 2, respectively. For simplicity, $s_{k,i}$ is substituted by s, and denote $sector_{js}$ the sector s in BS j ($\forall s_{k,i} \in S$, $j \in B$), where B is the set of base stations.

In hybrid F/CDMA scheme, the available wideband spectrum is divided into a number of sub-bands with smaller bandwidths. Each sub-band employs direct sequence (DS) spreading with reduced processing gain and is transmitted in one and only one sub-band. The capacity of this F/CDMA system is calculated as the sum of the capacities of the sub-band. Given the whole bandwidth BW_{WHOLE} 60MHZ in both uplink and downlink (in both $W_{js_{k,i}}^{DL}$ and $W_{js_{k,i}}^{UL}$), which is made up of N_{FU} frequency units (FU) with BW_{FU} =6MHZ, where $N_{FU} = BW_{WHOLE}/BW_{FU}$ =10. By integrating FU a number of frequency segments (FS), so-called sub-band, can be separated. FS instead of whole bandwidth is deployed in sector$_{js}$. The term frequency segment and sub-band will be used in turn throughout the paper.

We investigate the load balancing by jointly considering sectorization and hybrid F/CDMA scheme in the scenario of non-uniform environment. If there are four frequency segments (FS0, FS1, FS2, and FS3) to be assigned in a cell/sector, for each of traffic distributions in Fig. 1 where shadow cell means heavy loads, the probable assignment would be different. Since the nature of non-uniform traffic distribution, the bandwidth requirement in each cell to satisfying SIR would be varied. The proposed scheme is called "adaptive load balancing" to optimally assigning FS with respect to traffic loads, the interferences between cells/sectors can be mitigated by bandwidth segmentation. For example, if BW_{WHOLE} allocated in downlink connection is decomposed into five FUs, $N_{FU} = |FU| = BW_{WHOLE}/BW_{FU}$ =5, the frequency segments (FS) that combined consecutively from FU are categorized into five groups of FS length. Thus, the set FU ={ 1, 2, 3, 4, 5 }, the set FS ={ (1), (2), (3), (4), (5), (1,2), (2,3), (3,4), (4,5), (1,2,3), (2,3,4), (3,4,5), (1,2,3,4), (2,3,4,5), (1,2,3,4,5) }. The total number of FS, $N_{FS} = |FS| = N_{FU}^{DL} \times (N_{FU}^{DL} +1)/2 = 5 \times 6/2$ =15.

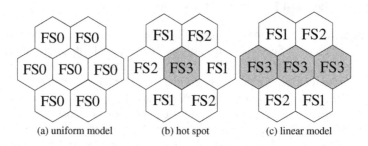

(a) uniform model   (b) hot spot   (c) linear model

Fig. 1. Scenarios of F/CDMA scheme applied in distribution diversity

2.2   Interference Model

Given sector configuration, the interference indicator functions $\Omega_{js\,j's'}^{UL}$ and $\Omega_{js\,j's'}^{DL}$ for uplink (UL) and downlink (DL) from sector$_{js}$ to sector$_{j's'}$, respectively, can be pre-calculated. Denote FS^{UL} and FS^{DL} the FS set in UL and DL, respectively. Applying bandwidth segmentation scheme, let x_{jsu} (y_{jsd}) be the decision variable which is 1 if sector$_{js}$ deploys frequency segment $u \in FS^{UL}$ ($d \in FS^{DL}$) or 0 otherwise. Thus, bandwidth allocated for uplink and downlink is calculated by w_{js}^{UL} $= \sum_{u \in FS^{UL}} x_{jsu} \cdot L_u \cdot BW_{FU}^{UL}$ and $w_{js}^{DL} = \sum_{d \in F^{DL}} y_{jsd} \cdot L_d \cdot BW_{FU}^{DL}$, respectively, where L_u (L_d) is the length of u (d). Frequency segments deployed in any two $sector_{js}$ probably exists overlapping of FUs. Furthermore, the interference indicator function $\Psi^{UL}_{u\,u'}$ from u to u' can be pre-calculated, and the indicator function $\Psi^{DL}_{d\,d'}$ is similar to $\Psi^{UL}_{u\,u'}$.

For traffic distribution, denote C the set of traffic classes and $c(t)$ ($c(t) \in C$) the traffic class of call request from mobile station (MS) t ($\forall t \in T$), where T is the set of mobile stations. If call type of MS t belongs to class-c, class-$c(t)$ is equivalent to class-c. Let $d^{UL}_{c(t)}$ ($d^{DL}_{c(t)}$) be the information rate in uplink (downlink). Denote z_{jst} the decision variable which is 1 if MS t is granted by $sector_{js}$ or 0 otherwise. Assuming both link powers are perfectly controlled, it ensures the received power at $sector_{js}$ from MS t with constant value in same traffic class-$c(t)$. Denote $P^{UL}_{c(t)}$ ($P^{DL}_{c(t)}$) the received uplink (downlink) power signal, the signal-to-interference ratio (SIR) $SIR^{UL}_{js,c(t)}$ in uplink, which consists of processing gain and intra-inter cell/sector interferences, is defined as $SIR^{UL}_{js,c(t)} = \frac{W^{UL}_{js}}{d^{UL}_{c(t)}} \cdot \frac{P^{UL}_{c(t)} + (1-z_{jst})V}{(1-\rho^{UL})I^{UL}_{jst,intra} + I^{UL}_{jst,inter}}$, where ρ^{UL} is the uplink orthogonality factor, $\alpha^{UL}_{c(t)}$ is uplink activity factor. A very large constant value V in numerator is to satisfy constraint requirement if MS t is rejected ($z_{jst}=0$). Denote D_{jt} the distance from MS t to $sector_{js}$, and given attenuation factor $\tau = 4$, the intra-cell interference is $I^{UL}_{jst,intra} = \sum_{\substack{t' \in T \\ t' \neq t}} \alpha^{UL}_{c(t')} P^{UL}_{c(t')} z_{jst'}$ and inter-cell $I^{UL}_{jst,inter}$ in uplink can be expressed by (1). Equation (1) jointly considers effect of inter-sector and inter-FS on inter-cell interference.

Linear model is a best form to efficiently solve the integer programming problem. Unfortunately, the coupling of decision variables (x_{jsu}, z_{jst}) and decision variables ($x_{j's'u'}$, x_{jsu}, z_{jst}) results to non-linear form in $SIR^{UL}_{js,c(t)}$ and (1), respectively. Here we introduce auxiliary variable $\gamma_{jsut} = x_{jsu} z_{jst}$ (s.t. $x_{jsu} + z_{jst} \geq 2 \cdot \gamma_{jsut}$, $x_{jsu} + z_{jst} - 1 \leq \gamma_{jsut}$) so that non-linear form can be reduced to linear form. Applying auxiliary variable γ_{jsut}, $SIR^{UL}_{js,c(t)}$ can be rewritten as (2). Based upon γ_{jsut}, another auxiliary variable $\zeta_{j's'u'jsut'} = \gamma_{j's'u't'} x_{jsu} = x_{j's'u'} z_{j's't'} x_{jsu}$ is also employed (s.t. $\gamma_{j's'u't'} + x_{jsu} \geq 2 \cdot \zeta_{j's'u't'jsu}$, $\gamma_{j's'u't'} + x_{jsu} - 1 \leq \zeta_{j's'u't'jsu}$). Applying auxiliary variable $\zeta_{j's'u'jsut'}$, (1) can be rewritten as (3).

$$I^{UL}_{jst,inter} = \sum_{\substack{j' \in B \\ j' \neq j}} \sum_{\substack{s' \in S \\ s' \neq s}} \sum_{\substack{t' \in T \\ t' \neq t}} \Omega^{UL}_{j's'js} \, \alpha^{UL}_{c(t')} P^{UL}_{c(t')} \left(\frac{D_{j't'}}{D_{jt'}}\right)^{\tau} z_{j's't'} \sum_{u \in FS^{UL}} \sum_{u' \in FS^{UL}} \left(\Psi^{UL}_{u'u} \, x_{j's'u'} x_{jsu}\right) \quad (1)$$

$$SIR^{UL}_{js,c(t)} = \frac{\sum_{u \in FS^{UL}} \left[L_u BW^{UL}_{FU} \left(x_{jsu} \left(P^{UL}_{c(t)} + V\right) - \gamma_{jsut} V\right)\right]}{d^{UL}_{c(t)} \left((1-\rho^{UL}) I^{UL}_{jst,intra} + I^{UL}_{jst,inter}\right)} \quad (2)$$

$$I^{UL}_{jst,inter} = \sum_{\substack{j' \in B \ s' \in S \ t' \in T \\ j' \neq j \ s' \neq s \ t' \neq t}} \Omega^{UL}_{j's'\,js} \left(\frac{D_{j't'}}{D_{jt'}}\right)^{\tau} \sum_{u \in FS^{UL}} \sum_{u' \in FS^{UL}} \Psi^{UL}_{u'u} \, \zeta_{j's'u't'\,jsu} \alpha^{UL}_{c(t')} P^{UL}_{c(t')} \tag{3}$$

In downlink connection, it is similar to uplink except that the interference of intra-cell is more complicated. Auxiliary variables $\eta_{jsdt} = y_{jsd} z_{jst}$ and $\xi_{j's'd'jsdt'} = \eta_{j's'd't'} y_{jsd} = y_{j's'd'} z_{j's't'} y_{jsd}$ are introduced that subject to ($y_{jsd} + z_{jst} \geq 2 \cdot \eta_{jsdt}$, $y_{jsd} + z_{jst} - 1 \leq \eta_{jsdt}$) and ($\eta_{j's'd't'} + y_{jsd} \geq 2 \cdot \xi_{j's'd't'jsd}$, $\eta_{j's'd't'} + y_{jsd} - 1 \leq \xi_{j's'd't'jsd}$), respectively. Associated models are expressed in (4)-(6).

$$I^{DL}_{jst,intra} = \sum_{\substack{t' \in T \\ t' \neq t}} \alpha^{DL}_{c(t')} P^{DL}_{c(t')} \left(\frac{D_{jt'}}{D_{jt}}\right)^{\tau} z_{jst'} \tag{4}$$

$$I^{DL}_{jst,inter} = \sum_{\substack{j' \in B \ s' \in S \ t' \in T \\ j' \neq j \ s' \neq s \ t' \neq t}} \Omega^{DL}_{j's'\,js} \left(\frac{D_{j't'}}{D_{j't}}\right)^{\tau} \sum_{d \in FS^{DL}} \sum_{d' \in FS^{DL}} \Psi^{DL}_{d'd} \xi_{j's'd't'\,jsd} \alpha^{DL}_{c(t')} P^{DL}_{c(t')} \tag{5}$$

$$SIR^{DL}_{js,c(t)} = \frac{\sum_{d \in FS^{DL}} \left[L_d BW^{DL}_{FU} \left(y_{jsd} \left(P^{DL}_{c(t)} + V \right) - \eta_{jsdt} V \right) \right]}{d^{DL}_{c(t)} \left((1 - \rho^{DL}) I^{DL}_{jst,intra} + I^{DL}_{jst,inter} \right)} \tag{6}$$

3 Adaptive Load Balancing Model

3.1 Performance Measure

In this paper, we consider multiple traffic classes in adaptive load balancing. Kaufman model [9] is used as a performance measure to effectively analyze blocking probability for each traffic class. Assuming M channels are shared by all traffic requirements. For each traffic class- c ($\forall c \in C$) with distinct resource requirements, the traffic arrival is a stationary Poisson process with mean rate λ. The channel requirement b is an arbitrary discrete random variable ($P\{b = b_c\} = q_c$, $\forall c \in C$). A call request with channel requirement b_c has holding time with mean μ_c. Thus, traffics with channel requirement b_c generate in Poisson arrival process with mean rate $\lambda_c = \lambda q_c$ and the class- c offered load $a_c = \lambda_c \mu_c$. The blocking probability of traffic class- c is defined as $B^c(a,b) = \sum_{i=0}^{b_c-1} q(|M|-i)$, where the distribution of $q(\bullet)$, the number of channels occupied for the complete sharing policy, satisfies the equation $\sum_{c \in C} a_c b_c q(j - b_c) = jq(j)$, $j = 0, 1, ..., M$, and $q(x) = 0$ for $x < 0$ and $\sum_{j=0}^{M} q(x) = 1$.

3.2 Problem Formulation and Solution Approach

The capacity of each cell/sector is calculated subject to SIR requirement. Probably, the cell that is lightly loaded is incurred more interference from the heavily loaded cell, it results to increasing blocking probability in lightly loaded cell. Appropriately allocating FS in a sector/cell to mitigate interference is considerable in the environment with heterogeneous traffics. The aim of the model is to investigate the load balancing among all cells/sectors in terms of blocking probability. Performance measure, B_{js}^c the call blocking probability of traffic class-c in sector$_{js}$, developed by Kaufman is applied. Denote $g_{js} = \sum_{c \in C} g_{js}^c$ the aggregate traffic (in Erlangs) in sector$_{js}$, where g_{js}^c is the aggregate intensity of class-c, and denote $m_{js} = \sum_{t \in T} z_{jst} m^{c(t)}$ the number of total channels allocated in sector$_{js}$, where $m^{c(t)}$ is the number of channels required for traffic class-$c(t)$. The weighted factor w_{js} is expressed by $w_{js} = \sum_{c \in C} g_{js}^c / \sum_{j \in B} \sum_{s \in S} \sum_{c \in C} g_{js}^c$. To objective function (IP) is to minimize weighted blocking probability, in which K^c is the balancing coefficient (BLC) where $\sum_{c \in C} K^c = 1$. If $K^{c1} > K^{c2}$, it claims that class-$c1$ is more concerned than class-$c2$ about traffic load balancing.

$$Z_{IP} = \min \sum_{c \in C} K^c \sum_{j \in B} \sum_{s \in S} w_{js} B_{js}^c \left(g_{js}, m_{js} \right) \qquad \text{(IP)}$$

$$\left(\frac{E_b}{N_{TOTAL}} \right)_{c(t)}^{UL} \leq \frac{\sum_{u \in F^{UL}} \left[L_u BW_{FU}^{UL} \left(x_{jsu} \left(P_{c(t)}^{UL} + V \right) - \gamma_{jsut} V \right) \right]}{d_{c(t)}^{UL} \left((1 - \rho^{UL}) I_{jst,intra}^{UL} + I_{jst,inter}^{UL} \right)} \qquad \forall j \in B, s \in S, t \in T \text{ (7)}$$

$$\left(\frac{E_b}{N_{TOTAL}} \right)_{c(t)}^{DL} \leq \frac{\sum_{d \in F^{DL}} \left[L_d BW_{FU}^{DL} \left(y_{jsd} \left(P_{c(t)}^{DL} + V \right) - \eta_{jsdt} V \right) \right]}{d_{c(t)}^{DL} \left((1 - \rho^{DL}) I_{jst,intra}^{DL} + I_{jst,inter}^{DL} \right)} \qquad \forall j \in B, s \in S, t \in T \text{ (8)}$$

$$\sum_{t \in T} z_{jst} \lambda_{c(t)} \mu_{c(t)} = g_{js}^c \qquad \forall j \in B, s \in S, c \in C \text{ (9)}$$

$$\sum_{t \in T} z_{jst} m^{c(t)} = m_{js} \qquad \forall j \in B, s \in S \text{ (10)}$$

$$z_{jst} D_{jt} \leq R_{js} \delta_{jst} \qquad \forall j \in B, s \in S, c \in C \text{ (11)}$$

$$\sum_{j \in B} \sum_{s \in S} z_{jst} \leq 1 \qquad \forall t \in T \text{ (12)}$$

$$2 \cdot \gamma_{jsut} \leq x_{jsu} + z_{jst} \qquad \forall j, \in B, s \in S, t \in T, u \in FS^{UL} \text{ (13)}$$

$$x_{jsu} + z_{jst} - 1 \leq \gamma_{jsut} \qquad \forall j, \in B, s \in S, t \in T, u \in FS^{UL} \text{ (14)}$$

$$2 \cdot \zeta_{j's'u'jsut} \leq \gamma_{j's'u't'} + x_{jsu} \qquad \forall j, j' \in B, j \neq j', s, s' \in S, s \neq s', t' \in T, u, u' \in FS^{UL} \text{ (15)}$$

$$\gamma_{j's'u't'} + x_{jsu} - 1 \leq \zeta_{j's'u'jsut} \qquad \forall j, j' \in B, j \neq j', s, s' \in S, s \neq s', t' \in T, u, u' \in FS^{UL} \text{ (16)}$$

$$2 \cdot \eta_{jsdt} \leq y_{jsd} + z_{jst} \qquad \forall j, \in B, s \in S, t \in T, d \in FS^{DL} \text{ (17)}$$

$$y_{jsd} + z_{jst} - 1 \leq \eta_{jsdt} \qquad \forall j, \in B, s \in S, t \in T, d \in FS^{DL} \text{ (18)}$$

$$2 \cdot \xi_{j's'd'jsdt'} \le \eta_{j's'd't'} + y_{jsd} \qquad \forall j, j' \in B, j \ne j', s, s' \in S, s \ne s', t' \in T, d, d' \in FS^{DL} \ (19)$$

$$\eta_{j's'd't'} + y_{jsd} - 1 \le \xi_{j's'd'jsdt'} \qquad \forall j, j' \in B, j \ne j', s, s' \in S, s \ne s', t' \in T, d, d' \in FS^{DL} \ (20)$$

$$\Phi_{js}^c \le \frac{\displaystyle\sum_{t \in T} z_{jsc(t)}}{\displaystyle\sum_{t \in T} \delta_{jsc(t)}} \qquad \forall j \in B, s \in S, c \in C \ (21)$$

$$\sum_{u \in FS^{UL}} x_{jsu} = 1 \qquad \forall j \in B, s \in S \ (22)$$

$$\sum_{d \in FS^{DL}} y_{jsd} = 1 \qquad \forall j \in B, s \in S \ (23)$$

$$z_{jst} = 0 \text{ or } 1 \qquad \forall j \in B, s \in S, t \in T \ (24)$$

$$x_{jsu} = 0 \text{ or } 1 \qquad u \in FS^{UL} \ (25)$$

$$y_{jsd} = 0 \text{ or } 1 \qquad d \in FS^{DL} \ (26)$$

$$\gamma_{jsut} = 0 \text{ or } 1 \qquad \forall j, \in B, s \in S, t \in T, u \in FS^{UL} \ (27)$$

$$\eta_{jsdt} = 0 \text{ or } 1 \qquad \forall j, \in B, s \in S, t \in T, d \in FS^{DL} \ (28)$$

$$\zeta_{j's'u'jsut'} = 0 \text{ or } 1 \qquad \forall j, j' \in B, j \ne j', s, s' \in S, s \ne s', t' \in T, u, u' \in FS^{UL} \ (29)$$

$$\xi_{j's'd'jsdt'} = 0 \text{ or } 1 \qquad \forall j, j' \in B, j \ne j', s, s' \in S, s \ne s', t' \in T, d, d' \in FS^{DL} \ (30)$$

Denote $(E_b/N_{TOTAL})_{c(t)}^{UL}$ and $(E_b/N_{TOTAL})_{c(t)}^{DL}$ the QoS requirement of admission control, then SIR constraint for uplink and downlink is expressed by (7) and (8), respectively. Traffic intensity of class- c in sector$_{js}$ is calculated in (9), where $\lambda_{c(t)}$ is mean arrival rate for class-$c(t)$. The number of channels allocated is constrained by (10). Denote δ_{jst} indication function which is 1 if MS t is covered by sector$_{js}$ or 0 otherwise. MS t can be serviced in the coverage of sector$_{js}$ by (11), where R_{js} is the power transmission radius and D_{jt} is the distance from sector$_{js}$ to MS t. Constraint (12) requires that each mobile user can be homed to only one base station. We do not take soft handoff into account. Since four auxiliary decision variables are introduced, a number of constraints are listed from (13) to (20). A pre-defined service rate Φ_{js}^c for class-c is given in (21). Using bandwidth segmentation scheme, only one FS be deployed in uplink and downlink by constraint (22) and (23), respectively. Constraints (24)-(30) are integer properties of the decision variables.

To solving the complicated optimization model, Lagrangean relaxation method is applied [10]. Problem (IP) is transferred to be a dual problem $Z_D = \max Z_D(V)$ by relaxing ten constraints (7), (8), (13)-(20), then multiply the relaxed constraints with corresponding Lagrangean multipliers vector $V=(v_1, v_2, v_3, v_4, v_5, v_6, v_7, v_8, v_9, v_{10})$, and add them to the primal objective function. We then get Lagrangean relaxation (LR) problem, which is further decomposed into three independent subproblems, including admission control subproblem related to z_{jst}, uplink bandwidth allocation subproblem related to x_{jsu}, γ_{jsut}, $\zeta_{j's'u'jsut'}$, and downlink bandwidth allocation subproblem related to y_{jsu}, η_{jsut}, $\xi_{j's'u'jsut'}$. All of them can be optimally solved efficiently by proposed algorithms[1]. According to the weak Lagrangean duality theorem, for any $V \geq 0$, the objective value of $Z_D(V)$ is a lower bound (LB) of Z_{IP}. Thus, the dual problem (D) $Z_D = \max Z_D(V)$ subject to $V \geq 0$ is constructed to calculate the tightest lower bound by adjusting multipliers. Subgradient method is used to solving the dual problem. Let the vector S be a subgradient of $Z_D(V)$ at $V \geq 0$. In iteration k of subgradient optimization procedure, the multiplier vector π is updated by $\pi^{k+1} = \pi^k + t^k S^k$, in which t^k is a step size determined by $t^k = \lambda\left(Z_{IP}^* - Z_D(\pi^k)\right)/\left\|S^k\right\|^2$, where Z_{IP}^* is an upper bound (UB) on the primal objective function value after iteration k, and λ is a constant where $0 \leq \lambda \leq 2$. To calculate upper bound of (IP), the algorithm of getting primal feasible solutions is also proposed [2].

4 Computational Experiments

4.1 Environment and Parameter

The structure of 5×5 two-dimensional array with hexagonal cells is deployed, and given $R_{js} =5.0$km. The required bit energy-to-noise density (QoS) for voice (v) and data (d) traffics $(E_b/N_{TOTAL})_v^{UL} = (E_b/N_{TOTAL})_v^{UL} =7$dB and $(E_b/N_{TOTAL})_d^{UL} = (E_b/N_{TOTAL})_d^{UL} =10$ dB, respectively, and the information rate $d_v^{UL} = d_v^{DL} =9.6$bps, $d_d^{UL} =19.2$bps, $d_d^{DL} =38.4$bps. Activity factor $\alpha_v^{UL} = \alpha_v^{DL} = \alpha_d^{UL} = \alpha_d^{UL} =0.5$. Number of channel required $m^v =1$, $m^d =4$. Orthogonality factor $\rho^{UL} =0.9$, $\rho^{DL} =0.7$. Power is perfectly controlled by $P_v^{UL} =10$dB, $P_v^{DL} =15$ dB, $P_d^{UL} =15$ dB, $P_d^{DL} =20$ dB. Service rate $\Phi_{js}^v = \Phi_{js}^d =0.1$. Call requests for voice and data are generated in Poisson arrival process with λ_v and λ_d, respectively. The mean call holding time is given $\mu_v =180$(sec), $\mu_d =600$(sec). Traffic intensity generated in heavy loaded cells of non-uniform distribution is multiple of five in uniform distribution.

For each case of distributions, Z_{IP} is solved with maximum number of 1000 iterations. The improvement counter is given 25. Time consumed in each case is up to 370 (sec). The error gap defined by (UB-LB)/LB*100% is calculated less than 30% in all cases. For the purpose of statistic analysis, 100 tests are experimented in each case. Performance analysis is based on the average of 100 tests.

[1, 2] Detailed algorithms are omitted due to the length limitation of the paper. A complete version of the paper is available upon request.

4.2   Performance Analysis

Traffic distributions including uniform (U), linear (L), as well as hop spot (H) models as shown in Fig. 1 are considered, performance analysis of adaptive load balancing is manipulated by proposed bandwidth segmentation scheme; it is denoted adaptive scheme (AS). For the comparison purpose, non-adaptive (NA) approach by common power control scheme [4][5] is implemented. Besides, sectorization is also taken into account. We analyze the weighted blocking

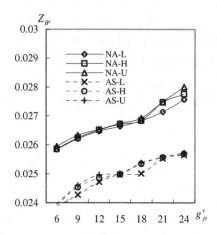

Fig. 2. Blocking as a function of voice traffics with $g_{js}^d = 30$ and $|S| = 1$

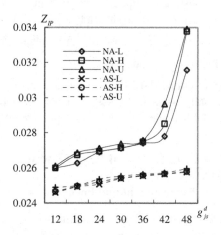

Fig. 3. Blocking as a function of data traffics with $g_{js}^v = 15$ and $|S| = 1$

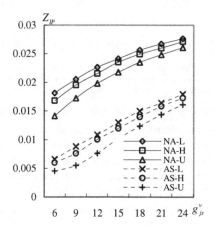

Fig. 4. Blocking as a function of voice traffics with $g_{js}^d = 30$ and $|S| = 3$

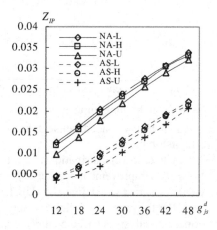

Fig. 5. Blocking as a function of data traffics with $g_{js}^v = 15$ and $|S| = 3$

probability (Z_{IP}) as a function of voice (data) traffic intensity with constant data (voice) traffic, in which load balancing scheme v.s. traffic distribution is compared. BLC ratio of K^v vs. K^d (0.5 vs. 0.5) is given.

First of all, without sectorization ($|S|=1$) and given $g^d_{js}=30$, Z_{IP} is a function of voice traffic in Fig. 2, while Z_{IP} is a function of data traffic in Fig. 3 with $g^v_{js}=15$. In the case of voice intensity $g^v_{js}=15$ in Fig. 2, no matter which distribution is considered, proposed AS scheme reduces blocking percentage of (0.0265-0.025)/0.0265*100% = 56.6%. In Fig. 3, much more blocking (up to 0.034) is incurred in case of data intensity $g^d_{js}=48$ with NA scheme. Considering sectorization with $|S|=3$, Z_{IP} is harmonically increasing function of data and voice traffics intensity in Fig. 4, and Fig. 5, respectively. This implies that combining sectorization and bandwidth segmentation approaches provides novel adaptive load balancing scheme. In summary, performance improvement that proposed adaptive scheme outperforms power control scheme is about 50%.

5 Conclusion

To maximize entire system capacity in ever-increasing non-uniform distribution, we propose the load balancing mechanism by jointly considering sectorization and hybrid F/CDMA scheme. Experiments show the mechanism is outstanding for performance management. For more practical, further experiments can be conducted, such as more generic cell planning rather than hexagonal cell structure, a great diversity of non-uniform distributions, a lot of BLC combinations.

References

1. W.-M. Tam and F.C.M. Lau, "Analysis of power control and its imperfections in CDMA cellular systems," *IEEE Trans. Veh. Technol.*, vol. 48, pp. 1706–1717, Sep. 1999.
2. C. U. Saraydar and A. Yener, "Adaptive cell sectorization for CDMA systems," *IEEE J. Select. Areas Commun.*, vol. 19, pp. 1041–1051, June 2001.
3. C. Y. Lee, H. G. Kang, and T. Park, "A dynamic sectorization of microcells for balanced traffic in CDMA: genetic algorithms approach," *IEEE Trans. Veh. Technol.*, vol. 51, pp. 63–72, Jan. 2002.
4. X.H. Chen and K.L. Lee, "A novel adaptive traffic load shedding scheme for CDMA cellular mobile systems," in *Proc. ICCS*, 1994, pp. 566–570.
5. X.H. Chen, "Adaptive traffic-load shedding and its capacity gain in CDMA cellular systems," in *Proc. IEE Communications*, 1995, pp. 186–192.
6. T. Eng and L. B. Milstein, "Comparison of hybrid FDMA/CDMA systems in frequency selective Rayleig fading," *IEEE J. Select. Areas Commun.*, vol. 12, pp. 938–951, June 1994.
7. L. Zhuge and V. O. K. Li, "Reverse-link capacity of multiband overlaid DS-CDMA systems," *Mobile Networks and Applications*, vol. 7, pp. 101–113, 2002.

8. L. Zhuge and V. O. K. Li, "Overlaying CDMA systems with interference differentials," *Mobile Networks and Applications*, vol. 8, pp. 269–278, 2003.

9. J. S. Kaufman, "Blocking in a shared resource environment," *IEEE Trans. Commun.*, vol. 29, pp. 1474–1481, 1981.

10. M. L. Fisher, "The Lagrangian Relaxation Method for Solving Integer Programming Problems," *Management Science*, vol. 27, pp. 1–18, 1981.

An Active Framework for a WLAN Access Point Using Intel's IXP1200 Network Processor

R. Sharmila, M.V. LakshmiPriya, and Ranjani Parthasarathi

Dept. of Computer Science and Engg., College of Engineering, Guindy,
Anna University, Tamil Nadu, India
sharmilaradhakrishnan@rediffmail.com,
{tkslp, rp}@cs.annauniv.edu

Abstract. Active Networks provide the user with the capability to inject customized programs into the network. The network nodes should interpret these programs and perform the desired operation on the data flowing through the network. The Network Processor (NP), a special purpose programmable device designed specifically to process packets at high speed with its concurrent packet processing model, forms an ideal platform for the implementation of active networks. This paper presents the development of an active framework using the IXP1200 network processor. This active framework is implemented for a Wireless LAN Access Point that bridges wired and wireless network segments. This framework allows the access point to use different classification, scheduling or queue management algorithms for different applications. The special features of IXP1200 such as multiprocessing, multithreading, block data movement, etc., are exploited to develop the access point and the active framework, with increased processing speed, scalability and flexibility.

1 Introduction

Active Networks (AN)[1] are networks that allow routing elements to be programmed by the packets passing through them. The network nodes interpret the programs carried in the packets and perform desired operation on the data flowing through the network. This allows computation previously possible only at endpoints to be carried out within the network itself, thus enabling optimizations. Although, this concept of active networks has been around for a long time, the deployment of these networks has not really taken off. The prime reason here is the additional processing that needs to be done at the intermediate routers, which has a negative impact on the efficiency in packet processing. Added to this is the lack of flexibility for this kind of processing in routers that are built using ASICs and FPGAs. With the advent of Network processors, however, the scenario has changed. Network Processors are well suited for most packet-processing tasks, ranging from content switching, load balancing, network security, and terminal mobility. With their high processing capability, they also provide an ideal platform for supporting the implementation of active networks [2].

L. Bougé and V.K. Prasanna (Eds.): HiPC 2004, LNCS 3296, pp. 71–80, 2004.
© Springer-Verlag Berlin Heidelberg 2004

This paper presents the development of an active framework on a network processor. The network environment chosen for the active framework is the wireless LAN. In a wireless LAN, the access point is the ideal location to implement the active framework, as it acts as a bridge connecting the wired and wireless segments, with vastly different bandwidth and reliability characteristics [3, 4]. For instance, in a wired network, having static packet classification rules may suffice, whereas, in a wireless environment, where nodes enter/leave the network, there is a need to change the classification rules dynamically. Hence, it would be reasonable to have an active framework on the access point, where, classification rules can be dynamically downloaded and used. Similarly, other network functions like scheduling, queue management, Qos support etc., can benefit from this flexibility.

The use of network processors for the access point itself is a recent approach. Access points are typically implemented entirely in software on a general-purpose processor or in hardware. Access points implemented in software [5,6,7] have the disadvantage of lower packet processing speed but the advantage of greater flexibility. On the other hand, access points implemented in hardware [8,9] have greater processing power but are not flexible. Thus there exists a gap in the access point domain. Network processor with its concurrent packet-processing model forms an ideal platform for the implementation of access points, by providing both flexibility and processing power. It is this flexibility that is further exploited to provide an active framework in our work. The active framework is used to choose the desired classification / scheduling / queue management algorithms in the access point for different applications. The network processor used here is the Intel IXP1200 [10]. Even though network processors like the Intel's IXP440, which have been specifically designed for developing VPNs, access points, etc., could be used, we have chosen the IXP1200 for its low cost. Thus we present a low cost solution to develop an active access point.

The rest of the paper is organized as follows. Section 2 gives an overview of IXP1200 network processor. Section 3 gives the design of the access point using IXP1200 NP. Section 4 gives the design of the active framework on IXP1200 NP. Section 5 presents the implementation and test results. Finally, section 6 gives the conclusions and future work.

2   Overview of IXP1200 Network Processor

The IXP1200 [10] is an integrated Network Processor, comprised of a single StrongARM processor, six Microengines, standard memory interfaces (SRAM, SDRAM and scratch pad memory), and high-speed bus interfaces. Each microengine can support four threads. The unique architecture of the IXP1200 provides the user a highly concurrent packet-processing model, while keeping the programming model simple.

The microengines are custom processors implemented specifically for networking applications. The microengines being fully programmable processors are able to examine packet contents at all levels of the networking stack. This makes them suitable not only for layer 2 and 3 switching and/or forwarding, but also for applications that require deeper inspection and manipulation of packet contents.

3   Design of the Access Point Using IXP1200 Network Processor

The access point sits between the wired and wireless LAN. The wired network considered is 802.3 Ethernet network and the wireless network is 802.11. The access point basically forwards packets from the wireless nodes (802.11) destined to nodes on the wired segment (Ethernet) and vice versa. Packets destined to the same network segment from which it originated are discarded. The packets that are selected for forwarding are mapped to the packet format of the destination segment.

To implement the above functions a Bridge Table (BT), a table that maps MAC address of the network nodes to the ports on which it arrived is maintained. An Active Station Table (AST) is also maintained which contains the MAC addresses of the wireless network nodes. The system also has a default classifier, a default queue manager and a default scheduler. The default queue management used is 'tail-drop' policy. Here packets in the tail of the queue are removed when overflow occurs. For classification, Class Based Queuing (CBQ)[11] is used. Till now CBQ has been used at the IP layer. In this access point, CBQ is used at the MAC layer i.e., MAC addresses are used for classification instead of IP addresses. In a wireless scenario, the IP addresses are not fixed. Hence classifying packets based on IP addresses is not reasonable, in this scenario. So, we classify packets based on their MAC addresses that are fixed. The default scheduling algorithm used is priority scheduling. Here all higher priority packets are serviced before lower priority packets.

3.1   Access Point Modules

The overall block diagram of the access point on IXP1200 NP is shown in Fig. 1. The various modules are Ingress, Default Classifier, Receive Handler, Default Scheduler, Send Handler, and Egress for Ethernet/802.11, Default Queuing Management Module and Table management Module. The block diagram also shows the mapping of the modules to various microengines of the IXP1200 network processor. The functionality of each of the modules is described below.

Ingress for Ethernet/802.11. This module gets the packet from Ethernet/802.11 port and copies it into an input buffer in SDRAM. The buffer is allocated from a buffer pool by name IBUFF1/IBUFF2 respectively. When a buffer is allocated, memory of the defined size is allocated both in SRAM and SDRAM. The incoming packet is stored in SDRAM and the packet descriptor is stored in SRAM. The fragmented 'mpackets' of a single incoming packet, that arrive at the Ethernet/802.11 ports, are defragmented by this module and copied into the corresponding buffer. The packet descriptor contains details of the total number of quad words in the packet and the total number of bytes in the packet. The send handler and egress modules use the packet descriptor. The packet descriptor is constructed as shown in Fig. 2 and placed

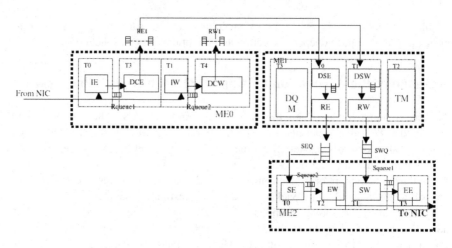

ME-n – Micro Engine *n*   TM –Table Manager (Bridge Table & Station Table)
DQM – Default Queue Manager Tn – Thread *n*
Wireless-802.11
DSW- Default Scheduler   RW-Receive Handler   IW-Ingress
DCW- Default Classifier   SW- Send Handler   EW – Egress
Ethernet
DSE-Default Scheduler   RE-Receive Handler   IE – Ingress
DCE-Default Classifier   SE-Send Handler   EE – Egress

Fig. 1. Overall block diagram of Access Point On IXP1200

in SRAM. The address of this packet descriptor is placed into a queue Rqueue1/Rqueue2 depending on whether it is an Ethernet/802.11 packet. The default classifier for the Ethernet/802.11 module handles this respectively.

Default Classifier for Ethernet/802.11. Four queues namely, RE1_Q1..RE2_Q4/ RW1_Q1.RW2_Q4 are maintained for Ethernet/802.11 packets. RE1_Q1/RW1_Q1 has the lowest priority and RE1_Q4/RW1_Q4 has the highest priority Default classifier for Ethernet/802.11 module gets the descriptor address of the next packet to be processed from Rqueue1/Rqueue2. It then places the incoming packet into any one of the four queues RE1_Q1..RE1_Q4/ RW1_Q1..RW2_Q4 respectively based on the default Class Based queuing (CBQ) tree. The source MAC address and TCP/UDP port number are taken into consideration while classifying using the CBQ tree.

Default Scheduler for Ethernet/802.11. The default scheduler uses Priority Queuing (PQ) algorithm. Here packets in the queues, RE1_Q1..RE1_Q4/ RW1_Q1..RW2_Q4, are scheduled. Here, the packets in the higher priority queues get serviced first. A packet in a queue is serviced only if there are no packets to be serviced in higher priority queues. The scheduled packets are sent to the Receive Handler for Ethernet/802.11 module.

Total no. of quad words in the packet 10:3 (bits)	Total no. of bytes in the packet 10:3 (bits)

Fig. 2. Packet descriptor Format

Destination MAC (6 Bytes)	Port Number (2 Bytes)	Timestamp (8 Bytes)

Fig. 3. Bridge Table Format

Destination MAC (6 Bytes)	Port Number (2 Bytes)	Time stamp (8 Bytes)

Fig. 4. Active Station Table Format

Destination MAC Address (6 Bytes)	Source MAC Address (6 Bytes)	Pointer to data packet in memory (4 Bytes)	Port number (2 Bytes)

Fig. 5. Internal Packet Format

Receive Handler for Ethernet/802.11 Module. This module gets the address of the packet descriptor of the Ethernet/802.11 packet to be processed next from the above module. It gets the source and destination address from Ethernet/802.11 header. It searches the bridge table (BT), which maps station addresses to ports. The format of BT entry is given in Fig. 3. If an entry is found in the BT, only the timestamp is updated, else a new entry is added. If the data packet is from 802.11 port then the active station table (AST), the table that contains the list of registered active wireless LAN terminals is also updated. The format of this table is shown in Fig. 5. It then searches the BT for destination address. If the port number in the BT for that address is equal to source port number, then the memory buffer held by that packet is freed. If the port number in the BT for that address is not equal to source port number, an internal packet whose format is given in Fig. 4 is, constructed and placed in a queue SEQ/SWQ. The Send handler for Ethernet/802.11 modules processes this queue respectively.

Send Handler for Ethernet/802.11 Module. This gets the internal packets from SEQ/SWQ. It then constructs the corresponding 802.11/Ethernet headers and calculates the CRC. It then copies the constructed 802.11/Ethernet packets into another buffer OBUFF2/OBUFF1. The packet descriptor is constructed according to the format as in Fig. 3. The constructed descriptor is written into the corresponding SRAM of the allocated buffer. It then places the address of this packet descriptor into a queue Squeue2/Squeue1, the queue that contains the descriptor address of 802.11/Ethernet packets to be handled by Egress for 802.11/Ethernet modules.

Egress for Ethernet/802.11. This module gets the next packet to be sent out from Squeue1/Squeue2, fragments it to a maximum size of 64b and sends it onto the corresponding Ethernet/802.11 interface.

Default Queuing Management Module. This module is responsible for removing packets from the queues if the access point's load is high and the queues approach overflow. Overflow is detected by checking a shared variable residing in scratch memory (FOVERFLOW). All the other modules, which send data to the queues, update this variable. FOVERFLOW is incremented by 1 by other modules if send operation fails because of the queue being full. Default queuing management module checks FOVERFLOW periodically (say every 2000 machine cycles), and if it exceeds specific threshold (Q_THRESHOLD), a single element is removed from all the queues. It also resets the value of FOVERFLOW to zero.

Table Management Module. This module is responsible for removing obsolete entries in the BT and the AST Obsolete entries are those in which the difference between timestamp in table for that entry and the current machine cycle is greater than a specific threshold value.

4   Design of Active Framework for a WLAN Access Point Using IXP1200 Network Processor

The access point discussed above has a fixed classifier, scheduler and queue manager. An active framework that enables the access point to use different scheduling, classification and queue management algorithms that can be dynamically downloaded is presented here. The key design issues that are considered for the development of the active framework are choice of active approach, interoperability, extensibility, and abstraction.

There are two possible approaches to building active networks - a discrete or out-of band approach and an integrated or in-band approach. In the discrete approach, programs are injected into the programmable active node separately from the actual data packets that traverses through the network. In an integrated approach, also termed as the encapsulation approach, the program is integrated into every packet of data sent to the network. In this system, discrete approach [3] is selected, as the active code is relatively large. Each data packet need not carry the classification/scheduling /queue management algorithms as they are flow dependent and not specific to a particular packet.

There are three well-known methods of achieving interoperability [3]. One is to express the programs in a high level source language that may be interpreted at the nodes. The second is to adopt a platform independent intermediate representation. The third is to have programs in platform-dependent format and arrange to carry multiple encoding of its program - one for each type of the platform it traverses. In this system, the third method is chosen, wherein, platform dependent binary code is passed. However, code for only one NP platform (IXP1200) has been developed.

The system has been developed using Micro-c, a high level type-safe language as a basis for extensibility. Moreover the system has the ability to avoid disruption of inflow of data packets to the maximum extent possible when switching over from one algorithm to another. This is one of the greatest advantages of the system.

Application Programming Interfaces (API) has been developed to provide an abstraction to the AP, which enable the implementation of various classification, scheduling and queue management algorithms. Active packets are encapsulated in Active Network Encapsulation Packet (ANEP) format.

4.1 Active Framework

The overall block diagram for the active framework for the WLAN access point system is given in Fig. 6. The various modules are Packet Forwarder, Active Code Handler, Active Classifier, Active Scheduler and Active Queue Manager. The figure also shows the mapping of the modules to the various microengines of the IXP1200 NP. The functionality of each of the module is described below.

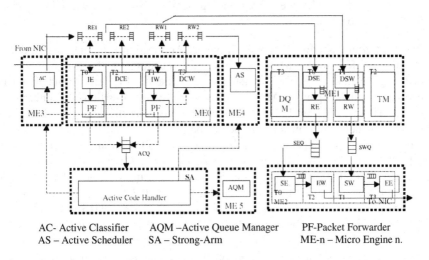

AC- Active Classifier   AQM –Active Queue Manager   PF-Packet Forwarder
AS – Active Scheduler   SA – Strong-Arm   ME-n – Micro Engine n.

Fig. 6. Active Framework For WLAN access point

Packet Forwarder. This module decides where to forward the incoming packet. It forwards it to the classifier if it is a data packet or to the active-code handler module of the strong -arm if it is an active code packet. If it is a data packet it sends it to the active classifier that has been enabled for that port. Otherwise it sends it to the default classifier.

Active Code Handler. This is built on the StrongArm Processor. The incoming active packets are in Active Network Encapsulation Packet (ANEP) format. The format of ANEP packet is shown in Fig. 7. The ANEP packets are embedded in the IP payload of the incoming packets. The type field of the ANEP packet header indicates whether the payload has code for the scheduler, classifier or queue manager. The active code handler module extracts the active code from the ANEP packet. It inserts

0	7	15	31
Version	Flags	TypeID	
ANEP Header Length		ANEP Packet Length	
Options			
Payload			

Fig. 7. ANEP Packet Format

the code into instruction store of the corresponding microengine depending on whether the code is for the classifier, scheduler or queue manager. It discards control packets of the same type if they arrive within the defined MINTHRESHOLD time. Two sets of queues (RE1/RE2 and RW1/RW2) are available for each port. After inserting the active code onto the instruction store of the microengines, the active code handler module finds the empty sets of queues and assigns it as the active set for that port. It then enables the corresponding microengine onto which the active code was inserted and disables the corresponding default classification/scheduling/queue management algorithm.

Active Classifier/Scheduler/Queue Manager. Application Programming Interfaces (API) are used for developing the classification, scheduling and queue management algorithms. Some of the functions of the API are send_to_queue, receive_from_queue, get_src_MAC and enable/disable classifier/ scheduler/queue manager. These API's enable development of any classification, scheduling and queue management algorithms for the access point.

For a proof of concept, we have developed the following using these APIs- an active CBQ tree, two scheduling algorithms namely Round Robin (RR) and Weighted Fair Queuing (WFQ), two queue management algorithms namely Partial Buffer Sharing (PBS) algorithm and Random Early Detection (RED).

If the active classifier is enabled, the incoming packets are classified based on the CBQ tree that has been actively downloaded and placed on the currently active set. In the Round Robin scheduling algorithm packets are classified and sent to 'n' queues. Queues are serviced in order 1..n. In the Weighted Fair Queuing-scheduling algorithm, the virtual finish time for each incoming packet in each of the queues are calculated. The virtual finish time of a packet is equal to the sum of the largest finish time of a packet in its queue and the size of the arriving packets (in bits). The scheduler will compare the virtual finish time of the first packets in the queues, and select the packet with the smallest virtual finishing time. The scheduled packets are sent to the corresponding receive handler modules.

PBS uses a threshold to determine whether an arriving packet should be allowed to enter the buffer or not. RED estimates the average queue size, using a simple exponentially weighted moving average method. Two RED parameters, minimum threshold (minth) and maximum threshold (maxth), are used in the packet drop decision process. As the average queue size varies from minth to maxth, packets will be dropped with a probability that varies linearly from 0 to maxp.

5   Implementation and Test Results

The system has been developed using MicroC [12,13]. The IXP Networking Library has also been used. MicroC compiler has been for compilation of source code to machine code. The BT is defined in SRAM starting at a specific memory location. The size of each entry of the BT is 4 long words. The AST is defined in SRAM starting at a specific memory location. The size of each entry of the AST is 4 long words. IBUFF1/ OBUFF1 are buffers that hold Ethernet packets. The size of each buffer is 190 quad words. IBUFF2/OBUFF2 are buffers that hold 802.11 packets. The size of each buffer is 192 quad words. Vxworks, an RTOS is used as the Operating System of the StrongArm. IXP1200 Developer Workbench - Version 2.01[14], has been used for the development of code for the microengines. Tornado 2.0 has been used for the development of code for the StrongArm. The system has been simulated using the workbench and implemented using the IXP1200 evaluation kit.

The system has been tested for different data rates ranging from 10 Mbits/sec to 100 Mbits/sec with transmit and receive buffer size of 256 bytes on both the wireless and wired sides. The system has also been tested for different packet sizes ranging from 64bytes to 1518 bytes on Ethernet side and from 80 bytes to 1534bytes on 802.11 side.

To test the active classifier API developed, four different classification algorithms have been developed using them and tested on the AP. To test the active scheduler API, scheduler algorithms such as Round Robin, Priority Scheduling and Weighted fair queuing algorithm have been developed using them and tested. Similarly, active queue management API has been tested by developing PBS and RED algorithms. Active classifier/scheduler/queue management code packets were sent to the AP. The access point used the corresponding active classifier/scheduler/queue management algorithms that were dynamically downloaded. In essence, a test-bed to evaluate various algorithms has been developed and tested. The actual evaluation of the algorithms per se in a wireless AP, using the active framework for different workloads is in progress.

6   Conclusions and Future Work

In this paper, we have presented the design of an active access point on the IXP1200 NP. The design has been verified using both simulator and actual hardware implementation. There are two key observations that we make based on this work. One is the use of a low cost NP for an access point, and the second is the successful development and deployment of an active framework based on NP.

Although the active framework has been designed for the wireless access point, the concept is applicable for any active node in any type of network. Thus we have shown that it is practically feasible to use the NPs capability and flexibility innovatively.

Future work that we plan to do include designs using multiple NPs to study the scalability of the approaches considered, and bring in additional functionalities such as security issues which are critical in a wireless network scenario.

References

1. H.Hasim: "Active Network Implementations", Research and development, 2002,Student Conference at MARA Univ. Of Technol., Selangor, Malaysia, 16-17 July 2002,Pages: 371–374.
2. Andreas Kind: "The Role Of Network Processors in Active Networks", ANTA 2002.

3. D. L. Tennenhouse and D. J. Wetherall: "Towards an Active Network Architecture," ACM Computer Communication Review, Vol. 26, No. 2, April 1996.
4. D.Scott Alexander, Marianne Shaw, Scott M. Nettles, and Jonathan M.Smith: "Active Bridging," ACM SIGCOMM 1997
5. Kuorilehto, M. et al.: "Implementation of wireless LAN access point with quality of service support", IECON 02 [Industrial Electronics Society, IEEE 2002 28th Annual Conference], Volume: 3, Nov 5-8, 2002 Page(s): 2333–2338.
6. M.Hännikäinen, et al.: "Windows NT Software Design and Implementation for a Wireless LAN Base Station", ACM International Workshop on Wireless Mobile Multimedia (WoWMoM'99), August 20, Seattle, USA, pp. 2–9.
7. M.Kuorilehto, et al.: "Design for a Wireless LAN Access Point Driver", International Conference on Telecommunications (ICT'2001), June 4-7, 2001, Bucharest, Romania, Vol. 3, pp. 167–173.
8. http://www.fujitsu-siemens.com/
9. http://www.pc-zubehoehelbling.ch/wireless/nokia/AccessPoint/ WirelessLANAccessPoint.html
10. Intel ® IXP1200 Network Processor Family Hardware Reference Manual, Version 1.0, December 2001.
11. http://www.icir.org/floyd/cbq.html
12. Intel ® Microengine C Networking Library for the IXP1200 Network Processor, Reference Guide, Version 1.0, December 2001.
13. Intel ® Microengine C Compiler Language Support, Reference Manual, Version 1.0, December 2001.
14. Intel ® IXP1200 Network Processor Family Development Tools User's Guide, Version 1.0, December 2001.

MuSeQoR: Multi-path Failure-Tolerant Security-Aware QoS Routing in Ad Hoc Wireless Networks*

S. Sriram, T. Bheemarjuna Reddy, B. S. Manoj, and C. Siva Ram Murthy**

Department of Computer Science and Engineering,
Indian Institute of Technology, Madras, India 600036
sriram@dcs.cs.iitm.ernet.in,
{arjun, bsmanoj}@cs.iitm.ernet.in,
murthy@iitm.ernet.in

Abstract. In this paper, we present MuSeQoR: a new multi-path routing protocol that tackles the twin issues of reliability (protection against failures of multiple paths) and security, while ensuring minimum data redundancy. Unlike in all the previous studies, reliability is addressed in the context of both erasure and corruption channels. The reliability and security requirements of a session are specified by a user and are related to the parameters of the protocol adaptively. In addition, by using optimal coding schemes and by dispersing the original data, we minimize the redundancy. Finally, extensive simulations were performed to assess the performance of the protocol under varying network conditions. The simulation studies clearly indicate the gains in using such a protocol and also highlight the enormous flexibility of the protocol.

1   Introduction

To ensure reliable communication in Ad hoc wireless networks, one of the common paradigms involves setting up multiple paths between the source and the destination. Due to the inherent broadcast nature of the medium, there is the additional concern of security of the transmitted data. For critical real-time applications, service disruptions cannot be tolerated and hence, the fault-handling method used is based on Forward-recovery (Hot Standby). Forward-recovery approaches are characterized by two parameters: *Redundancy* and *Dispersion*.

Diversity coding [1] is one of the approaches for reliable transmission. Assuming erasure channels, this approach encodes blocks to be sent on k paths as $(k + b)$ blocks, where $b < k$ is equal to the number of path failures to be protected against. Ref. [2] studies the allocation of blocks of data to multiple paths using diversity coding such that the probability of successful transmission is maximized under the erasure channel assumption. Ref. [3] links user reliability requirement to path availability by setting up more paths to the destination if a smaller number does not meet the requirement. But

---

* This work was supported by the iNautix Technologies India Private Limited, Chennai, India and the Department of Science and Technology, New Delhi, India.
** Author for correspondence.

L. Bougé and V.K. Prasanna (Eds.): HiPC 2004, LNCS 3296, pp. 81–90, 2004.
© Springer-Verlag Berlin Heidelberg 2004

this may involve high call setup times. In [4] Rabin's Information Dispersal Algorithm (IDA) is used to construct a framework for a reliable multi-path scheme.

The resource constraints of Ad hoc wireless networks place a premium on resource usage. Thus, a protocol that ensures reliable communication while at the same time minimizing resource usage is desirable. Reliability implies protection against faults such as link breakages and node failures as well as against corruption of data. Unlike wired networks, the local broadcast nature of the channel allows nodes that are not on the path of data transmission to listen to the data. Naturally, we would like to minimize the number of nodes that can listen to data not intended for them. We need to know how reliability and security can be related to the protocol parameters.

We outline a multi-path QoS routing protocol that ensures reliable communication in the event of multiple path failures and the presence of untrustworthy nodes, with minimum resource overhead. We estimate the overhead involved for a given failure model. Further, we define a security metric for the protocol. The two key issues of reliability and security are what the protocol attempts to simultaneously address. The scheme proposed is adaptive as the failure model and the number of paths to be setup are decided depending on the current state of the network and the reliability of the paths available between the source and the destination.

2   MuSeQoR: An Analytical Description

We provide an overview of the scheme followed by a description of the reliability and security aspects. We describe the protocol in detail in the subsequent section.

2.1   Overview

Given a set of n node-disjoint paths, let $P_i(\delta t)$ denote the probability of failure of i $(0 \le i \le n)$ of these n paths in a time-interval $(t, t + \delta t)$. We say that the set of n node-disjoint paths follows an f-path failure model $(f < n)$, if for a given $0 \le \epsilon \le 1$, $P_f(\delta t) > \epsilon$ and $P_{f+1}(\delta t) < \epsilon$ where ϵ is a path-failure metric. We assume that node-disjointedness is sufficient to ensure that the events of failure of the individual paths are independent. Thus, the probability of failure of the paths $p_0, \ldots, p_{i-1}$ in the interval $(t, t + \delta t)$ is given by $P_{(0,\ldots,i-1)}(\delta t) = \prod_{l=0}^{l=i-1} P_{p_l}(\delta t)$ where $P_{p_l}(\delta t)$ is the probability of failure of the path p_l in that interval. The path failure probability for a path p can be computed as $P_p(\delta t) = 1 - \prod_{(i,j) \in p} A_{(i,j)}(\delta t) \prod_{i \in p} B_i(\delta t)$ where $A_{(i,j)}(\delta t)$ is the link availability of the link connecting nodes i and j on path p as discussed in [5] and $B_i(\delta t)$ is the node availability of the node i on path p.

For a channel model (erasure or corruption) obeying a given failure model, we need to establish a session of bandwidth B between source S and destination D so that the paths can sustain communication. We do this by setting up n node-disjoint paths each of bandwidth $\frac{B}{k}, n \ge k$. This is done by splitting the message into k m-bit blocks. We transform the k blocks into n blocks using an (n, k)-code and then these n blocks are transmitted over the n paths setup. The (n, k)-code ensures that the destination can recover the original k blocks in the case of adversary events. A session that uses an (n, k)-code is referred to as an (n, k)-session.

We need to determine the minimum number of paths to be setup *i.e.*, the value of n so that the connection formed by the set of paths established is reliable for the given channel and fault model and a given value of k. We do this analysis for the cases of the erasure and corruption channels.

2.2   Erasure Channel

In erasure channels, faults are manifested in the non-arrival of packets (erasures). Packets that arrive can be assumed to be correct. This is the case when packet losses are caused by failures at links and nodes. We define an f-erasure channel as one in which at most f of the channels (node-disjoint paths) can fail simultaneously.

Consider an f-erasure channel. From the theory of erasure channel codes, the minimum value of n is given by $n = k + f$ provided $2^m > n$. The code vector $C = (C_1, \ldots, C_n)$ is derived from the data vector $M = (M_1, \ldots, M_k)$ as $C = MG$ with G being a $(k \times n)$ matrix, and M and C being m-bit blocks. All operations are performed in the field $\mathbb{GF}(2^m)$. G is constructed as $G = [g_{ij}] = [\alpha^{(i-1)j}]$ where α is the generator of $\mathbb{GF}(2^m)$.

At the destination, some subset of the n coded blocks arrive. For an f-erasure channel, at least $n - f$ blocks are expected to be received. If C' is the vector of some set of k blocks received and G' is the sub-matrix formed by the subset of columns of G corresponding to the blocks of C'. Then, we have $C' = MG'$ and $M = C'G'^{-1}$ where G'^{-1} is the inverse of G'. For the construction of G given, G' is always invertible provided $2^m > n$. Thus, in an f-erasure channel, it is sufficient to setup $k + f$ paths. The redundancy ratio RO defined as the ratio of the amount of data sent in excess to the original data, in this case, is $RO = \frac{f}{k}$.

2.3   Corruption Channel

In corruption channels, in addition to the erasures, packets delivered may be corrupted by malicious nodes. An (f, g)-corruption channel is one in which at most f erasures and g corruptions can occur simultaneously. Reliability in corruption channels is achieved by the use of codes such as the Reed-Solomon codes. The Reed-Solomon codes are based on the Fourier transform. An (n, k) Reed-Solomon code can detect and correct f erasures and g failures provided $n - k \geq 2g + f$ [6]. Thus, for a (f, g)-corruption channel, we need to setup a minimum of $k + f + 2g$ paths. The redundancy ratio in this case is $RO = \frac{2g+f}{k}$.

2.4   Security

The notion of security that we consider has to do with the fact that the data being transmitted can be overheard by nodes other than those on the $n = (k + b)$ paths. In this notion of security, we assume the adversaries are passive and non-colluding. Improving the security is done by reducing the number of nodes that can listen to the data. Our protocol enhances security by dispersity and coding. Dispersity results in the data being scattered over a wider area of the network, thereby reducing the probability that a node can overhear all the packets being transmitted. On the other hand, dispersity also increases the number of nodes that can access some portion of the data. The mapping that is used to code the packets is shared between the source and the destination by means of a Public-key cryptosystem. Since adversaries do not have the mapping information, even if they have access (by overhearing) to the packets transmitted on a certain fraction of the paths, it would be difficult to obtain the entire data.

We introduce a metric the Eavesdropping Ratio ER: a measure of the leakage of data due to the routing protocol relative to shortest-path routing. Consider a session between two nodes occurring over n paths $P_1, \ldots, P_n$. Define a batch of n packets as the set of n packets formed by encoding the k packets to be transmitted. For successful decoding of the packets at the destination, it is necessary that at least k packets in each batch must reach the destination. Let $T_k(P_1, \ldots, P_n)$ denote the set of nodes which can listen to at least $k < n$ of the packets of a batch being transmitted on the paths $P_1, \ldots, P_n$. Define the *Eavesdropping Ratio ER* as $ER = \frac{|T_k(P_1,\ldots,P_n)|}{|T_{uni-path}|}$, where $T_{uni-path}$ is the set of nodes that can listen to all packets on the shortest uni-path route between the source and the destination. A relation between ER and the protocol parameters has been derived in [8].

Our protocol is characterized by the three parameters (k, f, g) in the general case. The user requirements are specified by the parameters ϵ and ER. These requirements are then translated into values for k and f. Parameter g is determined by the state of the network. Thus, the protocol tackles the twin issues of reliability and security.

3   Description of the Protocol

The protocol that we propose is an on-demand protocol that is a modification of the Dynamic Source Routing (DSR) [7] protocol. Thus, no global topology information is maintained at the nodes. A session is divided into time frames Δt during which the network state is assumed to be fairly constant. The parameters computed and the paths setup during a time frame remain unaltered unless disruption of the existing paths occurs. To enhance security, each time frame is further divided into code sessions of length Δt_c. A code session is the period for which a single mapping is used by the source for encoding the packets. This mapping may be changed at the end of a code session by means of a Public-Key cryptosystem.

3.1   Route Setup and Parameter Selection

In this phase, paths between the source and the destination are probed. Depending on the reliability of the paths found, which is estimated during route discovery, the fault-model is chosen. This manner of choosing the fault-model ensures that the protocol adapts to the state of the network. Accordingly, the parameters k and b are estimated. The protocol then attempts to find these routes.

Route Discovery. Route request (REQUEST) packets are sent by the source to its neighbors. Each REQUEST packet carries a sequence number, the source and destination IDs, the connection ID, the path traversed so far, the reliability of the path traversed so far, the path bandwidth, the bandwidth required for this session, and the path failure metric ϵ and the ER specified for this session. Each intermediate node checks the route of the packet to ensure that there are no loops. The intermediate node forwards a maximum of MAX_FORWARD (a system parameter) route request packets. The higher the value of the system parameter, the greater the amount of flooding in the network while the number of routes discovered increases.

Before forwarding the REQUEST packet, the node multiplies the link availability of the link on which it received the packet to the reliability being carried in the packet. This link availability is calculated using information from the upstream node and the forwarding node (*i.e.*, the instantaneous velocity, the direction of motion, and the current position of nodes) and the former is included in the REQUEST packet. It also modifies the bandwidth available on the path and forwards the packet to its neighbors.

Determination of f, b, and k. The destination receives r REQUEST packets (r must be greater than the number of paths that need to be setup else no reply is sent back). It then sorts the paths according to their probabilities of failure p_f. Let $\{P_1, \ldots, P_r\}$ be the paths where $p_f(P_1) > \ldots > p_f(P_r)$. Find $i+1$ so that $p_f(P_1) \times p_f(P_2) \times \ldots \times p_f(P_{i+1}) < \epsilon$. Then an i-path failure model is used. This calculation of i is an approximation in that the paths $P_1, \ldots, P_r$ are not node-disjoint. The number of additional paths b to be setup is determined either as $b = f$ for the case of erasure channels and $b = f + 2g$ for the case of corruption channels. k must be chosen subject to the constraint that $k + b \leq min(|N(S)|, |N(D)|)$, where $N(S)$ and $N(D)$ are the set of nodes lying within the transmission range of source S and destination D, respectively. In addition, the ER can also be used to determine the value of k [8].

Route Calculation and Resource Reservation. Once k is determined, the destination finds out if there exists $(k + b)$ node-disjoint paths with bandwidth $\frac{B}{k}$. If it finds a set of paths, it sends a route reserve (RESERVE) packet on each of the paths and stores that set of paths in its route cache. The RESERVE packets carry the route, values k, f, and b and the mapping to be used for encoding for the remaining session. On receiving the RESERVE packets on the $(k + b)$ paths, the source begins data transmission. Some of the paths on which the RESERVE packets are being sent may not have the bandwidth as other calls may have reserved it in the meantime. The node which does not possess the required resources simply drops the RESERVE packet. Resources that have been reserved thus far are released when a timeout occurs due to non-arrival of packets of the connection on that path. The call is accepted only if n RESERVE packets are received at the source. Communication takes place only if the paths can meet the requirements of the session, and not merely if the source and the destination are connected.

If the destination is unable to select a set of node-disjoint paths that satisfy the bandwidth constraint, it does not send any RESERVE packets. The source on non-receipt of RESERVE packets re-initiates the route discovery. This process is repeated MAX_ATTEMPT times. Failure on all occasions results in the call being rejected at the source. When the destination is unable to find the set of node-disjoint paths when the REQUEST packets are sent for the MAX_ATTEMPT time (this is indicated in the REQUEST packet), all the paths to the source are removed from the route cache.

3.2   Route Maintenance

Route maintenance occurs in two scenarios.

Expiry of the Time Frame. At a time $\Delta t - 2RTT$, where RTT is the round trip time of the transmission, the data packets on the various paths carry the path failure probability. At the destination, this is used to verify that the paths still conform to the f-path fault model. If they do not, the destination sends back a REPAIR packet to the source. The source then initiates a route discovery to search for a set of paths disjoint with the current set of paths. The destination uses the newly discovered paths along with the currently used $(k + b)$ paths to recompute the parameter f. If f has increased, the destination sees if the newly discovered paths can be added to the already existing paths to meet the new reliability constraint. If this is possible, RESERVE packets are sent along the entire set of paths and the data transmission continues. On the other hand, failure to establish the required number of paths will be treated in the same way as in the route setup process. If the value of f decreases, on the other hand, the paths are maintained as such.

Route Break. If the route break occurs when the parameters are being re-estimated (during the time $\Delta t - 2RTT$ to Δt), no action is taken as anyway the routes may need to be reconfigured. Otherwise, failure of a link (u,v) where u is the upstream node results in node u sending ERROR packets to the source. On receiving these packets the source initiates a route discovery. The resources reserved on the downstream nodes are released on a timeout due to non-arrival of any more packets.

In the protocol, both the source and the destination need to know the set of paths used for communication. The encoding and decoding of the transmitted data requires this. Every node has a route cache of paths to a particular destination node for which it is the source (S-CACHE) and another route cache of paths from a source node for which it is the destination (D-CACHE). For the communication to be successful, the paths in the route caches of the source (S-CACHE) and the destination (D-CACHE) must be consistent. After the first route request by the source, the destination selects the set of node-disjoint paths if available and inserts these into its D-CACHE (*Note:* Even if the number of paths discovered is higher than required, the destination inserts only the required number into the D-CACHE. Thus, the set of paths represents the paths used for actual data transmission). When the RESERVE packets arrive on the paths at the source, the source adds these paths to its S-CACHE. A path break causes a ERROR packet to be sent to the source node which removes the path (and all paths of any other connection that may be passing through the failed link) from its S-CACHE. It then initiates a REQUEST to the destination. At the destination, paths which have not been delivering packets (either data or control) for a time exceeding a window are marked stale. When a REQUEST arrives at the destination, if this has arrived on a path marked stale, the path is re-included into the D-CACHE. Otherwise, a new set of paths to the source that includes the existing set of non-stale paths is computed and if this set is found to satisfy the bandwidth and the reliability requirements, the stale paths are expunged and RESERVE packets are sent along the newly-discovered paths.

4   Experiment and Simulation Results

The experiments are intended to study the following metrics and their response to changes in the mobility of nodes, load in the network, and terrain dimensions.

– Average Call Acceptance Rate:

$$ACAR = \frac{No.\ of\ calls\ successfully\ setup}{Total\ no.\ of\ call\ requests} \tag{1}$$

– Average Information Delivery Ratio:

$$IDR = \frac{No.\ of\ data\ packets\ successfully\ decoded}{Total\ no.\ of\ data\ packets\ sent\ by\ the\ source} \quad (2)$$

– Resource Consumption Ratio:

$$RCR = \frac{No.\ of\ data\ packets\ transmitted\ by\ nodes\ across\ the\ network}{Total\ no.\ of\ data\ packets\ sent\ by\ the\ source \times Average\ Hopcount} \quad (3)$$

– Control Overhead Ratio:

$$COR = \frac{No.\ of\ control\ bytes\ transmitted\ across\ the\ network}{Total\ no.\ of\ data\ bytes\ sent\ by\ the\ source} \quad (4)$$

– Eavesdropping Ratio:

$$ER = \frac{No.\ of\ nodes\ that\ can\ listen\ to\ data\ packets\ on\ at\ least\ k\ paths}{No.\ of\ nodes\ that\ can\ listen\ to\ data\ packets\ on\ the\ shortest\ uni-path} \quad (5)$$

4.1 Simulation Results

We simulated our protocol using Glomosim [9]. The network contains 75 nodes in a 1000 m × 1000 m terrain area. The channel capacity is fixed at 2 Mbps and the duration of the simulation is 10 minutes. The transmission range is 200 m. The traffic is generated in the form of Constant Bit Rate (CBR) sessions each of which lasts for 600 seconds. The load on the network is varied by varying the number of CBR sessions. Mobility is simulated according to the Random waypoint model. For all cases of mobility in the network, we set the pause time to 0 and set the minimum and maximum speeds to the same value to ensure that the nodes move at a constant speed. For the purpose of the simulation, we assume that the protocol parameters are known *a priori* and are fixed for the duration of the simulation for all sessions.

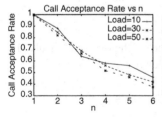 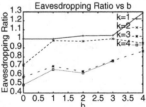 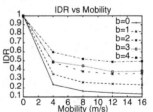

Fig. 1. Variation of Call Acceptance Rate vs n (total number of paths setup) for varying Load   **Fig. 2.** Variation of Eavesdropping Ratio vs protocol parameters k and b **Fig. 3.** Variation of Information Delivery Ratio vs Mobility for various b ($k=2$)

Average Call Acceptance Rate. From Figure 1, the call acceptance rate decreases with increasing n. With an increase in n, majority of the calls are rejected due to lack of n

node-disjoint paths. The call acceptance rate decreases with an increase in the load, yet the decrease is not considerable showing that the protocol can handle high loads.

Eavesdropping Ratio. From Figure 2, for a fixed value of b, ER decreases with increasing k. As b increases for a fixed k, the number of nodes that can potentially eavesdrop increases. This increases the value of ER.

Information Delivery Ratio. We have plotted the IDR for different node velocity values for different values of b in Figure 3 with $k = 2$. The higher fault-protection schemes perform better at higher mobility. We also studied the effect of k and b on the IDR, Figure 4, at a fixed mobility value of 12 m/s. While higher b increases IDR, higher k decreases IDR. The decrease in IDR due to increasing k is due to the higher fraction of packets in a batch that must now reach the destination for the original data packets to be successfully recovered.

Resource Consumption Ratio. We have plotted the variation of the RCR with node velocity for different k in Figure 5 at $b = 2$. By increasing k, the RCR decreases as b is constant for a given failure model. At high mobility, a decrease in the RCR is more likely a result of loss of data packets. We have also plotted the variation of the RCR with k and b at a fixed mobility value of 12 m/s (Figure 6). The values indicate an increase in this ratio with an increase in b due to an increase in the redundancy while a decrease in this ratio occurs with an increase in k since the redundancy is amortized over the increased number of paths.

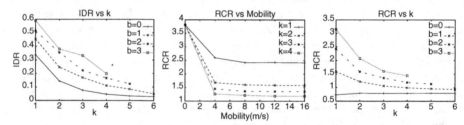

Fig. 4. Variation of Information Delivery Ratio vs k and b (Mobility=12 m/s)

Fig. 5. Variation of Resource Consumption Ratio vs Mobility for various k (b=2)

Fig. 6. Variation of Resource Consumption Ratio vs k and b (Mobility=12 m/s)

Control Overhead Ratio. Control overhead ratio, a measure of the overhead involved in route setup and maintenance, is seen to increase only slowly for $k = 1$ and $k = 2$ ($b = 2$) at high speeds indicating the suitability of the protocol to a mobile environment. As expected the control overhead is higher as a greater number of paths are setup as can be seen from the Figures 7 and 8.

Effect of Load. Figures 9 and 10 show the variation in IDR and RCR with changing load. IDR is seen to decrease with an increase in the load. With a higher load, IDR decreases though the value of b is increased. This shows that at high load, an increase in b may not imply an increase in the reliability. This is because, for a fixed k, the bandwidth requirement per path remains the same but each node may have to handle more data due to the setting up of higher number of paths. The RCR decreases with an increase in k as explained earlier.

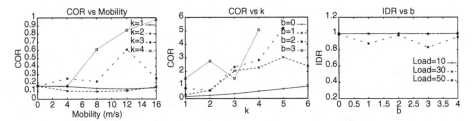

Fig. 7. Variation of Control Overhead vs Mobility for various k (b=2)

Fig. 8. Variation of Control Overhead vs protocol parameters k and b (Mobility=12 m/s)

Fig. 9. Variation of Information Delivery Ratio vs protocol parameter b (k=2)

Comparison of Block Allocation Strategies. In our protocol, the packets are uniformly distributed over the paths setup. In [2], allocation of packets to the paths is done so that the probability of successful reception is maximized. Transmission of a majority of packets of a batch on a single path, however, allows greater eavesdropping. We have compared the IDR and the ER of our protocol using uniform allocation to that which uses non-uniform allocation. The study was done by setting up 6 paths ($k = 3, b = f = 3$) between the source and the destination in a static network. The reliability of the paths was simulated by dropping packets at the destination. The numerical figures of the example in Section III-B of [10] were used. The reliability of all the paths except one were fixed at 0.8 while the remaining path's reliability (given by q) was varied between 0.8 and 1. The two allocation vectors compared were $(3, 1, 1, 1, 0, 0)$ and $(1, 1, 1, 1, 1, 1)$. The Figures 11 and 12 indicate that the two schemes have comparable IDR while uniform allocation has a lower ER.

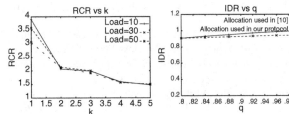
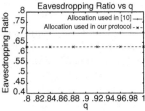

Fig. 10. Variation of Resource Consumption Ratio vs parameter k when parameter b is fixed at 2

Fig. 11. Comparison of Information Delivery Ratios of the allocation scheme used in [10] and our allocation scheme

Fig. 12. Comparison of Eavesdropping Ratios of the allocation scheme used in [10] and our allocation scheme

5   Conclusion

MuSeQoR attempts to address the twin issues of reliability and security while ensuring that the overhead involved is minimum. The protocol is characterized by the parameters

k, f, and g. A higher value of f and g indicates higher reliability while a higher value of k implies a lower ER and a lower redundancy ratio. The IDR is higher for high values of b while the redundancy ratio and the ER decrease for high values of k. In a real scenario, these values of k and b need to be chosen using the user specifications of ϵ and ER. For the static scenario, values of k and b between 2 and 3 offer a good compromise. Finally, we compared the block allocation strategy of [2] with the allocation strategy used in our scheme, which allocates packets equally on each path.

We are currently working on the translation mechanism that will translate the user requirements into the protocol parameters which would involve estimating f and k, and Δt based on the mobility of the nodes and the available bandwidth. Further work needs to be done to address the issues related to QoS protocols that do not require node-disjoint paths and those that allow differentiated service based on multiple QoS parameters.

References

1. E. Ayanoglu, I. Chih-Lin, R. D. Gitlin, and J. E. Mazo, "Diversity Coding for Transparent Self-Healing and Fault-Tolerant Communication Networks", *IEEE Transactions on Communications*, vol. 41, no. 11, pp. 1677-1686, November 1993.
2. A. Tsirigos and Z. J. Haas, "Multi-path Routing in the Presence of Frequent Topological Changes", *IEEE Communications Magazine*, vol. 39, no. 11, pp. 132-138, November 2001.
3. R. Leung, J. Liu, E. Poon, A. C. Chan, and B. Li, "MP-DSR: A QoS-Aware Multi-path Dynamic Source Routing Protocol for Wireless Ad hoc Networks", *in proceedings of IEEE LCN 2001*, pp. 132-141, November 2001.
4. L. Chou, C. Hsu, and F. Wu, "A Reliable Multi-path Routing Protocol for Ad hoc Network", *in proceeding of IEEE ICON 2002*, pp. 305-310, August 2002.
5. A. B. McDonald and T. Znati, "A Path Availability Model for Wireless Ad hoc Networks", *in proceedings of IEEE WCNC 1999*, vol. 1, pp. 35-40, September 1999.
6. R. E. Blahut, "Algebraic Coding for Data Transmission", *Cambridge University Press*, 1st edition, 2002.
7. D. B. Johnson and D. A. Maltz, "Dynamic Source Routing in Ad hoc Wireless Networks", *Mobile Computing*, edited by T. Imielinski and H. Korth, Kluwer Academic Publishers, Chapter 5, pp. 153-181, 1996.
8. S. Sriram, T. Bheemarjuna Reddy, B. S. Manoj, and C. Siva Ram Murthy, "MuSeQoR: Multi-path Failure-tolerant Security-aware QoS Routing in Ad hoc Wireless Networks", *Technical Report, HPCN Lab, Department of Computer Science and Engineering, IIT Madras*, December 2003.
9. X. Zheng, R. Bagrodia, and M. Gerla, "GloMoSim: A Library for Parallel Simulation of Large-Scale Wireless Networks", *in proceedings of the 12th Workshop on Parallel and Distributed Simulations (PADS 1998)*, pp. 154-161, May 1998.
10. A. Tsirigos and Z. J. Haas, "Analysis of Multi-path Routing- Part I: The Effect on the Packet Delivery Ratio", *IEEE Transactions on Wireless Communications*, vol. 3, no. 1, pp. 138-146, January 2004.

A Tunable Coarse-Grained Parallel Algorithm for Irregular Dynamic Programming Applications

Weiguo Liu and Bertil Schmidt

School of Computer Engineering, Nanyang Technological University, Singapore 639798
liuweiguo@pmail.ntu.edu.sg, asbschmidt@ntu.edu.sg

Abstract. Dynamic programming is a widely applied algorithm design technique in many areas such as computational biology and scientific computing. Typical applications using this technique are compute-intensive and suffer from long runtimes on sequential architectures. Therefore, many parallel algorithms for both fine-grained and coarse-grained architectures have been introduced. However, the commonly used data partitioning scheme can not be efficiently applied to irregular dynamic programming applications, i.e. dynamic programming applications with an uneven computational load density. In this paper we present an efficient coarse-grained parallel algorithm for such kind of applications. This new algorithm can balance the load among processors using a tunable block-cyclic data partitioning scheme. We present a theoretical analysis and experimentally show that it leads to significant runtime savings for several irregular dynamic programming applications on PC clusters.

1 Introduction

Dynamic programming (DP) is a popular algorithm design technique for optimization problems. Problems such as genome sequence alignment [8, 14], RNA and protein structure prediction [3, 10, 16], context-free grammar recognition [4, 11], and optimal static search tree construction [6] have efficient sequential DP solutions. In order to reduce the high computing cost of DP problems, many efficient parallel algorithms on different parallel architectures have been introduced [1].

On fine-grained architectures, the computation of each cell within an anti-diagonal is parallelized [12, 13]. However, this way is only efficient on architectures such as systolic arrays, which have an extremely fast inter-processor communication.

On coarse-grained architectures like PC clusters it is more convenient to assign an equal number of adjacent columns to each processor as shown in Fig. 1. In order to reduce the idle time further, matrix cells can be grouped into blocks. Processor i then computes all the cells within a block after receiving the required data from processor $i-1$. Fig. 1a shows an example of the computation for 4 processors, 8 columns and a block size of 2×2, the numbers 1 to 7 represent consecutive phases in which the cells are computed. We call this scheme *block-based*. It works efficiently for regular DP computations with an even load across matrix cells, i.e. each matrix cell is computed from the same number of other matrix cells.

L. Bougé and V.K. Prasanna (Eds.): HiPC 2004, LNCS 3296, pp. 91–100, 2004.
© Springer-Verlag Berlin Heidelberg 2004

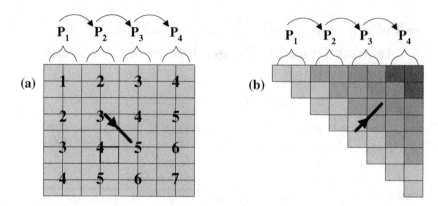

Fig. 1. (a) Parallel computation for 4 processors, 8 columns and a 2×2 block size. The complete 8×8 matrix can then be computed in 7 iteration steps; (b) Example of an irregular DP computation

In practice, there are many irregular DP applications. Fig. 1b shows an example of such an application. The load to compute one cell on the matrix will increase along the shift direction of the computation. We call this the *computational load density*. Fig. 1b shows the change of computational load density along the computation shift direction by using increasingly blacking shades. We can see that the computational load density at the top right-hand corner is much higher. The block-based partitioning scheme will therefore lead to a poor performance, since the load on processor P_i is much higher than the load on processor P_{i-1}.

In this paper, we propose a general parameterized parallel algorithm to solve this problem. By introducing two performance-related parameters, we can get the trade-off between load balancing and communication time by tuning these two parameters and thus obtain the maximum possible performance. We demonstrate how this algorithm can lead to substantial performance gains for irregular DP applications.

The rest of the paper is organized as follows: Section 2 gives an introduction to DP algorithms and describes the characters of irregular DP applications. Section 3 presents our new parallel algorithm. Section 4 evaluates the performance of our algorithm for several irregular DP applications on PC clusters. Section 5 concludes this paper.

2   Irregular DP Algorithms

DP views a problem as a set of interdependent sub-problems. It solves sub-problems and uses the result to solve larger sub-problems until the entire problem is solved. In general, the solution to a dynamic programming problem is expressed as a minimum (or maximum) of possible alternative solutions. Each of these alternative solutions is constructed by composing one of more sub-problems. If r represents the cost of a solution composed of sub-problems $x_1, x_2 \ldots x_l$, then r can be written as:

$$r = g \ (f(x_1), f(x_2) \ldots f(x_l))$$

(1)

The function $g()$ in Eq. (1) is called the composition function, and its nature depends on the problem described. If the optimal solution to each problem is determined by composing optimal solutions to the sub-problems and selecting the minimum (or maximum), Eq (1) is then said to be a dynamic programming formulation [9].

DP algorithms can be classified according to the matrix size and the dependency relationship of each cell on the matrix [7]: a DP algorithm is called a tD/eD algorithm if its matrix size is $t \times t$ and each matrix cell depends on $O(n^e)$ other cells. The DP formulation of a problem always yields an obvious algorithm whose time complexity is determined by the matrix size and the dependency relationship. If a DP algorithm is a tD/eD problem, it takes time $O(n^{t+e})$ provided that the computation of each term takes constant time. Two examples are given in Algorithms 1 and 2.

Algorithm 1 (2D/0D): Given $D[i, 0]$ and $D[0, j]$ for $1 \le i, j \le n$

$$D[i, j] = \min\{D[i-1, j] + x_i, D[i, j-1] + y_j, D[i-1, j-1] + z_{i,j}\} \text{ for } 1 \le i, j \le n \qquad (2)$$

where x_i, y_j and $z_{i,j}$ are computed in constant time.

Algorithm 2 (2D/1D): Given $w(i, j)$ for $1 \le i < j \le n$; $D[i, i] = 0$ for $1 \le i \le n$

$$D[i, j] = w(i, j) + \min_{i < k \le j} \{D[i, k-1] + D[k, j]\} \text{ for } 1 \le i, j \le n \qquad (3)$$

where $w(i, j)$ is computed in constant time.

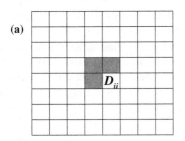
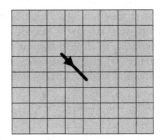
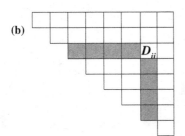
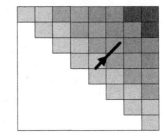

Fig. 2. Dependency relationship and distribution of computational load density along the computation shift direction for (a) Algorithm 1, (b) Algorithm 2

Fig. 2 shows the dependency relationships and the distributions of computational load density for Algorithm 1 and Algorithm 2. We can notice that the $2D/1D$ DP algorithms are irregular ones, i.e. the computational load density changes along the computation shift direction. Table 1 shows some $2D/1D$ DP algorithms in different areas.

Table 1. Some irregular DP algorithms

Algorithm	Application	Classification	Reference
Nussinov	RNA base pair maximization	2D/1D	[2, 4, 5, 14,15]
Matrix chain order	Scientific computing		
CYK	CYK parsing for context-free grammars		
Skyline matrix	Scientific computing		
Smith-Waterman with general gap penalty	Genome local alignment		

3   The Parameterized Parallel Algorithm

Fig. 3 shows the dependency relationship and the change of computational load density for the irregular algorithms in Table 1. The change is indicated by using increasingly blacking shades along the computation shift direction. For these algorithms, the block-based partitioning scheme of Fig. 1 leads to a poor load balancing. Thus, a new data partitioning scheme is needed.

The problem of determining an appropriate data partitioning scheme is to maximize algorithm performance by balancing the computational load among processors. Since the data partitioning scheme largely determines the performance and scalability of a parallel

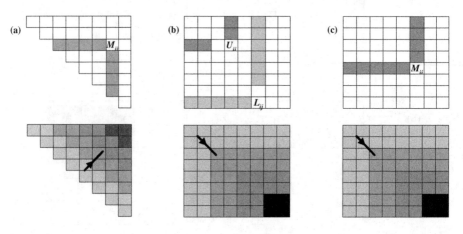

Fig. 3. Dependency relationship and distribution of computational load density along computation shift direction for (a) Nussinov, matrix chain and CYK algorithm, (b) Skyline matrix problem, (c) Smith-Waterman algorithm with the general gap penalty function

algorithm, a great deal of research has aimed at studying different data partitioning schemes.

As a result the block-cyclic partitioning has been suggested as a general-purpose basic scheme for parallel algorithms because of its scalability, load balancing and communication properties [9].

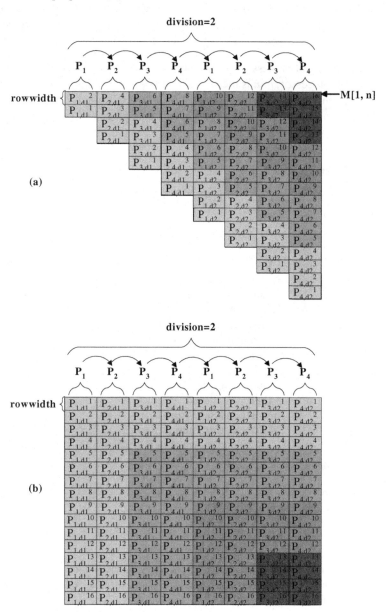

Fig. 4. The tunable block-cyclic partitioning scheme for (a) Nussinov, matrix chain and CYK algorithms, (b) Skyline matrix problem and SW with the general gap penalty function

In this section, we introduce a tunable block-cyclic based distribution of columns for irregular DP algorithms to balance the load among processors. The concept is illustrated in Fig. 4. The parameter *division* is used to implement a cyclic distribution of columns to processors. The parameter *rowwidth* is used to control the size of messages sent by processor P_i to processor P_{i+1} at a time. Increasing the number of cyclic divisions and decreasing the size of messages can lead to a better load balancing. Of course, doing this also increases the communication time. Thus, the choice of the parameter *division* and *rowwidth* is a trade-off between load balancing and communication time.

Input: The number of processors p, the value of *division* and *rowwidth*. (P_k denotes the k-th processor, $n{\times}n$ is the size of matrix M, d_t denotes the tth division).

Output: Depending on the requirements of the given applications, the output will be the optimal score $M[1, n]$ or the whole matrix M.

Begin
for P_k $(1 \leq k \leq p)$ in division d_t **do**

$\quad$ **for** $\left\{ \begin{array}{l} \text{(a) } i = Pk \times \dfrac{n}{p \times division} + (dt - 1) \times \dfrac{n}{division} \text{ to } 1 \\ \text{(b) } i = 1 \text{ to } n \end{array} \right\}$ **do**

$\qquad\qquad$ $\left\{ \begin{array}{l} \text{(a) after } i \text{ is reduced by } rowwidth \\ \text{(b) after } i \text{ is increased by } rowwidth \end{array} \right\}$ **do**

$\qquad\qquad\qquad$ **if** $k = 1$ **then**
$\qquad\qquad\qquad\qquad$ **if** $d_t = 1$ **do**
$\qquad\qquad\qquad\qquad\qquad$ send message to P_2;
$\qquad\qquad\qquad\qquad$ **if** $d_t > 1$ **do**
$\qquad\qquad\qquad\qquad\qquad$ receive message from P_p;
$\qquad\qquad\qquad\qquad\qquad$ send message to P_2;
$\qquad\qquad\qquad$ **if** $1 < k < p$ **then**
$\qquad\qquad\qquad\qquad$ receive message from P_{k-1};
$\qquad\qquad\qquad\qquad$ send message to P_{k+1};
$\qquad\qquad\qquad$ **if** $k = p$ **then**
$\qquad\qquad\qquad\qquad$ receive message from P_{k-1};
$\qquad\qquad\qquad\qquad$ **if** $d_t \neq division$ **then**
$\qquad\qquad\qquad\qquad\qquad$ send message to P_1;

$\quad$ **for** $\left\{ \begin{array}{l} \text{(a) } j = i \text{ to } Pk \times \dfrac{n}{p \times division} + (dt - 1) \times \dfrac{n}{division} \\ \text{(b) } j = (Pk - 1) \times \dfrac{n}{p \times division} + (dt - 1) \times \dfrac{n}{division} + 1 \text{ to } Pk \times \dfrac{n}{p \times division} + (dt - 1) \times \dfrac{n}{divison} \end{array} \right\}$ **do**

$\qquad\qquad$ compute $M[i, j]$;
End

Fig. 5. The general parameterized parallel algorithm for irregular DP problems. (a) For Nussinov, matrix chain and CYK algorithm, (b) For skyline matrix problem and SW algorithm with the general gap penalty function

Two scheduling schemes for irregular DP problems are illustrated in Fig. 4. $P_{i,dj}{}^k$ denotes the block of processor P_i at division j and step k. In Fig. 4a, all processors start computing at block 1 and block 2 in division 1. In the same division, after P_i completes computing block 1, it sends this part to P_{i+1} and then P_{i+1} can go on computing the block 3 in this division. Between two different divisions, the last processor will send message to P_1. This parallel procedure will continue until the final cell $M[1, n]$ is computed.

Fig. 4b shows the scheduling scheme for the skyline matrix problem and the Smith-Waterman algorithm with the general gap penalty function. Initially P_1 starts computing at block 1 in division 1. In the same division, after P_i completes computing block 1, it sends this part to P_{i+1} and then P_{i+1} can work at the block 1 in this division. Between two different divisions, the last processor will send message to P_1. This parallel procedure will continue until all cells on matrix M are computed.

The general parameterized parallel algorithm for these two scheduling schemes is presented in Fig. 5. We can get the following theorem according to Fig. 4 and Fig. 5.

Theorem. *The proposed algorithm uses* $\alpha\times(division\times p-1)\times\dfrac{n}{rowwidth}$ *communication steps with* $O(\frac{n^3}{p})$ *sequential computing time on each processor.* (α *is* $\frac{1}{2}$ *for the triangular matrix computation in Fig. 4a;* α *is 1 for the square matrix computation in Fig. 4b).*

Proof. Processor P_i sends $P_{i,dj}{}^k$ to P_{i+1} after the k-th step in division d_j. In the same division, after $\alpha\times\dfrac{n}{rowwidth}$ communication steps, processor P_i completes its work and moves to the next division to continue this loop until finish computing the sub matrix allocated to itself. Each moving to the next division will bring another computation and communication loop, thus, after $\alpha\times(division\times p-1)\times\dfrac{n}{rowwidth}$ communication steps, all the processors have completed their work.

As mentioned in Section 2, the time complexity of 2D/1D DP algorithms is $O(n^3)$. Increasing the number of *division* can balance the load among processors. For example, as to the Nussinov algorithm, the load on processor P_i is about $(2p^2\times division^3+6\times p\times pi\times division^2)\times\dfrac{n^3}{division^3 p^3}$. When the parameter *division* is increased, the lower power item $6\times P\times P_i\times division^2$ can be omitted. Thus, each processor needs almost the same $O(\frac{n^3}{p})$ sequential computing time.

4 Performance Evaluations

We have used the described algorithm to develop parallel applications for the algorithms in Table 1. The parallel programs are presently implemented using standard C++ and the MPI library provided by MPICH 1.2.5 [17]. We have executed the program on a PC cluster. The cluster comprises eight AlphaServer

ES45 nodes. Each node contains 4 Alpha-EV68 1GHz processors with 1GB RAM for each processor. All the nodes are connected with each other by a Gbit/sec Quadrics switch.

Tables 2 and 3 show the average speedups for parallel programs using different *division* and *rowwidth*.

Table 2. Speedup comparison using different *division* (*d*) and *rowwidth* (*r*) for Nussinov, matrix chain and CYK. The matrix size is 5000×5000. The number of processor is 32

	r = 5	10	15	20	30
d = 1	10	10.5	9.8	9.6	8.5
40	22.2	24.5	21.9	21.2	20.6
60	26.2	27.5	26.7	25.2	24.9
65	27.7	**29.8**	26.9	26.4	25.7
70	24.2	26.2	22.5	22.2	19.9
75	24.2	26.1	23.3	21.6	21.5

Table 3. Speedup comparison using different *division* (*d*) and *rowwidth* (*r*) for skyline matrix problem and Smith-Waterman algorithm with the general gap penalty function. The matrix size is 5000×5000. The number of processor is 32

	r = 5	10	15	20	30
d = 1	10.5	12.8	11.3	10.1	8.5
40	20.4	23.6	21.3	20.6	19.1
60	22.1	24.9	22.2	21.2	19.4
65	25.8	**27.9**	26.2	25	23.4
70	23.2	26.1	23.2	22.3	21
75	23	24.7	21.9	20.7	20.3

We have analyzed the behavior of *division* and *rowwidth* to estimate the optimal block size for these applications. The measurements show that the best speedups using 32 processors are obtained when *division* is set from 60 to 70 and *rowwidth* is set from 5 to 15. Fig. 6 shows the speedups for different number of processors. We compare the performance between the block-based partitioning scheme and the tunable block-cyclic partitioning scheme. Notice the super linear speedups are observed in several applications. This is because of the effects due to better caching.

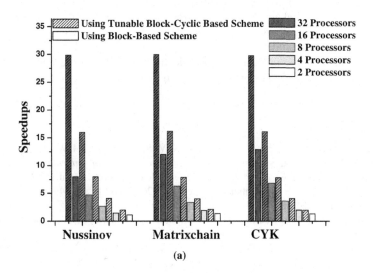

(a)

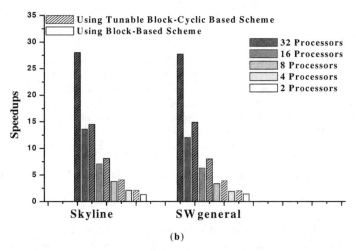

(b)

Fig. 6. Speedups for irregular DP applications using different partitioning schemes: (a) For Nussinov, matrix chain and CYK algorithms, (b) For Skyline matrix problem and SW algorithm with the general gap penalty function

5 Conclusions

Dynamic programming is a general problem-solving technique that has been widely used in many areas. In this paper, we have described a new parameterized parallel algorithm for irregular dynamic programming applications. By introducing the tunable block-cyclic partitioning scheme and tuning the block size, our algorithm can be used to develop parallel programs with significant speedups. We have presented the effectiveness of our algorithm for five irregular dynamic programming applications: Nussinov, matrix chain ordering, CYK, skyline matrix and Smith-Waterman with the general gap penalty function. Measurements show that the described algorithm can be used to develop parallel programs with substantial performance gains for irregular dynamic programming applications on coarse grained architectures.

Our future work includes designing efficient parallel algorithms for three dimensional irregular dynamic programming applications. It will also be interesting to develop the cost analysis and performance prediction tool for our algorithm.

References

1. C.E.R. Alves, E.N. Cáceres, F. Dehne, and S.W. Song: A Parallel Wavefront Algorithm for Efficient Biological Sequence Comparison, *Proceedings of. International Conference on Computational Science and Its Applications (ICCSA 2003)*, Montreal, Canada, 2003, Springer Lecture Notes in Computer Science, Vol. 2667, Part II, pp. 249–258.
2. D.S. Bouman: A Parallel Skyline Matrix Solver in Orca, *Proceedings of the 3rd Annual Conference of the Advanced School for Computing and Imaging*, Heijen, Netherlands, pp. 106–110, 1997.
3. J. Bowie, R. Luthy, D. Eisenberg: A Method to Identify Protein Sequences That Fold Into A Known Three-dimensional Structure, *Science*, 253, 164–170, 1991.
4. C. Ciressan, E. Sanchez, M. Rajman, J.C. Chappelier: An FPGA-based coprocessor for the parsing of context-free grammars, *IEEE Symposium on Field-Programmable Custom Computing Machines*, April, 2000.
5. R. Durbin, S. Eddy, A. Krogh, G. Mitchison: Biological Sequence Analysis - Probabilistic Models of Protein and Nucleic Acids, Cambridge University Press, 1998.
6. M. Farach and M. Thorup: Optimal evolutionary tree comparison by sparse dynamic programming, *35th Annual Symposium on Foundations of Computer Science*, pp. 770–779, Santa Fe, New Mexico, 20-22 November 1994.
7. Z. Galil, K. Park: Dynamic Programming with Convexity, Concavity and Sparsity, *Theoretical Computer Science*, 92, pp. 49–76, 1992.
8. X. Huang, K.M. Chao: A Generalized Global Alignment Algorithm, *Bioinformatics*, 19(2), pp. 228–233, 2003.
9. V. Kumar, A. Grama, A. Gupa and G. Karypis: Introduction to Parallel Computing, The Benjamin-Cummings Publishing Company Inc, 1994.
10. D.W. Mount: Bioinformatics-Sequence and Genome Analysis, Cold Spring Harbor Laboratory Press, 2001.
11. H. Ney: The Use of a One-Stage Dynamic Programming Algorithm for Connected Word Recognition, *IEEE Trans. on Acoustic, Speech and Signal Processing*, Vol. ASSP-32, num.2, pp.263–271, April 1984.
12. B. Schmidt, H. Schroder, M. Schimmler: Massively Parallel Solutions for Molecular Sequence Analysis, Proc. of IPDPS'02, 2002.
13. B. Schmidt, H. Schroder, M. Schimmler: A Hybrid Architecture for Bioinformatics, *Future Generation Computer System*, 18, 855–862, 2002.
14. T.F. Smith, M.S. Waterman: Identification of Common Subsequences, *Journal of Molecular Biology*, pp. 195–197, 1981.
15. G.V. Wilson: Assessing the Usability of Parallel Programming Systems: The Cowichan Problems", *Proceedings of the IFIP Working Conference on Programming Ecvironments for Massively Parallel Distributed Systems*, 1994.
16. M. Zuker, P. Stiegler: Optimal Computer Folding of Large RNA Sequences Using Thermodynamics and Auxiliary Information, *Nucleic Acids Research*, 9, 1981.
17. http://www.unix.mcs.anl.gov/mpi/ mpich

A Feedback-Based Adaptive Algorithm for Combined Scheduling with Fault-Tolerance in Real-Time Systems

Suzhen Lin and G. Manimaran

Dept. of Electrical and Computer Engineering, Iowa State University, Ames, IA 50011, USA
{linsz, gmani}@iastate.edu

Abstract. In this paper, we propose a feedback-based combined scheduling algorithm with fault tolerance for applications that have both periodic tasks and aperiodic tasks in real-time uniprocessor systems. Each periodic task is assumed to have a primary copy and a backup copy. By using the rate monotonic scheduling and deferrable server algorithm, we create two servers, one for serving aperiodic tasks and the other for executing backup copies of periodic tasks. The goal is to maximize the schedulability of aperiodic tasks while keeping the recovery rate of periodic tasks close to 100%. Our algorithm uses feedback control technique to balance the CPU allocation between the backup server and the aperiodic server. Our simulation studies show that the algorithm can adapt the parameters of the servers to recover the failed periodic tasks.

Keywords: Real-time systems, feedback-based scheduling, deferrable server algorithm, fault-tolerance.

1   Introduction

Real-time systems are defined as those systems in which the correctness of the system depends not only on the logical result of computation, but also on the time at which the results are produced [1]. Most real-time systems involve both periodic tasks and aperiodic tasks. Usually, periodic tasks are more important than aperiodic tasks. Due to the critical nature of tasks in a real-time system, it is essential that every periodic task admitted in the system completes its execution even in the presence of faults. Therefore, fault tolerance is an important requirement in such systems.

To address the fault tolerance problem, the primary/backup technique is used [2]. Each periodic task is assumed to have a primary copy and a backup copy. If the primary copy fails, the backup copy will be scheduled and then executed.

In order to schedule both periodic and aperiodic tasks in real-time systems, the simplest approach is to create a periodic server, with a certain computation time and period, whose purpose is to serve one or more aperiodic tasks each time it is invoked. In our paper, the concept of deferrable server algorithm [3] is used. We create two deferrable servers, one for serving aperiodic tasks and the other for executing backup copies of failed periodic tasks.

In scheduling systems, faults cause the performance of the system unpredictable. Control theory is one of the successful areas in addressing performance in the presence of uncertainty [4]. To the best of our knowledge, ours is the first work that uses feedback control theory to address the problem of combined scheduling with fault tolerance.

L. Bougé and V.K. Prasanna (Eds.): HiPC 2004, LNCS 3296, pp. 101–110, 2004.
© Springer-Verlag Berlin Heidelberg 2004

In this paper, we propose a feedback-based combined scheduling for uniprocessor real-time systems. By feeding back the performances, we adjust the utilization capacity for backup deferrable server. Specifically, we adjust the period of the backup deferrable server. By adjusting the period, the utilization capacity and priority are both adjusted. Since the CPU resource is limited, we cannot adjust the utilization for a server without changing the utilization of the other server. If the utilization capacity of the backup server is increased (decreased), then the utilization capacity of the aperiodic server must be decreased (increased). Our goal is to maximize the schedulability of aperiodic tasks while keeping the recovery rate of periodic tasks close to the desired value (100%).

The rest of the paper is organized as follows. In Section 2, the related work is discussed. In Section 3, the feedback-based fault-tolerant scheduling algorithm is proposed. In Section 4, we validate the result through simulation. Finally, Section 5 concludes the paper.

2   Background

In this section, we will introduce the system model for our algorithm, feedback control technique, and deferrable server [3] concept.

2.1   System Model

Figure 1 shows the system model. In our model, we make following assumptions.

- The system is uniprocessor real-time system.
- Tasks arrive at the system are periodic and aperiodic tasks. Each periodic task T_i is denoted by $T_i = \langle c_i, p_i \rangle$, where c_i and p_i are the computation time and period of T_i respectively. Each aperiodic task T_j is denoted by $T_j = \langle a_j, c_j, d_j \rangle$, where a_j, c_j and d_j are the arrival time, computation time and deadline of task T_j respectively.
- We assume a primary-backup scheduling for periodic tasks [6][7][8] wherein a backup copy of a task is executed if its primary fails the acceptance test.
- We assume transient faults for the periodic tasks and the faults are independent. The failure of aperiodic tasks is not considered as they are not very critical.
- There exists a fault-detection mechanism such as acceptance tests to detect both processor failures (transient) and software failures.
- We use the concept of deferrable server to serve aperiodic tasks and the backup copies. We have n periodic tasks: $T_1, T_2, ..., T_n$, and we also create an aperiodic deferrable server ($T_{as} = \langle c_{as}, p_{as} \rangle$) to serve aperiodic tasks and another backup deferrable server ($T_{bs} = \langle c_{bs}, p_{bs} \rangle$) to serve backup copies of periodic tasks. c_{as} and p_{as} are the computation time budget and period of the aperiodic deferrable server respectively. c_{bs} and p_{bs} are the computation time budget and period of the backup deferrable server respectively. When a failure is detected in a primary copy, the backup copy is put into the backup task queue of the backup deferrable server.
- The scheduling algorithm used is rate monotonic scheduling (RMS) [5] algorithm. The RMS algorithm schedules periodic tasks and the server instances.
- Feedback control technique is used to adjust the utilization allocated to the aperiodic server and the backup server. The adjustment is based on the fault rate and recovery rate of the periodic tasks.

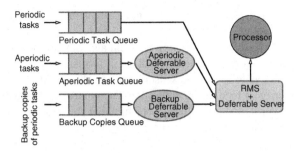

Fig. 1. System model

We define *miss ratio* (MR) of periodic tasks and aperiodic tasks and *recovery rate* (RR) of periodic tasks for the model. Miss ratio of periodic tasks is the ratio of the number of periodic tasks that miss their deadlines to the number of periodic tasks admitted into the system. Similarly, miss ratio of aperiodic tasks is defined for aperiodic tasks. Recovery rate of periodic tasks is the ratio of the number of recovered tasks (backup copies that meet their deadlines) to the number of failed primary copies. The desired MR of the periodic tasks is zero and the desired recovery rate is 100%.

Deferrable Server Algorithm: In the deferrable server algorithm, a periodic task known as a *deferrable server* [3] is created to serve aperiodic tasks. When the server is invoked but no aperiodic tasks are outstanding, the server does not execute but defers its assigned time slot. When an aperiodic task arrives, the server is invoked to execute aperiodic tasks and maintains its priority. The computation time budget for the server is replenished at the beginning of each period of the server.

For periodic tasks and deferrable server, we use the following schedulability checks. Assume that the tasks are ordered in non-increasing order of priority, that is, we have $T_1, T_2, ..., T_m, T_s, T_{m+1}, ..., T_n$, where T_s is the server. For schedulability check of each periodic task T_j that have higher priorities than the server, Equation 1 is used. For the server, Equation 2 is used, where c_s and p_s are computation time budget and period of the server respectively. For each task T_j that have lower priorities than the server, Equation 3 is used.

$$\sum_{i=1}^{j} \frac{c_i}{p_i} \leq j(2^{1/j} - 1) \tag{1}$$

$$\sum_{i=1}^{m} \frac{c_i}{p_i} + \frac{c_s}{P_s} \leq (m+1)(2^{1/(m+1)} - 1) \tag{2}$$

$$\sum_{i=1}^{j} \frac{c_i}{p_i} + \frac{c_s}{p_s} + \frac{c_s}{p_j} \leq (j+1)(2^{1/(j+1)} - 1) \tag{3}$$

2.2   Feedback Control

Figure 2 shows a typical control system, consisting of a controller, a plant to be controlled (controlled system), sensors, and actuators [9]. The system defines four variables: (1) exogenous variables are inputs from outside of the system, e.g., set points (desired values of the output values) and disturbance. (2) regulated variables are the output values that the system wants to regulate. (3) measured variables are values that the sensors measure. (4) control variables are the inputs of the actuators. The actuators will actuate the plant based on the control variables. Besides, The system also defines the error, which is the difference between the set points and the feedback information.

The system works as follows: The sensors periodically monitor the regulated variables and get the error to feed to the controller. The controller computes the required control, using the control function of the system, based on the error. The actuators change the control (manipulated) variables to control the system.

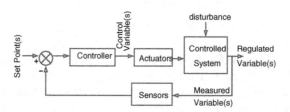

Fig. 2. Control system

3   Proposed Combined Scheduling Using Feedback

The goal of the proposed algorithm is to maximize the schedulability of aperiodic tasks while keeping the recovery rate of primary copies of periodic tasks close to 100%. Since aperiodic tasks are less important than periodic tasks, we give the aperiodic deferrable server a period larger than all periodic tasks, that is, the aperiodic deferrable server has lowest priority. According to Section 2.1, the CPU utilization allocated to the backup deferrable server is $u_{bs} = \frac{c_{bs}}{p_{bs}}$, the utilization that the periodic tasks need is $u_p = \sum_{i=1}^{n} \frac{c_i}{p_i}$, and the utilization allocated to aperiodic tasks is $u_{as} = \frac{c_{as}}{p_{as}}$.

The problem is how to allocate CPU utilization to the periodic tasks, aperiodic deferrable server and backup deferrable server. Since there is no way to know exactly how many tasks and which tasks will fail until the failures happen, we need to estimate u_{bs} to allocate resource. When an application starts, we guarantee enough capacity to the periodic tasks, allocate small capacity to the backup deferrable server, and the remaining capacity is allocated to the aperiodic deferrable server. When faults occur, we increase u_{bs}. u_{bs} can be increased by increasing c_{bs} or decreasing p_{bs}. Since decreasing p_{bs} can not only increase u_{bs} but also increase the priority of the backup deferrable server, we adjust the period of the backup deferrable server to change u_{bs}. The failure information can be obtained by measuring the recovery rate of failed periodic tasks in a past interval. The remaining utilization is assigned to the aperiodic deferrable server with lowest priority. When u_{bs} changes, we change the capacity of the aperiodic deferrable server to achieve suitable CPU utilization. The priority of the aperiodic deferrable server is fixed to be the lowest priority, thus p_{as} is fixed and c_{as} will be adjusted.

3.1   Admission Test

During the adjusting of the utilization of servers, the situation may happen. At the beginning, the period of the backup deferrable server is larger than the aperiodic deferrable server; Later, the period of the backup deferrable server may become smaller than the aperiodic deferrable server. In the admission test, when $p_{bs} > p_{as}$, Equation 4 is used, otherwise, Equation 5 is used.

$$\sum_{i=1}^{n} \frac{c_i}{p_i} + \frac{c_{as}}{p_{as}} + \frac{c_{bs}}{p_{bs}} + \frac{c_{as}}{p_{bs}} \leq (n+2)(2^{1/(n+2)} - 1) \tag{4}$$

$$\sum_{i=1}^{n} \frac{c_i}{p_i} + \frac{c_{bs}}{p_{bs}} + \frac{c_{as}}{p_{as}} + \frac{c_{bs}}{p_{as}} \leq (n+2)(2^{1/(n+2)} - 1) \tag{5}$$

All the utilization adjustments must satisfy Equation 4 or Equation 5. That is, every time we change the period of the backup deferrable server, we need to use Equation 4 or Equation 5 to decide the value of c_{as}.

3.2   Feedback Control Mechanism

The system architecture is shown in Figure 3. The measured variable is recovery rate of failed periodic tasks. The set point is desired value of the recovery rate. The control variable is the utilization of backup sever. The regulated variable is the recovery rate. However, we notice that if we assign the set point to a desired value of 100%, the period of the backup deferrable server will not change when the fault rate decreases. To avoid such undesired situation, we also measure the failure rate of the periodic tasks and feedback this information to the controller to adjust the period of the backup deferrable server. The controller algorithm is shown in Equation 6. In Equation 6, when the recovery rate at time instant $k - 1$ (RR_{k-1}) is less than the set point ($RR_s = 100\%$), the term $-k_r(RR_s - RR_{k-1})$ contributes a negative part to the period at time instant k (p_{bsk}) and hence the period of the backup deferrable server will decrease. This means the utilization allocated to the backup deferrable server increases. Therefore backup copies will get more chances to be executed, and the recovery rate will increase. In Equation 6, we also compare the measured failure rate (FR_{k-1}) with the average failure rate (FR_a) in the past t intervals. If the failure rate is less than FR_a, the term $k_f(FR_a - FR_{k-1})$ will contribute a positive part to the period. This will increase the period, and hence the utilization allocated to the backup deferrable server will decrease.

$$p_{bsk} = p_{bs(k-1)} - k_r(RR_s - RR_{k-1}) + k_f(FR_a - FR_{k-1}) \tag{6}$$

After getting p_{bs}, c_{as} can be calculated; when $p_{bsk} > p_{ask}$, we use Equation 7, otherwise, we use Equation 8. The subscript k is time instant and U_{rms} is the utilization bound of RMS.

$$c_{ask} = \frac{U_{rms} - u_p - u_{bsk}}{\frac{1}{p_{as}} + \frac{1}{p_{bsk}}} \tag{7}$$

$$c_{ask} = (U_{rms} - u_p - u_{bsk} - \frac{c_{bs}}{p_{as}}) \times p_{as} \tag{8}$$

Note that when we use controller to adjust the parameters of the two servers, negative values of p_{bsk} and c_{ask} should not appear. There may be two reasons of getting negative values: One is that the parameter of the controller is too large which results in the task parameters going to negative values; The other reason is that the fault rate is very high, and the system resource does not have enough capacity to execute all the tasks. We

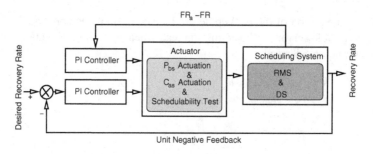

Fig. 3. System architecture

assume low fault rate, so the latter case will not appear. To deal with the formal case, we need to choose the controller parameters carefully such that the task parameters will not go out of their reasonable ranges. When faults occur, the period of backup deferrable server decreases. According to Equation 5, if the period decreases to a value that c_{as} has to be zero such that Equation 5 can still be satisfied, p_{bs} can not be decreased any more. From Equation 5, we get Equation 9.

$$\frac{c_{bs}}{p_{bs}} + \frac{c_{as}}{p_{as}} = U_{rms} - u_p - \frac{c_{bs}}{p_{as}} \tag{9}$$

We require that $c_{as} \geq 0$, then we have Equation 10.

$$p_{bs} \geq \frac{c_{bs}}{U_{rms} - u_p - \frac{c_{bs}}{p_{as}}} \tag{10}$$

In order to let p_{bs} satisfy Equation 10, we must be careful with selecting k_r and k_f. If these two values are too large, it will cause large fluctuation and unreasonable task parameters before the system becomes stable. Thus, we need to confine the value of k_r and k_f such that p_{bs} will be greater than $p_{bst} = \frac{c_{bs}}{U_{rms} - u_p - \frac{c_{bs}}{p_{as}}}$. The maximum change in p_{bs} is $p_{bsmax} = p_{bso} - p_{bst}$, where p_{bso} is the original period assigned to the backup deferrable server. Thus we have Equation 11.

$$-k_r \Delta RR + k_f \Delta FR \leq p_{bsmax} \tag{11}$$

ΔRR and ΔFR are the change in RR and FR respectively in the corresponding situation. Assume $\Delta RR = -1$ and $\Delta FR = 1$, we have a conservative restriction on k_r and k_f as shown in Equation 12, where $k_r > 0$ and $k_f > 0$.

$$k_r + k_f \leq p_{bsmax} \tag{12}$$

When we choose the parameters for the controller, Equation 12 must be satisfied.

4   Simulation Studies

The simulation studies were conducted in two parts. The first part shows the effect of feedback adjustment by injecting fault at the beginning. The second part shows the steady state performance for different fault rate.

The periodic tasks used for the simulation are generated as follows:

- The computation time (c_i) is uniformly chosen between 10 and 20.

- The period of task T_i is uniformly chosen between $(6 * c_i)$ and $(8 * c_i)$.
- The backup copies have identical characteristics of their primary copies.

The aperiodic tasks used for the simulation are generated as follows:

- The computation time (c_i) of task T_i is uniformly chosen between 10 and 20.
- The deadline of T_i is uniformly chosen between $(r_i + 30 * c_i)$ and $(r_i + 40 * c_i)$.
- The inter-arrival time of tasks is exponentially distributed with mean $\theta = 50$.

4.1   Part 1: Fault Injection at Time 0

In the first part of the simulation, we generate three periodic tasks and inject fault at time instant 0. Fault rate for task T_1, T_2 and T_3 are 0.01, 0.02 and 0.01 respectively.

Figure 4 shows that the period for the backup server decreases due to the fault injection. Then the period curve becomes flat after the fluctuation at the beginning.

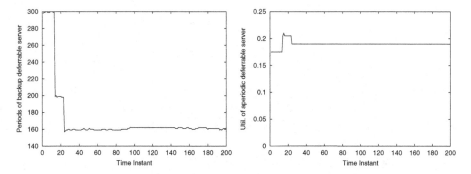

Fig. 4. Periods of backup deferrable server   **Fig. 5.** Utilization of aperiodic deferrable server

Figure 5 shows the utilization allocated for the aperiodic deferrable server. At the beginning, the utilization is lower, this is due to the preassigned budget. Then the utilization goes up since the utilization of the backup server has not gone up very much. Then, when the utilization of the backup server increases more, the utilization of the aperiodic deferrable server decreases.

Figure 6 shows the utilization allocated for the backup deferrable server. The curve increases at the beginning and then becomes stable due to the injection of fault.

Figure 7 shows the recovery rate of the failed periodic tasks. It reaches 100% soon. This means failed periodic tasks can be recovered after the curve reaches 100%.

4.2   Part 2: Steady State Performances

In the second part of the simulation, we measure the system performances for different average fault rate of tasks. In the steady state, the recovery rate is 100%. We plot the utilization of the aperiodic server instead of the schedulability of aperiodic tasks.

Figure 8 shows the periods of the backup deferrable server for different fault rates. The period of the backup server decreases when the fault rate increases.

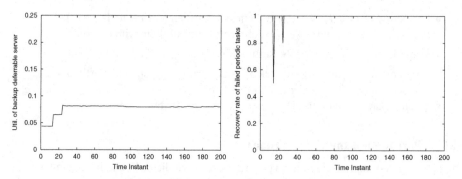

Fig. 6. Utilization of backup deferrable server   **Fig. 7.** Recovery rate of failed periodic tasks

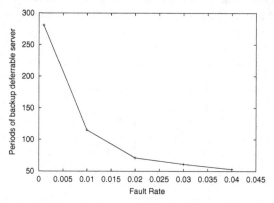

Fig. 8. Periods of backup deferrable server

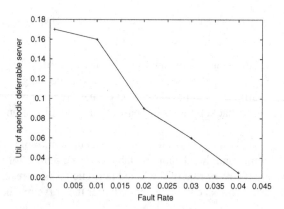

Fig. 9. Utilization of aperiodic deferrable server

Figure 9 shows the final values of the utilization allocated to the aperiodic deferrable server. The utilization decreases when the fault rate increase, since the aperiodic deferrable server needs to give some utilization to the backup server so that the recovery rate of the backup copies will reach 100%.

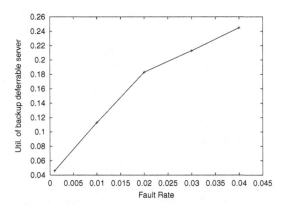

Fig. 10. Utilization of backup deferrable server

Figure 10 shows the final values of the utilization allocated to the backup deferrable server. The utilization increases when the fault rate increase since more utilization needs to be allocated to the backup server when the fault rate increases.

5   Related Work

In [10][11], authors proposed a new methodology for automatically adapting the rate of a periodic task set by using feedback control technique. The rate adaptation is good to achieve correct behavior of the system and high resource utilization. In [12][13], the authors present a feedback control EDF scheduling algorithm for real-time uniprocessor systems. The adaption is applied on tasks' service level. In [14][15][16], authors present a closed-loop scheduling algorithm based on execution time estimation. [17] assumes that each task consists of n sequential replicable sub-tasks. The controller adjusts the number of the replicas of sub-tasks to achieve low MR and high resource utilizations when the system is overloaded.

The above papers adopt the feedback control technique, but they do not address the issue of fault tolerance, which is important for periodic tasks in real-time systems. Our paper is the first one which addresses the combined scheduling with fault-tolerant issue.

There are several papers [6][7][8] which use PB approach to address fault tolerant scheduling problems in real-time systems, however, these papers do not use feedback technique, thus do not have the flexibility to allocate suitable utilization for tasks.

6   Conclusions

In this paper, we proposed a feedback-based fault-tolerant scheduling algorithm for real-time uniprocessor systems. The system has both periodic tasks and aperiodic tasks. Each periodic task can have a primary copy and a backup copy. The rate monotonic scheduling algorithm and deferrable server algorithm are used to schedule tasks. Two deferrable servers are created, one for aperiodic tasks and one for the backup copies of periodic tasks. The recovery rate and failure rate of periodic tasks are fed back to the controller, and the utilization capacity of the backup deferrable server is adjusted using feedback control theory. Suitable utilization capacity is allocated to backup deferrable server and the remaining utilization capacity is used for the aperiodic deferrable server. The simulation studies show that the algorithm can guarantee 100% recovery rate for periodic tasks with a balanced utilization for backup tasks.

References

1. K. Ramamritham and J. A. Stankovic, "Scheduling algorithms and operating systems support for real-time systems", in Proc. IEEE, vol.82, no.1, pp.55-67, Jan. 1994.
2. D. K. Pradhan, "Fault Tolerant Computing: Theory and Techniques", *Prentice Hall*, NJ, 1986.
3. J. K. Strosnider, J. P. Lehoczky and L. Sha, "The deferrable server algorithm for enhanced aperiodic responsiveness in hard real-time environments", *IEEE Trans. on Computers*, vol.44, no.1, pp.73-91, Jan. 1995.
4. Katsuhiko Ogata, "Modern Control Engineering", *Prentice Hall*, Upper Saddle River, New Jersey, 2002.
5. C. Liu and J. Layland, " Scheduling algorithms for multiprogramming in a hard real-time environment", *Journal of ACM*, vol.20, no.1, pp.45-61, Jan. 1973.
6. A. L. Liestman and R. H. Campbell, "A fault-tolerant scheduling problem", *IEEE Trans. Software Engineering*, vol.12, no.11, pp.1089-1095, Nov. 1988.
7. S. Ghosh, R. Melhem, and D. Mosse, "Fault-tolerance through scheduling of aperiodic tasks in hard real-time multiprocessor systems", *IEEE Trans. on Parallel and Distributed Systems*, vol.8, no.3, pp.272-284, Mar. 1997.
8. G. Manimaran and C. Siva Ram Murthy, "A fault-tolerant dynamic scheduling algorithm for multiprocessor real-time systems and its analysis", *IEEE Tran. on Parallel and Distributed Systems*, vol.9, no.11, pp.1137-1152, Nov. 1998.
9. C. Siva Ram Murthy and G. Manimaran,"Resource Management in Real-Time Systems and Networks", *MIT Press*, April 2001.
10. G. Buttzaao, G. Lipari and L. Abeni, "Elastic task model for adaptive rate control", in Proc. *IEEE Real-Time Systems Symposium*, pp.286-295, 1998.
11. G. Buttazzo and L. Abeni, " Adaptive workload management through elastic scheduling", *Real-Time Systems*, vol.23, no.1-2, pp.7-24, July-September, 2002.
12. C. Lu, J. A. Stankovic, G. Tao, and S.H. Son, "Design and evaluation of feedback control EDF scheduling algorithm", In Proc. *IEEE Real-Time System Symposium*, pp.56-67, 1999.
13. J. A. Stankovic, Chenyang Lu, S. H. Son, and G. Tao, "The case for feedback control real-time scheduling", in Proc. *Euromicro Conference on Real-Time Systems*, pp.11-20, 1999.
14. D. R. Sahoo, S. Swaminathan, R. Al-Omari, M. V. Salapaka, G. Manimaran, and A. K. Somani, "Feedback control for real-time scheduling", in Proc. *American Controls Conference*, vol.2, pp.1254-1259, 2002.
15. R. Al-Omari, G. Manimaran, M. V. Salapaka, and A. K. Somani, "Novel algorithms for open-loop and closed-loop scheduling of real-time tasks based on execution time estimation", in Proc. *IEEE Intl. Parallel and Distributed Processing Symposium*, pp.7-14, 2003.
16. S. Lin, S. Sai Sudhir and G. Manimaran, "ConFiRM-DRTS: A certification framework for dynamic resource management in distributed real-time systems," in Proc. *Intl. Workshop on Parallel and Distributed Real-Time Systems*, pp.110-117, 2003.
17. B. Ravindran, P. Kachroo, and T. Hegazy, "Adaptive resource management in asynchronous real-time distributed systems using feedback control functions", in Proc. *Intl. Symposium on Autonomous Decentralized Systems*, pp.39-46, 2001.

A Shared Memory Dispatching Approach for Partially Clairvoyant Schedulers

K. Subramani[1] and Kiran Yellajyosula[*,2]

[1] LCSEE,
West Virginia University,
Morgantown, WV
ksmani@csee.wvu.edu
[2] CSE,
University of Minnesota,
Minneapolis, MN
kiran@cs.umn.edu

Abstract. It is well known that in a typical real-time system, certain parameters, such as the execution time of a job, are not fixed numbers. In such systems, it is common to characterize the execution time as a range-bound interval, say, $[l, u]$, with l indicating the lower bound on the execution time and u indicating the upper bound on the same. Such intervals can be determined with a high degree of confidence in state of the art operating systems, such as MARUTI [7,5] and MARS [2]. Secondly, jobs within a real-time system are often constrained by complex timing relationships. In hard real-time applications, it is vital that all such constraints are satisfied at run time, regardless of the values assumed by environment-dependent parameters, such as job execution times. As described in [11], there are two fundamental issues associated with real-time scheduling, viz., the schedulability query and dispatchability. A positive answer to the schedulability query may not by itself guarantee that all the imposed constraints will be met at run-time; indeed the phenomenon in which the dispatcher fails to dispatch a schedulable job set, is called *Loss of Dispatchability*. This paper is concerned with techniques to address this phenomenon in Partially Clairvoyant schedulers; we primarily focus on distributing the dispatch computations across the processors of a shared-memory computer.

1 Introduction

Real-time systems have gained importance in a number of applications ranging from safety-critical systems such as nuclear reactors and automotive controllers, to gaming software and internet routing. There are two primary issues in real-time scheduling that distinguish this form of scheduling from the more traditional scheduling models, considered in [8] and [1], viz., the non-constant nature of job

---

[*] This work was conducted while the author was a graduate student at West Virginia University.

L. Bougé and V.K. Prasanna (Eds.): HiPC 2004, LNCS 3296, pp. 111–122, 2004.
© Springer-Verlag Berlin Heidelberg 2004

execution time and the existence of temporal constraints between jobs. When the execution times are non-constant and there are timing relationships between jobs, it is not obvious as to when a set of jobs can be declared schedulable. Accordingly, the E-T-C scheduling model was proposed in [11]; this model formally addresses the notion of schedulability in a real-time system. Within the framework of this model, we focus on Partially Clairvoyant scheduling with relative timing constraints [12].

As was pointed out in [11], schedulability and dispatchability are two distinct issues, with the former being conducted offline and latter being conducted online. It is the job of the scheduler to examine the constraints and check whether a feasible schedule exists, under the given schedulability query. If such a schedule exists, then it is the job of the dispatcher to compute the actual time at which a job is to be placed on the time line. In case of Partially Clairvoyant scheduling, computing the dispatch time of a job is a nontrivial task and it is possible that the time taken to perform this computation may cause constraint violation at run time. This phenomenon has been referred to as the *Loss of Dispatchability* phenomenon [10].

One technique to address Loss of Dispatchability in a scheduling system is to distribute the computation of the dispatch functions of jobs. This paper examines the benefits of this approach, by using a shared memory cluster to perform the dispatching and contrasting its performance with a sequential dispatcher. The detailed performance profile clearly demonstrates the superiority of the multiple processor approach.

2   Statement of Problem

Consider a set of ordered, hard real-time jobs $\mathcal{J} = \{J_1, J_2, \ldots, J_n\}$. These jobs are non-preemptive and occur exactly once in each scheduling window. Let $\vec{s} = [s_1, s_2, \ldots, s_n]^T$ denote the start time vector of the jobs and let $\vec{e} = [e_1, e_2, \ldots, e_n]^T$ denote the corresponding execution time vector. The execution time of the job J_i is known to vary in the interval $[l_i, u_i]$. There exist relative timing relationships between pairs of jobs, which can be represented by simple difference constraints between the start (or finish) times of jobs. For instance, the constraint: Start job J_2 six units after job J_1 finishes, can be expressed as: $s_1 + e_1 + 6 \leq s_2$. The set of all constraints can be thought of as a network, as described in [12]. Note that there is a strict ordering in job execution in that J_1 starts and finishes before J_2, which in turn starts and finishes before J_3 and so on.

In a Partially Clairvoyant schedule, the start time of a job can depend upon the execution times of jobs that are sequenced before it. Accordingly, the schedulability query is:

$$\exists s_1 \, \forall e_1 \in [l_1, u_1] \, \exists s_2 \, \forall e_2 \in [l_2, u_2], \ldots \exists s_n \, \forall e_n \in [l_n, u_n] \quad \mathbf{A} \cdot [\vec{s} \ \vec{e}]^\mathbf{T} \leq \vec{b}? \quad (1)$$

where $\mathbf{A} \cdot [\vec{s} \ \vec{e}]^\mathbf{T} \leq \vec{b}$ is the matrix representation of the constraint network. Familiarity with the contents of [12] is strongly recommended for a clear understanding of this paper.

Note that s_1 is numeric, since J_1 is the first job in the sequence, while s_i, $i > 1$, is a function of $e_1, e_2, \ldots, e_{i-1}$, in the above formulation.

The algorithm in [12] takes System (1) as input and proceeds by eliminating execution time variables and start time variables of jobs, beginning with J_n. For the purpose of saving space, we relegate the discussion of this algorithm to the journal version of the paper.

Consider a four job set J_1, J_2, J_3, J_4, with execution times $e_1 \in [3, 7]$, $e_2 \in [5, 6]$, $e_3 \in [2, 7]$ and $e_4 \in [8, 12]$ respectively, with the following set of temporal constraints imposed on them:

(a) J_1 finishes at least 2 units before J_2 starts: $s_1 + e_1 + 2 \leq s_2$.
(b) J_3 starts after the completion of J_2: $s_2 + e_2 \leq s_3$.
(c) J_3 starts at least 5 units after, but within 10 units of J_1 completing:
      $s_1 + e_1 + 5 \leq s_3$, $s_3 \leq s_1 + e_1 + 10$.
(d) J_3 finishes at least 5 units before J_4 starts: $s_3 + e_3 + 5 \leq s_4$.
(e) J_4 completes within 40 units of time: $s_4 + e_4 \leq 40$.

The Partially Clairvoyant schedule for the above example can be obtained by applying the algorithm in [12].

(i) $0 \leq s_1 \leq 1$
(ii) $s_1 + e_1 + 2 \leq s_2 \leq \min(10, s_1 + e_1 + 4)$
(iii) $\max(s_1 + e_1 + 5, s_2 + e_2) \leq s_3 \leq \min(s_1 + e_1 + 10, 16)$
(iv) $s_3 + e_3 + 5 \leq s_4 \leq 28$

In general, the dispatch function for s_i will have the following form:

$$\max(f_1, f_2, \ldots, f_{i-1}) \leq s_i \leq \min(f'_1, f'_2, \ldots, f'_{i-1}).$$

where f_j and f'_j are functions depending on the start and execution times of job J_j $(j < i)$. Further, each f_j is either a constant or a function of the form $s_j(+e_j) + k_1$, for some positive k_1.

Definition 1. *A safety interval for a job is the time interval during which the job can be started without violating any of the constraints imposed by the constraint system in (1).*

Thus, in order to determine the safety interval of J_i, we must know $e_j, \forall j < i$.

Definition 2. *A feasible Partially Clairvoyant schedule is said to be dispatchable on a machine M, if for every job J_i, M can start executing J_i in its safety interval.*

For the example above, assuming that $s_1 = 0$ and $e_1 = 6$, the safety interval for s_2 is $[8, 10]$.

The machine M computes the dispatch functions and obtains the safety interval during which each job can be dispatched without violating the constraints. The job J_i cannot be dispatched, if the time taken by M to perform this computation, exceeds its safety interval. Our goal in this paper is to determine safety intervals for jobs, such that *Loss of Dispatchability* does not occur.

3   Motivation and Related Work

The issues raised by execution time variability and complex timing constraints in real-time systems have been formalized in [11], where the E-T-C framework was proposed. Within this framework, we are interested in Partially Clairvoyant scheduling, as far as this paper is concerned. This problem was first discussed in [3], where an $O(n^3)$ algorithm was proposed for answering query (1). They also discuss a dispatching scheme, which could take as much as $\Omega(n)$ time on an n-job set, in the worst case. This computation cost suffered by the online dispatcher can cause constraint violation, i.e., the time after the computation of the safety interval of J_b, viz., $[l_b, r_b]$ could exceed r_b and hence J_b cannot be dispatched. For the example in Section §2, assume that the first two jobs take the maximum time to execute. Let J_1 start at time $t = 0$; it follows that the safety interval of J_3 is $[15, 16]$. If the dispatcher takes more than one unit of time to compute this safety interval, then J_3 cannot be dispatched.

[10] provides a theoretical discussion of a parallel online algorithm for eliminating Loss of Dispatchability. This algorithm provides $O(1)$ dispatch time per job and uses $O(n)$ space per processor; unfortunately, it uses one processor per job. We combine the design ideas in [10], with a comprehensive implementation profile in this paper; our approach uses a fixed number of processors, regardless of the number of jobs. One of the important consequences of our work is that we now have a smooth tradeoff between processors and quality of service, in that a greater level of quality can be guaranteed, by increasing the number of processors in the scheduling system.

4   Architecture and Algorithm

We use the Concurrent Read Exclusive Write (CREW) shared memory architecture, the details of which can be found in [4]. Each processor has a separate local memory in addition to the common shared memory and maintains a copy of the data it requires in its local memory. Changes to the shared data are made in the local memory first and then flushed to the shared memory, to achieve coherence. A shared data variable present in the memory of multiple processors is invalidated once a processor updates the shared data in the central memory. A processor requires far less time to access data from its memory than data in the memory of another processor. While reading a shared variable, the value resulting from the most recent write is loaded into the local memory.

4.1   Architecture

We assume that there is a central processor that runs the jobs and a number of satellite processors that are involved in the computation of dispatch functions. The processors share data with each other through the shared memory as indicated in Figure (1). The variables of interest, which are stored in the shared memory include (s_i, e_i) and $(l_{b_{i+1}}, r_{b_{i+1}})$. The central processor C executes Job J_i and stores (s_i, e_i) in the memory. It then updates a flag f_1 and waits on

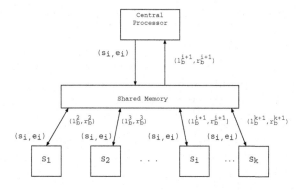

Fig. 1. Shared Memory Dispatcher Architecture

another flag f_2 to be updated by a satellite processor. Each satellite processor S_j updates and reports the safety intervals for the class of jobs C_j assigned to it. The satellite processor S_m, to which $J_{i+1} \in C_m$ has been assigned, writes the safety interval $(l_{b_{i+1}}, r_{b_{i+1}})$ in the memory and updates the flag f_2. After updating the safety intervals of all the remaining jobs in their class, the satellite processors wait on flag f_1 to be updated by C.

In this implementation, there are no communication costs as compared to a network distributed model but there is a cost for achieving memory coherence.

4.2   Algorithms

In [3], a sequential algorithm for dispatching was proposed; this algorithm has a dispatch complexity of $\Omega(n)$, since in the worst case, s_n may depend upon the execution times of all the jobs in the sequence.

Algorithm (4.2) describes the shared memory dispatcher. In this algorithm, each processor relaxes at most four constraints (see [12]), to determine the safety interval of s_{i+1}, after J_i has finished execution on the central processor. As stated in [10], relaxing 4 constraints takes at most four additions and four comparisons, i.e., $4 \cdot (T_{add} + T_{comp})$, where T_{add} and T_{comp} are the times taken to perform an addition and a comparison respectively. Let w_1 be the cost of writing a floating point number to the shared memory such that the data is coherent throughout the memory. C is required to flush the present values of (s_i, e_i, f_1) to the memory, while S_k writes $(l_{b_{i+1}}, r_{b_{i+1}}, f_2)$ into the shared memory. The time required to compute the safety interval is at most $4 \cdot (T_{add} + T_{comp}) + 6 \cdot w_1$ and hence Algorithm (4.2) takes at most $O(1)$ time, for computing the safety interval of a job.

Algorithm (4.2) updates the dispatch functions in parallel with the execution of the current job. Each processor updates constraints between the completed job and a fraction $(= \frac{1}{k})$ of the remaining jobs, where k is the number of satellite processors. When n is very large, the time required to update the constraints is larger than the execution time of a current job and this leads to loss of dispatchability on a sequential dispatcher. Increasing the number of processors helps in dispatching the schedule, as long as the memory coherence cost is not large.

Algorithm 4.1: Shared-Memory Dispatcher for a Partially Clairvoyant Scheduler

---

Function SHARED-ONLINE-DISPATCHER-FOR-J_a $(G =< V, E >)$
1: Let $[l_{b_i}, r_{b_i}], (l_{b_i} < r_{b_i})$ denote the current safety interval of J_i.
2: Let P denote the number of satellite processors.
3: **for** ($i = 1$ **to** n) **in parallel do**
4:   **if** (central processor) **then**
5:     **if** (current-time $< l_{b_i}$) **then**
6:       Sleep (l_{b_i}-current-time)
7:     **end if**
8:     **if** (current-time $\in [l_{b_i}, r_{b_i}]$) **then**
9:       Execute job J_i
10:       Save (s_i, e_i) to memory
11:       Update $flag_1$ and save to memory
12:       Wait till $flag_2$ is updated
13:       Read ($l_{b_{i+1}}, r_{b_{i+1}}$) from memory
14:     **else**
15:       Return (Schedule is not dispatchable)
16:     **end if**
17:   **end if**
18:   **if** (satellite processor S_m) **then**
19:     Compute S_k, the satellite processor to which the safety interval computation is to be reported
20:     Wait till $flag_1$ is updated
21:     Read (s_i, e_i)
22:     **if** ($S_k = S_m$) **then**
23:       UPDATE-CONSTRAINTS(i, $i + 1$)
24:       Write safety interval to memory
25:       Update $flag_2$ and write to memory
26:       UPDATE-CONSTRAINTS(i, q) for each Job $J_q \in C_k$
27:     **else**
28:       UPDATE-CONSTRAINTS(i, q) for each Job $J_q \in C_m$
29:     **end if**
30:   **end if**
31:   **if** ($i = n$) **then**
32:     **return**(schedule is dispatchable)
33:   **end if**
34: **end for**

Algorithm 4.2: Update function of Shared-Memory Dispatcher

Function UPDATE-CONSTRAINTS(i, q)
1: Relax constraints between J_i and J_q into absolute constraints of J_q.
2: Compare each absolute constraint with the existing safety interval for J_q
3: **if** (new constraint is not redundant) **then**
4:   Update Safety Interval ($[l_{b_q}, r_{b_q}]$)
5: **else**
6:   Leave the Safety Interval unchanged
7: **end if**

5   Empirical Analysis

5.1   Machine Description

We tested an implementation of the dispatcher on the SGI Origin2000 machine of the National Computational Science Alliance (NCSA). The hardware specification of the machine and environment are listed in Tables 1 and 2 respectively.

Table 1. Machine specifications of SGI Origin2000 of NCSA

Component	Description
Architecture	Distributed Shared Memory
Processors	MIPS R10000
Available number of processors	64 (or 128)
Clock Speed	250 MHz or 195 MHz
Instruction Cache Size	32Kbytes
Data Cache Size	32 Kbytes
User Virtual Address Space	4 GB
Interconnect between machines	Gigabit Ethernet

Table 2. Software

Component	Description
Operating System	Irix 6.5
Compiler	C
Programming Models	OpenMP
Floating Point Format	IEEE
Batch System	Load Sharing batch system

Due to space constraints, we omit the details of schedule generation in the experimental suites.

5.2   Results

In our experiments, we attempted to maximize the degree of parallelism, in that threads executed in parallel, whenever possible.

Our first focus was on measuring the update time of the sequential and the shared memory dispatchers. The sequential dispatcher updates all the existing constraints, depending on the start and execution time of a completed job before starting the next job. Figure (2) plots the update time in seconds against the number of jobs, for the sequential dispatcher and the shared dispatcher with two processors. The plot shows that the update time of the sequential dispatcher increases with size of the schedule, while that of the shared dispatcher with two processors is almost constant ($= 2.5 \times 10^{-5}s$).

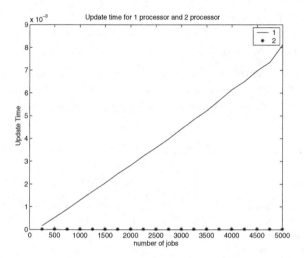

Fig. 2. Plot of update time of single processor dispatcher and 2-processor dispatcher versus number of jobs

It is not hard to see that after approximately 165 jobs, the shared memory dispatcher with two processors is superior to the uni-processor dispatcher.

We tested the shared-memory and sequential dispatcher with schedules of different sizes using three different random seeds. Similar results were observed in all three cases, i.e., schedules were not dispatchable using a single processor, but were dispatchable using multiple processors. Table 3 summarizes the results of the experiments performed.

Table 3. Results of dispatching job sets of different size using single and multiple processors. $\sqrt{}$ indicates that the schedule was successfully dispatched and $\times$ indicates that it was not

Processors	Number of Jobs							
	250	500	750	1000	2000	3000	4000	5000
1	$\sqrt{}$	$\sqrt{}$	$\sqrt{}$	$\sqrt{}$	$\times$	$\times$	$\times$	$\times$
2	$\sqrt{}$	$\sqrt{}$	$\sqrt{}$	$\sqrt{}$	$\sqrt{}$	$\sqrt{}$	$\sqrt{}$	$\sqrt{}$
3	$\sqrt{}$	$\sqrt{}$	$\sqrt{}$	$\sqrt{}$	$\sqrt{}$	$\sqrt{}$	$\sqrt{}$	$\sqrt{}$
4	$\sqrt{}$	$\sqrt{}$	$\sqrt{}$	$\sqrt{}$	$\sqrt{}$	$\sqrt{}$	$\sqrt{}$	$\sqrt{}$

5.3   Scalability

In this section, we observe the behaviours of dispatchers, as the size of the schedules (i.e., number of jobs) varies. We generated Partially Clairvoyant schedules with the number of jobs in each schedule increasing from 1000 to 9750. All the jobs in the schedules had execution time periods varying from one to five milliseconds and the spacing time between two adjacent jobs was between one-tenth and one-half of a millisecond.

Figure (3) shows the size of the largest schedule successfully dispatched with a certain number of processors and clearly demonstrates the scalability of the shared memory dispatcher. We observed that as the size of the schedule increased, additional processors were needed by the shared memory dispatcher to prevent loss of dispatchability. This is easily explained, since increasing the size of the schedule increases the update time required by the satellite processors. Consequently, the satellite processors need additional time to read the start and execution times of the current job, after it has been computed by the central processor. This increase in computing time in the satellite processors eventually results in the schedule breaking down. With 10 processors, all the generated schedules created were successfully dispatched.

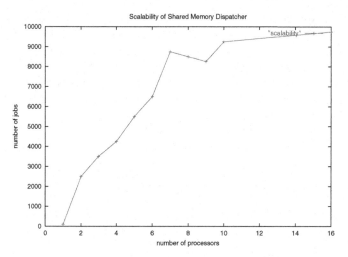

Fig. 3. The number of jobs that can be successfully dispatched by a given number of processors, where the job execution time was between 1 to 5 milliseconds and the spacing time was between 0.1 to 0.5 milliseconds. The area under the curve shows the schedules which can be successfully dispatched

In Figure (3), we observe that the slope of the curve decreases for large schedules, even as the number of processors is increased. As the size of the job set increases, the satellite processors need to access memory locations in the central memory for every constraint they relax. This causes frequent cache misses and page faults, thereby increasing the number of read and write operations to the main memory and slowing down the updating process. After a certain number of jobs, the time taken for constraint relaxation is significantly high and this causes delays in the reading of data, by the concerned satellite processor. We thus see that the memory to processor latency is a bottleneck for updating constraints and increasing the number of processors does not help when the latency and the spacing time are of the same order. A second reason for the increase in the latency is the increased number of read/write requests to the memory, as the number of processors increases. There is limited bandwidth between the shared memory and the processors; this limitation prevents all the processors from accessing and updating the memory at the same time, necessitating high speed connections and high memory bandwidth to ensure dispatchability of the schedules.

We also observed an interesting paradox, viz., that a given processor set successfully dispatched schedules of sizes larger than the size of the schedule on which it failed. In our opinion, this observation merits a serious investigation into the unpredictable nature of memory flushes; note that improving predictability is one of the fundamental concerns of real-time systems.

5.4   Effect of Execution Time

Increasing the execution times of the jobs gives the satellite processors additional time to compute their safety intervals. In most cases, this results in the satellite processors waiting for the central processor to update the flag f_1, which upholds our assumption in Section §4.1. On the other hand, decreasing the execution times of jobs voids our assumption, since the satellite processors would be updating safety intervals, even as the central processor completes a job and writes the corresponding values to memory. It follows that larger the job execution time, larger is the job set that can be dispatched.

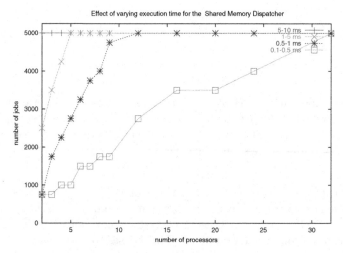

Fig. 4. The plot shows the effect of varying the execution time of jobs on the dispatchability of a job set by a certain number of processors. The spacing time was assumed to be between 0.1 to 0.5 milliseconds

In the test cases created, the spacing time between two adjacent jobs was set between one-tenth and one-half of a millisecond and job execution time was varied in the intervals $[0.1ms, 0.5ms]$, $[0.5ms, 1ms]$, $[1ms, 5ms]$ and $[5ms, 10ms]$.

The number of jobs in the schedule ranged from 250 to 5000. Experiments were conducted by varying the number of processors used by the dispatcher and finding the size of the largest schedule successfully dispatched. Figure (4) plots the size of the largest schedule successfully dispatched, with a given number of processors for the four different execution time intervals. From this figure, we conclude that greater the execution times of jobs, greater is the number of jobs that can be dispatched by the shared memory dispatcher. We also observed that schedules with higher execution time intervals can be dispatched using fewer processors, than the identical schedule with smaller execution time intervals. The sequential dispatcher was not able to dispatch any schedule used in these experiments.

6   Conclusion

In this paper, we implemented a shared memory dispatcher for a Partially Clairvoyant scheduler and empirically demonstrated its superiority over a sequential dispatcher. Whereas previous work had been confined to theoretical analysis, our work here conclusively establishes the need for a multi-processor approach to dispatching. We note that our strategy succeeds principally in situations when the time taken to compute the safety interval of a job exceeds the the time required to achieve memory coherence across a collection of satellite processors. On the whole, our approach is targeted to power up the controller by providing it with more processing power.

The principal conclusions that can be drawn from our work are as follows:

(a) Increasing the number of processors permits the dispatching of larger sized schedules and therefore the shared memory approach is highly scalable.
(b) When the execution times of the jobs are high, fewer processors are required to achieve successful dispatching.

From our perspective, the following issues merit further study:

(i) How does using non-blocking reads and writes in the dispatching algorithm (see [13]), affect its correctness and performance?
(ii) Can we construct a theoretical model to find the minimum number of processors that dispatches a given constraint set? We believe that this model and its solution, will have an immediate bearing on practical real-time systems.

Acknowledgements

The implementation effort involved in this project was supported by a grant from the National Computational Science Alliance (NCSA) under ASC30006N and utilized the account [kirany]. Their contributions are gratefully acknowledged.

References

1. P. Brucker. *Scheduling Algorithms*. Springer, 1998. 2nd edition.
2. A. Damm, J. Reisinger, W. Schwabl, and H. Kopetz. The Real-Time Operating System of MARS. *ACM Special Interest Group on Operating Systems*, 23(3):141–157, July 1989.
3. R. Gerber, W. Pugh, and M. Saksena. Parametric Dispatching of Hard Real-Time Tasks. *IEEE Transactions on Computers*, 1995.
4. Joseph Ja'Ja'. An introduction to parallel algorithms (contents). *SIGACTN: SIGACT News (ACM Special Interest Group on Automata and Computability Theory)*, 23, 1992.
5. S. T. Levi, S. K. Tripathi, S. D. Carson, and A. K. Agrawala. The `Maruti` Hard Real-Time Operating System. *ACM Special Interest Group on Operating Systems*, 23(3):90–106, July 1989.
6. Aloysius K. Mok, Chan-Gun Lee, Honguk Woo, and Konana. P. The monitoring of timing constraints on time intervals. In *Proceedings of the 23rd IEEE Real-Time Systems Symposium (RTSS'02)*, pages 191–200. IEEE Computer Society Press, 2002.
7. D. Mosse, Ashok K. Agrawala, and Satish K. Tripathi. Maruti a hard real-time operating system. In *Second IEEE Workshop on Experimental Distributed Systems*, pages 29–34. IEEE, 1990.
8. M. Pinedo. *Scheduling: theory, algorithms, and systems*. Prentice-Hall, Englewood Cliffs, 1995.
9. Manas Saksena. *Parametric Scheduling in Hard Real-Time Systems*. PhD thesis, University of Maryland, College Park, June 1994.
10. K. Subramani. *Duality in the Parametric Polytope and its Applications to a Scheduling Problem*. PhD thesis, University of Maryland, College Park, August 2000.
11. K. Subramani. A specification framework for real-time scheduling. In W.I. Grosky and F. Plasil, editors, *Proceedings of the 29th Annual Conference on Current Trends in Theory and Practice of Informatics (SOFSEM)*, volume 2540 of *Lecture Notes in Computer Science*, pages 195–207. Springer-Verlag, November 2002.
12. K. Subramani. An analysis of partially clairvoyant scheduling. *Journal of Mathematical Modelling and Algorithms*, 2(2):97–119, 2003.
13. Philippas Tsigas and Yi Zhang. Non-blocking data sharing in multiprocessor real-time systems. In *Proceedings of the Sixth International Conference on Real-Time Computing Systems and Applications*, page 247. IEEE Computer Society, 1999.

Data Redistribution Algorithms for Homogeneous and Heterogeneous Processor Rings

Hélène Renard, Yves Robert, and Frédéric Vivien

LIP, UMR CNRS-INRIA-UCBL 5668
ENS Lyon, France
{Helene.Renard, Yves.Robert, Frederic.Vivien}@ens-lyon.fr

Abstract. We consider the problem of redistributing data on homogeneous and heterogeneous processor rings. The problem arises in several applications, each time after a load-balancing mechanism is invoked (but we do not discuss the load-balancing mechanism itself). We provide algorithms that aim at optimizing the data redistribution, both for uni-directional and bi-directional rings. One major contribution of the paper is that we are able to prove the optimality of the proposed algorithms in all cases except that of a bi-directional heterogeneous ring, for which the problem remains open.

1 Introduction

In this paper, we consider the problem of redistributing data on homogeneous and heterogeneous rings of processors. The problem typically arises when a load balancing phase must be initiated. Because either of variations in the resource performances (CPU speed, communication bandwidth) or in the system/application requirements (completed tasks, new tasks, migrated tasks, etc.), data must be redistributed between participating processors so that the current (estimated) load is better balanced. We do not discuss the load-balancing mechanism itself: we take it as external, be it a system, an algorithm, an oracle, or whatever. Rather we aim at optimizing the data redistribution induced by the load-balancing mechanism.

We adopt the following abstract view of the problem. There are n participating processors P_1, P_2, ..., P_n. Each processor P_k initially holds L_k atomic data items. The load-balancing system/algorithm/oracle has decided that the new load of P_k should be $L_k - \delta_k$. If $\delta_k > 0$, this means that P_k now is overloaded and should send δ_k data items to other processors; if $\delta_k < 0$, P_k is under-loaded and should receive $-\delta_k$ items from other processors. Of course there is a conservation law: $\sum_{k=1}^{n} \delta_k = 0$. The goal is to determine the required communications and to organize them (what we call the data redistribution) in minimal time.

We assume that the participating processors are arranged along a ring, either unidirectional or bidirectional, and either with homogeneous or heterogeneous link bandwidths, hence a total of four different frameworks to deal with. There

L. Bougé and V.K. Prasanna (Eds.): HiPC 2004, LNCS 3296, pp. 123–132, 2004.
© Springer-Verlag Berlin Heidelberg 2004

are two main contexts in which processor rings are useful. The first context is those of many applications which operate on ordered data, and where the order needs to be preserved. Think of a large matrix whose columns are distributed among the processors, but with the condition that each processor operates on a slice of consecutive columns. An overloaded processor P_i can send its first columns to the processor P_j that is assigned the slice preceding its own slice; similarly, P_i can send its last columns to the processor which is assigned the next slice; obviously, these are the only possibilities. In other words, the ordered uni-dimensional data distribution calls for a uni-dimensional arrangement of the processors, i.e., along a ring.

The second context that may call for a ring is the simplicity of the programming. Using a ring, either uni- or bi-directional, allows for a simpler management of the data to be redistributed. Data intervals can be maintained and updated to characterize each processor load. Finally, we observe that parallel machines with a rich but fixed interconnection topology (hypercubes, fat trees, grids, to quote a few) are on the decline. Heterogeneous cluster architectures, which we target in this paper, have a largely unknown interconnection graph, which includes gateways, backbones, and switches, and modeling the communication graph as a ring is a reasonable, if conservative, choice.

As stated above, we discuss four cases for the redistribution algorithms. In the simplest case, that of a unidirectional homogeneous ring, we derive an optimal algorithm. Because the target architecture is quite simple, we are able to provide explicit (analytical) formulas for the number of data sent/received by each processor. The same holds true for the case of a bidirectional homogeneous ring, but the algorithm becomes more complicated. When assuming heterogeneous communication links, we still derive an optimal algorithm for the unidirectional case, but we have to use an asynchronous formulation. However, we have to resort to heuristics based upon linear programming relaxation for the bidirectional case. We point out that one major contribution of the paper is the design of optimal algorithms, together with their formal proof of correctness: to the best of our knowledge, this is the first time that optimal algorithms are introduced.

Due to the lack of space, the detailed proofs of correctness and optimality of the algorithms are not provided: please see the extended version [6]. Similarly, please refer to [6] for a survey of related work.

2   Framework

We consider a set of n processors $P_1, P_2, \ldots, P_n$ arranged along a ring. The successor of P_i in the ring is P_{i+1}, and its predecessor is P_{i-1}, where all indices are taken modulo n. For $1 \leq k, l \leq n$, $C_{k,l}$ denotes the *slice* of consecutive processors $C_{k,l} = P_k, P_{k+1}, \ldots, P_{l-1}, P_l$.

We denote by $c_{i,i+1}$ the capacity of the communication link from P_i to P_{i+1}. In other words, it takes $c_{i,i+1}$ time-units to send a data item from processor P_i to processor P_{i+1}. In the case of a bidirectional ring, $c_{i,i-1}$ is the capacity of the link from P_i to P_{i-1}. We use the one-port model for communications: at any given time, there are at most two communications involving a given processor, one sent and the other received. A given processor can simultaneously send and receive data, so there is no restriction in the unidirectional case; however, in the bidirectional case, a given processor cannot simultaneously send data to its successor and its predecessor; neither can it receive data from both sides. This is the only restriction induced by the model: any pair of communications that does not violate the one-port constraint can take place in parallel.

Each processor P_k initially holds L_k atomic data items. After redistribution, P_k will hold $L_k - \delta_k$ atomic data items. We call δ_k the *imbalance* of P_k. We denote by $\delta_{k,l}$ the total imbalance of the processor slice $C_{k,l}$: $\delta_{k,l} = \delta_k + \delta_{k+1} + \ldots + \delta_{l-1} + \delta_l$. Because of the conservation law of atomic data items, $\sum_{k=1}^{n} \delta_k = 0$. Obviously the imbalance cannot be larger than the initial load: $L_k \geq \delta_k$. In fact, we suppose that any processor holds at least one data, both initially ($L_k \geq 1$) and after the redistribution ($L_k \geq 1 + \delta_k$): otherwise we would have to build a new ring from the subset of resources still involved in the computation.

3   Homogeneous Unidirectional Ring

In this section, we consider a homogeneous unidirectional ring. Any processor P_i can only send data items to its successor P_{i+1}, and $c_{i,i+1} = c$ for all $i \in [1, n]$. We first derive a lower bound on the running time of any redistribution algorithm. Then, we present an algorithm achieving this bound (hence optimal), and we prove its correctness.

Lemma 1. *Let τ be the optimal redistribution time. Then:*

$$\tau \geq \left(\max_{1 \leq k \leq n,\, 0 \leq l \leq n-1} |\delta_{k,k+l}| \right) \times c.$$

Proof. The processor slice $C_{k,k+l} = P_k, P_{k+1}, \ldots, P_{k+l-1}, P_{k+l}$ has a total imbalance of $\delta_{k,k+l} = \delta_k + \delta_{k+1} + \ldots + \delta_{k+l-1} + \delta_{k+l}$. If $\delta_{k,k+l} > 0$, $\delta_{k,k+l}$ data items must be sent from $C_{k,k+l}$ to the other processors. The ring is unidirectional, so P_{k+l} is the only processor in $C_{k,k+l}$ with an outgoing link. Furthermore, P_{k+l} needs a time equal to $\delta_{k,k+l} \times c$ to send $\delta_{k,k+l}$ data items. Therefore, in any case, a redistribution scheme cannot take less than $\delta_{k,k+l} \times c$ to redistribute all data items. We have the same type of reasoning for the case $\delta_{k,k+l} < 0$.

Theorem 1. *Algorithm 1 is optimal.*

4   Heterogeneous Unidirectional Ring

In this section we still suppose that the ring is unidirectional but we no longer assume the communication paths to have the same capacities. We build on the results of the previous section to design an optimal algorithm (Algorithm 2 below). In this algorithm, the amount of data items sent by any processor P_i is

---
Algorithm 1 Redistribution algorithm for homogeneous unidirectional rings
---
1: Let $\delta_{\max} = (\max_{1 \leq k \leq n, 0 \leq l \leq n-1} |\delta_{k,k+l}|)$
2: Let **start** and **end** be two indices such that the slice $C_{\text{start,end}}$ is of maximal imbalance: $\delta_{\text{start,end}} = \delta_{\max}$.
3: **for** $s = 1$ to $\delta_{\max}$ **do**
4:    **for all** $l = 0$ to $n - 1$ **do**
5:       **if** $\delta_{\text{start,start}+l} \geq s$ **then**
6:          $P_{\text{start}+l}$ sends to $P_{\text{start}+l+1}$ a data item during the time interval $[(s - 1) \times c, s \times c[$

---

exactly the same as in Algorithm 1 (namely $\delta_{\text{start},i}$). However, as the communication links have different capabilities, we no longer have a synchronous behavior. A processor P_i sends its $\delta_{\text{start},i}$ data items as soon as possible, but we cannot express its completion time with a simple formula. Indeed, if P_i initially holds more data items than it has to send, we have the same behavior than previously: P_i can send its data items during the time interval $[0, \delta_{\text{start},i} \times c_{i,i+1}[$. On the contrary, if P_i holds less data items than it has to send ($L_i < \delta_{\text{start},i}$), P_i still starts to send some data items at time 0 but may have to wait to have received some other data items from P_{i-1} to be able to forward them to P_{i+1}.

Algorithm 2 Redistribution algorithm for heterogeneous unidirectional rings
---
1: Let $\delta_{\max} = (\max_{1 \leq k \leq n, 0 \leq l \leq n-1} |\delta_{k,k+l}|)$
2: Let **start** and **end** be two indices such that the slice $C_{\text{start,end}}$ is of maximal imbalance: $\delta_{\text{start,end}} = \delta_{\max}$.
3: **for all** $l = 0$ to $n - 1$ **do**
4:    $P_{\text{start}+l}$ sends $\delta_{\text{start,start}+l}$ data items one by one and as soon as possible to processor $P_{\text{start}+l+1}$

The asynchronousness of Algorithm 2 implies that it is correct by construction: we wait for receiving a data item before sending. Furthermore, when the algorithm terminates, the redistribution is complete.

Lemma 2. *The running time of Algorithm 2 is*

$$\max_{0 \leq l \leq n-1} \delta_{start,start+l} \times c_{start+l,start+l+1}.$$

The result of Lemma 2 is surprising. Intuitively, it says that the running time of Algorithm 2 is equal to the maximum of the communication times of all the processors, if each of them initially stored locally all the data items it will have to send throughout the execution of the algorithm. In other words, there is no forwarding delay, whatever the initial distribution.

Theorem 2. *Algorithm 2 is optimal.*

5   Homogeneous Bidirectional Ring

In this section, we consider a homogeneous bidirectional ring. All links have the same capacity but a processor can send data items to its two neighbors in the ring: there exists a constant c such that, for all $i \in [1, n]$, $c_{i,i+1} = c_{i,i-1} = c$. We proceed as for the homogeneous unidirectional case: we first derive a lower bound on the running time of any redistribution algorithm, and then we present an algorithm achieving this bound.

Lemma 3. *Let τ be the optimal redistribution time. Then:*

$$\tau \geq \max \left\{ \max_{1 \leq i \leq n} |\delta_i|, \max_{1 \leq i \leq n, 1 \leq l \leq n-1} \left\lceil \frac{|\delta_{i,i+l}|}{2} \right\rceil \right\} \times c. \tag{1}$$

The new (rightmost) term in this lower bound just states that a slice of processor can send (or receive) simultaneously at most two data items. Algorithm 3 is a recursive algorithm which defines communication patterns designed so as to decrease the value of $\delta_{\max}$ (computed at Step 1) by one from one recursive call to another. The intuition behind Algorithm 3 is the following:

1. Any non trivial slice $C_{k,l}$ such that $\lceil \frac{|\delta_{k,l}|}{2} \rceil = \delta_{\max}$ and $\delta_{k,l} \geq 0$ must send two data items per recursive call, one through each of its extremities.

2. Any non trivial slice $C_{k,l}$ such that $\lceil \frac{|\delta_{k,l}|}{2} \rceil = \delta_{\max}$ and $\delta_{k,l} \leq 0$ must receive two data items per recursive call, one through each of its extremities.

3. Once the mandatory communications specified by the two previous cases are defined, we take care of any processor P_i such that $|\delta_i| = \delta_{\max}$. If P_i is already involved in a communication due to the previous cases, everything is settled. Otherwise, we have the freedom to choose whom P_i will send a data item to (case $\delta_i > 0$) or whom P_i will receive a data item from (case $\delta_i < 0$). To simplify the algorithm we decide that all these communications will take place in the direction from P_i to P_{i+1}.

Algorithm 3 is initially called with the parameter $s = 1$. For any call to Algorithm 3, all the communications take place in parallel and exactly at the same time, because the communication paths are homogeneous by hypothesis. One very important point about Algorithm 3 is that this algorithm is a set of rules which *only* specify which processor P_i must send a data item to which processor P_j, one of its immediate neighbors. Therefore, whatever the number of rules deciding that there must be some data item sent from a processor P_i to one of its immediate neighbor P_j, only one data item is sent from P_i to P_j to satisfy all these rules.

Theorem 3. *Algorithm 3 is optimal.*

Algorithm 3 Redistribution algorithm for homogeneous bidirectional rings (for step s)

1: Let $\delta_{\max} = \max\{\max_{1 \leq i \leq n} |\delta_i|, \max_{1 \leq i \leq n, 1 \leq l \leq n-1} \lceil \frac{|\delta_{i,i+l}|}{2} \rceil\}$
2: **if** $\delta_{\max} \geq 1$ **then**
3:     **if** $\delta_{\max} \neq 2$ **then**
4:         **for all** slice $C_{k,l}$ such that $\delta_{k,l} > 1$ and $\lceil \frac{|\delta_{k,l}|}{2} \rceil = \delta_{\max}$ **do**
5:             P_k sends a data item to P_{k-1} during the time interval $[(s-1) \times c, s \times c[$.
6:             P_l sends a data item to P_{l+1} during the time interval $[(s-1) \times c, s \times c[$.
7:         **for all** slice $C_{k,l}$ such that $\delta_{k,l} < -1$ and $\lceil \frac{|\delta_{k,l}|}{2} \rceil = \delta_{\max}$ **do**
8:             P_{k-1} sends a data item to P_k during the time interval $[(s-1) \times c, s \times c[$.
9:             P_{l+1} sends a data item to P_l during the time interval $[(s-1) \times c, s \times c[$.
10:     **else if** $\delta_{\max} = 2$ **then**
11:         **for all** slice $C_{k,l}$ such that $\delta_{k,l} \geq 3$ **do**
12:             P_l sends a data item to P_{l+1} during the time interval $[(s-1) \times c, s \times c[$.
13:         **for all** slice $C_{k,l}$ such that $\delta_{k,l} = 4$ **do**
14:             P_k sends a data item to P_{k-1} during the time interval $[(s-1) \times c, s \times c[$.
15:         **for all** slice $C_{k,l}$ such that $\delta_{k,l} \leq -3$ **do**
16:             P_{k-1} sends a data item to P_k during the time interval $[(s-1) \times c, s \times c[$.
17:         **for all** slice $C_{k,l}$ such that $\delta_{k,l} = -4$ **do**
18:             P_{l+1} sends a data item to P_l during the time interval $[(s-1) \times c, s \times c[$.
19:     **for all** processor P_i such that $\delta_i = \delta_{\max}$ **do**
20:         **if** P_i is not already sending, due to one of the previous steps, a data item during the time interval $[(s-1) \times c, s \times c[$ **then**
21:             P_i sends a data item to P_{i+1} during the time interval $[(s-1) \times c, s \times c[$.
22:     **for all** processor P_i such that $\delta_i = -(\delta_{\max})$ **do**
23:         **if** P_i is not already receiving, due to one of the previous steps, a data item during the time interval $[(s-1) \times c, s \times c[$ **then**
24:             P_i receives a data item from P_{i-1} during the time interval $[(s-1) \times c, s \times c[$.
25:     **if** $\delta_{\max} = 1$ **then**
26:         **for all** processor P_i such that $\delta_i = 0$ **do**
27:             **if** P_{i-1} sends a data item to P_i during the time interval $[(s-1) \times c, s \times c[$ **then**
28:                 P_i sends a data item to P_{i+1} during the time interval $[(s-1) \times c, s \times c[$.
29:             **if** P_{i+1} sends a data item to P_i during the time interval $[(s-1) \times c, s \times c[$ **then**
30:                 P_i sends a data item to P_{i-1} during the time interval $[(s-1) \times c, s \times c[$.
31: Recursive call to Algorithm 3 $(s+1)$

---

6   Heterogeneous Bidirectional Ring

In this section, we consider the most general case, that of a heterogeneous bidirectional ring. We do not know any optimal redistribution algorithm in this case. However, if we assume that each processor initially holds more data than it needs to send during the whole execution of the redistribution (what we call a *light redistribution*), then we succeed in deriving an optimal solution.

Throughout this section, we suppose that we have a *light* redistribution: we assume that the number of data items sent by any processor throughout the redistribution algorithm is less than or equal to its original load. There are two reasons for a processor P_i to send data: (i) because it is overloaded ($\delta_i > 0$); (ii) because it has to forward some data to another processor located further in the ring. If P_i initially holds at least as many data items as it will send during the whole execution, then P_i can send at once all these data items. Otherwise, in the general case, some processors may wait to have received data items from a neighbor before being able to forward them to another neighbor.

Under the "light redistribution" assumption, we can build an integer linear program to solve our problem (see System 2). Let S be a solution, and denote by $S_{i,i+1}$ the number of data items that processor P_i sends to processor P_{i+1}. Similarly, $S_{i,i-1}$ is the number of data items that P_i sends to processor P_{i-1}. In order to ease the writing of the equations, we impose in the first two equations of System 2 that $S_{i,i+1}$ and $S_{i,i-1}$ are nonnegative for all i, which imposes to use other variables $S_{i+1,i}$ and $S_{i-1,i}$ for the symmetric communications. The third equation states that after the redistribution, there is no more imbalance. We denote by τ the execution time of the redistribution. For any processor P_i, due to the one-port constraints, τ must be greater than the time spent by P_i to send data items (fourth equation) or spent by P_i to receive data items (fifth equation). Our aim is to minimize τ, hence the system:

$$
\text{MINIMIZE } \tau, \text{ SUBJECT TO}
$$

$$
\begin{cases}
S_{i,i+1} \geq 0 & 1 \leq i \leq n \\
S_{i,i-1} \geq 0 & 1 \leq i \leq n \\
S_{i,i+1} + S_{i,i-1} - S_{i+1,i} - S_{i-1,i} = \delta_i & 1 \leq i \leq n \\
S_{i,i+1}c_{i,i+1} + S_{i,i-1}c_{i,i-1} \leq \tau & 1 \leq i \leq n \\
S_{i+1,i}c_{i+1,i} + S_{i-1,i}c_{i-1,i} \leq \tau & 1 \leq i \leq n
\end{cases}
\tag{2}
$$

Lemma 4. *Any optimal solution of System 2 is feasible, for example using the following schedule: for any $i \in [1, n]$, P_i starts sending data items to P_{i+1} at time 0 and, after the completion of this communication, starts sending data items to P_{i-1} as soon as possible under the one-port model.*

We use System 2 to find an optimal solution to the problem. If, in this optimal solution, for any processor P_i, the total number of data items sent is less than or equal to the initial load ($S_{i,i+1} + S_{i,i-1} \leq L_i$), we are under the "light redistribution" hypothesis and we can use the solution of System 2 safely. But even if the "light redistribution" hypothesis holds, one may wish to solve the redistribution problem with a technique less expensive than integer linear programming (which is potentially exponential). An idea would be to first solve System 2 to find an optimal *rational* solution, which can always be done in polynomial time, and then to round up the obtained solution to find a "good" integer solution. In fact, it turns out that one of the two natural ways of rounding always lead to an optimal (integer) solution [6]. The complexity of the light redistribution problem is therefore polynomial.

If we no longer assume the light redistribution hypothesis, we can still derive a lower bound, and use some heuristics [6]. However, we point out that the design of an optimal algorithm in the most general case remains open. Given the complexity of the lower bound, the problem looks very difficult to solve.

7   Experimental Results

To evaluate the impact of the redistributions, we used the SIMGRID [5] simulator to model an iterative application, implemented on a platform generated with the Tiers network generator [2, 3]. We use the platform represented in Figure 1. The capacities of the edges are assigned using the classification of the Tiers generator (local LAN link, LAN/MAN link, MAN/WAN link,...). For each link type, we use values measured using `pathchar` [4] between some machines in ENS Lyon and some other machines scattered in France (Strasbourg, Lille, Grenoble, and Orsay), in the USA (Knoxville, San Diego, and Argonne), and in Japan (Nagoya and Tokyo).

We randomly select p processors in the platform to build the execution ring. The communication speed is given by the slowest link in the route from a processor to its successor (or predecessor) in the ring. The processing powers (CPU speeds) of the nodes are first randomly chosen in a list of values corresponding to the processing powers (expressed in MFlops and evaluated thanks to a benchmark taken from LINPACK [1]) of a wide variety of machines. But we make these speeds vary during the execution of the application.

We model an iterative application which executes during 100 iterations. At each iteration, independent data are updated by the processors. We may think of a $m \times n$ data matrix whose columns are distributed to the processors (we use $n = m = 1000$ in the experiment). Ideally, each processor should be allocated a number of columns proportional to its CPU speed. This is how the distribution of columns to processors is initialized. To motivate the need for redistributions, we create an imbalance by letting the CPU speeds vary during the execution. The speed of each processor changes two times, first at some iteration randomly chosen between iterations number 20 and 40, and then at some iteration randomly chosen between iterations number 60 and 80 for each node (see Figure 2 for an illustration). We record the values of each CPU speed in a SIMGRID trace.

In the simulations, we use the heterogeneous bidirectional algorithm for light redistributions, and we test five different schemes, each with a given number of redistributions within the 100 iterations. The first scheme has no redistribution at all. The second scheme implements a redistribution after iteration number 50. The third scheme uses four redistributions, after iterations 20, 40, 60 and 80. The fourth scheme uses 9 redistributions, implemented every 10 iterations, and the last one uses 19 redistributions, implemented every 5 iterations. Given the shape of the CPU traces, some redistributions are likely to be beneficial during the execution. The last parameter to set is the computation-to-communication ratio, which amounts to set the relative (average) cost of a redistribution versus the cost of an iteration. When this parameter increases, iterations take more time, and the usefulness of a redistribution becomes more important.

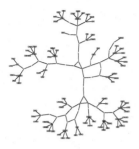

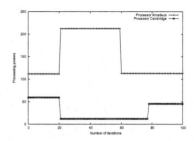

Fig. 1. The platform is composed of 90 machine nodes, connected through 192 communication links

Fig. 2. Processing power of 2 sample machine nodes

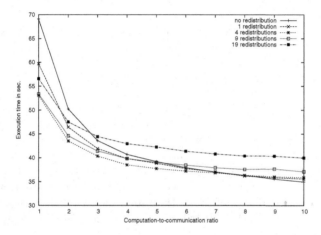

Fig. 3. Normalized execution time as a function of the computation-to-communication ratio, for a ring of 8 processors

In Figures 3 and 4, we plot the execution time of different computation schemes. Both figures report the same comparisons, but for different ring sizes: we use 8 processors in Figures 3, and 32 in Figures 4. As expected, when the processing power is high (ratio = 10 in the figures), the best strategy is to use no redistribution, as their cost is prohibitive. Conversely, when the processing power is low (ratio = 1 in the figures), it pays off to uses many redistributions, but not too many! As the ratio increases, all tradeoffs can be found.

8   Conclusion

In this paper, we have considered the problem of redistributing data on rings of processors. For homogeneous rings the problem has been completely solved. Indeed, we have designed optimal algorithms, and provided formal proofs [6] of correctness, both for unidirectional and bidirectional rings. The bidirectional algorithm turned out to be quite complex, and requires a lengthy proof [6].

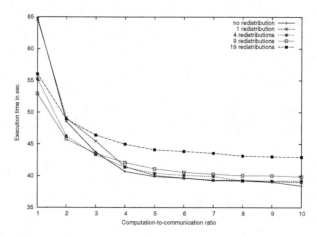

Fig. 4. Normalized execution time as a function of the ratio computation-to-communication, for a ring of 32 processors

For heterogeneous rings there remains further research to be conducted. The unidirectional case was easily solved, but the bidirectional case remains open. Still, we have derived an optimal solution for light redistributions, an important case in practice. The complexity of the bound provided for the general case shows that designing an optimal algorithm is likely to be a difficult task.

All our algorithms have been implemented and extensively tested. We have reported some simulation results for the most difficult combination, that of heterogeneous bi-directional rings. As expected, the cost of data redistributions may not pay off a little imbalance of the work in some cases. Further work will aim at investigating how frequently redistributions must occur in real-life applications.

References

1. R. P. Brent. The LINPACK Benchmark on the AP1000: Preliminary Report. In *CAP Workshop 91*. Australian National University, 1991. Website http://www.netlib.org/linpack/.
2. K. L. Calvert, M. B. Doar, and E. W. Zegura. Modeling internet topology. *IEEE Communications Magazine*, 35(6):160–163, June 1997. Available at http://citeseer.nj.nec.com/calvert97modeling.html.
3. M. Doar. A better model for generating test networks. In *Proceedings of Globecom '96*, Nov. 1996. Available at http://citeseer.nj.nec.com/doar96better.html.
4. A. B. Downey. Using pathchar to estimate internet link characteristics. In *Measurement and Modeling of Computer Systems*, pages 222–223, 1999. Available at http://citeseer.nj.nec.com/downey99using.html.
5. A. Legrand, L. Marchal, and H. Casanova. Scheduling Distributed Applications: The SimGrid Simulation Framework. In *Proceedings of the Third IEEE International Symposium on Cluster Computing and the Grid (CCGrid'03)*, May 2003.
6. H. Renard, Y. Robert, and F. Vivien. Data redistribution algorithms for heterogeneous processor rings. Research Report RR-2004-28, LIP, ENS Lyon, France, May 2004. Also available as INRIA Research Report RR-5207.

Effect of Optimizations on Performance of OpenMP Programs

Xinmin Tian and Milind Girkar

Intel Compiler Labs, Software and Solutions Group, Intel Corporation,
3600 Juliette Lane, Santa Clara, CA 95052, USA
{Xinmin.Tian, Milind.Girkar}@intel.com

Abstract. In this paper, we describe several compiler optimization techniques and their effect on the performance of OpenMP programs. We elaborate on the major design considerations in a high performance OpenMP compiler and present experimental data based on the implementation of the optimizations in the Intel® C++ and Fortran compilers for Intel platforms. Interactions of the OpenMP translation phase with other sequential optimizations in the compiler are discussed. The techniques in this paper are responsible for achieving significant performance improvements on the industry standard SPEC* OMPM2001 and SPEC* OMPL2001 benchmarks, and these results are presented for Intel® Pentium® and Itanium® processor based systems.

1 Introduction

The OpenMP* specification [7][11] for shared memory parallel programming has a rich set of features that allows the user to write parallel programs with a modest development effort using directives. These directives are translated by the compiler to generate threaded code that will usually show increased performance on shared memory multiprocessor systems. Performance can also be gained on processors that allow simultaneous execution of multiple threads (e.g. IBM* Power 5 [1] or Intel® processors with Hyper-Threading Technology [6]). The principle behind the OpenMP specification is to shift most of the complex tasks of thread management from the user to a compiler, freeing the user to concentrate on the expression of parallelism through the OpenMP directives. The Intel C++/Fortran95 compilers support the OpenMP 2.0 specification on Windows and Linux platforms on the IA-32 and Itanium® Processor Family (IPF) architectures [4][5].

There have been several papers discussing OpenMP parallelization in the compilers [2][3][4][5]. Various OpenMP implementations are possible. An OpenMP preprocessor [2][3] accepts C/C++ and Fortran95 OpenMP programs and translates them to C++/Fortran95 programs (without OpenMP directives) that are subsequently compiled with a native compiler (one that generates machine code). A more integrated approach is to have an internal OpenMP translation phase in the native compiler [5][10] itself eliminating the preprocessor. In most cases an auxiliary runtime library for thread management is used with the compiler generated code making many calls to this library. All implementations have to worry about the OpenMP translation

L. Bougé and V.K. Prasanna (Eds.): HiPC 2004, LNCS 3296, pp. 133–143, 2004.
© Springer-Verlag Berlin Heidelberg 2004

phase adversely affecting the other optimization phases in the native compiler. As a simple example, when implementing OpenMP through a preprocessor, the preprocessor may require the generation of calls to the OpenMP runtime library and passing of addresses of variables as parameters in such calls. However, taking the addresses of variable can significantly affect the ability of the compiler to determine accurately which variables are read (or written) at various points in the program. In this paper we study the interaction between other optimization phases and the OpenMP* translation phase and show the cooperation that is required between the two to generate optimized code.

The remainder of this paper is organized as follows. The Section 2 presents an overview of the Intel® C++/Fortran95 compiler. Section 3 describes the phase implemented in the Intel compiler for generating multithreaded code from OpenMP directives. Section 4 presents techniques to make the parallelizer interact tightly with other optimizations. In Section 5, we discuss the quantitative effect of optimizations on the industry standard SPEC OMPM2001 benchmarks [9][10] on Intel Pentium- and Itanium-processor based systems. In Section 6, we report industry leading performance results of SPEC OMPL2001 on the Itanium 2 (1.5GHz) processor based SGI* Altix* system with 128 processors obtained by an application of the ideas presented in this paper. Finally, concluding remarks can be found in Section 7.

2   Intel Compiler Architecture

A high-level overview of the Intel C++ and Fortran95 compilers is shown in Fig.1. The compiler incorporates many well-known and advanced optimization techniques [5][8][12] that are designed and extended to fully leverage Intel processor features for higher performance. The code transformations and optimizations in the compiler can be classified as below:

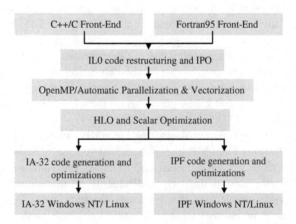

Fig. 1. Compiler Architecture Overview

- Code restructuring, inter-procedural optimizations (IPO), and OpenMP directive-guided parallelization, automatic parallelization and automatic vectorization
- High-level optimizations (HLO) and scalar optimizations including memory optimizations such as loop control and data transformations, partial redundancy elimination (PRE), and partial dead store elimination (PDSE).
- Low-level machine code generation and optimizations such as register allocation, instruction scheduling and software pipelining.

The Intel compiler has a common intermediate representation, called IL0, which has been extended to represent OpenMP directives and clauses, so that OpenMP directive guided parallelization and a majority of optimizations are applicable through *high-level* code transformations on IL0. Implementing the OpenMP translation phase at the IL0 level allows the same implementation to be used across languages (C++/C, Fortran95) and architectures (IA-32 and IPF). The Intel® compiler generated code has references to a high-level library API. The library implements this API in terms of the threading functionality provided by the OS; the use of such an API thus allows the compiler OpenMP* translation phase to be independent of the underlying operating systems. Our compiler architecture makes it possible to have one OpenMP implementation that covers differing language (C++ and Fortran95), differing architectures (IA-32 and IPF) and differing operating systems (Windows* and Linux*).

3   Threaded Code Generation

In order to support the OpenMP programming model, various components in the Intel compiler have been extended. First, the IL0 intermediate representation was extended to represent OpenMP directives/pragmas and clauses. The compiler front-end parses OpenMP directives in the source program to generate an IL0 representation of the OpenMP source code for the parallelizer and optimizer. The parallelizer generates multi-threaded code based on IL0 representation corresponding to OpenMP constructs.

The multithreaded code generator consists of many modules such as variable classification, privatization, array lowering, loop analysis, enclosing-while-loop generation for *runtime, dynamic* and *guided* scheduling, post-pass *threadprivate* handler and stack optimization. Essentially, it converts the OpenMP constructs to multithreaded code at the IL0 level. For the example shown in Fig. 2, with the *worksharing* loop in the routine *parwork* with the scheduling type *dynamic*, multithreaded code generation involves: (i) generating a runtime dispatch and initialization (*__kmpc_dispatch_init*) routine call to pass global loop lower-bound, upper-bound, stride, and all other necessary information to the runtime system; (ii) generating an enclosing while loop to dispatch *loop-chunk* at runtime through the *__kmpc_dispatch_next* routine supported in the library; (iii) localizing the loop lower-bound, upper-bound, and privatizing the loop control variable 'k' and local defined stack variable 'x'. With the *MET* technology [5], one threaded entry, or T-entry[1], is created within the *parwork*() routine for

---

[1] In [5], T-entry refers strictly to the entry point of a threaded region, or T-region, which is the section of code enclosed between a T-entry and its matching T-return. In this paper, we use T-entry to refer to the threaded entry or region, as this use is unambiguous from the context and often interchangeable.

each *parallel* region. The function call *__kmpc_fork_call* spawns a team of threads to execute the threaded codes in parallel.

```
void parwork() /* OpenMP C code sample */
{ double a[1000], b=1000;  int k;
#pragma omp parallel shared(a, b) private(k)
    {  int x = 7;
#pragma omp for schedule(dynamic)
        for (k=0; k<16; k++) {  x = x + b*b; a[k] = a[k] + b * x;  }
    }
}
entry extern void _parwork() /* IL0 pseudo-code after MT-code generation */
{  ... ...
    b = 1000.00 (F64)                    /* F64 denotes the 64-bit float type */
    __kmpc_fork_call(..., __parwork_par_region, &a, &b) ;   goto L46;
    T-entry __parwork_par_region(ap, bp)
    {  prv_x = 7;   prv_k = 0
       if  (1000 > prv_k) {
            t0 = (* F64)bp;  lower  = 0;   upper  = 999;   stride = 1;
            __kmpc_dispatch_init(..., lower, upper, stride, ...)
       L33:   t3 = __kmpc_dispatch_next(..., &lower, &upper, &stride)
            if ((t3 & upper>=lower) != 0(SI32)) {
                prv_k = lower
            L17:
                prv_x = prv_x + t0 * t0
                ((* F64)ap)[prv_k] = ((* F64)ap)[prv_k] + t0 * prv_x
                prv_k = prv_k + 1
                if (upper >= prv_k)    goto L17
                goto L33
            }
        }
        __kmpc_barrier(...) ;   T-return;
    }
```

Fig. 2. Pseudo-code after Threaded code Generation for a simple OpenMP program

4   Enabling Advanced Optimizations

The OpenMP* implementation in the Intel compiler strives to generate multithreaded code which gains speedup over optimized uniprocessor code by integrating parallelization tightly with advanced inter-procedural, scalar and loop optimizations such as vectorization [8] and memory hierarchy oriented optimizations [5][12] to achieve better cache locality and minimize the overhead of data-sharing among threads. In this section, we describe some of the techniques for generating efficient threaded code.

4.1   Effective Ordering of Optimization Phases

The phase ordering of optimizations in the compiler is critical for achieving optimal performance. It is difficult to architect an effective ordering to achieve speedups over well-optimized serial code through parallelization if significant sequential optimizations are affected adversely by the parallelization. Our design consideration includes:

- leverage classical peephole optimizations fully within basic-block, and perform inlining and OpenMP construct-aware constant propagation, and memory disambiguation before parallelization and multithreaded code generation.
- perform high-level optimizations (HLO) such as loop tiling, loop unroll, loop distribution, loop fusion, vectorization [8], software prefetching, scalar replacement, complex type lowering after parallelization and multithreaded code generation.
- enable advanced optimizations such as Partial Redundancy Elimination (PRE), Partial Dead-Store Elimination (PDSE), Dead Code Elimination (DCE) after high-level optimization (HLO).

4.2 Reducing Side-Effects of Privatization

Privatization is one of key components when generating threaded code. Privatizing a local stack variable, static variable or global variable is straight forward - the compiler can simply create its clone on the stack. Some Fortran95 arrays (unknown-size, assumed-size and assumed-shape) can be allocated on the stack or heap. Sometimes heap allocation is preferred for large objects in the sequential case as stack space is limited. However, in the case of parallel programs, heap allocation may cause performance slowdowns, as thread-safe memory allocation routines usually have critical sections guarded by a locking mechanism. Our solution takes advantage of the proper nesting structure of OpenMP directives to limit the lifetime of such allocations by judiciously allocating and freeing up stack space in a LIFO (last in first out) manner. The parallelization pass generates stack allocation and free intrinsics, _vla_alloc(size) and _vla_free(p, size). The _vla_alloc and _vla_free are lowered to stack adjustment instructions in the machine code generation. This is an efficient scheme for privatizing an object for each thread.

```
          subroutine foo(arr, n)                     threaded-entry par_region_foo()
            integer arr(n)                            ...
!$omp parallel private(arr, k)                          dv_clone_arr_size = n
            do k=1, 100                                 dv_clone_arr_baseaddr = vla_alloc(dv_clone_arr_size)
              arr(k) = omp_get_thread_num()             do clone_k = 1, 100
            end                                           dv_clone_arr_baseaddr(k) = ... ...
!$omp end parallel                                      end do
            arr = 10000                                 _vla_free(dv_clone_baseaddr, dv_clone_arr_size)
          end                                         threaded-return
   (a) An unknown size array example            (b) Pesudo threaded-code with _vla_alloc/_vla_free
```

In the above example, the size of array 'arr' is unknown, so the Front-End creates a structure (called the dope-vector) for it to fill the array size, base address, stride, array bounds information at runtime, During privatization phase, the compiler clones the original dope-vector by creating *dv_clone_ar,* and substituting the original array 'arr' with *dv_clone_arr_baseaddr* for each thread. The memory allocation/deallocation on stack is done by the intrinsic *_vla_alloc/_vla_free* through incrementing/decrementing the stack pointer of each thread for short lifetime private objects.

4.3 Preserving Memory Disambiguation Tokens

As we mentioned in Section 4.1, in the Intel® compiler, the memory disambiguation phase is invoked before the OpenMP* translation phase. The disambiguation phase annotates each memory reference with DISAM tokens that are used later by other optimizations.

```
common /ccc/ a(100), b(100), c(100)
common /eee/ d(100)
!$omp threadprivate(/ccc/, /eee/)
do ... ...
  do k = ... ...
    a(k) = d1 +b(k)                    ... ...
    b(k) = d2               tpv_ccc_base = _threadprivate_cached(tid, &a, size1, ccc_cache)
    c(k) = d3 + d(k)        tpv_eee_base = _threadprivate_cached(tid, &d, size2, eee_cache)
    d(k) = d4              ... ... ! DISAM tokens for 'a', 'b', 'c', 'd' are preserved in expr nodes
  end do                   do k = ... ...
end do                          *(P32 *)(tpv_ccc_base+0)(k)   = d1 + ... ...
                                *(P32 *)(tpv_ccc_base+400)(k) = d2
                                *(P32 *)(tpv_ccc_base+800)(k) = d3 + ... ...
                                *(P32 *)(tpv_eee_base+0)(k)   = d4
                         ... ...
```

In the simple kernel above from a real large application, with the array 'a', 'b', 'c', 'd' members of common block 'ccc' and 'eee', those optimizations relying on DISAM information such as the loop distribution and software pipelining are disabled if threaded code generation phase does not preserve DISAM token information in the new array referencing expression e.g. *(P32 *)(tpv_ccc_base+0)(k) of the threaded-code, since tpv_ccc_base and tpv_eee_base are allocated at runtime, it would be hard for compiler to figure out if those point to distinct memory areas.

By preserving the DISAM token for each expression during the parallelization, other optimizations know that there is no memory overlap among those memory reference expressions *(P32 *)(tpv_ccc_base+0)(k), *(P32 *)(tpv_ccc_base+400)(k), and *(P32 *)(tpv_ccc_base+800)(k), and *(P32 *)(tpv_eee_base+0)(k) by simply querying the DISAM tokens. In addition, the threaded code generation phase propagates the original attributes (e.g. address_taken, no_pointer_aliasing) of variable 'a' to '_tpv_ccc_base', and annotates the _threadprivate_cached call statements. For example, the annotation tells other optimizations that there is no aliasing between "a" and '_tpv_ccc_base', as thread local storage allocated by the call is disjoint from that with the original 'a'. Proper representation and propagation of such information is necessary for not disabling optimizations that happen later.

5   Effect of Optimizations on SPEC OMPM2001 Performance

SPEC* OMPM2001 suite consists of a set of OpenMP* based application programs [9][10]. The input data sets of the SPEC OMPM2001 suite (also referred to as the medium suite) are derived from state-of-the-art computations on modern medium-scale (4- to 16-way) shared-memory multiprocessor systems. This benchmark suite consists of 11 large application programs, which represent the type of software used in scientific technical computing. Table 1 provides an overview of SPEC* OMPM2001 benchmarks. Of the 11 application, 8 applications are written in FORTRAN, and 3 applications are written in C.

Table 1. Overview of SPEC OMPM2001 Benchmark Suite

Code	Applications	Language	# of Lines
310.wupwise_m	Quantum chromodynamics	Fortran	2200
312.swim_m	Shallow water modeling	Fortran	400
314.mgrid_m	Multigrid solver	Fortran	500
316.applu_m	Fluid dynamics and physics	Fortran	4000
318.galgel_m	Fluid dynamics	Fortran	15300
320.equake_m	Earthquake modeling and simulation	C	1500
324.apsi_m	Air pollution modeling and computation	Fortran	7500
326.gafort_m	Genetic algorithm	Fortran	1500
328.fma3d_m	Crash simulation	Fortran	60000
330.art_m	Image Recognition neural networks	C	1300
332.ammp_m	Chemistry and biology	C	13500

Those benchmarks require a virtual address space of about 2GB to run. The datasets are significantly larger than those of the SPEC CPU2000 benchmarks, while still fitting in a 32-bit address space. Our results show that the performance of single-thread run of the threaded code generated by the compiler with base options for each SPEC* OMP benchmark shows less than 2% overhead comparing with the serial code with same set of base options by disabling all OpenMP* directives. The next section shows the performance gain of SPEC OMPM2001 due to hyper-threading technology.

5.1 Effect of Hyper-Threading Technology

This performance study of SPEC OMPM2001 benchmarks is conducted on a single processor system with Hyper-Threading Technology enabled Intel Pentium® 4 processor built with 90nm technology running at 2.8GHz, with 2GB memory, an 8K L1-Cache, and 1M L2-Cache. For our performance measurement, all SPEC OMPM2001 benchmarks are compiled by the Intel 8.0 C++/Fortran compilers with the option set of our base performance run: –Qopenmp –Qipo –O3 –QxP (OMP w/ QxP) under Windows* XP on a Hyper-Threading enabled Pentium 4 processor. The -QxP switch enables the compiler to generate SSE3 instructions available on the more recent Intel® Pentium 4 processors.

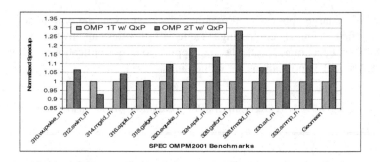

Fig. 3. Performance Gain from Hyper-Threading Technology

The normalized speedup of the SPEC OMPM2001 benchmarks is shown in the Fig. 3 that demonstrates the performance gain attributed to the Hyper-Threading Technology. The hyper-threading performance scaling is derived from the baseline performance of single thread binary with OMP 1T w/ QxP, and two threads execution under OMP 2T w/ QxP, respectively. As we see, Hyper-Threading Technology enabled Intel® Pentium® 4 processor to achieve a performance improvement of 4.3% to 28.3% (OMP 2T w/ QxP) on 9 out of 11 benchmarks except 316.applu_m (0.0%) and 312.swim_m (-7.4%). The 312.swim_m slowdown under two thread execution mode is due to the 312.swim_m being a memory bandwidth bound application. Overall, the improvement in geomean with OMP 2T w/ QxP is 9.1% due to Hyper-Threading. Considering that Hyper-Threading Technology does not add significant extra hardware execution (engine) resources, the gain of 9.1% illustrates that sequential optimizations are also triggered in the OpenMP threaded code.

5.2   Effect of Optimizations on SPEC OMPM2001 Performance

In this section, we examine the effect of compiler optimizations on generating multi-threaded code of SPEC OMPM2001 application programs with different optimization sets. Ideally, given a fixed number (we used 4 threads in our performance study as the 4-way system is the most common used system,) of threads or processors, it would be more interesting to study the effect of each compiler optimizations one-by-one to demonstrate their effectiveness on performance improvement. However, given the complexity of interaction among compiler optimizations and the length limitation of the paper, we decided to study the effect of a few optimizations. In our performance study, all SPEC* OMPM2001 benchmarks are compiled by the latest Intel 8.0 C++/Fortran compilers with four sets of base options: (i) -openmp -O2 (used as a baseline performance measurement); (ii) –openmp –O2 –ipo; (iii) –openmp –O3; and (iv) –openmp –O3 –ipo. The experiments were done on a 4-way 1.5GHz Itanium 2 based system with 6MB L3 cache. Fig.4 illustrates the performance gain with different higher level optimizations vs. the performance measured at a default optimization level –openmp –O2.

The O2 level optimization includes many traditional optimizations such as peephole optimization, constant propagation, copy propagation, dead code elimination,

partial-dead store elimination, partial redundancy elimination, etc.; the O3 level optimization includes advanced loop transformations (loop tiling, loop fusion, loop distribution, etc.), scalar replacement, software-prefetching, array contraction, etc.; the
IPO flag enables Inter-Procedural (IP) optimizations such as function inlining, IP
mod-ref analysis, etc. Fig. 4 provides the performance results at different optimization
levels. As is evident from the graph, the performance gain from OMP+O2 to
OMP+O2+IPO is 3% on Geomean, which is relatively small. This is because many
advanced optimizations that can exploit the inter-procedural information such as
mod-ref analysis are run only at O3. The results of OMP+O3 showed 22% performance gain vs. the performance of OMP+O2, and 19% performance gain over vs.
OMP+O2+IPO performance. This result reveals that the high-level optimizations are
effectively enabled for multithreaded-code generated for OpenMP programs.

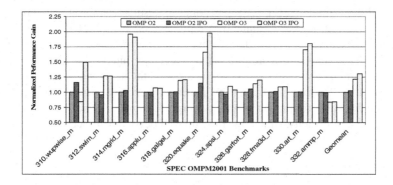

Fig. 4. Effect of Compiler Optimizations on SPEC OMPM Performance

As expected, the best performance is achieved with OMP+O3+IPO, Fig. 4 shows a
31% performance gain over the baseline performance (OMP+O2). For example, the
performance of 310.wupwise_m is dominated by a few hot loops with unknown trip
count, at OMP+O3 the compiler needs to be conservative without knowing those trip-
counts, in this case, OMP+O3 actually causes a 16% performance slow down on
310.wupwise. However, the addition of IPO provides a 49% performance gain with
known trip-counts through IPO constant propagation. Overall, 10 out of 11 benchmarks in SPEC* OMPM2001 benchmark suite achieved a performance gain ranging
from 7% to 98% with OMP+O3+IPO. An anomaly is 332.ammp which shows a
slowdown at OMP+O3+IPO; this needs to be investigated further.

6   SPEC OMPL2001 Performance Results

SPEC* OMPL2001 shares most of the application code base with SPEC OMPM2001,
and consists of 9 application programs from SPEC OMPM2001. However, the code
and the data sets are modified to achieve better scaling and also to reflect the class of
computation regularly performed on large-scale systems (32-way and larger).

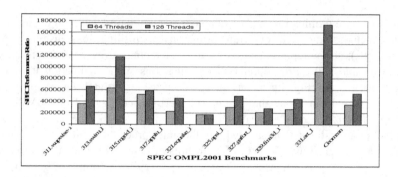

Fig. 5. SPEC OMPL2001 Performance on 64- and 128-CPU SGI Altix System

Fig.5 shows SPEC OMPL2001 performance results measured on SGI* Altix* 3000 (using Intel® 1500MHz Itanium® 2 processors) with 256KB L2 cache, 6MB L3 cache, 512GB memory (16*1024MB per core). Performance is measured on 64- and 128-CPU system configurations using the latest Intel® 8.0 C++ and Fortran compilers with OpenMP* support which incorporate the ideas and solutions discussed in this paper. At the time of writing this paper, these results were the best published SPEC* OMPL2001 results on www.spec.org. 315.mgrid_l and 321.equake_l are sparse matrix calculations, which do not scale well beyond 64 processors.

7 Conclusions

Exploiting effective parallelism for multithreaded processors and multiprocessor systems adds one more dimension of difficulty of the compiler development for generating optimized code. We tackled this performance challenge in our OpenMP design and implementation by developing techniques that produce well-defined and annotated IL to ensure that all classical optimizations are enabled seamlessly for Intel® Pentium® and Itanium® processor based systems. In this paper, we also studied the interaction between other optimization phases in the compiler and the OpenMP translation phase and show that cooperation that is required between the two to generate optimized threading code. The main contributions of this paper are:

– Several practical compiler techniques and solutions are proposed to ensure the generation of efficient threaded-code for OpenMP programs while interacting with other compiler optimizations phases. The implementation of these techniques in the Intel compilers is discussed.
– The effect of compiler optimizations on OpenMP programs is studied experimentally based on industry standard SPEC OMPM2001 and SPEC OMPL2001 benchmark suites on Intel® Pentium® 4 and Itanium® 2 processor based systems. This paper also reported the best performance results (published on www.spec.org) of SPEC* OMPL2001 delivered by the Intel® compilers on SGI* Altix* system.

---

*Other brands and names may be claimed as the property of others.

Acknowledgements

The authors thank all members of the Intel compiler team for developing the Intel C++/Fortran compilers. In particular, we thank Aart Bik, Ernesto Su, Hideki Saito, Dale Schouten for their contributions in PAROPT projects, Diana King and Michael Ross for OpenMP C++/Fortran FE support. Thanks also go to the library group at KSL for tuning OpenMP runtime library.

References

1. Ron Kalla, Balaram Sinharoy, Joel Tendler, "Simultaneous Multi-threading Implementation in POWER5 – IBM's Next Generation POWER Micorprocessor", *Hot Chips Conference 15*, August, 2003, http://www.hotchips.org/archive/hc15/pdf/11.ibm.pdf.
2. Feng Liu, Vipin Chaudhary, A Practical OpenMP Compiler for System on Chips, *in proc. of International Workshop on OpenMP Applications and Tools*, WOMPAT 2003, Toronto, Canada, June 26-27, 2003. LNCS 2716, pp.54-68.
3. Dan Quinlan, Markus Schordan, Qing Yi, Bronis R. de Supinski, "A C++ Infrastructure for Automatic Introduction and Translation of OpenMP Directives", *in proc. of International Workshop on OpenMP Applications and Tools*, WOMPAT 2003, Toronto, Canada, June 26-27, 2003, LNCS 2716, pp.13-25
4. Xinmin Tian, Yen-Kuang Chen, Milind Girkar, Steven Ge, Rainer Lienhart, and Sanjiv Shah, "Exploring the Use of Hyper-Threading Technology for Multimedia Applications with Intel OpenMP Compiler", In *Proc. of IEEE International Parallel and Distributed Processing Symposium, Nice, France,* April 22-26, 2003.
5. Xinmin Tian, Aart Bik, Milind Girkar, Paul Grey, Hideki Saito, Ernesto Su, "Intel OpenMP C++/Fortran Compiler for Hyper-Threading Technology: Implementation and Performance*", Intel Technology Journal,* http://www.intel.com/technology/itj, V6, Q1 issue, 2002.
6. Debbie Marr, Frank Binns, David L. Hill, Glenn Hinton, David A. Koufaty, J. Alan Miller, and Michael Upton, "Hyper-Threading Technology Microarchitecture and Architecture", http://www.intel.com/technology/itj, *Intel Technology Journal,* Vol. 6, Q1, 2002.
7. OpenMP Architecture Review Board, "OpenMP C and C++ Application Program Interface," Version 2.0, March 2002, http://www.openmp.org
8. Aart Bik, Milind Girkar, Paul Grey, and Xinmin Tian, "Automatic Intra-Register Vectorization for the Intel® Architecture", *International Journal of Parallel Programming*, Volume 30, page.65-98, April 2002.
9. Vishal Aslot, Max Domeika, Rudolf Eigenmann, Greg Gaertner, Wesley B Jones and Bodo Parady, "SPEComp: A New Benchmark Suite for Measuring Parallel Computer Performance", *in proc. of International Workshop on OpenMP Applications and Tools,* WOMPAT 2001, West Lafayette, Indiana, July 30-31, 2001, LNCS 2104, pp.1-10
10. Hidetoshi Iwashita, Eiji Yamanaka, Naoki Sueyasu, Matthijs van Waveren, Ken Miura, "SPEC OMP2001 Benchmark on the Fujitsu PRIMEPOWER System", *in proc. of Third European Workshop on OpenMP,* EWOMP'01 Barcelona, Spain, September 8-9th, 2001
11. OpenMP Architecture Review Board, "OpenMP Fortran Application Program Interface," Version 2.0, November 2000, http://www.openmp.org
12. Michael J. Wolfe, *High Performance Compilers for Parallel Computers,* Addison-Wesley Publishing Company, Redwood City, California, 1996.

®Intel is a registered trademark of Intel Corporation or its subsidiaries in the United States and other countries.

Sparse Matrices in MATLAB*P: Design and Implementation *

Viral Shah and John R. Gilbert

Department of Computer Science,
University of California, Santa Barbara
{viral, gilbert}@cs.ucsb.edu

Abstract. MATLAB*P is a flexible interactive system that enables computational scientists and engineers to use a high-level language to program cluster computers. The MATLAB*P user writes code in the MATLAB language. Parallelism is available via data-parallel operations on distributed objects and via task-parallel operations on multiple objects. MATLAB*P can store distributed matrices in either full or sparse format. As in MATLAB, most matrix operations apply equally to full or sparse operands. Here, we describe the design and implementation of MATLAB*P's sparse matrix support, and an application to a problem in computational fluid dynamics.

1   Introduction

MATLAB is a widely used tool in scientific computing. It began in the 1970s as an interactive interface to EISPACK, and LINPACK. Today, MATLAB encompasses several modern numerical libraries such as ATLAS, and FFTW, rich graphics capabilities for visualization, and several toolboxes for such domains as control theory, finance, and computational biology.

Almost all of today's supercomputers are based on parallel architectures. Companies such as IBM, Cray, SGI sell supercomputers with proprietary interconnects. Commodity clusters are omnipresent in research labs today. However, the tools used to program them are still predominantly Fortran and C with MPI or OpenMP.

MATLAB*P brings interactivity to supercomputing. There have been several efforts in the past to parallelize MATLAB. The parallel MATLAB survey [6] discusses most of these projects. Perhaps the most notable project that provides a large scale integration of parallel libraries with a MATLAB interface is NetSolve [2]. NetSolve provides an interfaces by invoking RPC calls through a

---

*This material is based on research sponsored by Air Force Research Laboratories under agreement number AFRL F30602-02-1-0181. The U.S. Government is authorized to reproduce and distribute reprints for governmental purposes not withstanding any copyright notation thereon.

L. Bougé and V.K. Prasanna (Eds.): HiPC 2004, LNCS 3296, pp. 144–155, 2004.
© Springer-Verlag Berlin Heidelberg 2004

special Matlab function, as opposed to Matlab*P which takes a unique approach to parallelization. The Matlab*P language is a superset of the Matlab language, and parallelism is propagated through programs using the the dlayout object – p. In the case where these systems use the same underlying packages, we believe that they can all achieve similar performance. However, we believe that the systems have different design goals otherwise and it is unfair to compare them.

Sparse matrices may have dimensions that are often in millions and enough non–zeros that they cannot fit on one workstation. Sometimes, the sparse matrices are themselves not too large, but due to the fill–in caused by intermediate operations (for eg. LU factorization), it becomes necessary to distribute the factors over several processors. Iterative methods maybe a better way to solve such large sparse systems. The goal of sparse matrix support in Matlab*P is to allow the user perform operations on sparse matrices in the same way as in Matlab.

2   User's View

In addition to Matlab's sparse and dense matrices, Matlab*P provides support for distributed sparse (dsparse) and distributed dense (ddense) matrices. The system design of Matlab*P and operations on ddense matrices are described elsewhere [12, 7].

The p operator provides for parallelism in Matlab*P. For example, a random parallel dense matrix (ddense) distributed by rows across processors is created as follows:

```
>> A = rand (100000*p, 100000)
```

Similarly, a random parallel sparse matrix (dsparse) also distributed across processors by rows is created as follows: (An extra argument is required to specify the density of non-zeros.)

```
>> S = sprand (1000000*p, 1000000, 0.001)
```

We use the overloading facilities in Matlab to define a *dsparse* object. The Matlab*P language requires that all operations that can be performed in Matlab be possible with Matlab*P. Our current implementation provides a working basis, but is not quite a drop–in replacement for existing Matlab programs.

Matlab*P achieves parallelism through polymorphism. Operations on ddense matrices produce ddense matrices. But once initiated, sparsity propagates. Operations on dsparse matrices produce dsparse matrices. An operation on a mixture of dsparse and ddense matrices produces a dsparse matrix unless the operator destroys sparsity. The user can explicitly convert a ddense matrix to a dsparse matrix using *sparse(A)*. Similarly a dsparse matrix can be converted to a ddense matrix using *full(S)*. A dsparse matrix can also be converted into a Matlab sparse matrix using S(:,:) or p2matlab(S). In addition to the data–parallel SIMD view of distributed data, Matlab*P also provides a task–parallel SPMD view through the so–called "MM–mode".

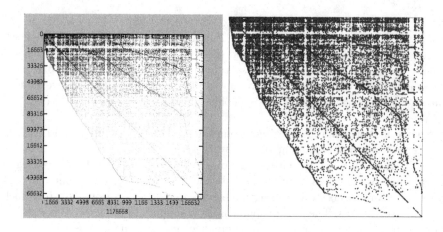

Fig. 1. MATLAB and MATLAB*P Spy plots of a web crawl dsparse matrix

MATLAB*P currently also offers some preliminary graphics capabilities to help users visualize dsparse matrices. This is based upon the parallel rendering for ddense matrices [5]. Again, this demonstrates the philosophy that MATLAB*P should feel like MATLAB. Figure 2 shows the spy plots (showing the non–zeros of a matrix) of a web crawl matrix in MATLAB*P and in MATLAB.

3   Data Structures and Storage

MATLAB stores sparse matrices on a single processor in a Compressed Sparse Column (CSC) data structure [10]. The MATLAB*P language allows for matrices to be distributed by block rows or block columns. This is already the case for ddense matrices [12, 7]. The current implementation supports only one distribution for dsparse matrices – by block rows. This is a design choice to prevent the combinatorial explosion of argument types. Block layout by rows makes the Compressed Sparse Row data structure a logical choice to store the sparse matrix slice on each processor. The choice to use a block row layout is not arbitrary, but based on the following observations:

- The iterative methods community largely uses row based storage. Since we believe that iterative methods will be the methods of choice for large sparse matrices, we want to ensure maximum compatibility with existing code.
- A row based data structure also allows efficient implementation of matvec (sparse matrix dense vector product) which is the workhorse of several iterative methods such as Conjugate Gradient and Generalized Minimal Residual.

By default, a dsparse matrix in MATLAB*P has the block row layout which would be obtained by ScaLAPACK [3] for a ddense matrix of the same dimensions. This allows for roughly the same number of rows on each processor. The user can override this block row layout in a couple of ways. The MATLAB sparse function takes arguments specifying a vector of row indices i, a vector of column indices j, a vector of non–zero values v, the number of rows m and the number of columns n as follows:

```
>> S = sparse (i, j, v, m, n)
```

By using a vector *layout* which specifies the number of rows on each processor instead of the scalar m which is simply the number of rows, the user can create a dsparse matrix with the desired layout:

```
>> S = sparse (i, j, v, layout, n)
```

The block row layout of a dsparse matrix can also be changed after creation with:

```
>> changelayout (S, newlayout)
```

The CSR data structure stores whole rows contiguously in a single array on each processor. If a processor has nnz non–zeros, CSR uses an array of length nnz to store the non–zeros and another array of length nnz to store column indices, as shown in Figure 2. Row boundaries are specified by an array of length $m+1$, where m is the number of rows on that processor.

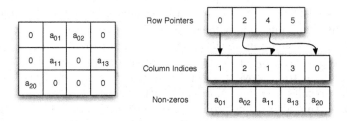

Fig. 2. Compressed Sparse Row (CSR) data structure

Assuming a 32–bit architecture and using double precision floating point values for the non–zeros, an $m \times n$ real sparse matrix with nnz non-zeros uses up $12nnz + 4m$ bytes of memory. Support for complex sparse matrices will be available very soon in MATLAB*P.

It would be simple to modify this data structure to allow some slack in each row so that element–wise insertion, for example, could be efficient. However, the current implementation uses the simplest possible data–structure for robustness and efficiency in matrix and vector operations.

4   Operations and Implementation

In this section, we describe the implementation of several sparse matrix operations in MATLAB*P. All experiments are performed on a cluster of 2.6GHz Pentium IV Xeon processors with 3GB RAM and a gigabit ethernet interconnect.

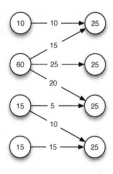

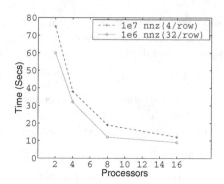

Fig. 3. Starching

Fig. 4. Scalability of **sparse** (Altix)

4.1   Parallel Sorting

Sorting of ddense vectors is an important building block for several parallel sparse matrix operations in MATLAB*P. Sorting is an extremely important primitive for parallel irregular data structures in general. Several parallel sorting algorithms have been surveyed in the literature [4, 8]. Cluster computers have distributed memory, and the interconnect typically has high latency and low bandwidth compared to shared memory computers. As a result, it is extremely important to minimize the communication while sorting. We are experimenting with Sample–Sort with different sampling techniques, and other median–based sorting ideas. Although we have an efficient parallel sorting algorithm already in MATLAB*P, we are still trying to improve it, given the importance of having fast sorting. We will describe results from our efforts in a future paper.

Several sorting algorithms produce a data distribution that is different from the input distribution. Hence a reshuffling of the sorted data is often required. We refer to this process as Starching, as shown in Figure 4.1. The distribution after sorting is on the left, whereas the desired distribution is on the right in the bipartite graph. The weights on the edges of the bipartite graph show the communication required between pairs of processors during starching. This step is required to ensure consistency of MATLAB*P's internal data structures.

4.2   Constructors

There are several ways to construct parallel sparse matrices in MATLAB*P:

1. matlab2pp converts a sequential MATLAB matrix to a distributed MATLAB*P matrix. If the input is a sparse matrix, the result is a dsparse matrix.
2. sparse – sparse works with $[i, j, v]$ triples, which specify the row value, the column value and the non–zero value respectively. If i, j, v are ddense vectors with nnz non–zeros, then sparse assembles a sparse matrix with nnz non–zeros. If there are duplicate $[i, j]$ indices, the corresponding values are summed. The pseudocode for sparse is shown in Figure 4.1. However, in our implementation, we implement this by sorting the vectors simultaneously

```
function s = sparse (i, j, v)

[j, perm] = sort(j);
i = i(perm); v = v(perm);

[i, perm] = sort(i);
j = j(perm); v = v(perm);

starch (i, j, v);
s = assemble (i, j, v);
```

Fig. 5. Implementation of `sparse`

using row numbers as the primary key, and column numbers as the secondary key.

The starch phase here is similar to the starching used in the parallel sort, except that it redistributes the vectors so that row boundaries do not overlap among processors and the required block row distribution for the sparse matrix is achieved. The assemble phase actually constructs a dsparse matrix and fills it with the non–zero values. Figure 4 shows the scalability of `sparse` on an SGI Altix 350. Although performance for commodity clusters cannot be as good as that of an Altix, our initial experiments do indicate good scalability on commodity clusters too. We will report more detailed performance comparisons in a future paper.

3. `spones`, `speye`, `spdiag`, `sprand` etc. – Some basic functions implicitly construct dsparse matrices.

4.3   Matrix Arithmetic

One of the goals in designing a sparse matrix data structure is that, wherever possible, it should support matrix operations in time proportional to flops. As a result, arithmetic on dsparse matrices is performed using a sparse accumulator (SPA). Gilbert, Moler and Schreiber [10] discuss the design of the SPA in detail. MATLAB*P uses a separate SPA for each processor.

4.4   Indexing, Assignment and Concatenation

The syntax of matrix indexing in MATLAB*P is the same as in MATLAB. It is of the form $A(p,q)$. p and q can each be either a range $(1:n)$, or a permutation vector or scalars. Depending on the context, however, this can mean different things.

```
>> B = A(p,q)
```

In this case, the indexing is done on the right side of "=" which specifies that B is a submatrix of A. This is the `subsref` operation in MATLAB.

```
>> B(p,q) = A
```

On the other hand, indexing on the left side of "=" specifies that A should be stored in a submatrix of B. This is the `subsasgn` operation in MATLAB.

If p and q are both integers, $A(p, q)$ directly accesses the dsparse data structure. If p or q are vectors or a range, $A(p, q)$ calls `find` and `sparse`. `find` is the reverse of `sparse` – it converts the matrix from CSR to $[i, j, v]$ format. In this format, it is very easy to find $[i, j]$ pairs which satisfy the indexing criteria. The resulting submatrix is then assembled by simply calling `sparse` .

MATLAB also supports horizontal and vertical concatenation of matrices. The following code, for example, concatenates A and B horizontally, C and D horizontally, and finally concatenates the results of these two operations vertically.

```
>> S = [ A B; C D ]
```

The basic primitives, `find` and `sparse` are used to provide support for concatenation operations in MATLAB*P.

4.5   Matvec

The matvec operation multiplies a dsparse matrix with a ddense column vector, producing a ddense column vector as a result. Matvec is the kernel for many iterative methods.

For the matvec, $y = Ax$, we have A and x distributed across processors by rows. The submatrix of A at each processor will need a piece of x depending upon its sparsity structure. When matvec is invoked for the first time on a dsparse matrix A, MATLAB*P computes a communication schedule for A and caches it. When more matvecs are performed using A, this communication schedule does not need to be recomputed, which saves some computing and communication overhead, at the cost of extra space required to save the schedule. MATLAB*P also overlaps the communication and computation during matvec. This way, each processor starts computing the result of the matvec whenever it receives a piece of the vector from any other processor. Figure 6 also shows how matvec scales in MATLAB*P, since it forms the main computational kernel for conjugate gradient.

Communication in matvec can be reduced by performing graph partitioning of the graph of the sparse matrix. If fewer edges cross processors, lesser communication is required during matvec. MATLAB*P can use several of the available tools for graph partitioning. However, by default, MATLAB*P does not perform graph partitioning during matvec. The philosophy behind this decision is similar to that in MATLAB, that reorganizing data to make later operations more efficient should be possible, but not automatic.

4.6   Solutions of Linear Systems

MATLAB solves the linear system $Ax = b$ with the matrix division operator, $x = A \backslash b$. In sequential MATLAB, $A \backslash b$ is implemented as a polyalgorithm [10], where every test in the polyalgorithm is cheaper than the next one.

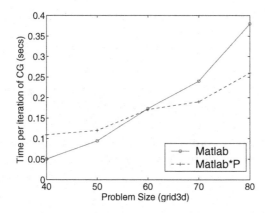

Fig. 6. Time per iteration of CG, scalability of matvec

1. If A is not square, solve the least squares problem.
2. Otherwise, if A is triangular, perform a triangular solve.
3. Otherwise, test whether A is a permutation of a triangular matrix (a "morally triangular" matrix), permute it, and solve it if so.
4. Otherwise, if A is Hermitian and has positive real diagonal elements, find a symmetric minimum degree ordering p of A, and perform the cholesky factorization of $A(p,p)$. If successful, finish with two sparse triangular solves.
5. Otherwise, find a column minimum degree order p, and perform the LU factorization of $A(:,p)$. Finish with two sparse triangular solves.

Different issues arise in parallel polyalgorithms. For example, morally triangular matrices and symmetric matrices are harder to detect in parallel. One also expects to be able to use iterative methods. Design for the right polyalgorithm for \ in parallel is an active research problem. For now, MATLAB*P uses a parallel general direct sparse solver for \, which is SuperLU_DIST [14] by default, although a user can choose to use MUMPS [1] too.

The open question at this point is, should MATLAB*P use preconditioned iterative methods to solve sparse linear systems instead of direct methods. Currently, iterative methods are not usable as a black box, and not yet suitable for MATLAB*P.

4.7   Iterative Methods - Conjugate Gradient

Conjugate Gradient is an iterative method used to solve a symmetric, positive definite system of equations. The same code is used for MATLAB*P and matlab, except that the input is dsparse in the MATLAB*P case. In Fig 6, grid3d(k) is a routine used from the meshpart [9] toolbox, which returns a $k^3 \times k^3$ symmetric positive definite matrix A with the structure of the $k \times k \times k$ 7–point grid.

5   An Application in Computational Fluid Dynamics

We are using a prototype version of MATLAB*P in collaboration with a number of domain scientists for applications in computational science and engineering. We describe an application here.

Goyal and Meiburg [11] are studying the influence of viscosity variations on the density–driven instability of two miscible fluids. The two fluids, of different density and viscosity are in a vertical Hele–Shaw cell as shown in figure 7. This problem is used to model porous media flows and finds applications in enhanced oil recovery, fixed bed regeneration and groundwater flows.

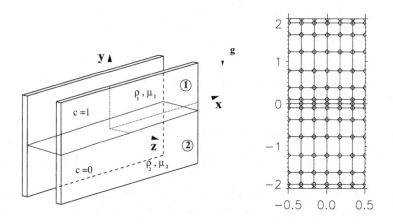

Fig. 7. Geometry of the Hele-Shaw cell (left). The heavier fluid is placed above the lighter one. Either one of the fluids can be the more viscous one. Mesh point distribution in the computational domain (right). A Chebyshev grid is employed in the y–direction, and compact finite differences in the z–direction

Fig. 7 shows the discretization of the problem, which yields an algebraic system of the form $A\phi = \sigma B\phi$. The eigenvalue σ represents the growth rate of the perturbations, while the eigenvector ϕ reflects the shape of the perturbations. A positive (negative) eigenvalue indicates unstable (stable) behavior. The system has a 5×5 block structure reflecting the 5 variables at each mesh point (3 velocity components u, v and w, relative concentration of the heavier fluid c, and pressure p).

A discretization of 165×25 points turns out to be sufficient for this problem. Since we solve for 5 variables at each grid point, the matrix A is of the size $20,625 \times 20,625$. The number of non–zeros is $3,812,450$. The matrix is unsymmetric, both in values and nonzero structure, as shown in the spy plots in Figure 5. In order to calculate the largest eigenvalue, we use the power method with shift and invert in MATLAB*P.

Fig. 8. Spy plots of the matrices A and B

```
function lambda = peigs (A, B, sigma, iter)

  [m n] = size (A);
  C = A - sigma * B;
  y = rand (n*p, 1);

  for k=1:iter
    q = y ./ norm (y);
    v = B * q;
    y = C \ v;

    theta = dot (q, y);
    res = norm (y - theta * q);
    if res <= 0.0001,  break;   end
  end

lambda = 1 / theta + sigma;
```

Fig. 9. MATLAB*P code for power method with shift and invert

The original non–MATLAB*P code used LAPACK with ARPACK [13], while the MATLAB*P code is using SuperLU_DIST with the power method as shown in figure 5. We use a guess of 0.1 to initialize the power method and it converges to 0.0194 which is enough precision for linear stability analysis. We use a cluster with 16 processors to solve the generalized eigenvalue problem. Each node has a 2.6GHz Pentium Xeon CPU, 3GB of RAM and a gigabit ethernet connection. Results are presented in Table 1.

As a next step we want to incorporate a variable viscosity net flow through the Hele–Shaw cell to incorporate the potentially destabilizing effects of viscous fingering into play, so that the possibility of complex interactions between density- and viscosity-driven instabilities arises. The existence of a more complex flow field necessitates a finer grid and a larger domain size for the linear stability calculations as compared to the previous case discussed where the two fluids were essentially at rest with respect to each other. We expect that we will require about 10 times more computing resources (CPU and memory) to tackle these challenges.

Table 1. Time to solve the generalized eigenvalue problem

No. of processors	Time (seconds)
4	90
8	39
16	33

6   Conclusion

The implementation of sparse matrices in MATLAB*P is work in progress. Current available functionality includes being able to construct sparse matrices, perform element–wise arithmetic and indexing operations on them, multiply a sparse matrix with a dense vector and solve linear systems. This level of functionality allows us to implement several algorithms such as conjugate gradient and the power method.

Much remains to be done. A complete implementation of sparse matrices requires matrix–matrix multiplication and several factorizations (Cholesky, QR, SVD etc). Improvements in the sorting code can lead to general improvements in many parts of MATLAB*P. It is also important to make existing graph partitioners available in MATLAB*P – Meshpart and ParMetis. Several preconditioning methods also need to be implemented for MATLAB*P, since iterative methods might possibly be the way to solve large linear systems.

The goal of sparse matrix support in MATLAB*P is to provide an interactive environment for users to perform operations on large sparse matrices in parallel, while being compatible with MATLAB. Our current implementation is ready to be used for simple real life problems.

Acknowledgements

David Cheng made useful contributions to the sorting code. We had several useful discussions with Parry Husbands, Per–Olof Persson, Ron Choy, Alan Edelman, Nisheet Goyal and Eckart Meiburg.

References

1. Patrick Amestoy, Iain S. Duff, and Jean-Yves L'Excellent. Multifrontal solvers within the PARASOL environment. In *PARA*, pages 7–11, 1998.
2. D. Arnold, S. Agrawal, S. Blackford, J. Dongarra, M. Miller, K. Seymour, K. Sagi, Z. Shi, and S. Vadhiyar. Users' Guide to NetSolve V1.4.1. Innovative Computing Dept. Technical Report ICL-UT-02-05, University of Tennessee, Knoxville, TN, June 2002.
3. L. S. Blackford, J. Choi, A. Cleary, E. D'Azevedo, J. Demmel, I. Dhillon, J. Dongarra, S. Hammarling, G. Henry, A. Petitet, K. Stanley, D. Walker, and R. C. Whaley. *ScaLAPACK Users' Guide*. SIAM, Philadelphia, PA, 1997.

4. Guy E. Blelloch, Charles E. Leiserson, Bruce M. Maggs, C. Greg Plaxton, Stephen J. Smith, and Marco Zagha. A comparison of sorting algorithms for the connection machine cm-2. In *Proceedings of the third annual ACM symposium on Parallel algorithms and architectures*, pages 3–16. ACM Press, 1991.

5. Oskar Bruning, Jack Holloway, and Adnan Sulejmanpasic. Matlab *p visualization package. 2002.

6. Long Yin Choy. Parallel Matlab survey. 2001. http://theory.csail.mit.edu/~cly/survey.html.

7. Long Yin Choy. MATLAB*P 2.0: Interactive supercomputing made practical. *M.S. Thesis, EECS*, 2002.

8. D. E. Culler, A. Dusseau, R. Martin, and K. E. Schauser. Fast parallel sorting under LogP: from theory to practice. In *Proceedings of the Workshop on Portability and Performance for Parallel Processing*, Southampton, England, July 1993. Wiley.

9. John. R. Gilbert, Gary L. Miller, and Shang-Hua Teng. Geometric mesh partitioning: Implementation and experiments. *SIAM Journal on Scientific Computing*, 19(6):2091–2110, 1998.

10. John R. Gilbert, Cleve Moler, and Robert Schreiber. Sparse matrices in MATLAB: Design and implementation. *SIAM Journal on Matrix Analysis and Applications*, 13(1):333–356, 1992.

11. Nisheet Goyal and Eckart Meiburg. Unstable density stratification of miscible fluids in a vertical hele-shaw cell: Influence of variable viscosity on the linear stability. *Journal of Fluid Mechanics*, (To appear), 2004.

12. P. Husbands and C. Isbell. MATLAB*P: A tool for interactive supercomputing. *The Ninth SIAM Conference on Parallel Processing for Scientific Computing*, 1999.

13. R. B. Lehoucq, D. C. Sorensen, and C. Yang. *ARPACK Users Guide: Solution of Large Scale Eigenvalue Problems with Implicitly Restarted Arnoldi Methods*. SIAM, Philadelphia, 1998.

14. Xiaoye S. Li and James W. Demmel. Superlu_dist: A scalable distributed–memory sparse direct solver for unsymmetric linear systems. *ACM Trans. Math. Softw.*, 29(2):110–140, 2003.

Architecture and Early Performance of the New IBM HPS Fabric and Adapter

Rama K Govindaraju, Peter Hochschild*, Don Grice, Kevin Gildea,
Robert Blackmore, Carl A Bender, Chulho Kim, Piyush Chaudhary,
Jason Goscinski, Jay Herring, Steven Martin, and John Houston

Server Development Lab, IBM Poughkeepsie, NY, 12533
*T. J. Watson Research Center, IBM Research, Hawthorne, 10954
{ramag, phoch}@us.ibm.com

Abstract. In this paper we describe the architecture, design, and performance of the new cluster switch fabric and adapter called HPS (High Performance Switch). HPS delivers very low latency and very high bandwidth. We demonstrate latency of less than 4.3us MPI library; 1.8GB/s of delivered unidirectional bandwidth and 2.9GB/s of bidirectional bandwidth between 2 MPI tasks running on 1.9GHz Power 4+ IH based nodes. HPS also supports RDMA (remote direct memory access capability). A unique capability of RDMA over HPS is that reliable RDMA is supported over an underlying unreliable transport (unlike Infiniband and other RDMA transport protocols which depend on the underlying transport being reliable). We profile the performance of RDMA and its impact on striping for systems in which multiple network adapters are available to tasks of parallel jobs.

1 Introduction

IBM has been one of the leaders in the development of High Performance Computing Systems and Supercomputers since 1993 [15]. At the heart of IBM's supercomputers have been the development of high speed switches along with an integrated and complete integrated suite of software.

Table 1 above shows a high level overview of the evolution of "IBM's High Performance Supercomputer products." In this paper we discuss the architecture, design and scaling elements (number of tasks in a single parallel job, number of end points supported by the HPS switch network) of the HPS based supercomputers. The rest of the paper is organized as follows: In Section 2, we provide an overview of the high level architecture and design of the HPS switch and adapter. In Section 3, we discuss the software architecture that drives communication over the HPS switch. In Section 4, we describe the design of the unreliable datagram mode (also called FIFO mode) and its performance at the MPI level. In Section 5, we highlight the importance of RDMA; describe the design of a reliable RDMA protocol over an unreliable datagram service and early performance data. We discuss some of the usage models of RDMA and cases where it can show improved application performance. In Section 6, we briefly touch on emerging technologies and the efforts in the standards body to exploit emerging networking technologies like RDMA and our thoughts on where the industry is headed, and the key questions that remain.

L. Bougé and V.K. Prasanna (Eds.): HiPC 2004, LNCS 3296, pp. 156–165, 2004.
© Springer-Verlag Berlin Heidelberg 2004

Table 1. Historical Performance of user space MPI on IBM interconnects

Year	Hardware Configuration	CPU MHz	MPI latency	Peak MPI Bandwidth
1994	590 uni-processor nodes/TB2 adapter, hPS switch	66	49us	34MB/s
1995	591 uni-processor nodes/TB2 adapter, hPS switch	77	43us	35MB/s
1996	P2SC uni-processor nodes/TB3 adapter, TBS switch	120	29us	102MB/s
1997	P2SC uni-processor nodes/TB3 adapter, TBS switch	160	28us	110MB/s
1998	Silver 4-way SMP nodes/TBMX adapter, TBS switch	332MHz	24us	83MB/s
1999	WH-2 way SMP nodes/TBMX adapter, TBS switch	375MHz	22us	139MB/s
2000	WH-4 way SMP nodes/TBMX adapter, TBS switch	375MHz	20us	140MB/s
2001	NH-16way SMP nodes/Colony adapter & switch	375MHz	15us	330MB/s
2004	Power4+ - 8 way SMP nodes/ HPS adapter & switch	1.9GHz	4.3us	1.8GB/s

2 Architecture and Design of the HPS Switch and Adapter

HPS is IBM's fourth generation switch and adapter technology. It incorporates new link technology with the ability to drive 2GB/s peak performance in each direction. The new switch is based on a building block of an 8 port switch chip. Key enhancements to the switch include higher bandwidth, lower latency capabilities, automatic link level retry, flit-level flow control, support for both copper and optical cables, and multiple virtual channels per link. Very low latency is achieved by using cut-through routing and source routing tables. Each switch chip adds only around 59ns of latency.

The HPS switch and adapters support Power 4 and Power 5 processor based servers. The HPS switch adapters attach to the nodes through the GX (IO bus for Power 4 and Power 5). The GX interface is the closest IO interface to the CPU and hence ideally suited for attachment of a high performance switch adapter like HPS. Most commodity switch adapters are PCI based but in order to exploit the performance of the GX interface, proprietary HPS like technology is necessary. The switch adapters then connect to the switch through the LDC (Link driver chips). The switch architecture remains a multi-stage Omega based network which has excellent scaling and bisection bandwidth properties and has proven to be a very robust architecture for scalable high performance supercomputing systems as evidenced by the success of such cluster based supercomputers in the last 10 years [2, 15]. 8 switch chips are typically packaged into a switch board providing a scaling of 16 adapter ports and 16 link ports per switch board. Two stages of the switch boards are capable of supporting up to 512 end points and three stages of cascading allow for a maximum of 4096 end points. Each additional cascading stage of switch board increases the end to end latency between the farthest points by approximately 0.4us. The switch provides multiple paths between any two end points. The adapter uses up to four different paths from among them.

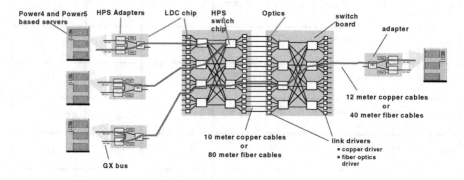

Fig. 1. Federation components: Nodes, the adapters and the switch elements

Figure 2 shows the high level HPS switch adapter hardware architecture. The Protocol processor is a sequencer engine supporting 16 task threads that can be programmed to activate the various hardware engines. The transport macro engines are used to move data from the host to the adapter and then to switch and vice versa. The link macros drive packets from the adapter into the switch and vice versa. Remaining engines are used for interrupts, user space and privileged MMIO (memory mapped IO) access to the adapter, error handling and other miscellaneous functions. The jtag hooks are used to service the switch and adapter and also set up the path and route tables necessary. The internal buffers and the SRAM are used by the protocol engine to maintain state to effect message passing. The multi-threaded protocol processor is

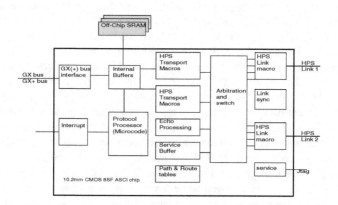

Fig. 2. Federation Adapter Hardware Architecture

programmed with microcode to effect efficient message passing. 4 tasks are reserved for send processing and 4 tasks for receive processing. Each of the 8 packet processing tasks is equipped with an on-chip packet buffer. This provides sufficient concurrency to achieve full utilization of the transport engines and IO buses. .

3   HPS Software Architecture

IBM provides a complete software stack that exploits the HPS switch. Figure 3 shows the high level software architecture. Application tasks executing on the node can make file system calls to IBM's GPFS (General Parallel File Systems) [4], or a socket call, or an MPI/LAPI based message passing [7] call over user space. All of these calls from the

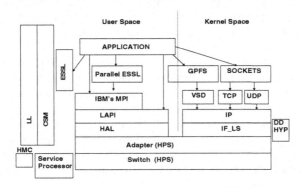

Fig. 3. HPC Software Architecture

end user application can result in communication over the HPS switch. The MPI library is supported over both user space and UDP. The architecture figure shows the layering for user space. MPI is layered on top of LAPI [9] which is IBM's proprietary interface for a low level communication API and presents a one-sided programming model to end users as well. LAPI supports reliable point to point message passing capability. Collective calls in MPI are broken down into point to point LAPI calls. LAPI ensures that the messages submitted to it are delivered to the target in a reliable fashion. LAPI maintains state to ensure packets which are not acknowledged are retransmitted by LAPI. LAPI also handles failure of adapters by re-driving pending messages through other adapters available on the node. LAPI is layered on top of HAL (the hardware abstraction layer) which is a packet interface. The HAL layer interfaces with the adapter microcode to exchange packets between the host side software and the adapter. HAL is stateless with respect to the upper layer protocols. The application can make file system calls which are fielded by GPFS (IBM's cluster and parallel file system). GPFS is built on shared disk architecture. For file data that is not on a disk that is locally attached to the node making the file system request, VSD (Virtual Shared Disk) traps those requests and gets the data shipped from the appropriate server. LL (Loadleveler)is the scheduler, CSM (cluster systems management), HMC (hardware management console), DD (device driver) and HYP (hyper-visor) are the other key components.

In Section 4 and Section 5, we discuss the details of how user space packet mode and RDMA capabilities are enabled and the performance of the various components. Details of IP performance, GPFS, and other components will be the subject of future papers currently in preparation.

4   HPS FIFO Mode

The implementation of LAPI [8] and MPI over LAPI [9] have been covered in earlier papers. Mapping a two-sided programming model like MPI on top of a one-sided programming model like LAPI requires efficient hooks to ensure an efficient implementation. The main data structures that allow the adapter and HAL to exchange packets are the send and receive FIFOs that reside in system memory and are mapped to the address space of the user process. These send and receive buffers are pinned and mapped onto AIX large pages (16M each). Depending on the size of the parallel job, the receive FIFO could be mapped onto multiple large pages. The mapping of these large pages is maintained by the adapter.

The protocol (LAPI) constructs the packet in the send FIFO and then issues an MMIO (memory mapped IO) request to the adapter. The adapter then DMAs the header, and the rest of the data from the FIFO slot (accomplished by a single DMA read operation with pre-fetch hints), adds the route header and injects the packet into the network. For multi-packet messages LAPI initially primes the FIFO with one packet, two packets and subsequently 4 or 8 packets. This allows the pipeline to be primed and started (with the first and second packets) and subsequent packets are bunched into groups of 8 before the adapter is notified to minimize the handshake between the adapter and host side software.

When a packet arrives from the network, the adapter first DMAs the data portion of the packet into the receive FIFO (which is all but the first cache line of the packet which contains the header). Once all the data portions of the packet have reached the point of coherence the first cache line containing the header is DMA-ed into the receive FIFO slot to signal completion of the receipt of the packet and that the protocol can now absorb the packet. This staging of first cache line is important since DMA transactions may complete out of order. A critical design goal was to minimize bus transactions in the transfer of data. The MMIO commands are also used by the adapter and HAL to ensure that the fill and available counts of packets in the FIFO are kept in synch. Packet arrival interrupts are controlled by the protocol by a simple thresholding mechanism.

4.1   Cache Effects and Bus Crossings

A typical send operation would result in the CPU loading from memory the data buffer, and storing it in the network send buffer. At first glance, it may appear that this results in 2 memory bus crossings. The CPU would then tap the adapter (via MMIO) to DMA the data from the network buffer resulting in an IO bus crossing. An important design choice we made greatly helped improve performance. The send FIFO network buffer was kept small (256K to hold 128 – 2K sized packets) so that it is usually in cache. Typically the user program prepares the buffers to be sent and it is reasonable to assume that the user buffer is likely in cache. So the copy from the user buffer to the network send buffer is often a cache to cache transfer not visible on the memory bus. Power architecture supports the satisfaction of DMA reads from cache. Therefore on the send side, typically the only bus crossing is the DMA across the IO bus. In the case of RDMA (discussed later) also there is one IO bus crossing.

The receive FIFOs are larger and scale with the number of user space tasks in the parallel job that the process must communicate with. This allows the receive FIFOs to absorb (buffer) packets from the network from multiple sources. The receive FIFO can be sized according to the number of tasks in the parallel job allowing the LAPI no loss of flow-control policy. This is crucial for performance so that the packets are not left clogging the network stalling packets belonging to different tasks, if the task for which the packets are designated for, is not actively pulling packets from the network (for e.g. it is busy in its compute loop or is de-scheduled for some reason). On the receive side the adapter DMAs packets into the receive FIFO. The CPU has to fetch the data from memory into its cache to absorb the newly arrived data into its on going computation. A future enhancement that is envisioned is to enable the adapter and the memory subsystem with smarts to allow cache injection for received messages into the processor cache running the relevant task. In most cases therefore, it is probably reasonable to assume that a well designed FIFO mode transport protocol can reduce unnecessary bus crossings which impact performance.

We have achieved less than 4.3us MPI level latency across 2 tasks that are communicating over the switch (across a single switch chip for the latency measurement). This implies that the MPI latency includes the locking overhead on both the 'send' and 'receive' sides. Currently, due to our layering, we incur two locks on the send side and two locks on the receive side in the critical path. To achieve this latency we designed an elegant mechanism to share locks between MPI and LAPI layers to remove two of the locks in the critical path. Simple smarts were put into the acknowledgement processing to remove it from the critical path to save up to 1us in our latency. The bandwidth profile for unidirectional bandwidth is shown in Figure 4. We have been able to achieve 1.8GB/s of unidirectional bandwidth between 2 tasks running on 2 different nodes across the switch. All these measurements were performed on 1.9GHz Power4+, 8way SMP nodes using HPS switch and adapter. The slight inflection at 64K message size is due to the switching over from eager mode of transport to

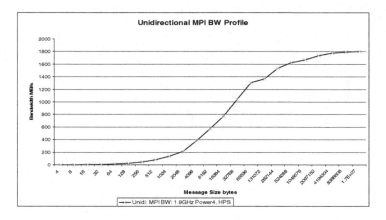

Fig. 4. MPI Unidirectional Performance

rendezvous protocol [MPIP]. Existing solutions typically use a PCI-X attachment which has a bandwidth limit of around 800MB/s.

5   HPS RDMA Mode

RDMA Definition and Terminology: RDMA is the transport capability that allows processes executing on one node to be able to "directly" access (execute reads or writes against) the memory of processes executing on a different node connected by an RDMA capable network. By "directly" we mean the capability of the network adapters at both ends to affect the transfer of data without any protocol processing by the CPU on the slave side of the transfer operation. "Master" is the one that initiates the operation and "slave" is the other target end point of the transport operation. RDMA is sometimes referred to as "memory semantics" for communication across a cluster network, or as "hardware put/get," or as "remote read/write."

5.1   Significance of RDMA Transport Protocols

We list some of the reasons why RDMA is an important emerging transport model: *a) Overlap value:* RDMA de-couples the CPU from the communication work of moving data. This allows for the computation on the CPU to be overlapped with communication. Many existing and emerging new applications are being written to exploit potential overlap between computation and communication [5]; *b) Memory subsystem bottleneck:* A critical bottleneck for improved sustained performance is memory system performance. RDMA can minimize the number of bus transactions and helps reduce the stress on the memory system for applications that can take advantage of overlap; *c) One-sided programming model:* Many applications are more ideally mapped on to a one-sided programming model. RDMA is a very natural and efficient mechanism to enable such a programming model [3]; *d) Striping:* Since the adapters are shared by multiple processors on the SMP node, and each node can have multiple adapters (typically one adapter for every 4 CPUs), striping of protocol messages may

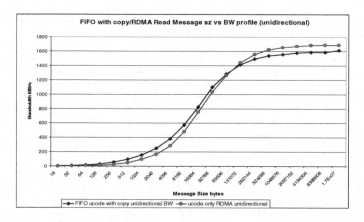

Fig. 5. FIFO Mode vs. RDMA performance (Power 1.7Ghz system with HPS)

provide some performance benefits. RDMA plays a critical role in how we do efficient striping of messages by a single task across multiple network adapters; *e) Interrupts:* RDMA offloads fragmentation and reassembly of the messages on to the adapter and this also helps reduce the number of per-packet interrupts that need to be processed by the protocol to absorb packets arriving from the network. This benefit can only be realized if the application is tuned to take advantage of the overlap possibilities with RDMA. Due to lack of space the details of the RDMA implementation will be the subject of another paper.

5.2 RDMA Performance

Figure 6 shows the performance of RDMA as compared to FIFO mode on 1.7GHz Power 4+ nodes. Note that the overheads associated with the setup cost (pinning and mapping of buffers) is not factored into the plots shown in Figure 5. Figure 6 shows the bidirectional performance using RDMA. Figure 6 and Figure 7 show the performance when striping using RDMA. Figure 6 shows that the unidirectional bandwidth using RDMA allows very close to linear scaling with multiple network interfaces. A single CPU can issue multiple RDMA requests to different network adapters in a pipelined fashion. This mechanism avoids unnecessary synchronization (since one CPU is submitting the requests) while still allowing the data transfer via RDMAs' across multiple network interfaces to proceed in parallel.

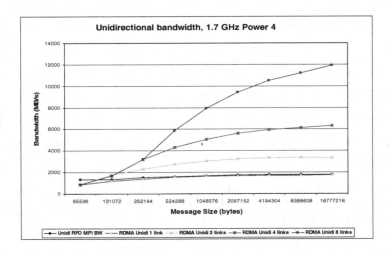

Fig. 6. Unidirectional bandwidth using RDMA and striping

Figure 7 shows the corresponding bidirectional data using RDMA and striping. Figures 6 and 7 show remarkable performance can be achieved using RDMA and efficient software striping models. Details of the implementation of the striping in software will be part of a different paper. The data in Figure 6 and 7 are plotted starting from 64K because that is the approximate cut-off point where RDMA crosses over the FIFO mode performance as seen in Figure 5. It is clear from Figures 6 and 7 that RDMA provides very good scaling of bandwidth across multiple network interfaces. With striping across 8 links between 2 tasks executing on different nodes, the unidirectional bandwidth was over 12GB/s (~83% striping efficiency since a single link is capable of 1.8GB/s unidirectional bandwidth) and the bidirectional bandwidth was over 20.5GB/s (~87% striping efficiency since a single link is capable of 2.9GB/s bidirectional bandwidth). RDMA is however not without significant challenges which will be addressed in a future paper.

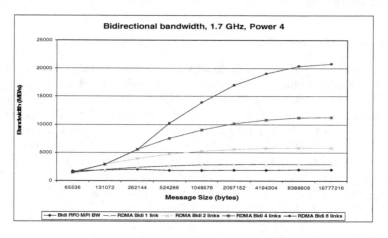

Fig. 7. Bidirectional bandwidth, using RDMA and multiple network interfaces

6   Conclusions and Future Work

HPS has demonstrated a significant increase in performance as compared to the previous network adapter and switch from IBM. Current work is underway to improve the bidirectional bandwidth, exploiting the use of the RDMA capability by MPI, LAPI and the IP transport protocols. Striping enablement is also currently underway. Future work also includes enabling HPS for Power 5 based systems, and adding additional capabilities into our high speed network hardware and software to ensure that a complete system offering is delivered. RDMA is an emerging technology that should spur more interest in new problems on how best to exploit it. RDMA also brings a lot of difficult problems that need further investigation to overcome many of the challenges discussed in Section 5. Addressing the challenges detailed with RDMA is a subject of active future research. IBM is engaged in the definition of API definitions in the standards body [5] and one of our focus areas has been to ensure we use our early experience with RDMA to help influence the standard so that the future API definitions lend themselves to efficient implementations and exploitation of RDMA capability.

Acknowledgements

A large number of people contributed to the architecture, design, development, tuning and testing of the HPS based system. Some of the many people who contributed to the development of this HPS based system are: Fu-Chung Chang, Amy Chen, Jennifer Doxtader, Bill Helmer, Su Huang, Patricia Heywood, Bin Jia, Doug Joseph, Chulho Kim, Eric Lais, Skip Lundin, Brett Patane, Aruna Ramanan, Nick Rash, Kevin J Reilly, Rajeev Sivaram, Carol Soto, Richard Treumann, William Tuel, Theodore Will, Hanhong Xue, and others we have surely missed.

References

1. [DGSMP] Richard Treumann, *"DGSM: Data Gather Scatter Machine,"* IBM Internal Report.
2. [FPGS] Daniel Frye, Kevin Gildea, Peter Hochschild, Marc Snir, *"The communication software and Parallel Environment for the IBM SP2,"* IBM Systems Journal, 34(2), pp. 205-221, 1995.
3. [GA] Jarek Nieplocha, Jialin Ju, Manoj Kumar Krishnan, Bruce Palmer, Vinod Tipparaju, *"The Global Arrays User Manual,"* http://www.emsl.pnl.gov/docs/global/user.html.
4. [GPFSP] GPFS White Paper: http://www-1.ibm.com/servers/eserver/pseries/software/sp/gpfs.html
5. [ITAPI] IT-API: Open group consortium on API definition for RDMA capable networks. http://www.opengroup.org
6. [IBTA] Infiniband Architecture, http://www.infinibandta.org/ibta/
7. [LAPIP] IBM's LAPI Documentation http://rs6ktech.dfw.ibm.com/sp/docs/pssp3.4/pssphtml/cmdsv2/am0trmst02.html.
8. [LAPIP2] Gautam Shah, Jarek Nieplocha, Jamshed Mirza, Chulho Kim, Robert J Harrison, Rama K. Govindaraju, Kevin Gildea, Paul DiNicola, Carl A Bender, *"Performance and Experience with LAPI – A New High Performance Communication Library for the IBM RS/6000 SP,* In Proceedings of IPPS (International Parallel Processing Symposium) 1998.
9. [MPILAPIP] Mohammad Banikazemi, Rama K Govindaraju, Robert Blackmore, D. B. Panda, MPI-LAPI: An Efficient implementation of MPI for RS/6000 SP Systems, in IEEE Transactions for Parallel and Distributed Computing, Vol. 12, Issue 10, pp. 1081-1093, Oct. 2001
10. [RDMAP] An Efficient reliable RDMA mechanism over an unreliable network transport protocol. IBM Patent submitted in April 2004.
11. [RDMAX] Brett M. Bode, Jason J. Hill, and Troy R. Benjegerdes, *"Cluster Interconnect Overview,"* USENIX 2004.
12. [REVIB] Troy R. Benjegerdes, Brett M. Bode, *"Infiniband Performance Review,"* USENIX 2004.
13. [VIBPCIEX] Jiuxing Liu, Amith Mamidala, Abhinav Vishnu, Dhabaleswar K. Panda, *"Performance Evaluation of Infiniband with PCI Express,"* Hot Interconnect 12, August, 2004
14. [VMPI] J. Liu, J. Wu, S. P. Kini, P. Wyckoff, and D. K. Panda, *"High Performance RDMA-Based MPI implementation over Infiniband,"* 17th International Conference on Supercomputing, June 2003.
15. [YNET] C. Bell, D. Bonachea, Y. Cote, J. Duell, P. Hargrove, P. Husbands, C. Iancu, M. Welcome, and K. Yelick, *"An evaluation of current high-performance networks,"* International Parallel and Distributed Processing Symposium, April 2003.

Scheduling Many-Body Short Range MD Simulations on a Cluster of Workstations and Custom VLSI Hardware

J.V. Sumanth, David R. Swanson, and Hong Jiang

Department of Computer Science and Engineering,
University of Nebraska-Lincoln, Lincoln, NE, USA
{sumanth, dswanson, jiang}@cse.unl.edu

Abstract. Molecular dynamics is a powerful technique used to obtain static or dynamic properties of liquids and solids. The sheer computational intensity of many of these simulations demands more computational power than what any uniprocessor system can provide. Fortunately, these simulations can be parallelized, allowing faster execution times on a cluster of workstations. Of late, custom VLSI chips have been designed to provide an alternative to parallel techniques. The MD-GRAPE 2 is one such solution, offering a peak performance of 64 Gflops. We evaluate the performance and cost-effectiveness of various methods used in sequential and parallel molecular dynamics and the MD-GRAPE 2. We then illustrate how MD simulations involving more complex potential functions can be scheduled on parallel machines and the MD-GRAPE 2 simultaneously.

1 Introduction

The complexity of MD simulations are overcome either by using parallel supercomputers [1] [2] or employing special purpose hardware such as the MD-GRAPE 2 [3]. Clusters of workstations are more general purpose architectures since it is possible to implement any kind of potential on them. On the other hand, the MD-GRAPE 2 board is more restrictive since it can perform simulations based only on selected 2-body potentials. We combine the advantages of both these techniques and perform simulations involving a composite 3-body potential on a combination of the two architectures.

In the next section, the computational aspects of a typical MD simulation are discussed, in section 3, we briefly explain and evaluate the performance[1] of basic MD techniques and in sections 4 and 5, parallel MD techniques and special purpose hardware like the MD-GRAPE 2 [4] to speed-up MD simulations are outlined. In section 6, we evaluate the performance of combining a cluster of workstations and special purpose hardware.

---

[*] The parallel code was executed on Prairiefire, a 256 processor cluster. Each processor was an 1800MHz Athlon MP. Each node contains 2 processors and 2GB of PC 2300 RAM. The nodes were interconnected by a 2 Gbit/sec Myrinet network. The peak performance of this cluster was 512 Gflops.

L. Bougé and V.K. Prasanna (Eds.): HiPC 2004, LNCS 3296, pp. 166–175, 2004.
© Springer-Verlag Berlin Heidelberg 2004

2   Computational Aspects of MD Simulations

2.1   Basic Equations

The computational task of a MD simulation [5] is to perform the time integration of the differential equation 1 with given initial atom positions and velocities i.e. $\{\vec{r}_i(0), \vec{v}_i(0) | i = 1, 2, \ldots, N\}$ and obtain the positions and velocities at a later time i.e. $\{\vec{r}_i(t), \vec{v}_i(t) | i = 1, 2, \ldots, N\}$.

$$\frac{\partial^2 \vec{r}_k(t)}{\partial t^2} = \vec{a}_k(t) = \sum_{i<j} \vec{r}_{ij}(t) \left(-\frac{1}{r} \frac{\partial u(r)}{\partial r} \right) \Bigg|_{r=r_{ij}(t)} \cdot (\delta_{ik} - \delta_{jk}) \tag{1}$$

where,

$$\delta_{ik} = \begin{cases} 1, i = k \\ 0, i \neq k \end{cases} \tag{2}$$

is the kronecker delta function and $u(r)$ is the potential function.

Forces are computed as the negative gradient of the potential as

$$F_k = -\frac{\partial V(\vec{r}^N)}{\partial \vec{r}_k} = -\left(\frac{\partial V}{\partial x_k}, \frac{\partial V}{\partial y_k}, \frac{\partial V}{\partial z_k} \right) \tag{3}$$

where $V(\vec{r}_k) = \sum_{i<j} u(\vec{r}_{ij})$ and $\vec{r}_{ij} = \vec{r}_i - \vec{r}_j$.

We use the velocity-verlet integration algorithm [5].

2.2   Lennard-Jones Potential - A 2-Body Potential

The Lennard-Jones(LJ) 2-body potential is commonly used to model weak, long range interactions. The LJ potential and force equations are computed by equations 4 and 5, where σ and ε are experimentally determined constants, related to atom size and potential strength, respectively.

$$V(r) = 4\varepsilon \left[\left(\frac{\sigma}{r} \right)^{12} - \left(\frac{\sigma}{r} \right)^6 \right] \tag{4}$$

$$\vec{F}_k = 24\varepsilon \left[2 \left(\frac{\sigma}{r} \right)^{12} - \left(\frac{\sigma}{r} \right)^6 \right] \left(\frac{1}{r} \right) \tag{5}$$

2.3   REBO Potential - A 3-Body Potential

The Reactive Empirical Bond Order (REBO) Potential computes the potential energy of a system of N atoms by equation 6 [6]. The functions called from this expression are detailed in table 1. r_{ij} is the scalar interatomic separation between atoms i and j. The function $f_c(r_{ij})$ restricts the range of the potential. The attractive forces are modelled by $V_A(r_{ij})$ and the repulsive forces are modelled by $V_R(r_{ij})$. The weak van der Waals forces are modelled by a LJ potential $V_{vdW}(r_{ij})$. The bond order function $\overline{B}_{ij} \equiv \frac{(B_{ij}+B_{ji})}{2}$, is a many-body term that models bond breaking and formation.

$$E_b = \sum_i^N \sum_{j>i}^N \{f_c(r_{ij}) [V_R(r_{ij}) - \overline{B}_{ij}V_A(r_{ij})] + V_{vdW}(r_{ij})\} \tag{6}$$

Table 1. REBO components and parameters used

$$V_R(r) = \frac{D_e}{S-1} exp\left[-\alpha\sqrt{2S}(r-r_e)\right]$$

$$V_A(r) = \frac{SD_e}{S-1} exp\left[-\alpha\sqrt{\frac{2}{S}}(r-r_e)\right]$$

$$B_{ij} = \left\{1 + G\sum_{k \neq i,j} f_c(r_{ik})exp\left[m(r_{ij}-r_{ik})\right]\right\}^{-n}$$

$$f_c(r) = \begin{cases} 1 & r < r_e + \delta \\ \frac{1}{2}\left\{1 + cos\left[\beta\pi(r-(r_e+\delta))\right]\right\} & r_e + \delta \leq r < r_e + \delta + \beta^{-1} \\ 0 & r_e + \delta + \beta^{-1} \leq r \end{cases}$$

$Mass_{A,B} = 14.0$ amu; $D_e^{AB} = 2.0eV$; $D_e^{AA} = D_e^{BB} = 5.0eV$; $S = 1.8$; $\alpha = 2.7\mathring{A}^{-1}$; $G = 5.0$; $m = 2.25\mathring{A}^{-1}$; $n = 0.5$

AB Model I: $r_e = 1.2\mathring{A}, \delta = 0.467\mathring{A}; \beta = 1.2\mathring{A}^{-1}$
AB Model II: $r_e = 1.0\mathring{A}, \delta = 0.400\mathring{A}; \beta = 1.0\mathring{A}^{-1}$
AB Model III: $r_e = 1.0\mathring{A}, \delta = 1.000\mathring{A}; \beta = 1.0\mathring{A}^{-1}$

3   Basic MD Techniques

The all-pairs method [5] and the link-cell method are the two most commonly used techniques. For a 2-body potential, the all-pairs method involves computing all N^2 interatomic interactions at every time-step and updating the new atomic-positions. For a 3-body potential the time complexity is $O(N^3)$.

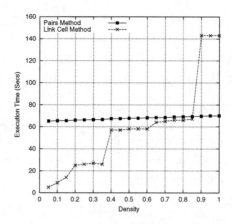

Fig. 1. Effect of density on all-pairs method and link-cell method

The link cell method [5] makes the complexity of the MD simulation nearly linear provided the system being simulated is not dense. The effect of density on the performance is illustrated in Fig. 1. The simulation space is divided into cubic cells each dimension. A link list is maintained for each cell containing the atoms in that cell. To determine the cell a particular atom belongs to, all that needs to be done is to inspect its position co-ordinates and no distances need to be computed.

4   Parallel MD Methods and Their Performance

The key to parallel MD algorithms is that the force computations and the integration algorithm can operate independently on each atom. The atom and force decomposition methods can be used for long range as well as short range potentials, while the spatial-decomposition algorithm can be used effectively only for short range potentials.

4.1   Atom Decomposition

The atom-decomposition method divides the N atoms into sets of N/P atoms and assigns each set to one of P processors. Each processor computes the forces on, and updates the positions and velocities of, its N/P atoms and hence the name atom-decomposition. At every time-step, two global communication operations [7] are required. The first is to ensure that all the processors have the updated positions of all the atoms in the system. The second is to ensure that every processor receives the forces computed on its atoms by other processors. These partial forces are summed to obtain the total force on each atom. When simulating a very dense system incorporating long-range potentials, this method is as effective as other sophisticated algorithms.

Fig. 2 illustrates the performance of this method when implemented on a cluster of workstations and on the MD-GRAPE 2. Clearly, the overall performance scales as $O(N^2)$. Fig. 3 illustrates the efficiency of the atom-decomposition

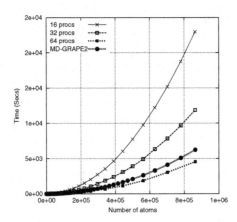

Fig. 2. Performance of atom-decomposition

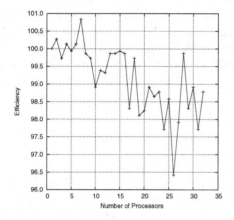

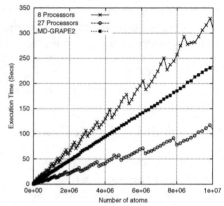

Fig. 3. Efficiency of atom-decomposition

Fig. 4. Performance of Spatial Decomposition and MD-GRAPE link cell method

method on half a million atoms with up to 32 processors. The efficiency gradually falls off from 100%. The reason for the perfect efficiency despite the presence of global communication operations twice at every time-step is the computation time is more dominant.

4.2 Spatial Decomposition

This method [2] the most effective for short-range molecular dynamics, involves dividing the simulation box into domains. Each processor is assigned a domain and computes the forces and updates the positions and velocities of the atoms in its domain. Each processor is also responsible for informing other processors if an atom leaves its domain and enters that of the other processors. This technique is most effective if the distribution of the atoms in the system is uniform. For highly localized distributions, this method can perform many times worse than the previous method. Fig.4 illustrates the performance of the spatial-decomposition algorithm on 8 and 27 processors.

Fig. 5 illustrates the efficiency of the spatial-decomposition algorithm using 4 million atoms. It can be seen that the efficiency does not scale as well as in the atom-decomposition algorithm, but the execution time of the spatial-decomposition for a given system of atoms is far superior. The reason for the fluctuations in this plot and the plot in Fig.4 is due to the fact that the actual number of processors per dimension is not always a perfect cube leading to imperfect load balancing.

5   MD-GRAPE 2 for MD Simulations

The MD-GRAPE(GRAvity PipE) 2 is a parallel pipelined special purpose hardware [3] designed to compute non-bonding forces.The bond-forces and time-integration of atom accelerations are performed on the host machine. The forces

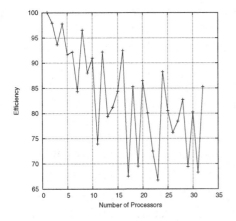

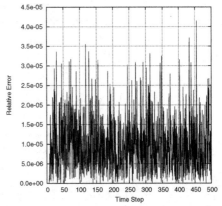

Fig. 5. Plot of efficiency of spatial-decomposition algorithm using 4M atoms

Fig. 6. Relative Error in computing Total Energy using MD-GRAPE 2

are computed using a function evaluator whose relative accuracy is of the order of 10^{-7}. The MD-GRAPE 2 chip computes forces exerted by all atoms on 24 different atoms in 6 clock cycles. In force mode, the force on atom i is computed by equation 7 and in the potential mode, the potential contribution by atom i is computed as in equation 8.

$$F_i = \sum_j b_j G(a_j r^2)(\vec{r_j} - \vec{r_i}) \tag{7}$$

$$\phi_i = \sum_j b_j G(a_j r^2) \tag{8}$$

where, $G(x)$ is any arbitrary smooth function that is evaluated by a segmented 4-th order polynomial, $\vec{r_i}$ is the vector position of the i−atom, $\vec{r_j}$ is the vector position of the j−atom and a_j and b_j are scaling factors. r^2 is the square of the distance between the two atoms.

The function $G(x)$ is evaluated using a segmented 4th order polynomial interpolation. Polynomial coefficients are stored in the RAM onboard. The range $[x_{min}, x_{max})$ is divided into 1024 segments. If $x \in [x_k, x_{k+1})$, the function $f(x)$ is approximated by equation 9.

$$f(x) \approx \sum_{i=0}^{4} c_i^{(k)} (\Delta x)^i \tag{9}$$

where, Δx is the difference of x from the center of the segment.

5.1   All Pairs Method

The input to the MD-GRAPE 2 is the atom positions and the output is the forces on the atoms or the potentials on each atom. Initially, the positions of the j-atoms (atoms that exert forces on other atoms) are sent to the MD-GRAPE 2 board and stored in RAM. The maximum number of j-atoms that can be stored at once is roughly half a million. If the simulation involves more than half a million atoms,

1: Divide simulation space into cells of side r_c.
2: Create a Link List for each cell containing the atoms belonging to the cell.
3: **for** each cell i **do**
4:     Place atoms belonging to cell i into a permuted position vector $r_p[\,]$.
5: **end for**
6: Send $r_p[\,]$ to the MD-GRAPE 2 board as j atoms.
7: **for** each cell i **do**
8:     Determine the 27 neighboring cells of i
9:     Send the base and size of the neighboring cells to the MD-GRAPE 2 board.
10:     Send the positions of the atoms in cell i as the i atoms.
11:     Compute the forces on the i atoms.
12:     Apply an inverse permutation to determine the atoms in the original force vector and add in the computed forces.
13: **end for**

Fig. 7. Link Cell Method on the MD-GRAPE 2

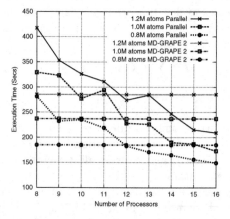

Fig. 8. Plot of number of processors vs. Execution time for MD-GRAPE 2 link-cell method and domain decomposition method

then the atom positions must be sent in batches of half a million each. Once the j-atoms are sent to the board, the i-atoms (atoms on which forces are exerted) are sent to the board. There is no effective limit on the number of i-atoms. Once the above positions are sent to the board, forces that the j-atoms exert on the i-atoms are computed. Fig. 2 illustrates the performance of this method.

5.2   Link Cell Method

While programming the MD-GRAPE 2 board to work in the link-cell method, the host machine has to perform more computation than in the all-pairs method. The host machine has to place all the atoms into cubic cells each of side at least

r_c. Finally, the host machine has to sort the atoms by their cell numbers. Further, the host machine has to remember the permutation created by the sorting and apply this to the forces returned by the MD-GRAPE 2 board. The algorithm is illustrated in Fig. 7.

Fig. 8 illustrates the number of processors it takes to equal the performance of the MD-GRAPE 2 board using the link-cell method by plotting the MD-GRAPE 2 execution time along with the execution times of the spatial-decomposition algorithm using the same simulation parameters and varying the number of processors. From the graph, we can conclude that it takes 12 processors to equal the performance of the MD-GRAPE 2 board.

Fig. 6 illustrates that the error in computing the total energy using the MD-GRAPE 2 board is 10^{-6}.

6   Scheduling MD on a Parallel Architecture and the MD-GRAPE 2 Simultaneously

It is possible to schedule a MD simulation involving a REBO potential on a cluster of workstations and the MD-GRAPE 2 board simultaneously. We choose to use the atom-decomposition method in parallel and the all-pairs method on the GRAPE board, since many applications involving non-uniform atom distributions will perform better using these techniques.

In table 1, we saw that the REBO potential comprises of three 2-body components namely V_R, V_A and V_{vdW} and a 3-body component $\overline{B}_{ij}$. There is no way of implementing any 3-body potential on the MD-GRAPE 2 architecture. Further, the custom function evaluation table does not allow for conditional statements to be placed in the function. A conditional statement is required to evaluate $f_c(r_{ij})$, which is to be multiplied into V_R and V_A. Hence, we cannot compute the V_R and V_A components on the MD-GRAPE 2 board, either.

We determined a second degree polynomial using data like that shown in 2 to predict the execution time of the all-pairs method on the MD-GRAPE 2 board. A second-degree polynomial can also be used to estimate the execution time of the atom-decomposition method. We compute the 3-body component of the REBO potential and the 2-body V_R and V_A components on a cluster of workstations using the atom-decomposition method. The V_{vdW} component is computed on the MD-GRAPE 2 board using a custom function evaluation table.

To allow the MD-GRAPE 2 board to communicate with the cluster of workstations, we implement a server on the MD-GRAPE 2 board's host machine that accepts a position vector and outputs a partial force vector and a partial potential energy. The forces and potential energies are called partial as they are only due to the V_{vdW} component and need to be added in with the forces and potential energy of the other components which are computed on the cluster.

At every time-step, before the parallel code begins computing the forces and potential energies, it sends a copy of the position vector to the MD-GRAPE 2 host machine. Now the cluster of workstations and the MD-GRAPE 2 board compute the forces and potential energies simultaneously. At the end of the

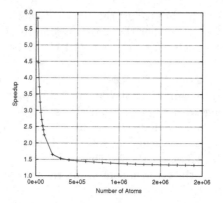

Fig. 9. Plot of additional speedup over 35 processors by including an MD-GRAPE 2 board

force computation, the MD-GRAPE 2 host machine returns the partial forces and partial potential energy. These partial components are summed to give the actual force vector and total potential energy.

Let the previously determined polynomials to estimate the execution time on the cluster using a single processor be $t_c(N)$ and on the MD-GRAPE 2 board be $t_g(N)$. Since the atom-decomposition has a nearly perfect efficiency, we can assume the execution time of this method using p processors to be $\frac{1}{p} \cdot t_c(N)$. Hence the total execution time using both the cluster and the MD-GRAPE 2 board T_{total} is given by equation 10.

$$T_{total}(N) = max \left[\frac{1}{p} \cdot t_c(N), t_g(N) \right] \tag{10}$$

To optimally schedule the MD simulation in parallel, we need to solve equation 11 for a sufficiently large N to determine an optimal number of processors p on which to run the atom-decomposition method.

$$p = \frac{t_c(N)}{t_g(N)} \tag{11}$$

Experimentally, we have determined the two polynomials and used them to estimate an optimal p to be 35 which agrees with the value of p predicted by equation 11. Fig. 9 illustrates the additional speedup obtained when combining the cluster and MD-GRAPE 2. The atom-decomposition component was run on 35 processors.

The speedup with respect to the cluster alone gradually reaches approximately 1.4 and remains constant from there on with N increasing.

7   Conclusion

We review the performance and trade-offs involved in the various techniques of sequential and parallel classical MD. We also study the performance and trade-offs involved in using the MD-GRAPE 2 board for MD simulations and determine it to equal the performance of 12 high end processors working in parallel using the link-cell method. But in the all-pairs method, the MD-GRAPE 2 performs as well as 61 processors working in parallel.

At the time of writing, the cost per cluster processor including the network was around 1500 US dollars and the MD-GRAPE 2 board around 15000 US dollars. The MD-GRAPE 2 is a very cost effective way to compute long-range potentials such as the coulomb potential since the MD-GRAPE 2 board performs as well as 61 processors. But for short-range potentials like the Lennard-Jones potential, it may be more cost-effective to use a cluster of workstations since the MD-GRAPE 2 board performs only as well as 12 processors in the link cell method.

We find that the relative accuracy of the MD-GRAPE 2 board for a typical MD simulation board is of the order of $10E - 5$ despite the function evaluator having a relative accuracy of the order of $10E - 7$. For most applications, this accuracy is acceptable.

We then show that using a combination of parallel techniques and custom hardware, we can perform MD simulations with significantly better execution times when compared to using either technique in isolation. We choose to assume a highly non-uniform distribution of atoms and hence choose to work with the atom-decomposition and all-pairs technique. We determine that using this hybrid technique, using 35 processors in parallel and a single MD-GRAPE 2 board is optimal, performing 40% better than a cluster alone.

Acknowledgements

We thank ONR, RCF and SDI (NSF 0091900) for their support of this project. We are grateful to Dr. Kenji Yasuoka, Dr. Takahiro Koishi and Mako Furukawa for the useful discussions we had about the MD-GRAPE 2 board.

References

1. Sumanth J.V, D. R. Swanson, and H. Jiang. Performance and cost effectiveness of a cluster of workstations and MD-GRAPE 2 for MD simulations. In *2nd International Symposium on Parallel and Distributed Computing , IEEE, ACM, TFC*, 2003.
2. S. Plimpton. Fast parallel algorithms for short-range molecular dynamics, Sandia Report, SAND91-1144 (1993)., 1993.
3. Tetsu Narumi. *Special-purpose computer for molecular dynamics simulations*. PhD thesis, Department of General Systems Studies, College of Arts and Sciences, University of Tokyo, 1998.
4. Tetsu Narumi, Atsushi Kawai, and Takahiro Koishi. An 8.61 tflop/s molecular dynamics simulation for nacl with a special-purpose computer: MDM. *SC2001*, 2001.
5. M. P. Allen and D. J. Tildesley. *Computer simulation of liquids*. Oxford University Press, 1987.
6. C.T.White, D. R. Swanson, and D. H. Robertson. Molecular dynamics simulations of detonations. In *Chemical Dynamics in Extreme Environments , Ed. Rainer A.*, pages 546–592. Dressler, World Scientific, 2001.
7. G.C.Fox, M.A.Johnson, G.A.Lyzenga, S.W.Otto, J.K.Salmon, and D.W.Walker. *Solving Problems on Concurrent Processors*. Prentice Hall, Englewood Cliffs, NJ, 1988.

Performance Characteristics of a Cosmology Package on Leading HPC Architectures

Jonathan Carter, Julian Borrill, and Leonid Oliker

CRD/NERSC,
Lawrence Berkeley National Laboratory,
Berkeley, CA 94720
{jtcarter, jdborrill, loliker}@lbl.gov

Abstract. The Cosmic Microwave Background (CMB) is a snapshot of the Universe some 400,000 years after the Big Bang. The pattern of anisotropies in the CMB carries a wealth of information about the fundamental parameters of cosmology. Extracting this information is an extremely computationally expensive endeavor, requiring massively parallel computers and software packages capable of exploiting them. One such package is the Microwave Anisotropy Dataset Computational Analysis Package (MADCAP) which has been used to analyze data from a number of CMB experiments. In this work, we compare MADCAP performance on the vector-based Earth Simulator (ES) and Cray X1 architectures and two leading superscalar systems, the IBM Power3 and Power4. Our results highlight the complex interplay between the problem size, architectural paradigm, interconnect, and vendor-supplied numerical libraries, while isolating the I/O filesystem as the key bottleneck across all the platforms.

1   Introduction

About 400,000 years after the Big Bang the expansion of space had cooled the Universe sufficiently for the charged electrons and protons to combine into neutral hydrogen atoms. At this point the primordial photons, which had been scattering off the free electrons, were suddenly able to propagate undisturbed through space, carrying with them a record of this moment which we call the Cosmic Microwave Background. The details of this snapshot — tiny variations in the photons' temperatures and polarizations — are an exquisitely sensitive probe of the fundamental parameters of cosmology, and measuring the detailed statistical properties of the CMB has been a high priority ever since its serendipitous discovery in 1965. The challenge lies in the fact that the continued expansion of the Universe has reduced the mean temperature of the CMB from around 3000K at last-scattering to only 3K today, and the anisotropies whose statistics we want to determine are at the 10^{-5} level in temperature, and anticipated to be at the 10^{-6}–10^{-8} level in polarization.

Realizing the extraordinary scientific potential of the CMB requires making precise measurements of the microwave sky temperature over a significant fraction of the sky at very high resolution. Such measurements are made by scanning the sky for as long as possible with a cryogenically cooled telescope and as many microwave detectors as possible. The reduction of the resulting datasets—first to a pixelized sky map, and then to an angular power spectrum—is a serious computational challenge, and one which is only getting

L. Bougé and V.K. Prasanna (Eds.): HiPC 2004, LNCS 3296, pp. 176–188, 2004.
© Springer-Verlag Berlin Heidelberg 2004

worse with increasing dataset sizes, as we try to make ever more precise measurements. It is therefore critical to choose the optimal algorithmic approach and supercomputing platform; one approach is the Microwave Anisotropy Dataset Computational Analysis Package (MADCAP) [1], which has been widely used on a variety of supercomputers.

Until recently, CMB analyses were performed almost exclusively on superscalar cache-based microprocessors, due to their generality, scalability, and cost-effectiveness. However, for many classes of applications, these architectural platforms suffer from a growing gap between their sustained performance and claimed peak capabilities. Recently, two innovative parallel-vector architectures have become available to the supercomputing community: the Japanese Earth Simulator (ES) and the Cray X1. In order to quantify what these modern vector capabilities offer to scientists that rely on numerical simulation and data analysis, it is critical to evaluate this architectural approach in the context of demanding scientific computing algorithms [2–6]. Our research team was the first international group to conduct a performance evaluation study of the Earth Simulator, currently the world's most powerful supercomputer [7]. As remote ES access is not available, the study was performed during the authors' visit to the Earth Simulator Center located in Kanazawa-ku, Yokohama, Japan in December 2003.

In this work, we compare MADCAP performance on the vector-based ES and X1 architectures and two leading superscalar systems, the IBM Power3 and Power4. Two of the architectures studied, the X1 and Power4, were only available as relatively small systems. This restricted the size of problem that we were able to use for the comparison to the correspondingly small (15,000 pixel) CMB dataset from the MAXIMA balloon-borne experiment [8] on at most 64 processors. However, MADCAP's algorithmic development has been targeted at analyzing much larger datasets on many more processors. In particular, the recent introduction of gang-parallelism has enabled the dominant component of MADCAP—a set of independent dense matrix-matrix multiplications—to achieve near perfect scaling for a 100,000 pixel dataset on 1024, 2048, 3072 and 4096 Power3 processors. The results here show that the overheads associated with implementing this optimization negate most of the performance benefits for our experimental data set. Our analysis highlights the complex interplay between the problem size, architectural paradigms, interconnect fabric, and vendor-supplied numerical libraries, while isolating the I/O filesystem as the key bottleneck across the suite of HPC platforms.

2   Architectural Platforms

Table 1 presents a summary of the architectural characteristics of the four supercomputers examined in our study. Observe that the vector systems are designed with higher absolute performance and better architectural balance than the superscalar platforms. The ES and X1 have high memory bandwidth relative to peak CPU (bytes/flop), allowing them to continuously feed the arithmetic units with operands more effectively than the superscalar systems in our study. Additionally, the custom vector interconnects show superior characteristics in terms of measured latency [9, 10], point-to-point messaging (bandwidth per CPU), and all-to-all communication (bisection bandwidth) — in both raw performance (GB/s) and as a ratio of peak processing speed (bytes/flop).

Table 1. Architectural highlights of the Power3, Power4, ES, and X1 platforms

Platform	CPU/ Node	Clock (MHz)	Peak (GF/s)	Mem BW (GB/s)	Peak bytes/flop	MPI Lat (μsec)	Netwk BW (GB/s/CPU)	Bisect BW bytes/s/flop	Network Topology
Power3	16	375	1.5	0.7	0.47	16.3	0.13	0.087	Fat-tree
Power4	32	1300	5.2	2.3	0.44	12.0	0.06	0.012	Fat-tree
ES	8	500	8.0	32.0	4.0	5.6	1.5	0.19	Crossbar
X1	4	800	12.8	34.1	2.7	7.3	6.3	0.088[1]	2D-torus

The Power3 experiments reported here were conducted on the 380-node IBM pSeries system running AIX 5.1 and located at Lawrence Berkeley National Laboratory. Each 375 MHz processor contains two floating-point units (FPUs) that can issue a multiply-add (MADD) per cycle for a peak performance of 1.5 Gflop/s. Each SMP node consists of 16 processors connected to main memory via a crossbar. Multi-node configurations are networked via the SP Switch2 (Colony) switch using an omega-type topology. The IBM distributed filesystem, GPFS, was used for all benchmarks. The filesystem was configured with 16 GPFS servers (each 16 processor SMP nodes), each with 32GB of main memory that can be used to cache files and metadata. The total size of the filesystem was 30TB, with a block size of 256KB. In this model disk I/O uses the switch fabric, sharing bandwidth with message-passing traffic.

The Power4 experiments were performed on the 27-node IBM pSeries 690 system running AIX 5.2 and operated by Oak Ridge National Laboratory (ORNL). Each 32-way SMP consists of 16 Power4 chips (organized as 4 MCMs), where a chip contains two 1.3 GHz processor cores. Each core has two FPUs capable of a fused MADD per cycle, for a peak performance of 5.2 Gflop/s. Our benchmarks were run on a system employing the Colony interconnect. As in the Power3 case, GPFS was used for all benchmarks. The filesystem was configured with 8 GPFS servers (each a 4 CPU 1.7GHz Power4+) with 32GB of main memory. These servers support two 2TB filesystems, both with a block size of 256KB. The benchmarks utilized only one of these filesystems.

The 640 node ES runs enhanced Super-UX, a 64-bit Unix-based operating system. Each SMP node contains eight processors with 16 GB of memory, and are connected through a custom single-stage crossbar. The 500 MHz ES processor contains an 8-way replicated vector pipe (vector length = 256) capable of issuing a MADD each cycle, for a peak performance of 8.0 Gflop/s per CPU. For scalar instructions, the ES contains a 500 MHz scalar processor. Like traditional vector systems, the ES vector unit is a cache-less architecture; memory latencies are masked by overlapping pipelined vector operations with memory fetches. Each group of 16 nodes has a pool of RAID disk (720GB per node) attached via fiber channel switch. The filesystem used for our experiments is NEC's Supercomputer Filesystem (SFS), with a block size of 4MB. Each node has a separate filesystem, in contrast to the other architectures studied.

All Cray X1 benchmarks were performed on a 256-MSP system (several reserved for OS services) running UNICOS/mp 2.4 and operated by ORNL. The computational core, called the single-streaming processor (SSP), contains two vector (vector length =

---

[1] X1 bisection bandwidth is based on a 2048 MSP configuration.

64) pipes running at 800 MHz, giving a 3.2 Gflop/s peak for 64-bit data. The SSP also contains a superscalar processor running at 400 MHz. The multi-streaming processor (MSP) combines four SSPs into one logical computational unit, sharing a 2MB data Ecache, that allows extremely high bandwidth (25–51 GB/s) for computations with temporal data locality. An X1 node consists of four MSPs sharing a flat memory, and large system configurations are networked through a modified 2D torus interconnect. The X1 at ORNL has four nodes available for I/O processing; each node is connected to a RAID array using fiber channel arbitrated loop protocol. Data transfer from a batch MSP must travel over the interconnect to one of the I/O nodes. The filesystem used in our study is a 4TB XFS filesystem, with a block size of 64KB.

3   MADCAP Overview

The analysis of a CMB dataset typically starts from the noise-dominated time-ordered data, constructs a pixelized map of the observed region (typically with signal-to-noise of around unity), and finally extracts the signal-dominated two-point angular correlation function, or power spectrum, of the CMB signal together with the errors on this spectral estimate (see Figure 1). The MADCAP approach is first to calculate the analytic maximum likelihood map and its residual pixel-pixel noise correlations, and then iteratively estimates the maximum likelihood power spectrum and its fisher information matrix. In this work we concentrate on the second step, which dominates the computational costs.

3.1   Methodology

The angular power spectrum is both a complete characterization of the CMB if its fluctuations are Gaussian, and is the statistic which can most readily be predicted for candidate cosmological models. MADCAP recasts the extraction of a CMB power spectrum from a map of the sky into a problem in dense linear algebra, and exploits the ScaLAPACK [11] libraries for its efficient parallel solution. The goal is to maximize the Gaussian log-

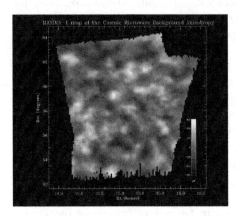
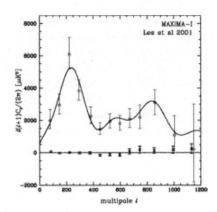

Fig. 1. The map and associated angular power spectrum of the part of the CMB sky measured by the MAXIMA experiment, as calculated by MADCAP

likelihood of the data d (a pixelized sky map of dimension $\mathcal{N}_p$) over all possible power spectrum multipole coefficients C_l where $\mathcal{L}(d|C_l) = -\frac{1}{2}\left(d^T\,D^{-1}\,d - \text{Tr}\ln D\right)$ and D is the data correlation matrix $\langle dd^T \rangle$. In an ideal experiment, there would be an independent coefficient C_l for each multipole in the angular power spectrum. However, because of finite beam size and incomplete sky coverage, the accessible multipoles are instead grouped into $\mathcal{N}_b$ bins and a single coefficient C_b is associated with each bin.

Using Newton-Raphson iteration to locate the peak of this log-likelihood requires the evaluation of its first two derivatives with respect to the binned power spectrum coefficients. First, the data correlation matrix D is constructed as the sum of the experiment-specific noise correlations N and the theory-specific signal correlations $S(C_b)$ – then a square linear system $W_b = D^{-1}\frac{\partial S}{\partial C_b}$ is solved for each of the $\mathcal{N}_b$ spectral coefficients. This is accomplished by inversion of D and direct matrix-matrix multiplication. These operations scale as $\mathcal{N}_b\,\mathcal{N}_p^3$ for a map with $\mathcal{N}_p$ pixels. This number of pixels has progressively increased from $O(10^3)$ for the initial detection of CMB anisotropies by the COBE satellite to $O(10^4) - O(10^5)$ for the ensuing ground- and balloon-based experiments, to $O(10^6) - O(10^7)$ for the current WMAP and forthcoming Planck satellite missions.

MADCAP achieves its highest performance when the data are dense on the processors so that the communication overhead is minimized. With the advent of supercomputers with thousands of processors this was becoming harder to achieve for all but the largest datasets. MADCAP has therefore recently been rewritten to exploit the parallelism inherent in performing $\mathcal{N}_b$ independent matrix-matrix multiplications. The analysis is split into two steps: first, all of the processors collectively build and invert D; then, the processors are divided into independent gangs, each of which performs a subset of the multiplications. Since the matrices involved are block-cyclically distributed over the processors, this incurs the additional overhead of redistributing the matrices between the two steps. Our results compare these single- and multi-gang approaches.

3.2   Major Components

Each iteration of MADCAP's power-spectrum extraction algorithm is divided into seven steps. Table 2 presents an overview of the resource requirements. To maximize its ability to handle large datasets and many bins, MADCAP works with at most $3\mathcal{N}_p^2$ double words of memory, which corresponds to supporting 3 matrices in memory concurrently. The out-of-core disk-based storage for the other matrices in the calculation is the only practical choice given the number of bins, but comes at the cost of heavy I/O. All matrices are block-cyclic distributed across the processors; when a matrix is stored to disk due to memory limitations, MADCAP generates one file per processor over which the matrix is distributed. This results in matrix I/O operations that are independent, however the simultaneity of multi-processor disk accesses can create contention within the I/O infrastructure, thus degrading overall performance.

(i) **dSdC** calculates each of the pixel-pixel signal correlation derivative matrices dS/dC_b. The elements of these matrices are weighted sums of the Legendre functions P_l for each multipole l in bin b, evaluated for each pixel-pixel pair. Here $\mathcal{N}_b$ distributed matrices are output to disk. This step has high computational intensity on the superscalar architectures (over 7 flops/byte on the Power3/4), and medium on the vector machines, see Section 3.3 for details.

Table 2. Computational requirements for each iteration of MADCAP's power spectrum algorithm, in terms of pixels (N_p), bins (N_b), and multipoles (N_l)

Phase	Disk	RAM	Flops
dSdC	$8N_bN_p^{\cdot}$	$16N_p^{\cdot}$	$O(N_lN_p^{\cdot})$
invD	$8N_p^{\cdot}$	$16N_p^{\cdot}$	$O(N_p^{\cdot})$
redist	—	$16N_p^{\cdot}$	—
W	$8N_p^{\cdot}$	$24N_p^{\cdot}$	$O(N_bN_p^{\cdot})$
dLdC	$8N_b$	$8N_p^{\cdot}$	$O(N_bN_p^{\cdot})$
fisher	$8N_b^{\cdot}$	$16N_p^{\cdot}$	$O(N_b^{\cdot}N_p^{\cdot})$
dC	$8N_b^{\cdot}$	$8N_b^{\cdot}$	$O(N_b^{\cdot})$
Total	$8(N_b+2)N_p^{\cdot}$	$24N_p^{\cdot}$	$O((N_b+1)N_p^{\cdot})$

(ii) **invD** calculates the full (symmetric positive definite) pixel-pixel data correlation matrix as $D = N + \sum_b C_b dS/dC_b$, explicitly inverts it using the ScaLA-PACK pdpotrf and pdpotri routines, and performs the matrix-vector multiplication $z = D^{-1}d$ (where d is the data vector, the pixelized sky map) using the ScaLAPACK pdsymv routine. This step has an intermediate computational intensity of around 1.7 flops/byte. For our benchmarks we perform only the initial iteration in which $C_b = 0$, so that the $dSdC_b$ matrices do not actually need to be read in; this routine then reads and writes one matrix.

(iii) **redist**, one by one, reads each of the dS/dC_b matrices and the D^{-1} matrix, which are block-cyclic distributed across all of the processors, block-cyclic redistributes them using the psgemr2d routine, and rewrites them over one gang's worth of processors. This step performs no calculations per se, but is a set of data gathers by the group of processors in one gang from all other processors, stressing both memory bandwidth and system interconnect.

So far, all the processors have worked on the same step of the code (*single-gang* mode), since these operations are inherently sequential. The next steps may be performed in *multi-gang* mode. Note that *redist* step re-maps the data from single- to multi-gang mode and is not performed for single-gang MADCAP calculations.

(iv) **W** performs the multiplication $W_b = D^{-1}dS/dC_b$ for a given bin using the pdgemm routine. (Although D^{-1} and dS/dC_b are both positive definite symmetric matrices, dS/dC_b may have significant block structure which we can exploit by using pdgemm to multiply just the appropriate non-zero blocks, rather than using pdsymm on the whole matrices.) Depending on the amount of data per processor, intermediate to high computational intensity is required. This step requires each gang to read D^{-1} and its subset of the dS/dC_b matrices, and write the resulting W_b matrices.

(v) **dLdC** calculates the bth element of the log-likelihood derivative vector using pdgemv, $dL/dC_b = z^T W_b d - Trace(W_b)$, since the W_b matrices are not symmetric. Each gang performs this multiplication for its subset of the bins. Matrix-vector multiply is of relatively low computational intensity, requiring an architectural balance between memory subsystem and peak arithmetic speed to achieve high performance. *dLdC* requires matrix input (W_b) but has no matrix output requirements.

(vi) **fisher** computes the bth column of the bin-bin fisher matrix by first reading and transposing W_b, followed by reading W_b' for all $b' > b$ and calculating the trace as the sum over all matrix element pair products: $F_{bb'} = Trace(W_b W_b')$. For the case where the number of bins exceeds the number of gangs, this step is load-balanced by giving each gang both a low and high numbered bin. In the case where number of gangs equal to the number of bins, there is an inherent load imbalance. The gang that processes the first bin will take on the order of $\mathcal{N}_b$ longer than the gang that processes the last bin. This step has low computational intensity, the main computational work being BLAS1; it has heavy I/O requirements, reading $\mathcal{N}_b(\mathcal{N}_b + 1)/2$ matrices.

(vii) **dC** calculates the correction $dC_b = F_{bb'}^{-1} dL/dC_b'$ using Cholesky decomposition of F and triangular solve. The number of bins is small enough that this is a simple serial code using the `dposv` and `dpotri` routines. We do not present an analysis of the dC phase, as it requires a trivial amount of runtime.

3.3   Vectorization

Most of the MADCAP routines utilize the ScaLAPACK library, making code migration a relatively simple task. Performance for both scalar and vector systems depends heavily on an efficient implementation of the vendor-supplied linear-algebra libraries. However, explicit vectorization was required for the hand-coded $dSdC$ routine. The basic structure of the $dSdC$ routine loops over all pixel pairs, calculating the value of Legendre polynomials up to some preset degree for the angular separation between these pixels. On superscalar architectures this constituted a largely insignificant amount of work, but due to the recursive computation, vectorization was prevented—resulting in significant overheads on the ES and X1. This routine was therefore rewritten so that at each iteration of the recursion a large batch of angular separations was computed in an inner loop. Compiler directives were required to ensure vectorization for both vector architectures. For our test case a speedup of approximately 10X and 30X were recorded on the ES and X1 respectively, bringing back a performance balance similar to the superscalar architectures.

3.4   Experimental Data

The data used in our experiments was collected by MAXIMA [8] (Millimeter Anisotropy eXperiment Imaging Array): a balloon-borne millimeter-wave telescope designed to measure the angular power spectrum of fluctuations in the CMB over a wide range of angular scales. MAXIMA has an unprecedented combination of sensitivity, angular resolution, and control of systematic effects. The experiment consists of a 1.3 m diameter off-axis Gregorian telescope and a receiver with a 16 element array of bolometers cooled to 100 mK. The high sensitivity of this receiver allows accurate measurements of the CMB power spectrum in a single overnight balloon flight. Each of the detectors in the array is sensitive to a single frequency band centered at 150, 240, or 410 GHz. The 150 GHz band is the most sensitive to the CMB and is close in frequency to the predicted minimum in galactic foregrounds. The higher frequency channels monitor emission from the atmosphere and galactic foregrounds such as dust.

4   Results

Our experiment used a dataset of 14996 pixels (N_p), 16 bins (N_b), and 1200 multipoles (N_l). We explore both single-gang (SG) runs, where all processors participate in each step of the calculation, and multi-gang (MG) runs, where gangs of processors carry out the N_b W, $dLdC$, and *fisher* steps concurrently. For each architecture we perform SG calculations using both 16 and 64 processors (SG processor counts are restricted to squared integers). The MG implementation depends on fast file-level synchronization across tasks. As the ES architecture provides this via MPI-IO or vendor-specific API (currently not utilized in MADCAP) and our short stay at the ES Center prevented code re-engineering, no MG experiments were performed. For all other architectures we performed MG calculations using 16, 32, and 64 processors with 4, 8, and 16 gangs of 4 processors respectively.

Tables 3-8 present a performance breakdown of MADCAP's five key steps. To distinguish between computational overhead and I/O requirements, we present two sets of runtime data (RT) for each experiments: the overall time (in wall-clock seconds); and the computational costs, without accounting for I/O operations and barrier wait times caused by I/O imbalance. For the ES, we were unable to measure the I/O and barrier times, so only the overall time is shown; we plan to gather these measurements on our next visit to the ES center. In addition, the parallel efficiencies (PE) of scaling from 16 to 64 processors are shown (P=32 is also presented for the MG case). Finally, we show the percentage of time each step accounts for in the overall MADCAP simulation (OT).

Recall that for *dSdC*, the SG and MG configurations are equivalent (exactly the same code is executed in both cases). We expect good scaling for this step, since it is embarrassingly parallel: The pixel map is divided amongst the processors and correlation matrix is computed independently, pixel by pixel. As the per-processor data density decreases at higher concurrencies, we expect to see a slight performance degradation due to loop overhead and decreased vector lengths.

Table 3 presents *dSdC* results, showing that the ES achieves the fastest raw performance, approximately 5x, 3.5x, and 2x faster than the Power3, Power4, and X1 respectively. For the Power3 and ES systems we see excellent speedup, but for the Power4 and X1 this is not the case. On further investigation, we determined that the main cause

Table 3. Performance of *dSdC* using single-gang (SG) and multi-gang (MG): in runtime seconds (RT), parallel efficiency (PE), and as percentage of MADCAP's overall runtime (OT)

dSdC		Total overhead							Computation only					
P	metric	Power3		Power4		X1		ES	Power3		Power4		X1	
		SG	MG	SG	MG	SG	MG	SG	SG	MG	SG	MG	SG	MG
16	RT	743	746	375	371	179	190	156	716	714	339	339	171	174
32	RT	—	373	—	199	—	97	—	—	359	—	172	—	93
	PE	—	100%	—	93%	—	98%	—	—	100%	—	98%	—	94%
64	RT	188	187	130	131	72	72	37	180	180	86	86	49	49
	PE	99%	100%	72%	70%	63%	66%	105%	100%	99%	98%	98%	88%	88%
	OT	7%	7%	6%	6%	8%	9%	4%	7%	7%	10%	11%	13%	12%

Table 4. Performance of *invD* using single-gang (SG) and multi-gang (MG): in runtime seconds (RT), parallel efficiency (PE), and as percentage of MADCAP's overall runtime (OT)

invD		Total overhead							Computation only					
		Power3		Power4		X1		ES	Power3		Power4		X1	
P	metric	SG	MG	SG	MG	SG	MG	SG	SG	MG	SG	MG	SG	MG
16	RT	395	389	141	141	71	78	74	378	373	130	132	69	72
32	RT	—	213	—	90	—	52	—	—	204	—	81	—	44
	PE	—	91%	—	78%	—	75%	—	—	91%	—	81%	—	82%
64	RT	131	131	77	70	81	80	22	126	125	67	62	71	75
	PE	75%	74%	46%	50%	22%	24%	84%	75%	74%	49%	53%	24%	24%
	OT	5%	4%	4%	3%	9%	10%	2%	5%	5%	6%	6%	14%	13%

of slowdown was increased I/O time for writing out the matrix. Both the Power4 and X1 systems have global filesystems with compute nodes having no direct connections to the I/O subsystem. We would expect to see some contention when increasing the number of I/O streams above a certain level, and this effect is most likely the reason for the slowdown. On removing the I/O time from the comparison the efficiency increases markedly, in line with our expectations.

For *invD*, we expect some slowdown on increasing the processor count, due the the decreased ratio of computation to communication; performance is presented in Table 4. As in *dSdC*, we note that the ES has the best parallel efficiency and absolute performance. Similarly, the Power3 has the second best efficiency, and is also the architecture where I/O scaling has the least impact. The Power4 and X1 both scale poorly, and removing I/O does not improve performance. For the Power4, we are comparing results between 16 processors (a half-populated SMP node) and 64 processors (two full SMP nodes), where the former case involves no intra-node communication and has twice the memory bandwidth per processor than in the latter case. For the X1, the scaling does not seem related to I/O, nor to the vector length (which remains similar going from 16 to 64 processors) — this issue is currently under investigation.

Table 5. Performance of *redist* using single-gang (SG) and multi-gang (MG): in runtime seconds (RT), parallel efficiency (PE), and as percentage of MADCAP's overall runtime (OT)

redist		Total overhead							Computation only					
		Power3		Power4		X1		ES	Power3		Power4		X1	
P	metric	SG	MG	SG	MG	SG	MG	SG	SG	MG	SG	MG	SG	MG
16	RT	0	461	0	280	0	254	0	0	456	0	273	0	252
32	RT	—	366	—	409	—	185	—	—	360	—	400	—	184
	PE	—	63%	—	34%	—	69%	—	—	63%	—	34%	—	69%
64	RT	0	222	0	296	0	186	0	0	217	0	289	0	185
	PE	—	52%	—	24%	—	26%	—	—	53%	—	24%	—	34%
	OT	0%	6%	0%	13%	0%	23%	0%	0%	9%	0%	23%	0%	30%

Table 6. Performance of W using single-gang (SG) and multi-gang (MG): in runtime seconds (RT), parallel efficiency (PE), and as percentage of MADCAP's overall runtime (OT)

P	metric	Power3 SG	MG	Power4 SG	MG	X1 SG	MG	ES SG	Power3 SG	MG	Power4 SG	MG	X1 SG	MG
		Total overhead							Computation only					
16	RT	8501	7847	3176	2818	958	906	1345	7928	7474	2865	2582	791	693
32	RT	—	4052	—	1811	—	537	—	—	3753	—	1531	—	413
	PE	—	97%	—	78%	—	84%	—	—	100%	—	84%	—	84%
64	RT	2204	2029	1173	930	421	280	357	2066	1873	972	727	323	207
	PE	96%	97%	68%	76%	57%	81%	94%	96%	100%	74%	89%	61%	84%
	OT	79%	56%	53%	41%	48%	34%	37%	79%	81%	89%	72%	74%	41%

The *redist* step, presented in Table 5 taxes both memory bandwidth and interconnect efficacy. In addition, any low-level support for strided gathers and the ability of the ScaLAPACK to effectively use these features will also affect performance. Observe that this operation is fastest on the X1, while slowest on the Power4. However, since the X1 is has significantly higher peak performance than the superscalar systems in our study, *redist* accounts for a significant fraction of the X1's overall time (23%), and shows little parallel efficiency (26%). The cost of this step is vital to the success or failure of the multi-gang strategy: If it is too high, we will not recoup the loss via faster matrix-matrix multiplies in step W.

Before discussing the following set of results, we note that close to perfect parallel efficiency is expected for the multi-gang experiments. In both the 16 and 64 processor experiments, the computations and matrix distributions are identical — the only difference being that in the 64-way run, four matrix-matrix multiplies are performed concurrently. Table 6 shows the performance of W on our suite of architectural platforms. Observe that, as expected, the MG strategy reduces the overhead of W in all test cases when compared with the SG approach. The Power3 performs the closest to ideal scalability, particularly when I/O times are removed. This is followed by the Power4 and the X1. Note, however, that significant variability was seen in the X1's I/O performance, sometime by up to a factor of 4x. Observe that the X1 has the faster MG raw performance, achieving a 4.8x

Table 7. Performance of *dLdC* using single-gang (SG) and multi-gang (MG): in runtime seconds (RT), parallel efficiency (PE), and as percentage of MADCAP's overall runtime (OT)

P	metric	Power3 SG	MG	Power4 SG	MG	X1 SG	MG	ES SG	Power3 SG	MG	Power4 SG	MG	X1 SG	MG
		Total overhead							Computation only					
16	RT	284	276	111	132	95	86	166	15	19	7	9	5	1
32	RT	—	141	—	81	—	51	—	—	10	—	9	—	2
	PE	—	98%	—	81%	—	85%	—	—	98%	—	48%	—	37%
64	RT	73	71	62	78	87	52	48	8	4	5	5	6	9
	PE	98%	98%	45%	42%	27%	42%	86%	50%	121%	34%	46%	20%	4%
	OT	3%	2%	3%	3%	10%	6%	5%	3%	3%	5%	6%	15%	8%

Table 8. Performance of *fisher* using single-gang (SG) and multi-gang (MG): in runtime seconds (RT), parallel efficiency (PE), and as percentage of MADCAP's overall runtime (OT)

fisher		Total overhead							Computation only					
P	metric	Power3		Power4		X1		ES	Power3		Power4		X1	
		SG	MG	SG	MG	SG	MG	SG	SG	MG	SG	MG	SG	MG
16	RT	1361	1719	936	853	291	47	1442	591	489	125	104	63	28
32	RT	—	863	—	699	—	21	—	—	223	—	61	—	15
	PE	—	100%	—	61%	—	109%	—	—	110%	—	86%	—	93%
64	RT	589	880	653	694	85	16	416	383	72	149	38	47	10
	PE	58%	49%	36%	31%	86%	73%	87%	39%	170%	21%	69%	33%	71%
	OT	21%	24%	30%	31%	10%	2%	44%	21%	16%	49%	38%	15%	2%

speedup over the Power3; however, it is also important to recall that the X1 is 8.5x faster in peak compared with the Power3.

For the *dLdC* step, shown in Table 7, we again expect high efficiency in MG mode. However, unlike step *W*, *dLdC* is dominated by I/O processing. The resulting computation-only runtimes are too small to clearly see the benefits of multi-gang parallelism, except perhaps for the Power3. The total overhead times are influenced strongly by the ability of the filesystem to handle concurrent streams of I/O without loss of efficiency.

Results for the *fisher* step are shown in Table 8. As previously mentioned in Section 3.2, the 64-way MG experiment is inherently load imbalanced, due to the equal number of gangs and bins (16); thus, we expect poor scalability when comparing with the 16 processors simulation. For the Power3 and Power4, I/O accounts for a significant fraction of the runtime, while the X1 shows negligible I/O effects. In terms of runtime, the ES achieves surprising poor performance, almost five times slower than the X1 for the SG case — we plan to investigate this issue during our next visit to the ES Center.

5   Summary and Conclusions

Table 9 summarizes our findings by putting together all of MADCAP's components. We find that the X1 has the best runtimes: 1.1x, 2.8x, and 4.4x faster than the ES, Power4, and Power3 respectively; however, it suffers the lowest parallel efficiency. The ES and Power3 demonstrate the best scalability, significantly higher than the Power4 and X1. The Power3 shows the highest percentage of peak, followed by the ES, X1, and Power4; it is also the only architecture where the multi-gang strategy pays off for this dataset.

Our in-depth analysis of the performance of the MADCAP package demonstrates the complex interplay between the architectural paradigms, interconnect technology, and I/O filesystem. These design tradeoffs play a key role in algorithmic design and system acquisitions. Preliminary multi-gang parallel optimization has previously demonstrated high sustained performance for large problem sizes at extremely high concurrencies. However, for our experimental data set and limited processor count, little or no benefit was attained on a broad spectrum of supercomputers when using this optimized approach. Additionally, all evaluated architectural platforms sustained a relatively low

Table 9. Overall *MADCAP* performance using single-gang (SG) and multi-gang (MG): in runtime seconds (RT), parallel efficiency (PE), MFlop/s per CPU (MF/s/P), and percentage of peak

MADCAP		Total overhead						Computation only					
P	metric	Power3		Power4		X1		ES	Power3		Power4		X1	
		SG	MG	SG	MG	SG	MG	SG	SG	MG	SG	MG	SG	MG
16	RT	11400	11522	4814	4646	1710	1672	3269	9688	9547	3506	3455	1171	1285
32	RT	—	6088	—	3344	—	1058	—	—	4924	—	2272	—	816
	PE	—	95%	—	69%	—	79%	—	—	97%	—	76%	—	79%
64	RT	3266	3606	2196	2264	873	823	954	2795	2492	1319	1240	567	624
	PE	87%	80%	55%	51%	49%	51%	86%	87%	96%	66%	70%	52%	52%
	MF/s/P	542	491	807	782	2029	2153	1857	634	711	1343	1429	3127	2840
	% peak	36%	33%	16%	15%	16%	17%	23%	42%	47%	26%	27%	24%	22%

overall fraction of peak, considering MADCAP's extensive use of computationally intensive dense linear-algebra calculations. Future work will examine higher-scalability simulations across a broad range of supercomputing systems, where we expect to cross the break-even point where multi-gang parallelism confers a clear performance advantage. We also plan investigate MADCAP's data transpositions and I/O transfer requirements in more detail, with the goal of reducing the impact of these overheads.

Acknowledgments

The authors would like to gratefully thank: the staff of the Earth Simulator Center, especially Dr. T. Sato, S. Kitawaki and Y. Tsuda, for their assistance during our visit; D. Parks and J. Snyder of NEC America for their help in porting applications to the ES; the MAXIMA team; and the NASA Advanced Information Systems Research Program which supported the development of MADCAP. IBM Power4 and Cray X1 access was graciously provided by ORNL; This research used resources of the NERSC at LBNL and CCS at ORNL supported by the DOE under Contract No DE-AC03-76SF00098 and DE-AC05-00OR22725 respectively. All authors from LBNL were supported by the Office of Advanced Scientific Computing Research in the DOE Office of Science under contract number DE-AC03-76SF00098.

References

1. Borrill, J.: MADCAP: The Microwave Anisotropy Dataset Computational Analysis Package. In: 5th European SGI/Cray MPP Workshop, Bologna, Italy (1999)
2. Agarwal, P.A., et al.: Cray X1 evaluation status report. In: Proc. of the 46th Cray User Group Conference, Knoxville, TN (2004)
3. Dunigan Jr., T.H., Fahey, M.R., III, J.B.W., Worley, P.H.: Early evaluation of the Cray X1. In: Proc. SC2003, Phoenix, AZ (2003)
4. Nakajima, K.: Three-level hybrid vs. flat mpi on the earth simulator: Parallel iterative solvers for finite-element method. In: Proc. 6th IMACS Symposium Iterative Methods in Scientific Computing. Volume 6. (2003)

5. Oliker, L., et al.: Evaluation of cache-based superscalar and cacheless vector architectures for scientific computations. In: Proc. SC2003, Phoenix, AZ (2003)
6. Oliker, L., et al.: A performance evaluation of the Cray X1 for scientific applications. In: 6th International Meeting on High Performance Computing for Computational Science. (2004)
7. Top500 Supercomputer Sites: (http://www.top500.org)
8. Hanany, S., et al.: Maxima-1: A measurement of the cosmic microwave background anisotropy on angular scales of $10'$–$5°$. The Astrophysical Journal **545** (2000) L5–L9
9. ORNL Cray X1 Evaluation: (http://www.csm.ornl.gov/~dunigan/cray)
10. Uehara, H., Tamura, M., Yokokawa, M.: MPI performance measurement on the Earth Simulator. Technical Report # 15, NEC Research and Development (2003/1)
11. The ScaLAPACK Project: (http://www.netlib.org/scalapack/)

A Dynamic Geometry-Based Shared Space Interaction Framework for Parallel Scientific Applications*

Li Zhang and Manish Parashar

The Applied Software Systems Laboratory, Rutgers University,
94 Brett Road, Piscataway, NJ 08854
{emmalily, parashar}@caip.rutgers.edu

Abstract. While large-scale parallel/distributed simulations are rapidly becoming critical research modalities in academia and industry, their efficient and scalable parallel implementations present many challenges. A key challenge is the dynamic and complex communication/coordination patterns required by these applications, which depend on states of the phenomenon being modeled and are determined by the specific numerical formulation, the domain decomposition and/or sub-domain refinement algorithms used, and are known only at runtime. In this paper, we present a dynamic geometry-based shared-space interaction framework for scientific applications. The framework provides the flexibility of shared-space coordination models while enabling scalable implementations. The design, prototype implementation and experimental evaluation using an adaptive multi-block oil reservoir simulation are presented.

Keywords: parallel scientific applications, dynamic geometry-based shared space, communication locality, scalability, tuple space, Hilbert space filling curve.

1 Introduction

Large-scale parallel/distributed simulations are playing an increasingly important role in science and engineering and are rapidly becoming critical research modalities in academia and industry. With the increasing scale of parallel systems and sophistication of application formulations and numerical techniques, emerging applications offer the potential for accurately simulating physically realistic models of complex phenomena and providing dramatic insights into complex applications such as interacting black holes and neutron stars, formations of galaxies, subsurface flows in oil reservoirs and aquifers, and dynamic response

---

* The research presented in this paper is supported in part by the National Science Foundation via grants numbers ACI 9984357 (CAREERS), EIA 0103674 (NGS), EIA-0120934 (ITR), ANI-0335244 (NRT), CNS-0305495 (NGS) and by DOE ASCI/ASAP via grant number 82-1052856.

L. Bougé and V.K. Prasanna (Eds.): HiPC 2004, LNCS 3296, pp. 189–199, 2004.
© Springer-Verlag Berlin Heidelberg 2004

of materials to detonation. However, the phenomena being modeled by these
applications and their implementations are inherently multi-phased, dynamic,
and heterogeneous in time, space, and state. Combined with the complexity and
scale of the underlying parallel/distributed system, efficient and scalable implementations of these applications present many challenges.

A key challenge is the dynamic and complex communication/coordination
patterns required by these applications. These communication/coordination patterns depend on states of the phenomenon being modeled and are determined
by the specific numerical formulation, domain decomposition and/or sub-domain
refinement algorithms used, and are known only at runtime. Implementing these
communication/coordination patterns using commonly used parallel programming frameworks is non-trivial. Message passing frameworks such as MPI
require matching sends and receives to be explicitly programmed for each interaction. Frameworks based on shared address spaces provide higher-level abstractions that can support dynamic interactions. However scalable implementation
of global shared address spaces remains a challenge.

Tuple spaces provide a very flexible and powerful mechanism for extremely
dynamic communication and coordination patterns [1]. In the model, processes
interact using an associative shared tuple space. A tuple is a sequence of fields,
each of which has a type and contains a value. The producer of a message
formulates the message as a tuple and places it into the tuple space. The con-
sumer(s) can associatively look up relevant tuples using pattern matching on
the tuple fields. The tuple space model provides two fundamental advantages:
simplicity and flexibility. The communicating nodes need not care about who produced or will consume a tuple. Furthermore, the communicating processes do not
have to be temporally or spatially synchronized. This decoupling feature automatically supports for dynamic communication/coordination. However, scalable
implementation of tuple spaces remains a challenge. In a pure tuple space environment, all the communication passes through a central tuple space with relatively slow associative lookup mechanisms [2], which is an inherent bottleneck
impeding scalability and efficiency.

In this paper, we present the design, implementation and evaluation of an
interaction framework for scientific applications that address the challenges outlined above. The proposed framework supports the flexibility and dynamism of
a tuple-based environment while enabling scalable implementations. It builds on
two key observations: (a) formulations of most scientific and engineering applications are based on geometric multi-dimensional domains (e.g., grid or mesh)
and (b) interactions in these applications are typically between entities that are
geometrically close in this domain (e.g., neighboring cells, nodes or elements).
Rather than implementing a general and global associative space, we enable the
dynamic creation of transient geometry-based interaction spaces, each of which
is localized to a sub-region of the overall geometric domain. The interaction
space is defined to cover a closed region of the application domain described
by an interval of coordinates in each dimension, and can be identified by any
set of coordinates contained in the region. It can then be used to share objects
between nodes corresponding to that region. Nodes do not have to know of or

synchronize with each other. The semantics of sharing is similar to traditional tuple space models.

The prototype implementation of the proposed model complements existing interaction frameworks (e.g., MPI, OpenMP) and provides a scalable geometry-based shared-space for dynamic runtime coordination and localized communication. It uses the Hilbert Space Filling Curve (SFC), a locality preserving recursive mapping from a multi-dimensional coordinate space to a 1-dimensional index space, to construct a distributed directory structure that enables efficient registration and lookup of objects in the shared-space. The prototype is evaluated using a parallel adaptive multi-block oil reservoir simulation [3]. Experimental results demonstrate system scalability, low space operation overheads, and that the performance is comparable to a pure message passing system.

The rest of the paper is organized as following. Section 2 presents a driving application and its interaction requirements. Section 3 presents the dynamic geometry-based shared space model. Section 4 presents design of the interaction framework. Section 5 presents the prototype implementation and experimental evaluation. Section 6 discusses related work and Section 7 draws a conclusion for our work.

2   A Driving Application: Parallel Adaptive Multi-block Oil Reservoir Simulation

In this section we use the parallel multi-block oil reservoir simulation as the driving application to motivate the interaction framework presented in this paper. In these simulations, the oil reservoir is discretized as a series of blocks and interfaces between blocks. The target domain consists of a coupled system of highly nonlinear transient partial differential equations. Its geometrical and geological features induce a multi-block decomposition so that each block is discretized by cell-centered finite differences on logically rectangular grids. Flux

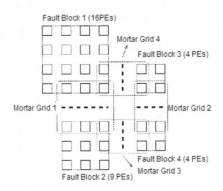

Fig. 1. 2-D view of decomposed domains with interface sharing [3]

matching conditions are imposed on the interfaces and a non-overlapping domain decomposition algorithm is exploited so that solving the interface problem only requires in-block solves and an exchange of interface values between neighboring blocks [3]. We refer to the face sharing described above as neighbor-neighbor relationship. Fig. 1 presents a 2-D view of decomposed domains in a 2-dimensional coordinate system. From the figure we can observe that communication between blocks in this particular environment is highly localized and is on the interfaces of neighboring blocks. The challenge presented in this application is that when the decomposed sub-blocks are distributed across nodes in the system, locating the processor assigned to a neighboring block is non trivial especially when dynamic load-balancing is used.

3   Dynamic Geometry-Based Shared Space(DGSS) Model

DGSS builds on the tuple spaces model. Communication entities interact with each other by sharing objects using a virtual shared space. However there is conceptual difference between the DGSS model and the general tuple space model. A general tuple space spans the entire problem domain, is accessible to all nodes in computing environments, and is associated with a generic tuple matching scheme. DGSS defines a dynamic shared space that is based on geometric regions within the application domain. It enables interactions that are localized to a geometric region by sharing objects in the DGSS based on geometry-associative semantics. DGSS supports for dynamic and flexible interaction/coordination while enabling scalable realizations based on the geometric nature of the computational domain and the local nature of communications, which are typical of most scientific applications. The geometric nature is due to the observation that formulations of scientific applications are based on a geometric descretization of the physical domain. Communication locality is due to the observation that interaction and coordination are defined by problem domains, and are typically local to sub-regions of the domain. Consequently, operations on an object shared in a DGSS only require communication within the DGSS. Coordinates from the geometric domain define the geometry-associative semantics for retrieving/storing objects from/to the spaces. DGSS is dynamic in the sense that it is created/destroyed at runtime and is constructed on top of a dynamic set of nodes that may change as the communication-integrated sub-domain changes. Through the model we automate the communication setup procedure among partitioned tasks, thus releasing programmers from the complicated and tiresome work of manually arranging all coordination patterns for every node during application development, facilitate localized communication, thus ensuring scalability of the model and benefit most scientific applications through its support for geometry-associative semantics.

4   Design of the DGSS Framework

4.1   DGSS Architecture

The DGSS architecture consists of two components: a distributed directory structure that enables locating shared spaces based on geometric relationships, and the dynamic shared-spaces that are associated with geometric regions. The distributed directory is constructed as a distributed table. The index of the table is generated by mapping the multi-dimensional problem domain to a 1-dimensional index space using Hilbert space-filling curves, which is then partitioned and distributed across the nodes in the system. A space-filling curve (SFC) is a continuous mapping from a d-dimensional space to a 1-dimensional space. The d-dimensional space is viewed as a d-dimensional cube, which is mapped onto a line such that the line passes once through each point in the volume of the cube, entering and exiting the cube only once [4]. Using this mapping, a point in the cube can be described by its spatial or d-dimensional coordinates, or by the length along the 1-dimensional index measured from one of its ends. The construction of SFCs is recursive. An important property of SFCs is locality preserving. Points that are close together in the 1-dimensional space are mapped from points that are close together in the d-dimensional space. As the index space is partitioned across the nodes in the system, each node is responsible for an interval of the index space and the region of the computational domain corresponding to this interval. The node manages information regarding the creation, deletion and memberships of any DGSS in this region. Note that the hash table will be typically sparsely populated and that the shared spaces are not uniformly distributed across the index space. As a result, load-balancing is used while partitioning the index space across the nodes. The mapping of a 2-dimensional domain using the Hilbert SFC and the creation of a distributed directory are illustrated in Fig. 2.

To create or access a shared space corresponding to a region in the computational domain, the region is first translated to interval(s) in the 1-d index space and these intervals are used to locate the processor where information about the space is maintained. The process of locating corresponding directory node is efficient, requiring only local computation. An interval tree is used to store index intervals corresponding to already created and registered shared spaces at each

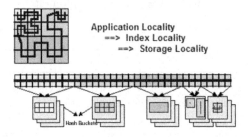

Fig. 2. Directory structure using Hilbert SFC [3]

node, and to detect geometric relationships between new and already registered spaces.

4.2   DGSS-Based Shared Storage

The DGSS-based shared storage is created to span a dynamic set of nodes based on the geometric relationship of their regions of interaction. Multiple DGSS-based shared spaces can co-exist in the system and each node can be part of more than one space. Physically, the shared space is replicated on each of the participating nodes and consistency is maintained using a combination of update propagation and multiple-versioned objects. Update propagation refers to propagating changes to a shared object to every node caching the object within the space. Since each DGSS-based shared space only spans a geometrically localized communication zone, it typically spans a small number of nodes and update propagation does not result in significant overheads. Multiple-versioned objects allows shared objects to have multiple co-existing versions, which can improve parallelism by enabling nodes to access and update different versions of the same object without synchronization.

4.3   DGSS Interface

The DGSS framework interface defines operators to allow nodes to join/leave a space and to access the space. The creation/destruction of a space is a non-collective operation and nodes can join and leave a space at runtime. A node joins a space by registering its interaction region described using geometric coordinates. If the region overlaps with an existing region, the querying node joins the existing space and the region covered by the space is redefined to be a union of the two regions, and the membership of the space is updated. If it does not overlap with any of the existing regions, a new shared space is dynamically created. A node leaves a space by de-registering itself. When the last node associated with a space de-registers, the space is destroyed.

The space access operators are similar to those provided by tuple space systems such as Linda [1], with the exception of the "eval" function, which is not supported. The space access operators are listed in Table 1. Given the geometry-based access semantic defined by DGSS, the search process for a finite region should uniquely return zero or one object from the shared space. This is unlike a generalized tuple space, which may have multiple matches.

Table 1. DGSS Interface

Interface Operators	Function Description	Linda Correspondence
get	A "get" operation moves an object from a DGSS to requesting node. Further "get" requests on the object are blocked until it is "put" back to DGSS	in
put	A "put" operation moves an object from requesting node to a DGSS.	out
read	A "read" operation copies an object to requesting node without removing it from DGSS. Multiple "read" operations can occur simultaneously.	rd
register	"register" is provided to register an object with DGSS. Based on registered geometric information, a pointer pointed at an existing DGSS or a new DGSS will be returned.	n/a

5   Implementation and Performance Evaluation of a Prototype System

5.1   Prototype Implementation

We have developed a prototype DGSS interaction framework. The implementation uses multi-threading. At application startup, a *DGSS-daemon* thread is created within the user application process on each node. This daemon handles registration requests by retrieving and updating local directory entries, and object access requests if the node is part of a DGSS. Besides the *DGSS-daemon*, the other key component is *DGSS-storage*, which is created to store shared objects. To create a DGSS at runtime, nodes in a sub-domain will register a geometric interaction region of interest with the underlying distributed directory layer. On receiving the registration request, the *DGSS-daemon* retrieves its local directory to determine whether the region of interest intersects with an existing DGSS or if a new DGSS should be created. The *DGSS-daemon* returns a pointer to the existing/new DGSS, which is then used for further space interactions. Current implementation needs to statically define a startup server, which is known a priori to all nodes in the computing environment. Table 2 lists a code sample showing how a node starts up a space daemon, joins a DGSS by registering an object and shares the object with other nodes through the DGSS within a computation loop.

Table 2. Pseudo Code Series Calling DGSS Interface

```
/*In the pseudo code series we first create a runtime DGSS by calling space-initiation function,
register an object with the DGSS and insert an object into it. Then a loop that takes the object from
the DGSS, performs local computation and updates the object, and puts the object back to the DGSS
is executed until loop condition becomes invalid. After that the object is de-registered from DGSS*/
/*Create DGSS by calling space-initiation function*/
SPACE* space=system_init(node id, shared space bootstrap server ip);
Initiate a local object;
/*Register an object with the DGSS*/
space->register(object geometric description);
/*Insert the object into the DGSS*/
space->put(object geometric description, object, object version number);
while(number of iterations<maximum number of iterations){
/*Get the object which has been updated by other nodes sharing it from the DGSS*/
space->get(object geometric description, object, object version number);
perform local computation, update object and its version number;
/*After performing local computation and updating the object, put it back to the DGSS*/
space->put(object geometric description, object, object version number);
}
/*De-register the object from the DGSS*/
space->deregister(object geometric description );
```

5.2   Experimental Evaluation

We have constructed a simulated oil reservoir environment to evaluate performance and scalability of the prototype system. In the simulation, we assumed

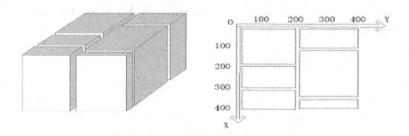

Fig. 3. 3-D/2-D view of the simulated oil reservoir experiment environment

the whole problem domain is mapped to a geometry model of a series of 6 blocks and 5 interfaces in a 3-dimensional coordinate system as shown in Fig. 3.

Four of shared interfaces have a size of 200*400 grid points and the fifth has 400*400 grid points. The data attached to each point is of type double. Blocks are decomposed at runtime into smaller blocks and assigned to nodes across a cluster of workstations. Thus possibly a node owns only a small partition of one block and its associated interface or possibly no associated interfaces, e.g., it is a central part of a block. The simulation is run on a 64-node Beowulf cluster connected by a high speed 100 MB LAN.

a) Performance Evaluation

The execution times for "register" "get" and "put" operations are measured for a range of system sizes, upto 64 nodes. Two observations result: First, as the system size increases, application grid blocks were partitioned into a larger number of partitions of smaller sizes. Consequently, the corresponding shared interfaces were also smaller in size. Second, not all nodes were assigned shared interfaces and so, not all nodes were part of a DGSS. Fig. 4 shows the execution time of each primitive. Lines with different colors represent experiments on different system sizes. The figure shows that the time for the "register" operation varies from 0.06428 second to 3.10842 seconds. This is because in the experiment all nodes that share an interface register that interface almost at the same time. Thus these registration requests nearly simultaneously reach the directory nodes that should handle them and are processed sequentially to guarantee consistency of the registration process. As a result, the execution time for "register" operation includes the time that a request blocks waiting for response, which can increase as system size increases. This potential bottleneck can be removed using dynamic load balancing. The times required for "get" and "put" operations are much smaller and comparatively stable, as seen from the figure. Further, the times required for these operations decrease as system size grows due to the two observations mentioned above. Of the three operations, "register" has the highest cost. However, note that each shared interface is registered only once in the application.

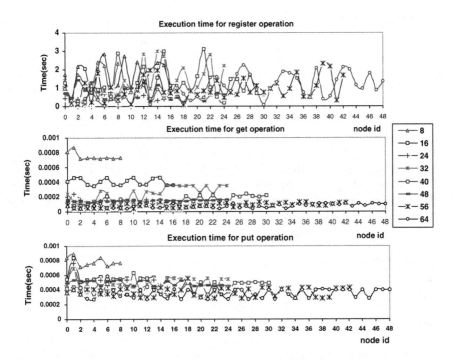

Fig. 4. Execution time for "register","get" and "put" operations

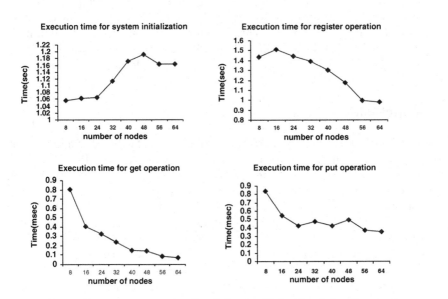

Fig. 5. Execution times for "init","register","get"and "put" operations

b) Scalability Evaluation

To evaluate system scalability, we averaged the execution time for each operation on different nodes for different system sizes. These results are plotted in Fig. 5. The figure shows that only system startup time increases as system size increases. Execution times for "register", "get" and "put" operations decrease as system size increases. The reason for increasing startup time is that the startup phase uses server-client communication to collect necessary network information from all nodes. As system size increases, the server becomes a communication bottleneck causing the startup time to increase. The other three operations scale well, which is because (a) they operate with DGSS which includes only a small number of processors, and (b) as the system size increases, the size of shared interface corresponding to a DGSS reduces and thus the "register", "get" and "put" operations operate on objects of smaller sizes.

6   Related Work

Several other projects also base their frameworks on tuple space concept such as Sun's JavaSpaces and IBM's Tspace. Sun's JavaSpaces combines Java with tuple spaces while IBM's Tspace emphasizes the integration of tuple space with database systems. These systems are quite complex and over-weighted for High Performance Computing. A lightweight Java Taskspaces framework for scientific computing on computational Grids [2] is a work similar to ours. The framework constructs lightweight shared taskspaces for node pairs that are assigned with tasks in problem sub-domains with neighbor-neighbor relationship. Because of the particular type of applications targeted, space sharing mechanism is supported by building direct communication channels between two nodes in a node pair. However the framework is limited to only one specific type of communication locality while our model addresses communication locality in general.

7   Conclusion

This paper presented a DGSS interaction framework to facilitate dynamic interaction/coordination in large-scale parallel scientific applications. The framework exploits geometric structure of the application domain and its communication locality to provide the flexibility and dynamic of shared space interaction models while enabling their scalable implementations. A DGSS is virtually shared among a group of nodes in a problem sub-domain and provides a powerful mechanism for dynamic complex interaction/coordination. The framework complements (and can co-exist with) existing interaction infrastructures (e.g. MPI). A prototype implementation and experimental evaluations were presented. Experimental results using a multi-block adaptive oil reservoir simulation show system scalability and small overheads of interface operations.

References

1. Nicholas Carriero, David Gelernter. Linda in context. Communications of the ACM, Volume 32, Issue 4, pp.444-458, April 1989, ISSN:0001-0782.
2. H. De Sterck, R.S.Markel, T.Pohl, U.Rude. A lightweight Java Taskspaces framework for scientific computing on computational grids. The eighteenth annual ACM symposium on applied computing, March 2003, Melbourne, Florida, USA. pp.1024-1030, 2003, ISBN:1-58113-624-2
3. M. Parashar and I. Yotov. An Environment for Parallel Multi-Block, Multi-Resolution Reservoir Simulations. Proceedings of the 11th International Conference on Parallel and Distributed Computing Systems (PDCS 98), Chicago, IL, International Society for Computers and their Applications (ISCA), pp.230-235, September 1998.
4. Cristina Schmidt, Manish Parashar. Flexible Information Discovery in Decentralized Distributed Systems. 12th IEEE International Symposium on High Performance Distributed Computing (HPDC'03), June 22-24,2003, Seattle, Washington, pp.226, 2003, ISBN:0-7695-1965-2.

Earthquake Engineering Problems in Parallel Neuro Environment

Sanjay Singh[1] and S V Barai[2]

[1] IBM Global Services India Pvt. Ltd., 2nd Floor, Subramanya Arcade-2,
12 Bannerghatta Main Road, Bangalore 560 029, India
sanjaysingh@in.ibm.com
[2] Department of Civil Engineering, IIT Kharagpur, Kharagpur 721 302, India
skbarai@civil.iitkgp.ernet.in
http://barai.sudhir.tripod.com

Abstract. The aim of the paper is to explore the application of Parallel Neuro Simulator for the generation of artificial earthquake. Parallel Neuro Simulator is a neural network code developed on PARAM 10000 using 'C' language and MPI library subroutines. In this study, two artificial neural network (ANN) models have been proposed to replace the auto-regressive moving average (ARMA) model. First ANN model substitutes the polynomial model that represents the relation of initial site information and coefficients of polynomial and the second ANN based model substitutes the estimated parameters of the ARMA model. Several Indian earthquake records have been used for present study on PARAM 10000. The variation in computational time with increasing number of processors has also been studied.

1 Introduction

Earthquake ground acceleration is a typical example of nonstationary process. For a nonstationary process both the frequency and the intensity changes with time. In the past a wide variety of models have been used to simulate the nonstationary processes. Earlier, most of the models were based on the spectral representation of the nonstationary process. After that, autoregressive moving average ARMA models had been applied to the analysis of earthquake acceleration time series with stationary and nonstationary characteristics [3,11]. Recently, the attempts have been made to exploit the learning capabilities of Artificial Neural Networks [5] to develop the methods for simulating ground motion [2,4,6,7,13].

Lin and Young [8] proposed the concept of random pulse train for the modeling of hypothetical ground acceleration and investigated an evolutionary Kanai-Tajimi model, a one-dimensional elastic model, and a one-dimensional Maxwell model. Artificial accelerograms were generated from these models and results were compared.

Wen and Yeh [14] and Yeh and Wen [15] proposed a stochastic process model for generating artificial earthquake ground motions using three different functions such as intensity, frequency modulation (FM) and power spectral density (PSD) function.

L. Bougé and V.K. Prasanna (Eds.): HiPC 2004, LNCS 3296, pp. 200–210, 2004.
© Springer-Verlag Berlin Heidelberg 2004

Polhemus and Cakmak [11] obtained earthquake acceleration time series by fitting stationary autoregressive moving average (ARMA) models after a variance-stabilization transformation.

Ghaboussi and Lin [4] recently proposed a new method of generating artificial earthquake accelerograms using neural networks. In the proposed method the neural networks learned the inverse mapping directly from the actual recorded earthquake accelerograms and their response spectra.

Lee and Han [7] developed efficient neural network based models for the generation of artificial earthquake and response spectra. Five neural network based models were proposed for replacing traditional processes.

In the past few years, efforts have been made to develop efficient neural network models. It is observed that while developing a reliable neural network model one has to try with the different combinations of neural network parameters and it is a time consuming process. In addition to this, as the size of data set increases the computational time also increases. In such cases, use of parallel computing environment offers an attractive solution to both the problems. Based on the brief review, following objectives are identified.

(i) Develop and explore neural network models for simulating earthquake acceleration. (ii) To expedite the process of training neural networks, perform simulation on parallel computing environment using Indian earthquake data.

2 Autoregressive Moving Average (ARMA) Models

Due to non-stationary characteristics of recorded earthquake ground motion, direct application of ARMA models to the recorded strong ground motion is not possible. A non-stationary process has a change in variance as a function of time. In constructing ARMA models with time dependent variance, it is common practice to multiply either the white noise a_t or the filtered noise $\Psi(B)a_t$ by non negative time varying function $g(t)$. In the first case, the non-stationary model Z_t takes the form

$$Z_t = \Psi(B)[g(t)a_t] \tag{1}$$

So that the variance $\sigma^2_z(t)$ of Z_t varies with time according to

$$\sigma_z^2(t) = \left[g(t)^2 + \sum_{j=1}^{\infty} \psi_j^2 g(t-j)^2 \right] \sigma_a^2 \tag{2}$$

in the latter,

$$Z_t = g(t)[\Psi(B)a_t] \tag{3}$$

With the variance changing as

$$\sigma_z^2(t) = g(t)^2 \left[1 + \sum_{j=1}^{\infty} \psi_j^2 \right] \sigma_a^2 \tag{4}$$

While the first case does allow for changes in the variance to be affected by the filter, the later case is more tractable for many analytical procedures and is the form assumed in this study. The procedure for construction such a model from an observed time series requires the following steps:

Step 1: Estimation of the variance function $\sigma_Z^2(t)$.

Step 2: Construction of a stationary series by transforming the non-stationary series according to the various functions determined in step 1.

Step 3: Estimation of an ARMA model from the stationary series obtained in step 2.

Step 4: Validation of the modeling procedure.

In the present analysis, an effort has been made to replace the ARMA model by ANN. Since earthquake is a non-stationary process, so first it is converted to a stationary process using a polynomial model because ARMA models can only be applicable to stationary time series. So from the given earthquake records polynomial model parameters and ARMA model parameters are determined. In the present approach, the obtained polynomial model parameters and ARMA model parameters have been related to the recorded site history using Neural Network. The advantage of this approach is that given the site condition, directly accelerorgams can be generated and these can be used in earthquake resistant design.

The following steps will ensure a reliable ANN model. (i) Data collection (ii) Development of neuro models (iii) Performance of neuro models.

3 Data Collection

The simulation procedure is carried out on the various earthquakes occurred in the North-Eastern part of India, Himanchal Pradesh and some parts of Uttaranchal during last two decades.The analysis is carried out for 100 strong earthquake records (provided by Department of Earthquake Engineering (DEQ), Indian Institute of Technology Roorkee) starting with North-east India earthquake on September 10, 1986 and ending with India-Burma border earthquake on May 6, 1995. Brief details are given below:

(a) September 10, 1986 earthquake in North-east India (b) April 26, 1986 earthquake in Himachal Pradesh (c) May 18, 1987 earthquake in India-Burma border (d) February 6, 1988 earthquake in India-Bangladesh border (e) August 6, 1988 earthquake in India-Burma border (f) January 10, 1990 earthquake in India-Burma border (g) October 20, 1991 earthquake at Uttarkashi (h) March 24, 1995 earthquake at Chamba, Himachal Pradesh (i) May 6, 1995 earthquake in India-Burma border.

4 Developments of Neuro Models

4.1 Neural Network Architecture

Inputs to the neural network are four site parameters namely *magnitude, focal depth, epicentral distance* and *ground motion components* of earthquake. The output parameters for both the neural network models are different and model specific. The information about the model output parameters is given in the Figure 1. The inputs and the outputs are normalized in order to have a significant number representing the vector between 0 and 1. The data samples were normalized during pre-processing and renormalized during post-processing. On the basis of experience and compromise between the accuracy and the computational time and going through different hit and trial with one and two hidden layers, it is decided to have two hidden layers. The transfer function between hidden layer and output layer has to be such that the output nodes contain the values in the range 0 to 1. So to achieve the above objective it was decided to use sigmoidal function which gives output value in the range of 0 to 1 and is also very commonly used in neural networks. Back-propagation algorithm is selected for training purpose. The various neural network parameters for both the models are summarized here in Tables 1 and 2.

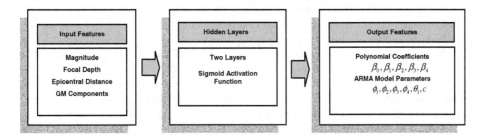

Fig. 1. Neural network architecture

Table 1. Neural network architecture data for polynomial coefficients prediction

Neural Network Architecture	Input neurons: 4
	Hidden layers: 2
	No. of neurons in 1st layer: 10
	No. of neurons in 2nd layer: 15
	Output neurons: 5
Neural Network Parameters	Learning rate parameter: 0.250
	Momentum rate parameter: 0.10
	Maximum normalized error: 0.001
	No. of processors used: 8
Input Parameter	Magnitude, Focal Depth, Epicentral Distance, G.M. Component
Output Parameter	Coefficients of Polynomial ($\beta_0, \beta_1, \beta_2, \beta_3, \beta_4$)

4.2 Implementation of Neural Network Model on PARAM 10000

Parallel execution occurs in terms of different neural network models. There are total 8 processors available on PARAM 10000. The loosely coupled shared memory multiprocessors approach is used. Processor with rank 0 acts as Master processor and rest as Slave processors (with rank 1,2, 3…7). Processor with rank 0 exchanges the messages (information) with rest of the processors and assigns various tasks to them.

Table 2. Neural network architecture data for ARMA model parameters prediction

Neural Network Architecture	Input neurons: 3
	Hidden layers: 2
	No. of neurons in 1st layer: 8
	No. of neurons in 2nd layer: 10
	Output neurons: 6
Neural Network Parameters	Learning rate parameter: 0.40
	Momentum rate parameter: 0.10
	Maximum normalized error: 0.001
	No. of processors used: 8
Input Parameter	Magnitude, Focal Depth, Epicentral Distance, G.M. Component
Output Parameter	ARMA Model Parameters ($\phi_1, \phi_2, \phi_3, \phi_4, \theta_1, c$)

Neural network models are evenly distributed among processors with rank 0 to rank 7 and each model is trained independently on assigned processor. For each neural network model input file contains training and testing examples. As soon as processor finishes the computations it sends messages that work has been done and new task is assigned to that processor. Finally, outputs generated by different processors are ensembled.

Several researchers in the field of neural networks have investigated and proposed various techniques for combining the predictions of multiple networks to produce a single prediction. The resulting model (referred to as an *ensemble*) is generally more accurate than any of the original models selected on the basis of statistical tests, tends to be more robust to overfitting phenomena, and avoids the instability problems [12].

In an ensemble, each neural network model is generally trained separately, and the predicted output of each neural network model is then combined to produce the output of the ensemble. However, combining the output of several models is more useful if there is disagreement. Obviously, the combination of identical models produces no gain. So it is desired that an ideal ensemble consist of highly correct prediction that disagree as much as possible, and that an effective combining scheme is to simply average the predictions of each neural network model. As a result, it is necessary to ensembles those models that disagree in their predictions. Neural network techniques that can be employed include different reliable methods for training with different topologies, different initial weights, different parameters, and different training sets. In case numbers of neural network models are less than the processors available then in that case work is assigned to processors equal to the number of neural network models.

For each model networks have been independently trained on N different processors and finally ensembling has been done. The output generated by different model are tested and validated to estimate the reliability and performance of the model.

For the present study, a standard Message Passing Interface (MPI), which was originally designed for writing applications and libraries for distributed memory environment, was used. More details about the MPI can be found elsewhere [9,10]. The programming and data structures details are not included due to space restrictions.

5   Performances of Neuro Models

In the present study, as discussed earlier total one hundred recorded accelerograms are selected from different recording stations. Training is carried out with ninety recorded accelerograms. Ten records are used for testing. The data set is randomized before carrying out the training. For training and testing purpose, Neuro Simulator developed on PARAM 10000 is used. The whole exercise is carried out on 8 different processors, independent to each other. For each processor the different randomized data set is used keeping the testing data same every time. Each set is trained and tested on 8 different processors and finally the average error for training sets is calculated. Performance of the ANN is shown for typical testing example of Baithalangso recording station.

5.1   Part 1: Construction of ARMA Models Using Neural Networks

Due to non-stationary nature of earthquake ground motion, Figure 2 shows an obvious change in variance as a function of time.

To estimate the variance function $\sigma_Z^2(t)$, a natural approach would be to fit a model to the observed $\{z_t^2\}$. However, the distribution of Z_t^2 is positively skewed with a variance that increases with its mean, so that some transformation must first be applied. Box and Cox [1] transformation has been applied and it is postulated that the expected values of the transformed squares accelerations follow a polynomial model $h\,(t)$ of the form with the coefficients $\beta_1, \beta_2, \beta_3 \ldots , \beta_k$.

$$h(t) = \beta_0 + \beta_1 t + \beta_2 t^2 + \ldots\ldots + \beta_k t^k \tag{5}$$

A fourth order polynomial model appears to describe the data well. Figure 2 shows the estimated model for $h\,(t)$ together with the transformed data.

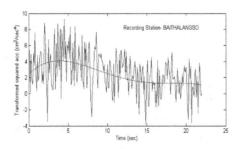

Fig. 2. Transformed square acceleration with estimated variance polynomial

(i) ANN for variance function estimation: In developing ANN for variance estimation, initial information is used as input variables and the coefficients of the polynomial model are chosen as the output parameters. Total 100 examples are taken, out of
which 90% used for training and rest 10% for testing purpose. Both input and output
variables are normalized. Here, two-layered model is considered with 10 and 15 hidden neurons in first and second layer respectively. Training is successfully completed
over about 400000 epochs. The results indicate that the predicted outputs are satisfac-
tory for all test patterns. The Table 3 shows the average error of all training patterns.
Table 4 shows numerical result for a typical testing pattern.

Table 3. Average error in training patterns

Parameters	β_0	β_1	β_2	β_3	β_4
% Average error	2.29	2.45	3.92	5.16	6.19

Table 4. Typical testing pattern result

No.	Parameter	Estimated Value	Predicted	% Error
Example 1	β_0	2.500	2.584	3.36
	β_1	0.963	1.027	6.65
	β_2	-0.18	-0.17	1.59
	β_3	0.0102	0.0114	11.76
	β_4	-0.0002	-0.00018	10.00

To obtain an estimate for $\sigma_z^2(t)$, the variance function, the reverse Box-Cox [1]
transformation is applied. The square root of $\sigma_z^2(t)$ provides an estimate of the standard deviation of acceleration versus time. The comparison of standard deviation
obtained from actual data and predicted by neural networks showed good agreement.
To obtain a series with a variance, which is constant over time, z_t is scaled by dividing it by a local estimate of its standard deviation according to

$$z_t^* = z_t / \hat{\sigma}_z(t) \tag{6}$$

The result of this transformation is illustrated in Figure 3, showing that the variance has been reasonably well stabilized.

(ii) ANN for ARMA model estimation: Having transformed the acceleration series
so that the variance is constant throughout, an ARMA model is then constructed to
represent the dynamic behavior of the transformed series. The ARMA (4, 1) appeared
to give a reasonable fit to the data.

$$Z_t = \phi_1 Z_{t-1} + \phi_2 Z_{t-2} + \phi_3 Z_{t-3} + \phi_4 Z_{t-4} + a_t - \theta_1 a_{t-1} + c \tag{7}$$

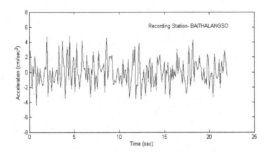

Fig. 3. Acceleration series after variance stabilization

In developing ANN for ARMA model estimation, initial information is used as input variables and the ARMA model parameters $(\phi_1, \phi_2, \phi_3, \phi_4, \theta_1, c)$ are chosen as the output variables. The same methodology is followed in this model as described for previous one. The Table 5 shows the average error for 10 randomized training patterns. Table 6 shows numerical result for typical testing pattern.

Table 5. Average error in training patterns

ARMA Parameters	ϕ_1	ϕ_2	ϕ_3	ϕ_4	θ_1	c
% Average Error	1.052	1.23	1.217	1.77	1.81	2.16

Table 6. Typical testing pattern result

No.	Parameter	Desired value	Predicted value	% Error
Example 1	ϕ_1	0.655	0.678	3.51
	ϕ_2	-0.212	-0.226	6.60
	ϕ_3	0.117	0.125	6.84
	ϕ_4	0.024	0.027	12.5
	θ_1	0.608	0.616	1.316
	c	0.019	0.017	10.53

5.2 Part 2: Simulation of Earthquake Acceleration

To produce the simulated ground motion acceleration series, stationary series are first generated from the fitted ARMA models and then multiplied by the estimated standard deviation function. Two ANN models are used to predict the ARMA model coefficients and the standard deviation function. So artificial accelerogram can directly be reproduced by the multiplication of two. Figure 4 shows the model ARMA (4, 1) being replaced by ANN.

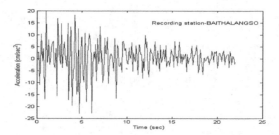

Fig. 4. Simulated Acceleration from ANN

6   Computation Time Comparisons

In parallel computing environment, there are many components of computational time. The total time is time taken in whole execution from the invocation till the termination of the program. This includes the time of communication among the processors and the execution of program on individual processor. Figure 5 shows the variation in computational time with the number of processors for both the neural networks models.

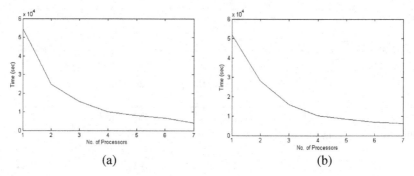

        (a)   (b)

Fig. 5. Comparison between computation time vs. number of processors. (a) ANN for variance estimation (b) ANN for ARMA parameters estimation

    Figure gives an indication that computational time decreases as number of processors increases. The performance of 5 processor-based models is near flattened because of its loosely coupled nature. The results are transferred to master node infrequently with a very small period of independent computation in between. Hence five, six and seven processor-based models do not show much improvement in terms of computation time. Other reasons could be that size of the data is too small in present context.

7   Conclusions

In the present study, neural networks simulations were carried out using Parallel Neuro Simulator developed on PARAM 10000 for various earthquake engineering problems. The problems were - replacing the ARMA model by ANN and simulating earthquake acceleration using ANN. It is found that computational time reduces significantly with increasing number of processors. It is not always true, as the number of processors increases the communication time also increases. At some point of time the communication time may take a substantial part of total computational time. In that case the total computational time keeps on increasing with the increase in processors. So one need to select the optimum number of processors for minimum time.

Acknowledgement

Authors gratefully acknowledge the facility of PARAM 10000 installed by Center for Development of Advanced Computing (C-DAC) at IIT Kharagpur and provided to them for the reported research work.

References

1. Box, G. E. P. and Cox, D. R.: An analysis of transformations', J. R. Statist. Soc. B 26, (1964) 211-252.
2. Cheng, M. and Popplewell, N.: Neural Network for Earthquake Selection in Structural Time History Analysis, Earthquake Engineering and Structural Dynamics, 23, (1994) 303-319.
3. Chang, M.K., Kwaitkowski J.W., Nau R.F., Oliver R.M. and Pister K.S. : ARMA models for earthquake ground motions, Earthquake Eng. Struct. Dyn., 10, (1982) 651-662.
4. Ghaboussi, J. and Lin, C.: New method of generating spectrum compatible accelerograms using neural networks, Earthquake Eng Struct Dyn., 27, (1998) 377–396.
5. Haykin S.: Neural networks: a comprehensive foundation. Englewood Cliffs, NJ: Prentice-Hall International, Inc; 1994.
6. Kerh, T. and Chu, D.: Neural networks approach and microtremor measurements in estimating peak ground acceleration due to strong motion, Advances in Engineering Software, 33, (2002) 733-742.
7. Lee, S. C. and Han, S.W.: Neural network based models for generating artificial earthquakes and response spectra, Computers and Structures 80, (2002) 1627-1638.
8. Lin, Y. K. and Young, Y.: Evolutionary Kanai-Tajimi earthquake models. J Eng Mech, ASCE, 113(8), (1987) 1119–37.
9. MSMPI, The Mississippi State MPI web page, http://www.erc.msstate.edu/mpi
10. Pacheco, P. S. :A User Guide to MPI, (Dept. of Mathematics, University of San Francisco, San Francisco, March 30, 1998), http://lavica.fesb.hr/~slap/upute/mpi.guide.ps
11. Polhemus, N.W. and Cakmak A. S.: Simulation of earthquake ground motions using autoregressive moving average (ARMA) models. Earthquake Engg. Struct. Dyn., 9, (1981) 343-354.
12. Roverso D.: Neural Ensembles for Event Identification, in Proceedings of Safeprocess'2000, the 4th IFAC Symposium on Fault Detection, Supervision and Safety for Technical Processes.(2000).
13. Tung, A. T. Y. , Wang, Y. Y. and Wong, F. S.: Prediction of the Spatial Distribution of the Modified Mercalli Intensity using Neural Networks, Earthquake Engineering and Structural Dynamics, 23(1), (1994) 49-62.

14. Wen, Y. K. and Yeh, C-H.: Biaxial and torsional responses of inelastic structures under random excitations, Structural Safety, 6 (2-4), (1989) 137-152.
15. Yeh, C-H and Wen, Y. K.: Modeling of nonstationary ground motion and analysis of inelastic structural response, Structural Safety, 8(1-4), (1990) 281-298.

Parallel Simulation of Carbon Nanotube Based Composites*

Jyoti Kolhe[1], Usha Chandra[2], Sirish Namilae[3],
Ashok Srinivasan[1], and Namas Chandra[3]

* Computer Science, Florida State University, Tallahassee FL 32306-4530, USA
asriniva@cs.fsu.edu
* Computer and Information Sciences, Florida A&M University,
Tallahassee FL 32307-5100
uchandra@cis.famu.edu
* Mechanical Engineering, FAMU-FSU College of Engineering, Tallahassee FL 32312
chandra@eng.fsu.edu

Abstract. Computational simulation plays a vital role in nanotechnology. Molecular dynamics (MD) is an important computational method to understand the fundamental behavior of nanoscale systems, and to transform that understanding into useful products. MD computations, however, are severely restricted by the spatial and temporal scales of simulations. This paper describes the methods used to achieve effective spatial parallelization of a MD code that is based on a multi-body bond order potential. The material system studied here is a carbon nanotube (CNT). We discuss the scientific and computational issues in the development and implementation of parallel algorithms, when the domain needs to be discretized with fine granularity. Specific issues in terms of neighbor-list computation, communication reduction, and cache awareness are delineated, with corresponding benefit in terms of speed up. Important practical problems relevant to CNT based composites are studied, and the effectiveness of various strategies reported. Our implementation achieves efficient parallelization at a finer granularity compared with published works on CNTs with complex configurations.

1 Introduction

Molecular dynamics simulation involves numerical solution of Newton's equation of motion on a set of atoms interacting through an interatomic potential. The position and velocity vectors for N atoms (6 N variables) are computed for each time step, with time steps of the order of femto seconds (10^{-15} s). Sequential computing naturally consumes significant effort even with limited system sizes, for meaningful time periods of simulation. For example, to understand a system

* We wish to acknowledge the School of Computational Science and Information Technology, Florida State University, for permitting use of their Teragold (IBM SP3) supercomputer. This work is partly funded by NSF grant # CMS-0403746.

L. Bougé and V.K. Prasanna (Eds.): HiPC 2004, LNCS 3296, pp. 211–221, 2004.
© Springer-Verlag Berlin Heidelberg 2004

behavior for even a 100nm^3 of material for one nanosecond, it requires five million iterations (time step 0.2 femto seconds) for about a million atoms. The computational effort is further increased when chemical interactions are required to be modeled accurately using complex multi-body potentials. For example, a problem involving the pull-out of a CNT with 3000 atoms for 800,000 time steps takes about two days of computational time on a 2GHz single processor PC running Linux. Each run is often just a component of a larger multi-scale simulation. It is therefore untenable to perform these simulations without the use of effective parallelization. Thus, for these classes of problems, parallelization is not a luxury but a necessity.

The specific class of problems addressed here involves the evaluation of mechanical properties, and should be clearly distinguished from a host of MD problems, where physical and thermodynamics properties, such as entropy and specific heat, are evaluated. The latter set of problems is usually formulated with periodic boundary conditions on all sides, representing samples of a microcanonical ensemble. Several independent simulations on the same set of atoms with differing starting states are then performed and the results are averaged in some sense to determine the physical properties. This leads to a trivially parallelizable scheme quite different from those cases when mechanical loads are applied, or when fracture can occur with no a-priori knowledge. Further, many mechanical property simulations involve a low number of atoms (e.g. 3000-10000) for long simulation times. In such cases, conventional spatial decomposition techniques often lead to fine granularity resulting in high communication overhead. Thus more care needs to be taken in the implementation to ensure that communication does not become a bottleneck.

The specific problem solved in this paper involves the determination of mechanical behavior of a CNT, either as a stand-alone unit or when connected to a matrix using chemical attachments; the latter case represents an important industrial application pertaining to CNT based composites. This is shown in Fig. 1. The CNT is pulled at a constant rate, while the polymer matrix remains fixed. One end of the attachment (which connects the CNT to the matrix) moves along with the CNT, while the other end is fixed to the matrix. The mechanical force vs. displacement response of these attachments then represents the mechanical behavior of the interface that we seek to understand. This figure also demonstrates that the physical domain is not homogeneous (it has an uneven distribution of attachments), and even for a case of 64 processors, the number of atoms per processor is of the order of only 50 for a 3000-atom CNT. Such fine granularity leads to communication cost being an important bottleneck to effective parallelization.

For the sake of completeness, and also to highlight the specifics of computational issues in MD, we present in Sect. 2, the problem description of the pull-out test, which should be carefully considered in the formulation and implementation of the parallelization scheme. In Sect. 3, we present the computational structure used, both in the serial and parallel versions of the code. In this section, we also present the details of the various schemes invoked to achieve parallelism.

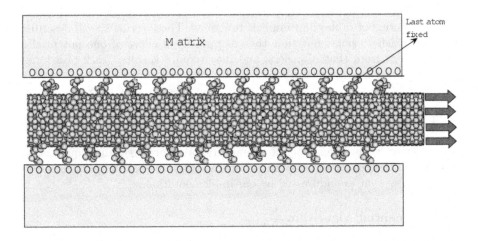

Fig. 1. Schematic of the boundary conditions applied in the pullout test simulation

In Sect. 4, we present the timing studies that illustrate the effectiveness of the various implementation schemes, and then summarize our conclusions in Sect. 5.

2   Description of the MD Problem

Molecular dynamics is a computational simulation method that determines the position r_i and velocity v_i of every atom $i = 1, ..., N$, that is contained in a computational cell subjected to external boundary conditions (force, pressure, temperature or velocities). The basic equation of motion that solves for these $6N$ variables is given by,

$$F_i = \frac{\partial v_i}{\partial t}, \quad v_i = \frac{\partial r_i}{\partial t}, \tag{1}$$

where F_i is the force on an atom i having mass m_i. In a numerical scheme, ∂t is approximated by Δt. A variety of schemes is available to numerically integrate the above equation. Irrespective of the scheme used, the incremental time step is restricted to a very low value based on physical considerations of atomic level vibrations.

The only input needed to solve the above equation is the inter-atomic potential $E = E(r)$, which is related to force in the above equation through $F = \nabla E$. The potential energy function can be as simple as a pair-potential (forces between two atoms) or as complex as a four-body potential (three neighbor interactions) with chemical bonding effect. Truly, this function is what determines the accuracy of the MD solution. The simple Lennard-Jones potential defines a potential energy function that is based only on two-body interaction. However, in the case of CNTs considered in this work, a more complex but accurate reaction bond order potential, the Brenner potential, is required. It considers two, three, and four-body interactions. These terms are very complex in nature and are expressed in terms of tables and implicit functions. These terms are all described in [1]. We wish to point out that the computational effort of one potential is quite different from that of others, and may require keeping track of a larger neighborhood (three neighbors compared to only one).

3   Computational Structure

In this section, we first describe the computational structure of the sequential algorithm and then discuss our optimizations. We then discuss the parallelization, and the issues that are addressed by our implementation.

3.1   Sequential Algorithm

The sequential algorithm for the simulation is given in Fig. 2. We examine the computational efforts required in each of the following components: (i) force, (ii) neighbor list, (iii) integration scheme, and (iv) thermostat computations.

Force and Neighbor List Computations. Brenner's potential can, in principle, require $O(N^2)$ time for two-body forces, $O(N^3)$ time for three-body forces, and $O(N^4)$ time for four-body forces. However, in practice, the number of neighbors of each atom is a constant, and so these take a constant time for each atom. This leads to O(N) time for the force computations, if we can determine all neighbors of an atom in constant time. This is typically accomplished through a neighbor list, which is an array that stores the neighbors of each atom. Though the list can theoretically change every time step, the physics of the problem dictates otherwise. Also, a neighbor list is re-computed only when some atom in the system has moved farther than a given threshold.

- Initialization
- Loop over n iterations

    • Determine the neighbor list, if necessary
    • For each atom

        * Compute two-body forces, three-body forces, and four-body forces
        * Determine new positions, velocities, and its higher derivatives

    • Compute potential and kinetic energies
    • Apply thermostat: that is, change velocities to keep temperature constant.
    • Determine if neighbor list needs to be updated

- End loop

Fig. 2. Sequential algorithm for MD simulation of CNT interface

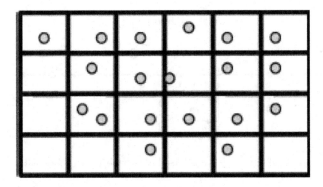

Fig. 3. Schematic of the cell-based approach to neighbor list computation

Numerical Integration Scheme and Thermostating. Once the forces are known, acceleration can be computed from Newton's law. Positions of the atoms can be predicted by integrating the acceleration twice in succession using a numerical scheme. We use a third order Nordsieck predictor-corrector scheme. Themostating is the generic name given to maintaining a constant temperature. While generic MD uses a microcanonical ensemble (Number of atoms N, volume V, and total energy E fixed), we use a canonical ensemble (fixed N, V, and temperature T). The temperature is a linear function of the kinetic energy, which is determined by summing the squares of the velocity component of each atom. In order to maintain constant temperature, we keep the kinetic energy constant by modifying the velocities of atoms.

The force and the neighbor list computations consume more than 95% of the time taken by the algorithm. For a 3000-atom simulation, the neighbor list computation accounts for around 25% of the time taken. Computations other than the neighbor list take $O(N)$ time. The neighbor list in the original Brenner algorithm is determined by a simple scheme, which compares all pairs of atoms and determines if they are sufficiently close to be considered neighbors, leading to $O(N^2)$ time complexity. So, for a larger number of atoms, the proportion of time taken by neighbor list computations is even greater. We discuss below some of the approaches that have been taken to improve the neighbor list computations.

Srivastava [2] and Caglar [3] improve the neighbor list computation by using a cell-based approach, as shown in Fig. 3. The spatial domain studied is divided into a number of cells, with the length of a cell equal to the cutoff distance. Atoms are placed into cells based on their positions in space. A linked list is used in each cell to store the atoms located in that cell. In order to determine neighbors of an atom, while creating the neighbor list, one needs to search only neighboring cells. The worst case time complexity of $O(N^2)$ occurs if all the atoms are in the same cell. This does not happen in practice, because repulsive forces prevent such an agglomeration. In practice, the cells have only a few atoms at most, and so the time complexity of computing the neighbor list is $O(N)$.

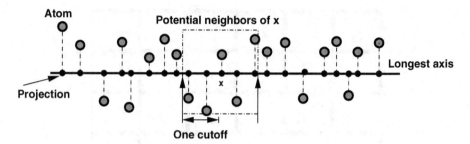

Fig. 4. Schematic of the projection-based approach to neighbor list computation

In the problem studied here, the spatial region containing the functional groups (which connect the CNT to the polymer matrix) is sparsely populated with atoms. So, memory is often wasted, with several cells being empty. We observe that the neighbor list computation is just a range-search calculation, and so the use of a data structure such as k-d tree or range tree appears promising. But we found that in practice they are not as effective as the cell-based approach, as they actually attain their worst-case time complexities. For example, we implemented a k-d tree and found it to be much worse (by a factor of four for a 1000 atom CNT) than the cell-based approach, though it was still much better than the crude algorithm (by a factor of three). There is one other disadvantage to the cell-based algorithm, apart from memory usage; it uses indirection, due to access through pointers, and is thus less efficient than an array-based representation.

We address both these problems by observing that the CNT calculations involve long, thin regions in space. So we project the coordinates of the atoms along the axis of longest dimension (z-axis), and then sort the atoms using insertion sort based on this projection, as shown in Fig. 4. Neighbors of an atom can be determined by traversing this sorted array on both sides for a length of the cut-off distance. This not only avoids waste of memory, but also facilitates an array representation. While the worst-case time complexity is $O(N^2)$, in practice it only takes $O(N)$ time, since atoms change positions infrequently, and have only a small number of neighbors falling within the cutoff. We found that the average time for sorting and computing the neighbor list is approximately half that of the cell-based approach. There is an additional advantage to sorting. In irregular computations, data for neighboring atoms may be stored at unrelated memory locations, and so cache misses occur frequently. When we sort atoms, we move the entire data of the atoms to correspond to the sorted order, and so data of neighbors tend to be close to each other. Consequently, when data for some atom is brought into cache, it is very likely that data of its neighbors too are brought into cache, thereby reducing the likelihood of additional cache misses.

3.2   Parallelization

There has been considerable effort in parallelizing MD codes for CNT applications. In general, these efforts use domain decomposition techniques and divide the overall computational effort to multiple processors. The methods differ in the manner in which the domain is divided, with the overall objective of balancing the load, while minimizing the communication cost.

Srivastava [2] used a lexical decomposition method to parallelize the computations on a shared memory machine. Here, the atoms are divided equally amongst the processors based on their indices, in an attempt to balance the load. Auxiliary arrays were used to avoid cache misses due to false sharing. (False sharing implies that data for different atoms fall in the same cache line, and update of data for one atom invalidates that of others.) However, this random assignment of atoms does not make good use of memory bandwidth on NUMA architectures. The speedup reached in these computations was around 16 on 32 processors, with over 2500 atoms per processor.

Caglar [3] performed parallelization using a cell-based decomposition. Atoms are placed in cells, as in Fig. 3, and blocks containing equal numbers of adjacent cells are then assigned to different processors. Boundary cells on a processor require data from neighboring processors, which are obtained by message passing. Caglar obtained good speedups with granularity of less than 1000 atoms per processor. However, the geometry simulated was simple, not involving functional groups. With the use of a functionalized CNT, as in our case, the disparity in the number of atoms per cell will lead to greater load imbalance in their algorithm. In addition, the speedup results reported were for simple tensile tests. Also, the performance of the baseline sequential code used to measure speedup was quite slow. For example, their baseline sequential code took about thirty times that of our baseline sequential code. Even accounting for the difference in computing platforms (Cray T3E, versus IBM SP3 for us), the difference in speeds is substantial.

Furthermore, we solved a more complex science problem. Our MD simulations are performed at a constant temperature, which implies that the thermostat needs to be applied at every time step. Maintaining a constant temperature requires computing the global energies, which normally needs a reduction in the parallel code. Reduction is expensive in terms of communication cost. The design of the parallel code aims to achieve a good speedup with a fine granularity, even for a complex geometric domain. The parallel algorithm is shown in Fig. 5.

We project the coordinates of the atoms along the longest axis, as shown in Fig. 4. The processors can be logically considered as a linear array having two neighbors each, except at the two ends. The data of the sorted atoms is divided equally into P blocks of N/P atoms each, and each block is assigned to a different processor, where P is the number of processors. Though the initial sorting takes significant time, subsequent sorting consumes much less time, since the operations are local. Each processor needs data for not only atoms that it owns, but also for a buffer region from adjacent processors. The buffer region comprises atoms that are within three times the cut-off distance (third neighbor) for the border atoms, and every processor exchanges this buffer data with its immediate neighbor.

- Initialization: Sort data for atoms; each processor decides its own domain and buffer
- Loop over n iterations

  - Send and receive buffer data
  - Determine the neighbor list for local and buffer atoms, if necessary
  - For each local atom

    * Compute two-body forces, three-body forces, and four-body forces
    * Determine new positions, velocities, and its higher derivatives

  - Compute potential and kinetic energies (may need a reduction)
  - Apply thermostat: that is, change velocities to keep temperature constant.
  - Determine if neighbor list needs to be updated

- End loop

Fig. 5. Parallel algorithm for MD simulation of CNT interface

The neighbor list is recomputed for the local and buffer atoms on a processor if any local atoms moves more than a specified threshold. Also, if a neighboring processor performs a neighbor list computation, then the atoms in the buffer regions are sorted, and the neighbor list is recomputed. This is needed to ensure that neighboring processors have a consistent view of atoms they own. When a neighbor list computation is required due to local changes, a processor informs the neighbors of this, along with the message that sends the boundary data, so that the neighbors too can recompute their lists. This scheme assumes that atoms stored in each processor (other than the end processors) span a range of at least three cut-offs, in order to ensure correctness.

Apart from the computations of neighbor list updates, our algorithm has to cater to certain global operations involving total and kinetic energy of the system. Energy needs to be computed for two different reasons: (i) Energy is a system level quantity that needs to be monitored and reported. However, this information is needed only periodically, say once in 100 iterations. Therefore the reduction is performed only occasionally, and the communication overhead is small. (ii) The kinetic energy evaluation is also needed for thermostating, and this needs to be performed every iteration, which requires a reduction. In our most optimized implementation, we changed the thermostat to use the average kinetic energy of only the local atoms. This is acceptable, since thermostating is performed just to ensure that the set of velocities for a large number of atoms has the required distribution. If there are enough atoms per processor, then this leads to a scientifically acceptable result, even though the answers may not be identical to that of the sequential code.

We discuss a few more implementation issues below.

(i) *Replicating force computations:* There are symmetries in the forces being computed that can be used to reduce the computational effort. In two-body forces, for example, the force on one atom is the negative of the force on the other atom. Thus one computation is sufficient for each pair of atoms. However, if different processors own the atoms of a pair, then the processor that performed the force computation will need to communicate the value to the neighboring processor. This applies to multi-body forces also. In order to avoid this additional communication cost, we do not use symmetries for atoms close to the boundary, but make use of these symmetries for only local atoms.

(ii) *MPI Communication mode:* Use of non-blocking communication (MPI ISend/IRecv) improved communication cost by a factor of two over the blocking (MPI Send/Recv) version.

(iii) *Packing and unpacking data:* The code is mixed Fortran/C, with all the communication being performed in C routines. Instead of sending/receiving several arrays, we send a single message by packing and unpacking several arrays in each communication step. The arrays that contain the original data are defined in Fortran code, which is inherently stored in column-major order. Packing and unpacking arrays in this order resulted in a four-fold improvement in performance for this computational component, compared with our original code, which accessed the arrays in row-major order, as is natural for C code.

(iv) *Redundant computations:* One way to save on message startup cost is to perform redundant computations. Processors do not send messages to neighbors each time step, but only every k time steps. When they do send a message, they send not only the buffer, but also a buffer k times larger. After this step, a processor can perform accurate computations for its local atoms, and for atoms in $k-1$ buffer zones. In the next time step, without any additional communication, the same processor can perform accurate computations for all local atoms and $k-2$ buffer zones. Following this reasoning, communication needs to be performed only every k iterations. If message startup cost is significant, then this can lead to significant improvements in efficiency. However, this was not effective in our implementation for the following reason. Four-body forces imply that the buffer is fairly large – three cutoffs – instead of just one with two-body forces. The extra computation is not compensated for by the decrease in message startup cost on the IBM SP3.

4   Results

Pullout simulations are performed as shown in Fig. 1, while tensile tests do not include attachments to the matrix. A displacement of 0.05 Åis applied to the atoms at one end of the CNT, in a region about 15 Åin length. A similar region in the other end is held fixed. After each displacement, the system is equilibrated for 1500 time steps. The simulations are carried out until some of the hydrocarbon chains fail. Typically, a simulation lasts for 500,000 to 800,000 time steps. They were performed on an IBM SP3 with 375 MHz IBM Power3 processors. Each node consists of four processors, and has 2GB RAM.

We first summarize improvements due to the optimizations in Fig. 6. Substantial improvements are obtained, especially for the neighbor list computations,

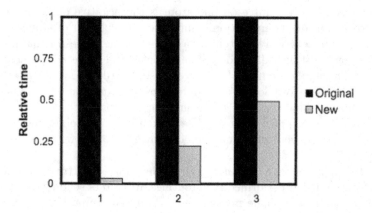

Fig. 6. Improvements in performance, relative to the original scheme. 1. Time for neighbor list computation on 3200 atoms, compared with Brenner's algorithm. The ratio (our time/Brenner's) decreases by a factor of two when the number of atoms doubles. The time relative to a cell-based algorithm is a little less than 0.5, and fairly independent of the number of atoms. 2. Time for packing and unpacking, using the cache-aware scheme, compared with a non-cache-aware one. The relative performance is independent of the number of atoms, for large numbers of atoms. 3. Time using non-blocking sends and receives, compared with blocking calls. The relative performance is independent of the number of calls and the number of atoms on an IBM SP3

which now ceases to be a bottleneck. The relative time for neighbor list computations is directly related to the number of atoms, and influences the speed of both the sequential and the parallel algorithms. The relative times for the other two techniques are independent of the number of atoms, and are relevant only to the parallel algorithm.

The speedup results are plotted below, for a 10000-atom simulation. The results shown are for tensile tests under isothermal conditions; pullout tests also show similar speedup with equivalent granularity. We can see that good efficiency is maintained up to 20 processors with a granularity of 500 atoms per processor. The efficiency reduces for a larger number of processors. An important reason for this is the replication of force computations on boundary atoms. This factor limits speedup to $P/(1 + 2bP/N)$, where $b = 50$ is the number of atoms in each buffer zone. This limits the maximum speedup on 20 processors for a 10000-atom CNT to 16.7, and on 32 processors to around 24.3. The loss in speedup beyond this is caused almost solely by the communication overhead; the load imbalance is negligible. As seen above, the code achieves a large fraction of the efficiency possible. In comparison, [2] requires a granularity greater than 2500 atoms per processor for good efficiency. Caglar [3] achieves good speedups at granularity less than 1000 atoms per processor too, but for simpler geometries and applications. Furthermore, the speedups in [3] are relative to a sequential code that are much slower.

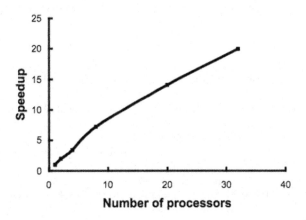

Fig. 7. Speedup results with 10000 atoms on an IBM SP3 for a tensile test

5   Conclusions

We demonstrate significant improvements in sequential and parallel performance, and also explain the relative effectiveness of different optimization techniques.

One future work is to investigate the frequency with which different (two-, three-, and four-body) force components are recomputed. For example, if four-body forces change less frequently, then they need not be updated every time step. This can improve the performance of both, the sequential and the parallel codes. Such optimizations have been used by other groups, in other applications. The speedup of the parallel code can also be improved on the IBM SP3 by avoiding replication of force computations, at the expense of increased communication cost, since communication is quite fast on this machine when the MPI implementation uses the shared memory.

References

1. Brenner, D.W.: Empirical potential for hydrocarbon for use in simulating the chemical vapor deposition of diamond films. Physical Review B **42** (1991) 9458
2. Srivastava, D., Bernard, S.T.: Molecular dynamics simulation of large-scale carbon nanotubes on a shared-memory architecture. In: Proceedings of the IEEE/ACM SC1997 Conference, IEEE Computer Society (1997)
3. Caglar, A., Griebel, M.: On the numerical simulation of Fullerene nanotubes: $C_{...........}$ and beyond! In et al., R.E., ed.: Molecular Dynamics on Parallel Computers, World Scientific (2000)

Design of a Robust Search Algorithm for P2P Networks[*]

Niloy Ganguly[1,2], Geoff Canright[3], and Andreas Deutsch[2]

[*]Indian Institute of Social Welfare and Business Management, Management House, Kolkata, India
n_ganguly@hotmail.com
[*]Center for High Performance Computing, Dresden University of Technology, Dresden, Germany
deutsch@zhr.tu-dresden.de
[*]Telenor Research and Development, 1331 Fornebu, Norway
geoffrey.canright@telenor.com

Abstract. In this paper, we report a decentralized algorithm, termed *ImmuneSearch*, for searching *p2p* networks. *ImmuneSearch* avoids query message flooding; instead it uses an immune-systems-inspired concept of proliferation and mutation for message movement. In addition, a protocol is formulated to change the neighborhoods of the peers based upon their proximity with the queried item. This results in topology evolution of the network whereby similar contents cluster together. The topology evolution help the *p2p* network to develop 'memory', as a result of which the search efficiency of the network improves as more and more individual peers perform searches. Moreover, the algorithm is extremely robust and its performance is stable even when peers are transient.

1   Introduction

Due to their flexibility, reliability and adaptivity, *p2p* solutions like Gnutella [6], Napster [5], and Freenet [2] are becoming hugely popular. However, especially due to the unreliability of the peers, the development of an efficient search algorithm poses a fundamental challenge to researchers. The algorithm for search in *p2p* networks proposed by us in this paper is termed *ImmuneSearch*. It has been inspired by the simple and well known concept of the humoral immune system where B cells undergo mutation and proliferation to generate antibodies which track the antigens (foreign objects). *ImmuneSearch* uses proliferation and mutation to spread query message packets across the network. In addition, it evolves the topology of the *p2p* network in terms of adjusting the neighborhood of the participating peers. This gives rise to a loosely structured network where the overlay topology [1] roughly corresponds to the content in the network. Consequently, the algorithm ensures better quality of service (in terms of the number

---

[*] This work was partially supported by the Future & Emerging Technologies unit of the European Commission through Project BISON (IST-2001-38923).

L. Bougé and V.K. Prasanna (Eds.): HiPC 2004, LNCS 3296, pp. 222–231, 2004.
© Springer-Verlag Berlin Heidelberg 2004

of search items found within a specified number of steps), and greater efficiency (in terms of the network congestion arising from the query packets) compared to the conventional schemes of random walk and message flooding [3]. The algorithm ensures robustness, that is, stability of performance in face of the transient nature of the network. It also guarantees autonomy to the users, who are not required to store any replicated files on their own machine.

The next section describes the *ImmuneSearch(IS)* algorithm in detail. *Section 3* details the different simulations performed based upon the algorithm *Immune-Search*. The simulation results reflect the potential of *ImmuneSearch* to perform fast and accurate search as well as point to its adaptability to continuously changing situations of *p2p* networks. The concluding section summarizes important insights from our simulation studies, and presents an outline for further work.

2   Simulation Model

In this section, we describe the framework chosen to model the *p2p* environment and the *ImmuneSearch* algorithm.

2.1   Abstraction

The factors which are important for simulating *p2p* environments are the overlay topology, the profile management of each individual peer, the nature of distribution of these profiles and the affinity measure based upon which the search algorithm is developed. Each of these factors is discussed one by one.

Topology: The overlay topology responsible for maintaining the neighborhood connections between the peers in the *p2p* network is considered to be a (100 × 100) toroidal grid where each node in the grid is conceived to be hosting a member (peer) of the *p2p* network. Each node has a fixed set of eight neighbors. A peer[1] residing in a particular node has correspondingly eight neighbors. Each peer carries two profiles - the *informational profile* and the *search profile*.

Profile: The *informational profile* (P_I) of the peer may be thought of as a description of the information stored by the user. The *search profile* (P_S) of a peer is built from the informational interest of the user. In general, the search profile may differ from the information stored on the peer. For simplicity we assume that there are 1024 coarse-grained profiles, and let each of these profiles be represented by a unique $(d =)$ 10-bit binary token. The query message packet (M) is also a 10-bit binary token. From now on we interchangeably use the term profile and token. Similarity between a profile P and a query message packet

---

[1] Although, in standard literature, 'peer' and 'node' are synonymous terms, the terms have been differentiated in the paper for ease of understanding. Node here means a position in the grid and essentially indicates a neighborhood configuration. A peer entering the network is assigned a node by the overlay management protocol. During topology evolution (discussed next) peers occupy new nodes and acquire new sets of neighbors.

(M) that is, $sim(P, M) = d - HD(P, M)$, where HD is the *Hamming distance* between P and M. The frequency of the profiles follows Zipf's distribution [4]. The ranking of tokens in terms of frequency is the same for both information and search profiles—for instance, the most popular information profile is also the most popular search profile.

On the basis of the above discussed model, we now present the search algorithm *ImmuneSearch*.

2.2   ImmuneSearch Algorithm

The *ImmuneSearch* algorithm defines the movement of the query message packets through the network and the topology evolution initiated as a result of search.

Packet Movement: The search in our *p2p* network is initiated from the user peer. The user (U in *Fig. 1*) emanates message packets (M) to its neighbors - the packets are thereby forwarded to the surroundings. The message packets (M) are formed from the search profile P_S of U. The message packets spread in the network by undergoing random walk on the grid, but when they come across a matching profile (information profile of any arbitrary peer), that is, the similarity between a message packet and informational profile is above a threshold, the message packet undergoes proliferation (as around peer A of *Fig. 1*), so as to find more peers with similar information profile around the neighborhood. Some of the proliferated packets are also mutated. (Cf. the differently colored message packets around A in *Fig. 1*). Mutation has two-fold consequences. First of all, due to mutation the chance of message packets meeting similar items increases, which in turn helps in packet proliferation. Secondly, the concept of mutation can be used in the future to help the user peer to find a wider variety of search items. The mutation results in finding new items which may not have been exactly queried. But these suggestive new results can be helpful for the user[2]. (This aspect of mutation is not dealt with in this paper).

Proliferation and mutation will initiate an intensified search around the neighbors of the peers which are already found to be similar to the queried profile. This implicitly points to the importance of topology evolution of the network, which should ensure that peers which have similar profiles come close to each other. Due to clustering, packets after proliferation will immediately begin to find peers with similar information profiles, thus enhancing the efficiency of search.

Topology Evolution: In the topology evolution scheme, the individual peers change their neighborhood configuration during search so as to place them 'closer' to U. *Fig. 1* illustrates the exact mechanism of this movement. In the figure, peer A moves (changes its neighborhood configuration) from node 7 to node 13 to place itself 'closer' to U. Correspondingly, other peers adjust their positions. We now explain the factors based upon which a peer decides to change position as well as the rules guiding the degree of change.

[.] An example of this is the Amazon.com criterion of statistical correlation: "Users who sought [query Q] have also often been interested in [...]."

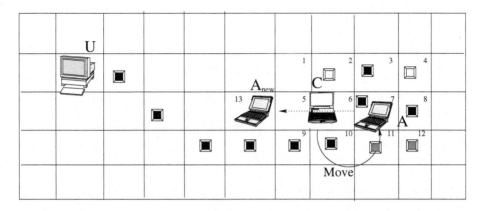

To facilitate A's movement from node 7 to 13; peers in node 6, 5, 13, respectively, move to node 7, 6, 5. For example, peer C which initially was residing in node 6 having peers at node numbers $\{1, 2, 3, 5, 7\ (= A), 9, 10, 11\}$ as neighbors, after topology evolution resides in node number 7 and has peers at node numbers $\{2, 3, 4, 6, 8, 10, 11, 12\}$ as neighbors. In effect, peer C changes three neighbors [Neighbors previously residing in $(1,9,7=(A))$ for $(4,8,12)$. (The peer which was earlier residing in 5 now resides in 6, hence there is no change in this case.)]

Fig. 1. The Search Mechanism (packets passing and topology evolution)

A peer (say A) decides to change its neighborhood configuration and places itself 'closer' to the user, when similarity between the profile of peer (P) and the message (M) sent by user peer (say U) is above a threshold level. The similarity can be of two types: (i) Similarity between *information profile* (P_I) of the peer (A) and the message packet (M), and (ii) similarity between *search profile* (P_S) of the peer (A) and the message packet (M).

The amount of movement of A towards the user peer (U) is proportional to (a) the similarity between them $(P$ and $M)$ either in terms of (i) P_I or (ii) P_S; and (b) the distance between node U and node A. (Each message packet carries the node number of the user peer U which initiated the query, so that each peer can estimate the distance between it and U). (c) The movement of the peer is also controlled by another important process which is inspired by natural immune systems - *aging*. The movement of a peer gets restricted as it ages. The age of a peer is determined in terms of the number of times it undergoes movement as a result of encountering similar message packets. That is, the longer it stays in the environment (*p2p* network), the more it is assumed that the peer has found its correct node position, and hence the less it responds to any call for change in neighbors towards any user peer U. If the search profile (P_S) of peer A matches M, but the peer (A) has performed the search operation more times than U, then there is no movement of the peer A towards user peer U. The aging concept lends stability to the system; thus a peer entering the *p2p* network, after undergoing some initial changes in neighborhood, finds its correct position.

3   Simulation Results

The experimental results, besides illustrating the efficiency of the *ImmuneSearch* algorithm, also show the self-organizing capacity of the algorithm in face of heavy unreliability of the peers participating in a *p2p* network. We also simulate experiments with a random walk, two schemes of proliferation/mutation termed *proliferation*$_1$ and *proliferation*$_2$, as well as a simple flooding technique. In *proliferation*$_1$ and *proliferation*$_2$, peers basically execute the *ImmuneSearch* algorithm without the *Topology_Evolution* step. The threshold conditions applied to the two schemes differ; this point will be discussed later.

3.1   Experimental Setup

To understand the effect of proliferation and mutation rates, experiments with different rates and different threshold values have been performed. From these, we report two cases which represent two main trends observed by us. In both cases, the proliferation and mutation rate is the same; however, the value of Threshold(Pro/Mut) differs. For the first case, Threshold(Pro/Mut) is $(d - 1)$; while in the second case, it is $(d - 2)$ (d is the length ($= 10$) of the token). *ImmuneSearch* and *proliferation*$_1$ represents the first case, while *proliferation*$_2$ represents the second case. The number of packets proliferated (NR) in the neighborhood is given by the following equation - $NR = 8 \cdot S$, where $S = \frac{sim(P_I,M)}{d}$; while the probability of each packet undergoing one bit mutation (MP) is 0.05. The threshold value required for topology evolution is set to d.

Each search is initiated by a peer residing at a randomly chosen node and the number of search items (n_s) found within 50 time steps from the commencement of the search is calculated. The search output (n_s) is averaged over 100 different searches (a generation), whereby we obtain N_s, where $N_s = \frac{\sum_{i=1}^{100} n_s}{100}$.

In the graphs (*Fig. 2 & 5*) we plot this average value N_s against generation number to illustrate the efficiency of different models. We perform two types of experiments within the above mentioned experimental setup. In the first experiment, no peers leave the system, while the second experiment represents a more transient situation where peers leave/join the network at random.

3.2   Expt. I: Search in Stable Conditions

This experiment is carried out with the assumption that no peer leaves the system. We have initiated experiments with random walk, two types of proliferation/mutation schemes (*proliferation*$_1$ and *proliferation*$_2$), limited flooding, and *ImmuneSearch*. The graph of *Fig. 2* displays the performance of the five different models. The x-axis of the graph shows the generation number while the y-axis represents the average number of search items (N_s) found in the last 100 searches. The performance comparison of the above mentioned five methods obeys fairness criteria which are discussed next.

Fairness in Power: To provide fairness in 'power', two different approaches are taken. The first approach defines fairness among *ImmuneSearch*, *proliferation*$_2$, random walk, and limited flooding, while the second approach defines fairness between *ImmuneSearch* and *proliferation₁*. The initial conditions (number of message packets) for *ImmuneSearch*, *proliferation₂*, and random walk, are chosen in a way such that the total number of packets used over 50 time steps of each individual search is roughly the same. In the case of flooding, we have allowed the process to run for x number of steps where x (< 50) steps uses the same number of packets as the aforesaid three cases used in 50 time steps. *Proliferation₁* and *ImmuneSearch* have the same threshold level for proliferation, and the same proliferation/mutation rate. But due to topology evolution, the message packets during *ImmuneSearch* pass through thickly populated areas with similar information profile and are able to produce more message packets.

Search Efficiency: In *Fig. 2*, it is seen that the number of search items (N_s) found is progressively higher in limited flooding, random walk, *proliferation₂*, *proliferation₁*, and *ImmuneSearch*, respectively. The *proliferation₁*, *proliferation₂*, random walk, and limited flooding maintain a steady search output of around 50, 40, 30, and 15 hits respectively.

In the *ImmuneSearch* algorithm, it is observed that after it starts at an initial output of around 55 items per search, it steadily increases to 80 within the 25th generation, and then maintains a steady output of about 80 per search. Therefore, the first 25 generations can be termed as 'learning' phase. During this time, similar to natural

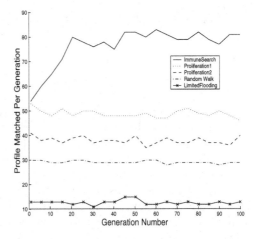

Fig. 2. Efficiency of different techniques of search namely *ImmuneSearch*, *proliferation.*, *proliferation.*, random walk and limited flooding. (Search results are averaged over 20 simulation runs)

immune systems, the *p2p* network *develops memory* by repositioning the peers. The repositioning results in clustering of peers with similar profiles which is discussed next.

Clustering Impact: The series of snapshots in *Fig. 3* demonstrates the clustering effect in the p2p network as a result of *ImmuneSearch*. Each figure represents the configuration on the 100 × 100 overlay grid taken to host the 10,000 peers. In *Fig. 3a. & b.*, each peer displays its two profiles P_I and P_S. (The big dots represent the search profile of a peer (P_S) while the small dots are the informational profile (P_I)). In *Fig. 3c*, we show only the informational profile represented as dots.

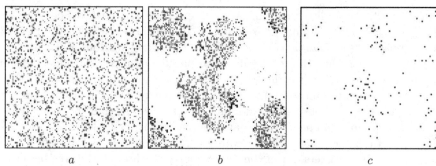

a	*b*	*c*

Clustering of information (small dots) and search (big dots) profile of peers possessing most frequent tokens at generation no. 0, 24 respectively.

Information profile of peers hosting 11th most frequent tokens at generation 100.

Fig. 3. Snapshots showing clustering of similar peers in the p2p network

The second snapshot (*Fig. 3b*) exhibits the clustering of the most frequently occurring profile at generation number 24 (the generation around which 'learning' is more or less complete) from the initial scattered setting (*Fig. 3a*). The snapshot of generation 24 shows that peers with search profile P_S intermingle with the peers with information profile P_I. That is, execution of the algorithm results in the peers with search profile (P_S) positioning themselves in 'favorable' positions whereby, when these peers initiate search, the message packets emanated by them immediately begin to find peers with similar profiles. The clustering of the peers, as seen in *Fig. 3b*, is roughly divided into three major clusters; and it is notable that the clusters are porous. The porous and separated clusters are a result of ongoing competition among the differently frequent tokens and as a result of it also less frequent tokens obtain space to form clusters. Subsequently, their search output is also enhanced. (As an example, *Fig. 3c* shows clusters of peers hosting the 11th most frequent token).

The next important aspect of the experimental results which needs to be discussed is the *cost* incurred during search.

Cost and Self-Regulation: *Cost* is defined as the number of message packets spent per successful item searched. Since, already, in order to be 'fair', we have assigned each process the same 'power', intuitively, we can say that cost will vary inversely to performance. However, while experimentally demonstrating this fact in the following paragraphs, we also illustrate another important self- organizing property displayed by the *ImmuneSearch* algorithm. In order to illustrate the self-organizing property, we next present the result of a single experiment.

The graph of *Fig. 4(a)* displays the performance analysis of the five different models based upon a single experiment. Similar to *Fig. 2*, the x-axis of the graph shows the generation number while the y-axis represents the average number of search items (N_s) found in the last 100 searches. However, unlike *Fig. 2*, in this figure we see that the search results of all the models are oscillating in proportion to their average output. The oscillations occur due to the sampling differences at

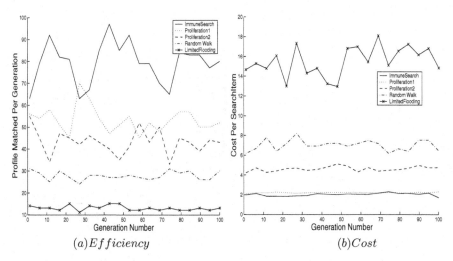

$(a) Efficiency$ $(b) Cost$

Fig. 4. Efficiency and cost of different techniques of search namely *ImmuneSearch*, *proliferation.* , *proliferation.* , random walk and limited flooding. (Search results are based on one experiment)

each generation. However, these oscillations help us to understand the cost regulation mechanism inbuilt within proliferation/mutation schemes, explained next.

Fig. 4(b) displays the cost each scheme incurs (y - axis) to generate the performance of *Fig. 4(a)*. The packets are assumed to be present in the network throughout the 50 time steps executed during a search instance. It is seen that for limited flooding, it is around 16 packets per item searched, while it is just around 2 in case of *ImmuneSearch* and $proliferation_1$.

Two interesting observations are worth mentioning here. First of all, the cost is almost the same in $proliferation_1$ and *ImmuneSearch*. It is seen that keeping the same constraints for proliferation/mutation implies almost the same cost in the network. Only the number of search items differs according to the topology of the network. The interesting feature is further demonstrated through the second observation. We see that the cost of all the three schemes using proliferation and mutation for packet movement ($proliferation_1$, $proliferation_2$, *ImmuneSearch*) almost remains constant throughout the total period, although the number of items found through search (*Fig. 4 (a)*) varies considerably during this period. The above two observations point to the fact that *proliferation and mutation have a self-regulatory quality inherent within them.* As a result of this, in the *ImmuneSearch* algorithm, the packets are not generated blindly, as with flooding, but are instead regulated by the availability of the searched item.

3.3   Expt II : Search in Transient Conditions

The robustness of the algorithm is demonstrated by the following experiment. In this experiment, 0.5%, 1%, 5%, or 50% of the population, respectively, is replenished after every generation. This mimics the transient nature of *p2p* networks where peers regularly join and leave the system.

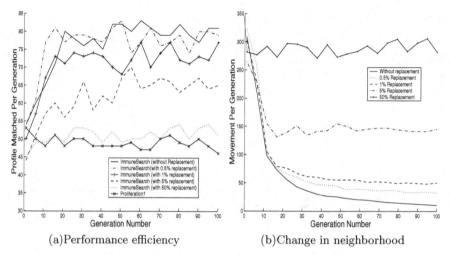

(a)Performance efficiency   (b)Change in neighborhood

Fig. 5. Performance efficiency and amount of change in neighborhood undergone by peers when there are 0%, 0.5%, 1%, 5% and 50% replacement of peers after each generation respectively. (Results are average of 20 simulation runs)

Fig. 5(a) showing the performance of the *ImmuneSearch*, under various degrees of replacement, illustrate two important results. (i) First of all, even in face of dynamic change, the *ImmuneSearch* algorithm 'learns', in that, after some initial generations, the efficiency increases. The amount of increase in search efficiency, are generally dependent on the percentage of replacement the *p2p* network undergoes after each generation. We find that the performance of *proliferation*$_1$ (which has been plotted as a reference point) is roughly the same when there is 50% replacement. But replacement of 50% of all peers within 100 searches is likely far higher than any realistic turnover rate. (ii) However, the more important point to be noted is that at 0.5% replacement, we observe that the performance is in fact at par and sometimes slightly better than *ImmuneSearch* without replacement! The result establishes one important positive point about the algorithm - that is, a little transience is helpful, rather than detrimental, to the performance of the algorithm. This happens because the problem of developing a search algorithm is in fact a multi-objective optimization problem, and due to the enormous complexity, we are obtaining a 'good', however not optimal solution. So a little change in peers is facilitating the system to move quickly towards a better solution.

We next discuss the extent of neighborhood changes the peers have to undergo.

Change in Neighborhood: *Fig. 5(b)* shows the amount of change in neighborhood of the constituent peers (*y*-axis) . One unit of movement implies the movement of a peer from one node to a neighboring node. We can also refer to a unit of movement as a neighborhood change.

It is seen that in all the cases (0%, 0.5%, 1%, 5% & 50%), initially, there are around 300 neighborhood changes, per search. That is, when a search goes on for

50 time steps, at each time step there are around 6 peers changing their neighborhood (or 1 peer changing neighborhood 6 times). However, the movement of the peers in all the cases, except the first case, gradually reaches a steady level. In the first case, plotting the movement over many generations produces a monotonically decreasing curve, which implies that the system will eventually reach a state in which there will be no further movement. In the other cases, the amount of movement at steady state increases monotonically (but sublinearly) with the percentage of peers leaving the network. For example, in cases of 0.5% and 1% replacement, the change in neighborhood becomes quite insignificant and are around 30 and 50 neighborhood changes per search respectively.

The monotonically decreasing graph for the case where no peers leave the system is the result of the concept of aging, whereby, after some time, the system stops rearranging the peers. The steady level of movement maintained by other cases can be directly attributed to the dynamic nature of the system. As new peers are joining the system, the system as a whole tries to adjust to the changing conditions. This shows that *the system on the whole tries to learn*, while at the same time does not unnecessarily undergo neighborhood changes.

4   Conclusion

This paper has presented a search algorithm which derives its inspiration from natural immune systems and whose underlying guiding rules are generally also very simple. We find that as a result of the algorithm, the *p2p* network 'learns' and subsequently develops memory, whereby the search efficiency improves dramatically after some initial learning/training phase. The system also gains capability to decide upon the number of message packets to be generated during the search for a particular item, according to the availability of that item. Thus the cost of search remains virtually constant irrespective of the item's availability and the nature of the topology. The system also can withstand the transient nature of the peers. The basic strengths displayed by the *ImmuneSearch* algorithm need to be further explored and developed, by applying it in more realistic circumstances in the near future.

References

1. G Canright, A Deutsch, M Jelasity, and F Ducatelle. Structures and functions of dynamic networks. Bison Deliverable, www.cs.unibo.it/bison/deliverables/D01.pdf, 2003.
2. Freenet. http://freenet.sourceforge.net/.
3. Q. Lv, P. Cao, E. Cohen, and S. Shenker. Search and Replication in Unstructured Peer-to-Peer Networks. In *Proceedings of the 16th ACM International Conference on Supercomputing*, June 2002.
4. G. K. Zipf. *Psycho-Biology of Languages*. Houghton-Mifflin, 1935.
5. Napster. *http://www.napster.com*, 2000.
6. Gnutella. *http://www.gnutellanews.com*, 2001.

Efficient Immunization Algorithm for
Peer-to-Peer Networks*

Hao Chen, Hai Jin, Jianhua Sun, and Zongfen Han

Cluster and Grid Computing Lab,
Huazhong University of Science and Technology, Wuhan, 430074, China
{haochen, hjin, jhsun, zfhan}@hust.edu.cn

Abstract. In this paper, we present a detail study about the immunization of
viruses in Peer-to-Peer networks exhibiting power-law degree distributions. By
comparing two different immunization strategies (randomized and degree-based),
we conclude that it is efficient to immunize the highly connected nodes in order to
eradicate viruses from the network. Furthermore, we propose an efficient updating
algorithm for global virus database according to the degree-based immunization
strategy.

1   Introduction

In the past several years, Peer-to-Peer (P2P) networks have emerged as effective ways
for communication and cooperation among geographically distributed computers. P2P
systems depend on voluntary participation of peers without any centralized control and
hierarchial organization, from which the underlying infrastructure is constructed. In
P2P networks (e.g. SETI@Home, Freenet, Gnutella, Napster), through cooperation of
all peers, tremendous computation and storage resources unoccupied on individual computers can be utilized to accomplish some kinds of tasks jointly. Individual computers
communicate with each other directly without a central point of coordination. P2P systems are often built at the application level and use their own communication protocols to
form a virtual network over the underlying physical network. The topology of the virtual
network shares some common properties of complex networks in other disciplines of
science, and has a significant impact on performance, scalability and robustness of P2P
systems.

Recently, a large proportion of research effort has been devoted to the study and
modeling of a wide range of natural systems that can be regarded as networks, focusing
on large scale statistical properties of networks other than single small networks. Some
reviews on complex networks can be found in [11]. From biology to social science to
computer science, systems such as the Internet [8], the World-Wide-Web [5], social communities, food web and biological networks can be represented as graphs, where nodes
represent individuals and links represent interactions among them. Despite this simple
definition, these networks often exhibit high degree of complexity due to the wiring

---

* This paper is supported by National Science Foundation of China under grant 60125208 and
60273076.

L. Bougé and V.K. Prasanna (Eds.): HiPC 2004, LNCS 3296, pp. 232–241, 2004.
© Springer-Verlag Berlin Heidelberg 2004

entanglement during their growth. Researches on these networks have revealed some commonalities. Specially, many of these networks have complex topological properties and dynamical features that can not be explained by the classical graph model of random networks, the Erdos-Renyi model [4].

These diverse networks can be characterized more accurately by small world phenomenon and power-law degree distributions. The first demonstration of small world effect was introduced by the classic experiment by Stanley Milgram [10], which showed that people could find a short sequence of acquaintances in order to deliver a message to each other, and are often referred to as "six degrees of separation". In networks with power-law degree distributions, the probability distribution of the degree of the node is approximately proportional to $k^{-\gamma}$, where k is the node degree and γ is a constant. Such networks are often called scale-free networks. However, the degree distribution in random graph networks follows a Poisson distribution.

One important characteristic of P2P networks, like some other complex networks, is that they often show high degree of tolerance against random failures, while they are vulnerable under intentional attacks [6]. Such property has motivated us to carry out a study about the virus spreading phenomena and hacker behaviors in P2P networks from a topological point of view. In our study, we choose Gnutella as our testbed, due to its large user community and open architecture. Some previous works have been done on the measurement and analysis of Gnutella network [12, 14, 9], such as bottleneck bandwidth [14] and search algorithms [1]. But few work has been devoted to investigate the behaviors of virus spreading and intrusion from a topological view. The main contributions of this paper are two-fold: firstly, an optimal immunization strategy is given; secondly, we propose an efficient information updating algorithm for P2P networks based on the immunization strategy.

The rest of this paper is organized as follows. Section 2 describes the immunization model of P2P networks. In section 3, we propose an information updating algorithm for P2P networks. In section 4, we give our conclusions and point out some directions for future work.

2   Immunization Model of P2P Networks

Some previous works [1, 6] indicate that P2P networks often display small-world phenomenon and power-law degree distributions. Many topological properties of such power-law networks are much different from those of networks modelled by random graphs [4]. One of the most important property is the network resilience, which measures the network robustness and weakness by random removal or targeted deletion of vertices in the network. In this section, we first review some materials about the resilience of P2P network, from which implications for designing immunization strategy of P2P network will be introduced.

2.1   Network Resilience and Its Implication for Immunization of P2P Networks

There are a variety of different strategies of removing nodes form a network, and different networks may show varying degrees of resilience to these strategies. For example, one

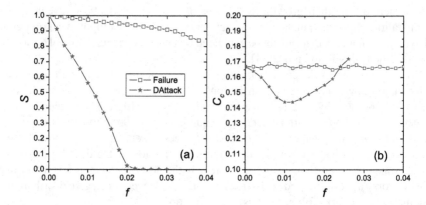

Fig. 1. Results for random failures (open square) and degree-based (star) attacks of nodes measured by the relative size of largest cluster S, the average closeness centrality C_c as functions of the fraction of removed nodes f in Gnutella network

could remove some nodes randomly in a network, or other nodes with high degrees. Removal of important nodes may affect the network significantly. With the removing of nodes from a network, some paths between pairs of nodes is broken. The average length of these paths increases. Eventually nodes are isolated in different clusters, and communications between them become impossible. Some real networks display high degree of robustness against random failures of nodes, but they are also very vulnerable under attacks of the high degree nodes.

In the following, we illustrate the network resilience of Gnutella network based on our previous work [6]. To explain the damages caused by attacks and random failures, we measure two parameters: the relative size of the largest cluster S (defined as the ratio between the size of the largest cluster and the size of original network) and the average closeness centrality[1] C_c (defined as the average of the closeness centralities of all nodes in the largest cluster).

As shown in Fig.1, Gnutella network shows high degree of tolerance against random failures. However, the fault tolerance comes at the expense of attack vulnerability: rapid decreasing of the relative size of the largest cluster and the average closeness centrality in early stage. In Fig.1 (b), after the critical point, the largest cluster becomes much smaller than the initial size of the network, which causes the fallback of average path length in such clusters and the increasing of C_c correspondingly. A more detailed description can be referred to [6].

It is intuitive that the attacks on high degree nodes are analogous to the malicious behavior of hackers in reality. They often make hosts malfunctioning by brute attacks. In addition, there are other dangers caused by computer viruses or backdoor programs (programs left by hackers that reside in hosts and can be used to intrude into other hosts

---

[1] Closeness centrality is the measurement of the shortest path length of one node to all others in the network. See [6] for a more detailed description.

without breaking down systems). If they are not controlled properly, they will spread to the whole system. Hence, the problems remained to us are: how efficiently can we stop the spreading of viruses (in the rest of this paper, viruses stand for both computer viruses and backdoor-like programs unless explicitly stated)? How can one node inform other nodes when it detects a virus? Imagine that if one successfully intrude into one host and spread in a P2P network, DDOS(Distributed Denial of Service) attacks could be easily performed. These are exactly the topics to be discussed in the following sections.

2.2 Modeling Immunization of P2P Network

One model of the spread of a virus over a network is the SIR (*susceptible-infective-recovered*) model [7]. This model assumes that the nodes in the network can be in three states: *susceptible* (one node is healthy but could be infected by others), *infective* (one node has the virus, and can spread it to others), or *recovered* (one node has recovered from the virus and has permanent immunity, so that it can never be infected again or spread it).

Another widely used model of virus spreading is called SIS (*susceptible-infective-susceptible*) model [7]. The main difference with SIR model is that one node can be infected again without permanent immunity, even though it once recovered from the virus. Comparing the two models, we know that the SIS model is more suitable for modeling the spread of computer virus or intrusion in P2P networks, since viruses or intrusions in the network can be cured by antivirus software or be blocked by intrusion detection system. But without a permanent virus-checking or intrusion-detecting program, they have no way to defend the subsequent attacks by the same virus or intrusion. Hence, we use SIS model to investigate the effect of virus spreading in P2P networks.

In SIS model, regarding P2P networks as graphs, we represent individuals by nodes, which can be either "healthy" or "infected", and represent connections between individuals by links, along which the infection can spread. Each node (susceptible) is infected with rate ν if it is connected to one or more infected nodes. At the same time, an infected node is cured with rate δ, defining an effective spreading rate $\lambda = \nu/\delta$ for the virus. Without lose of generality, we set $\delta = 1$. Viruses whose spreading rate exceeds a critical threshold λ_c will persist, while those under the threshold will die out shortly. This model can be used to investigate epidemic states of viruses of P2P networks, in which a stationary proportion of nodes is infected.

P2P networks often exhibit power-law degree distributions [1, 6], similar to other complex networks. A widely used theoretical model for such power-law networks is the Barabasi and Albert (BA) model [3], which describes the growth of complex networks by two basic features: the growing nature of the networks and a preferential attachment rule. The algorithm of BA model is as following: Staring with a small number (m_0) of nodes, at every step we add a new node with m edges that link the new node to m different nodes already in the system. The probability that a new node will be connected to node i depends on the degree k_i of node i, such that $\prod(k_i) = k_i / \sum_j k_j$. After n steps, we obtain a network with degree distribution $p_k(k) = 2m^2 k^{-3}$. In the following, we use the BA model to deduce a theoretical framework of the prevalence of virus, and then compare with the real data obtained from Gnutella network [6].

In order to take into account the different connectivity of all the nodes, we denote the density of infected nodes with degree k by $\rho_k(t)$, where the parameter t indicates the time evolution, and denote the average density of all infected nodes in the network by $\rho = \Sigma_k p(k)\rho_k$. According to the mean-field theory as in [13], we have the following equation:

$$\frac{d\rho_k(t)}{dt} = -\rho_k(t) + \lambda k[1 - \rho_k(t)]\Theta(\lambda). \tag{1}$$

The first term in the right-hand side describes the probability that an infected node is cured. The second term is the probability that a healthy node with degree k is infected, proportional to the infection rate λ, the probability $1 - \rho_k(t)$ that a node with degree k is healthy and the probability $\Theta(\lambda)$ that a given link point to an infected node. The probability $\Theta(\lambda)$ is proportional to the average degree $\langle k \rangle = \sum_k kp(k)$ of all the nodes, and it can be written as:

$$\Theta(\lambda) = \frac{\sum_k kp(k)\rho_k(t)}{\langle k \rangle}. \tag{2}$$

Imposing the stationary condition $\frac{d\rho_k(t)}{dt} = 0$ when the system is at large times such that the number of infected nodes are balanced with the number of healthy nodes, we find the stationary density as:

$$\rho_k = \frac{k\lambda\Theta(\lambda)}{1 + k\lambda\Theta(\lambda)}. \tag{3}$$

Using a continuous k approximation, we calculate $\Theta(\lambda)$ for BA model, whose average degree is $\langle k \rangle = \sum_k kp(k) = \int_m^\infty k2m^2k^{-3}dk = 2m$, as:

$$\Theta(\lambda) \simeq m \int_m^\infty \frac{\lambda\Theta(\lambda)}{k(1 + k\lambda\Theta(\lambda))}dk = \frac{e^{-1/m\lambda}}{(1 - e^{-1/m\lambda})m\lambda}. \tag{4}$$

By combining equations (3) and (4) we have:

$$\rho \simeq 2m^2 \int_m^\infty \frac{k^{-2}\lambda\Theta(\lambda)}{1 + k\lambda\Theta(\lambda)}dk = \frac{2e^{-1/m\lambda}}{1 - e^{-1/m\lambda}}. \tag{5}$$

The ρ is the stationary density of all infected nodes after time evolution of the stochastic cycle of SIS model. Equation (5) shows an explicit relationship between the infection density ρ and the effective infection rate λ, which can be used to evaluate different immunization strategies in the following section. The detailed calculations of $\Theta(\lambda)$ and ρ are shown in the appendix.

2.3   Immunization Strategies of P2P Networks

As discussed in section 2.1, the power-law networks exhibit different behaviors under random failures and intentional attacks, from which two intuitive immunization strategies may be regarded as randomized and degree-based immunizations. In the randomized immunization strategy, a proportion of nodes randomly chosen in the network are immunized, and these immune nodes will not be infected and do not spread the virus to their

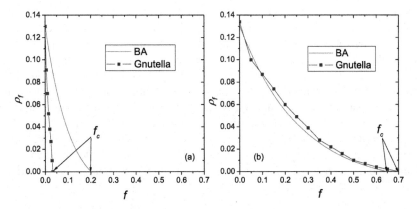

Fig. 2. Results for randomized and degree-based immunization measured by the density of infected nodes ρ_f as a function of the fraction of immune nodes f

neighbors. Accordingly, in the degree-based strategy, nodes are chosen for immunization if their degrees are greater than a predefined value.

Let us illustrate an example of installing a distributed intrusion detection system or a distributed firewall in a P2P system according to the two strategies. In both of the immunization strategies, the spreading dynamical properties can be considered as follows: suppose that a proportion of nodes are infected in the initial state, the immune nodes (intentionally protected by the IDS or firewall) in the system do not transmit viruses as if the links to their neighbors were eliminated, and the non-immune nodes spread viruses to their neighbors, but at the same time these nodes are cured with probability δ such as some nodes may install their personal antivirus program (not the same as the IDS or firewall mentioned above) or update the operating system. After a long time evolution, according to the mean-field theory, the system comes into balance between the infected and healthy nodes. In such a process, the two strategies have remarkably different impact on the density of infected nodes at the critical point of balance.

In the randomized case, for a fixed spreading rate λ, defining the fraction of immunized nodes in the network as f, we get the effective spreading rate $\lambda(1 - f)$, and substitute it into equation (5), we obtain

$$\rho_f = \frac{2e^{-1/m\lambda(1-f)}}{1 - e^{-1/m\lambda(1-f)}}. \tag{6}$$

Clearly, in the case of degree-based immunization, we can not use equation (5) to deduce an explicit formula as in the randomized case, but we will use simulations to compare the difference between the theoretical BA model and the real data of Gnutella network.

Our simulations are implemented with a fixed spreading rate $\lambda = 0.15$, the smallest node degree $m = 3$ and the number of nodes $N = 34206$ the same as the real data of the topology collected from Gnutella network [6]. Initially a proportion of healthy nodes are infected in the network. In Fig.2 (a), we plot the simulation results of degree-based immunization for BA network (line) and Gnutella network (square-line). As the increasing of f, ρ_f decays much faster in Gnutella network than in BA model, and the linear regression from the largest values of f yields the estimated thresholds $f_c \simeq 0.03$ in Gnutella network, $f_c \simeq 0.2$ in BA network. The value of f_c in Gnutella network indicates that the Gnutella network is very sensitive to the degree-based immunization, and the immunization of just a very small fraction (3%) of nodes will eradicate the spreading of virus. On the other hand, in Fig.2 (b), the simulation results of randomized immunization are plotted for Gnutella Network (square-line), which is in good agreement with the theoretical prediction (line) by equation (5), except for a larger value of $f_c \simeq 0.7$ compared with the value $f_c \simeq 0.64$ of BA network. Based on the analysis above, it is evident that the degree-based immunization is really better than randomized immunization, which also inspires us designing an efficient immunization algorithm for P2P networks in the next section.

3   Efficient Immunization Algorithm for P2P Networks

Since the degree-based immunization is more effective than randomized immunization, one problem arises naturally that how efficiently we can inform other nodes when one node finds a virus in real networks. Returning to the example discussed in section 2.3, we consider that in a distributed IDS or firewall system for P2P network, if one node detects an intrusion or a virus, how can it transfer the information to others to update their local database in an efficient way? An intuitive solution is that it transfers the update information by visiting the neighbor with the highest degree, followed by a node with the next highest degree, since the immunized nodes are always with high degrees. In such a way, one can walk down a degree sequence all having high degrees. First, we formulate the highest degree in the network as a function of the network size.

Generally, the highest degree k_{max} of a node in a network depends on the size of the network. In [2], Aiello et al. assumed that the highest degree was approximately the value above which there was less than one node of that degree in the network on average, i.e, $np_k = 1$. This implies that for the power-law degree distribution $p_k \sim k^{-\gamma}$, $k_{max} \sim n^{1/\gamma}$. However, this assumption is not accurate in many real networks, where there are nodes with significantly higher degrees than this in the network.

Given a specific degree distribution p_k, the probability that there are m nodes with degree k and no nodes with higher degree is

$$p_{k_{max}} = \binom{n}{m} p_k^m (1 - P_k)^{n-m}, \tag{7}$$

where $P_k = \sum_{k'=k}^{\infty} p_{k'}$ is the cumulative probability distribution and n is the number of nodes in the network. Hence, the probability Π_k that the highest degree in the network is k is

$$\Pi_k = \sum_{m=1}^{n} \binom{n}{m} p_k^m (1 - P_k)^{n-m} = (p_k + 1 - P_k)^n - (1 - P_k)^n, \tag{8}$$

and the expected value of the highest degree is $k_{max} = \sum_k k \Pi_k$.

The probability Π_k tends to zero for both small and large values of k. Thus, in most case, a good approximation to the mean value of the highest degree is given by the modal value. Based on equation (8), we find that the maximum of Π_k occurs when

$$\frac{d\Pi_k}{dk} = (\frac{dp_k}{dk} - 1)(p_k + 1 - P_k)^{n-1} + p_k(1 - P_k)^{n-1} = 0, \qquad (9)$$

where $\frac{dP_k}{dk} = p_k$. Assuming that p_k is sufficiently small for $k \geq k_{max}$ that $np_k \ll 1$ and $P_k \ll 1$, we can write equation (9) as

$$\frac{dp_k}{dk} = -p_k[(\frac{1 - P_k}{p_k + 1 - P_k})^{n-1} - 1]$$

$$= -p_k[(\frac{1}{p_k + 1})^{n-1} - 1] \simeq -np_k^2, \qquad (10)$$

where $0 < (p_k + 1 - P_k)^{n-1} < (1 - P_k)^{n-1} < 1$.

For BA model, the probability distribution of degree is $p_k = 2m^2k^{-3}$. Substituting it into equation (10), we have

$$k_{max} \simeq \sqrt{\frac{2m^2n}{3}}. \qquad (11)$$

As described above, the node that detected an intrusion or a virus transfers the information to other nodes through a degree sequence in which all the nodes have the highest degree. For simplicity, suppose that the degrees of the nodes in the sequence all approximate to k_{max}, then the number of steps needed to transfer the information in the network of size n is

$$s = \frac{n}{k_{max}} \simeq \sqrt{\frac{3n}{2m^2}}. \qquad (12)$$

We performed simulations of the real data of Gnutella network with a power-law exponent $\gamma = 2.0$ [6], and compared the simulation results with the theoretical prediction of BA network in equation (12). The number of nodes range from $N = 10^3$ to $N = 10^4$.

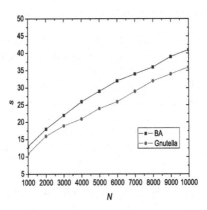

Fig. 3. The number s of steps needed to transfer information through high degree nodes as a function of the network size N

Fig.3 shows that the algorithm of transferring update information based on high degrees in Gnutella network is as efficient as the prediction of the theoretical BA model. We need only $s = 11$ steps to update all high degree nodes in Gnutella network with $N = 1000$ nodes, and $s = 36$ steps in Gnutella network with $N = 10000$ nodes. The discrepancy between the BA network and the Gnutella network is mainly due to the difference of power-law exponent of $\gamma = 3.0$ in BA network and $\gamma = 2.0$ in Gnutella network. Hence, in implementing a real distributed IDS or firewall system, we can update the global information effectively utilizing the highly connected nodes.

4   Conclusions

In this paper, based on the simple SIS model, we analyze the influence of virus spreading on P2P networks with two different immunization strategies namely randomized and degree-based immunization, and performe theoretical modeling and real data simulations. The results show that the degree-based strategy is more efficient than the randomized strategy, which also motivate us to design an effective information transferring algorithm for updating global virus databases. These methods are highly valuable in the implementation of real systems such as distributed IDSs or firewalls. As mentioned in the text, the immunization model is not flexible enough to analyze both randomized and degree-based strategies, hence, our future work is to improve the model to make it suitable for the analysis of both strategies.

References

1. L. Adamic, R. Lukose, A. Puniyani and B. Huberman, "Search in Power-Law Networks", *Phys. Rev. E*, Vol.64, 2001.
2. W. Aiello, F. Chung, and L. Lu, "A random graph model for massive graphs", *Proceedings of the thirty-second annual acm symposium on Theory of computing*, pp.171-180, 2000.
3. A. L. Barabasi and R. Albert, "Emergence of scaling in random networks", *Science*, Vol.286, pp.509, 1999.
4. B. Bollobas, *Random Graphs*, Academic Press, New York, 2nd ed, 2001.
5. A. Broder, R. Kumar, F. Maghoul, P. Raghavan, and R. Stata, "Graph structure in the web", *Computer Networks*, Vol.33, pp.309-320, 2000.
6. H. Chen, H. Jin, and J. H. Sun, "Analysis of Large-Scale Topological Properties for Peer-to-Peer Networks", *Proceedings of International Symposium on Cluster Computing and the Grid*, 2004.
7. O. Diekmann and J. A. P Heesterbeek, *Mathematical epidemiology of infectious diseases: model building, analysis and interpretation*, JohnWiley & Sons, New York, 2000.
8. M. Faloutsos, P. Faloutos, and C. Faloutsos, "On Power-law Relationships of the Internet Topology", *Computer Communications Review*, Vol.29, pp.251-262, 1999.
9. M. A. Jovanovic, "Modeling Large-scale Peer-to-Peer Networks and a Case Study of Gnutella", Master thesis, Department of Electrical and Computer Engineering, University of Cincinnati, 2000.
10. S. Milgram, "The small-world problem", *Psychology Today*, Vol.1, pp.62-67, 1967.

11. M. E. J. Newman, "The structure and function of complex networks", *SIAM Review* , Vol.45, pp.167-256, 2003.
12. M. Ripeanu, I. Foster and A. Lamnitchi, "Mapping the Gnutella Network: Properties of Large-Scale Peer-to-Peer Systems and Implications for System", *J. Internet Computing*, 2002.
13. R. P. Satorras and A. Vespignani, "Epidemic Spreading in Scale-Free Networks", *Phys. Rev. Lett*, Vol.86, pp.3200-3203, 2001.
14. S. Saroiu, K. P. Gummadi and S. D. Gribble, "Measuring and analyzing the characteristics of Napster and Gnutella hosts", *Multimedia Systems*, Vol.9, pp.170-184, 2003.

A   Appendix. Calculations of $\Theta(\lambda)$ and ρ

$$\Theta(\lambda) = \frac{\sum_k kp(k)\rho_k(t)}{\langle k \rangle} \simeq \int_m^\infty \frac{2m^2k^{-3}k^2\lambda\Theta(\lambda)}{2m(1+k\lambda\Theta(\lambda))}$$

$$= m\int_m^\infty \frac{\lambda\Theta(\lambda)}{k(1+k\lambda\Theta(\lambda))}dk$$

$$= m\lambda\Theta(\lambda)\int_m^\infty(\frac{1}{k} - \frac{\lambda\Theta(\lambda)}{1+k\lambda\Theta(\lambda)})dk$$

$$= m\lambda\Theta(\lambda)([\ln k]_m^\infty - [\ln(1+k\lambda\Theta(\lambda))]_m^\infty)$$

$$= m\lambda\Theta(\lambda)(\lim_{M\to\infty}\ln\frac{M}{1+M\lambda\Theta(\lambda)} - \ln\frac{m}{1+m\lambda\Theta(\lambda)})$$

$$= -m\lambda\Theta(\lambda)\ln\frac{m\lambda\Theta(\lambda)}{1+m\lambda\Theta(\lambda)} \Rightarrow$$

$$-\frac{1}{m\lambda} = \ln\frac{m\lambda\Theta(\lambda)}{1+m\lambda\Theta(\lambda)} \Rightarrow \Theta(\lambda) \simeq \frac{e^{-1/m\lambda}}{(1-e^{-1/m\lambda})m\lambda}.$$

$$\rho = \Sigma_k p(k)\rho_k \simeq \int_m^\infty \frac{2m^2k^{-3}k\lambda\Theta(\lambda)}{1+k\lambda\Theta(\lambda)}dk$$

$$= 2m^2\lambda\Theta(\lambda)\int_m^\infty(\frac{1}{k^2} - \frac{\lambda\Theta(\lambda)}{k} + \frac{(\lambda\Theta(\lambda))^2}{1+k\lambda\Theta(\lambda)})dk$$

$$= 2m^2\lambda\Theta(\lambda)([-\frac{1}{k}]_m^\infty - \lambda\Theta(\lambda)[\ln k]_m^\infty + \lambda\Theta(\lambda)[\ln(1+k\lambda\Theta(\lambda))]_m^\infty)$$

$$= 2m^2\lambda\Theta(\lambda)(\frac{1}{m} + \lambda\Theta(\lambda)\lim_{M\to\infty}\ln\frac{1+M\lambda\Theta(\lambda)}{M} + \lambda\Theta(\lambda)\ln\frac{m}{1+m\lambda\Theta(\lambda)})$$

$$= 2m^2\lambda\Theta(\lambda)(\frac{1}{m} + \lambda\Theta(\lambda)\ln\frac{m\lambda\Theta(\lambda)}{1+m\lambda\Theta(\lambda)})$$

$$= 2m^2\lambda\Theta(\lambda)(\frac{1}{m} - \lambda\Theta(\lambda)\frac{1}{m\lambda}) \quad \text{(by substituting }\Theta(\lambda)\text{ into the above equation)}$$

$$\simeq 2m\lambda\Theta(\lambda) \quad \text{(the lowest order of }\lambda\text{ is remained)}$$

$$= \frac{2e^{-1/m\lambda}}{(1-e^{-1/m\lambda})}.$$

Leveraging Public Resource Pools to Improve the Service Compliances of Computing Utilities

Shah Asaduzzaman and Muthucumaru Maheswaran

McGill University, Montreal QC H3A 2A7, Canada
{asad, maheswar}@cs.mcgill.ca,
http://www.cs.mcgill.ca/~anrl/

Abstract. Computing utilities are emerging as an important part of the infrastructure for outsourcing computer services. One of the major objectives of computing utilities is to maximize their net profit while maintaining customer loyalty in accordance with the service level agreements (SLAs). Defining the SLAs conservatively might be one easy way to achieve SLA compliance, but this results in underutilization of resources and loss of revenue in turn. In this paper, we show that inducting unreliable public resources into a computing utility enables more competetive SLAs while maintaining higher level of runtime compliance as well as maximizing profit.

1 Introduction

Constant improvements in computer communications and microprocessor technologies are driving the development of new classes of network computing systems. One such system is the *computing utility* (CU) that brings large number of resources and services together in a virtual system to serve its clients. Typically, CUs are built by connecting the resources or services to a *resource management system* (RMS) that itself is implemented either centrally or federally. The RMS allocates resources to the client requests such that some measure of delivered performance is maximized subject to fairness constraints. The organization of the RMS, which impacts the scalability, extensibility, and fault tolerance of the CU is a key consideration in CU design. Support for services with *quality of service* (QoS) assurances is another important design issue in order to attract business critical applications.

This paper is concerned about augmenting CUs using "public" resources (i.e., resources that wish to contribute their computing, storage, and network capacities without subjecting themselves to any contractual agreements). Several large-scale network computing systems such as Gnutella, SETI@home have demonstrated the tremendous potential of using public resources. Our proposed CU architecture augments the deployed dedicated resources with public resource for additional capacity and we refer to it as a *public computing utility* (PCU).

Although different applications can potentially use a PCU, here we consider only high-throughput computing applications. In this situation, job requests belonging to different clients arrive at the PCU at arbitrary times. In a practical

L. Bougé and V.K. Prasanna (Eds.): HiPC 2004, LNCS 3296, pp. 242–251, 2004.
© Springer-Verlag Berlin Heidelberg 2004

PCU setting, the RMS has to take the allocation decision as soon as the jobs arrive.

In this paper, we devise an online scheduling heuristic for the RMS of the PCU. The PCU online heuristic needs to decide what class of resources (public or private) should be used for servicing a given request. Because the PCU is bound by the SLAs when delivering services to the clients, we need to consider the SLAs in the resource allocation process as well. Section 2 of the paper discusses the related results found in the literature. Section 3 explains the proposed system architecture in detail. Section 4 defines the resource scheduling problem being dealt with in the PCU. Section 6 discusses the results from the simulations performed to evaluate the resource allocation alternatives.

2 Related Work

Multiprocessor job scheduling is a well-studied problem in operations research and computer science. Although several optimal algorithms are available [1] for simpler scheduling problems, most of the interesting and practical scheduling problems are computationally intractable. Scheduling jobs with arrival time and deadline constraints is proven to be a NP-hard problem for more than two processors [2]. In fact [3] proved that optimal scheduling of jobs in multiple processors is impossible if any of the 3 parameters - arrival time, execution time or deadline is unknown. Because in an online scheduling scenario, resource allocations have to be carried out with incomplete information regarding jobs, heuristic solutions are appropriate for this situation. A good survey of online scheduling heuristics can be found in [4].

One major goal of the RMS of a PCU is to enforce QoS according to the SLAs signed up with its clients. Architectures of SLA compliant resource management for cluster of dedicated machines has been studied in several research projects like Oceano [5], Globus Grid [6][7], etc. However, study of scheduling algorithms with detailed performance evaluations were not carried in the above works. Performance evaluation of scheduling heuristics for cluster based hosting centers are found in [8][9][10] with different optimization goals in different cases.

The Condor project [11] focuses on harvesting unused resources from heterogeneous public machines, but their resource management mainly emphasizes on discovery and co-allocation of resources through matchmaking and gangmatching. They do not support SLA driven QoS aware resource management on the public resource pool.

One work that is very close to our work is [12], which examines stochastic QoS on a similar architecture using dedicated private resource and stochastic public resources. Nevertheless, our work is significantly different from theirs in several dimensions. Instead of modeling the public resources with homogeneous performance and stochastic idle times, we have modeled their throughput to be stochastic which captures the real behavior more closely. They have assumed the QoS requirement (cycle length) of applications has distribution identical to that of underlying public resources, but we have relaxed that assumption.

Furthermore, their scheme does not have any long term SLA with the clients, whereas Our scheduling heuristic is devised to simultaneously maximize the net-profit of the service provider and the level of compliance with the long term SLAs. Our investigation also includes job streams arriving from multiple clients that have different SLAs with the PCU.

3   The PCU System Model and Assumptions

The underlying substrate of the PCU is a proximity aware planetary scale P2P network such as the Pastry [13] that connects all the resources that participate in the system. The public resources are expected to be dispersed throughout the network and the private resources can be concentrated as clusters at certain locations. The P2P network enables efficient discovery of the public resources.

Several issues like resource co-allocation, trust and incentive management, load-balancing, etc. should be addressed in developing the resource allocation process in a PCU system. As a first cut at the problem, we consider allocation of only one resource – the processors. The allocation decisions of the RMS are influenced by several parameters including: (a) current utilization of the private resource pool administered by the PCU, (b) current load offered by the different clients, (c) current value of the expected performance of the best-effort resources, and (d) throughput guaranteed to the particular client by the PCU in its SLA. In our current PCU RMS design, there is no progress monitoring of public resources, only process completions are notified. Inclusion of progress indicators can improve the contribution from public resources towards overall throughput, albeit at the cost of high communication overhead.

4   The Resource Management Problem

Computational jobs arrive from each client of the PCU service provider at arbitrary points in time with each job consisting of arbitrary number of mutually independent parallel components of possibly different but known sizes. An overall deadline is defined for the job before which all the components must finish their execution.

The SLA that is signed off-line between the provider and a client reserves a throughput guarantee for the corresponding client. The SLA defines various parameters including:

- ρ, the ratio of the client-offered workload that is guaranteed to be carried out by the PCU service provider.
- V, the maximum limit on the workload that can be offered by the client.

From these parameters it can be deduced that when the offered load is $v \leq V$, the delivered throughput should be $\geq \rho v$ to be compliant with the SLA. If offered load v is greater than V, it is sufficient for the PCU to deliver ρV amount of throughput.

The PCU provider earns revenue in proportion to the total delivered computational work for the jobs that finish completely within their deadline (with all of its components). There is penalty for violation of the SLA terms and the penalty is proportional to amount of deviation of the delivered throughput from guaranteed throughput, measured over a specified time window. The optimization goal of the job scheduler is to maximize the net revenue (i.e., revenue − penalty) of the PCU service provider.

5   Heuristic Solutions for Resource Management

In this section, we present three heuristic solutions to resource management in a PCU environment. The first solution, the PCU heuristic, is proposed as part of this work. The next two solutions are adopted from the scheduling literature for the PCU environment for comparison purposes.

5.1   PCU Heuristic: An Online Resource Allocator

The scheduler of the RMS uses an online heuristic to take decisions about allocating available resources to incoming jobs. To reduce the scheduling overhead, the RMS executes the scheduling rules at discrete points of time (i.e., at the end of each scheduling epoch δ). Another component of the RMS, the SLA monitor measures the current deviation D_c of delivered throughput from required throughput for each client c, according to the SLA specified time-window τ_c and moving average factor α_{sla}. Say the total arrived workload in a time-window is W_a and total completed and delivered workload is W_d, both W_a and W_d being smoothed by moving average with the past values. Then,

$$D_c = \max(V_c, W_a \rho_c) - W_d$$

In the above equation, V_c and ρ_c are SLA defined maximum load and acceptance ratio for client c. The current value of D_c is available to the scheduler at the end of every epoch. There are two parts of the decision taken by the scheduler at the end of every epoch (i) accept newly arrived jobs and start them on public and/or private resources, and (ii) relocate and restart the deadline vulnerable jobs from public resource to the private resource pool (in absence of checkpointing and progress monitoring, it is impossible to migrate without restarting).

Acceptance of Jobs. For each client, the scheduler maintains a priority queue for newly arrived jobs, ordered by highest contributing job first. For a job with total workload W and total available time T_a before deadline, the throughput contribution is $\frac{W}{T_a}$. Every time the foremost job from the queue of the client having highest $D_c - W_c$ value is chosen, where W_c is is the amount of workload so far accepted for client c in current SLA window.

All the jobs are ultimately accepted, and each of them are assigned one of the two different levels of *launch-time-priority*, which is used for restarting decisions. The jobs are accepted according to the following rules:

1. As long as available dedicated resources allow, schedule jobs with high launch-time-priority with *critical* components on dedicated and the rest on public resources. The components that are expected to violate deadline if scheduled on a public resource according to its currently estimated expected through-put μ, are identified as critical components. Among the M private resources, M_r are reserved for restarting phase (the ratio $\frac{M_r}{M}$ is a design parameter). If M_o resources are already occupied and the selected job has m critical components, this phase continues as long as $M_o + m \leq M - M_r$,

2. For the rest of the enqueued jobs all components are scheduled on public resources. For any client c, as long as total accepted workload in the current SLA window is below $\rho_c V_c$, the launch-time-priority of the accepted job is high, otherwise it is low.

Restart Jobs. At the end of every epoch, the scheduler restarts some dead-line vulnerable job-components from public resources. The job-components that have reached a point where it can be completed before deadline only if run on a dedicated machine, is identified as vulnerable. A priority queue is maintained for all the vulnerable components. The queue is ordered descending primarily by launch-time-priority (explained earlier) and secondly by violation probability (p_v). p_v is computed at the job-launch time from the available information (distribution of the public resource throughput, component size and the deadline). From the queue, high launch-time-priority components are restarted as long as any dedicated resource is available. Low launch-time-priority are restarted as long as available dedicated resource is $> M_r$. The rest of components are left on public resource.

5.2   Least Laxity First and Greedy Heuristics

For performance evaluation we compare our PCU heuristic with the well known *Least Laxity First* (LLF) [4] heuristic and a Greedy heuristic. We use the LLF heuristic to schedule the jobs only in the private pool of resources. The laxity is the slack between possible execution finish time and deadline. New jobs form each client enter a separate priority queue ordered by laxity and at every epoch jobs popped from the queue that fits in available dedicated resources are started there, otherwise the job is deferred until it becomes infeasible to execute before deadline. As a fairness scheme the queue of the client with highest deviation from SLA is favored when choosing every job.

The Greedy heuristic, another one that we used for comparison, works on the same PCU architecture with a combination of private and public resource pools. The greedy scheduling policy chooses jobs from the arrival queues in every scheduling epoch in the order of highest contributing job of the highest deviating client first. It schedules all components of incoming jobs on private resources in the order of longer component first, as long as there is spare capacity in the private resource pool. All the remaining job-components are scheduled on public resources until all the arrival queues are exhausted.

6   Simulation Results

Here we evaluate the performance of the PCU heuristic through a simulator written in Parsec [14] by changing different parameters and comparing it with the Greedy and LLF heuristics. In our simulation setup, the service provider had a pool of 100 dedicated machines and an infinite pool of public machines. There were five independent clients each feeding a stream of parallel jobs that should be completed within the given deadlines and having its own SLA. Jobs arrival is a Poisson process, with each job having a random number (k) of parallel components (geometrically distributed). Each component of a job also has a random workload that is from a geometric distribution. Each job has a feasible deadline, i.e., it can always be completed if all the parallel components run on dedicated machines. Unless stated otherwise, the deadline was computed with a uniform random laxity between 0.5 and 2 times the mean component length, from the longest component. This tight deadline allows one trial on the public pool and failing that it should be restarted on a private resource.

All private machines have homogeneous throughput, completing 1 unit of workload of a component per second. The public resource throughput is sampled from Lognormal distribution with standard deviation 1.0 and mean less than 1.0. Justification behind using lognormal distribution is that being left skewed it closely resembles the behavior of the resources in a PCU setting, where most of the public resources may have very low or even 0 throughput.

In the first set of experiments the PCU heuristic is compared with LLF and Greedy using throughput (Figure 1), SLA compliance (measured using penalty per unit revenue in Figure 2). The PCU heuristic delivers better throughput than LLF, which implies it useful to augment public resource in a CU. Also the PCU-heuristic is superior in performance to the greedy heuristic in similar setting.

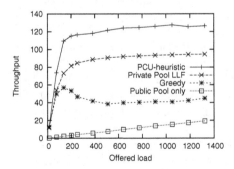
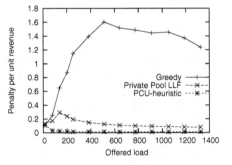

Fig. 1. Variation of mean throughput with offered load values for mean public resource throughput $\mu = 0.80$, mean number of parallel components $P = 25$, total number of private resources $M = 100$, and total SLA booking, $\sum \rho V = 100$

Fig. 2. Variation of penalty per unit revenue with offered load for $\mu = 0.80$, $P = 25$, $M = 100$, and $\sum \rho V = 100$

Figure 3 shows that a much higher gain in throughput is achievable, if the exact knowledge of throughput of each public machine is available at schedule time, because then there is no need for restarting jobs. How far of this gain can be achieved without apriori knowledge remains a problem for future research.

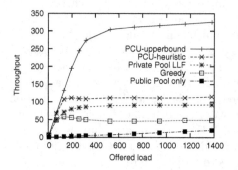 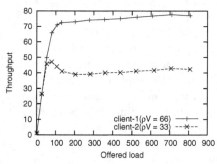

Fig. 3. Upper bound on PCU throughput assuming future behavior of public resources is known for $\mu = 0.80$, $P = 25$, $M = 100$, and $\sum \rho V = 100$

Fig. 4. Comparing delivered throughput to 2 clients having different max-load defined in SLA for $\mu = 0.80$, $P = 25$, and $M = 100$

Demonstrating the fairness of PCU heuristic figure 4 shows that for 2 different clients, who offers load at the same rate, but has SLA maxload (V) defined at $2 : 1$ ratio, the delivered throughput is proportional to the maxload of the clients for overloaded situations.

The penalty is higher with the LLF algorithm on private pool only system than the PCU heuristic, because jobs are not deprioritized when the client is offering more workload than the SLA upper bound. In case of the greedy algorithm, penalty grows even higher when the client is overloading, because the dedicated pool gets fully occupied and most of the newly arriving jobs are put on public resources. Consequently, only a small portion of the newly arriving jobs can finish before their deadlines.

As Figure 5 shows, the utilization of dedicated resources is higher for the greedy policy. This is because Greedy uses the dedicated resources exhaustively. The PCU heuristic tries to execute a job-component primarily using public resources unless it becomes vulnerable for deadline violation. Also, in PCU, to allow the restarting of vulnerable components, it reserves a portion of the dedicated resources (25%) as contingency resources. These factors lower the utilization of dedicated resources in the PCU heuristic. Greedy's utilization is even more than LLF, because, in LLF jobs are not allocated unless the all the components fit in the private resources, whereas, Greedy may put part of a job in private pool and rest in public pool.

To consider the flexibility in SLA overbooking, if and total agreed upon deliverable throughput (ρV) is higher than the maximum system capacity, the SLA deviation goes very high leading to correspondingly high penalties. This in turn

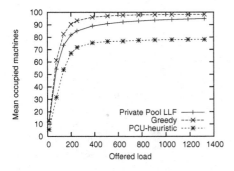
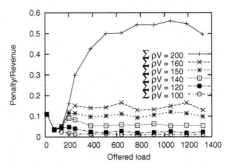

Fig. 5. Utilization of dedicated resources versus offered load for $\mu = 0.80$, $P = 25$, $M = 100$, and $\sum \rho V = 100$

Fig. 6. Penalty per unit revenue earned at different levels of SLA booking for $\mu = 0.80$, $P = 25$, and $M = 100$

reduces the net profit earned by the service provider. From Figures 6 it can be observed that SLA booking should be at 140% of the dedicated pool capacity to maximize the performance for the given PCU configuration.

Figure 7 shows that use of PCU-heuristic brings gain in delivered throughput in most region of the spectrum of public resource behavior. It should be noted that with lognormal distribution, even if the mean throughput is equal to that of a dedicated machine, 62% of the public resources have throughput less than that of a dedicated machine. For very low public resource throughput, almost all of the jobs scheduled there needs restart, and since restart is subject to availability in the limited capacity private pool, many jobs get discarded. This explains the less than one throughput-gain with poor quality of public resources. Figure 8 shows that PCU-heuristic outperforms the greedy heuristic across the whole spectrum.

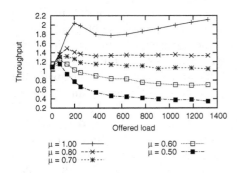
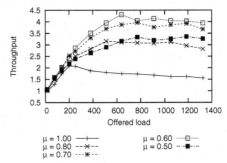

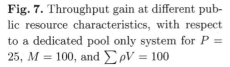

Fig. 7. Throughput gain at different public resource characteristics, with respect to a dedicated pool only system for $P = 25$, $M = 100$, and $\sum \rho V = 100$

Fig. 8. Throughput gain at different public resource characteristics, with respect to the greedy resource allocation policy on combined pools for $P = 25$, $M = 100$, and $\sum \rho V = 100$

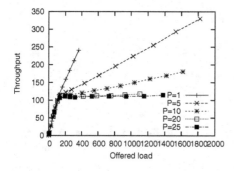
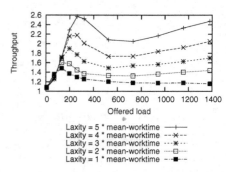

Fig. 9. Mean throughput at varying degree of parallelism for $\mu = 0.80$, $M = 100$, and $\sum \rho V = 100$

Fig. 10. Throughput gain at different amount of laxity in deadline, with respect to a dedicated pool only system for $\mu = 0.80$, $P = 25$, $M = 100$, and $\sum \rho V = 100$

Studying the effect of parallelism figure 9 shows that the effect is insignificant in underloaded situations, but when the system is overloaded, high number of parallel components increase the probability of failure of a whole job due to failure of only one or few components which could not be restarted when necessary. Hence, the total delivered throughput becomes low.

Study on the effect of laxity before deadline (Figure 10)f shows that throughput gain is much higher with relaxed laxity jobs. This is because with relaxed laxity the probability of getting a job component completed before deadline on a public resource increases, which incurs less restarts and better contribution from public resources.

7   Conclusion

In this paper, we presented the idea of creating a public computing utility by augmenting computing utilities of dedicated resources with public resources. A resource management strategy for such an augmented system was presented. We proposed a resource allocation heuristic that uses the public and private (dedicated) pools of resources in an efficient manner. We carried out extensive simulations to evaluate the performance of the proposed heuristic and compare it with two other heuristics.

The results indicate that the use of public resources can lead to significant performance improvements both in terms of obtainable throughput and the compliance with client SLAs. Further, the results indicate that the performance gain from PCU increases if the job has fewer components or relaxed deadlines. The performance of the PCU heuristic may be further improved by incorporating these parameters in the decision process.

One of the significant features of our PCU architecture is the minimal monitoring on the public resources. Because public resources are plenty this helps to keep the overhead low. It might be possible to selectively enable performance monitoring for high capacity public resources and increase the delivered performance levels even further.

References

1. Lawler, E.L., Lenstra, J.K., Kan, A.H.G.R., Shmoys, D.B.: 9. In: Handbooks in Operations Research and Management Science. Volume 4. Elsevier Science Publishers (1993) 445–522
2. Garey, M., Johnson, D.: Computers and Intractability: A Guide to the theory of NP-Completeness. W H Freeman and Company, New York (1979)
3. Dertouzos, M.L., Mok, A.K.L.: Multiprocessor on-line scheduling of hard-real-time tasks. IEEE Transactions on Software Engineering **15** (1989) 1497–1506
4. Sgall, J. In: On-Line Scheduling – A Survey. Springer Verlag (1997) 196–231
5. Appleby, K., Fakhouri, S., Fong, L., Goldszmidt, G., Kalantar, M., Krishnakumar, S., Pazel, D., Pershing, J., Rochwerger, B.: Oceano – SLA based management of a computing utility. In: Proceedings of the 7th IFIP/IEEE International Symposium on Integrated Network Management. (2001)
6. Foster, I., Kesselman, C., Tuecke, S.: The anatomy of the grid: Enabling scalable virtual organizations. International J. Supercomputer Applications **15** (2001)
7. Czajkowski, K., Foster, I., Kesselman, C., Sander, V., Tuecke, S.: SNAP: A protocol for negotiating service level agreements and coordinating resource management in distributed systems. Lecture Notes in Computer Science **2537** (2002) 153–183
8. Chase, J.S., Anderson, D.C., Thakar, P.N., Vahdat, A.M., Doyle, R.P.: Managing energy and server resources in hosting centers. In: 18th ACM Symposium on Operating Systems Principles. (2001)
9. Ranjan, S., Rolia, J., Knightly, E.: QoS driven server migraion for internet data centers. In: Proceedings of IWQoS 2002. (2002)
10. Aron, M., Druschel, P., Zwaenepoel, W.: Cluster reserves: A mechanism for resource management in cluster-based network servers. In: Proceedings of ACM SIGMETRICS. (2000)
11. Thain, D., Tannenbaum, T., Livny, M.: Distributed computing in practice: The condor experience. Concurrency and Computation: Practice and Experience (2004)
12. Kenyon, C., Cheliotis, G.: Creating services with hard guarantees from cycle harvesting resources. In: Proceeings of the 3rd IEEE/ACM International Symposium on Cluster Computing and the Grid (CCGRID'03). (2003)
13. Rowstron, A., Druschel, P.: Pastry: Scalable, distributed object location and routing for large-scale peer-to-peer systems. In: IFIP/ACM International Conference on Distributed Systems Platforms (Middleware), Heidelberg, Germany (2001) 329–350
14. Bagrodia, R., Meyer, R., Takai, M., Chen, Y., Zeng, X., Martin, J., Park, B., Song, H.: Parsec: A parallel simulation environment for complex systems. Computer **31** (1998) 77–85

Plethora: An Efficient Wide-Area Storage System*

Ronaldo A. Ferreira, Ananth Grama, and Suresh Jagannathan

Department of Computer Sciences,
Purdue University,
West Lafayette, IN 47907
{rf, ayg, suresh}@cs.purdue.edu
Phone: (765) 494 0971, Fax: (765) 494 0739

Abstract. Trends in conventional storage infrastructure motivate the development of foundational technologies for building a wide-area read-write storage repository capable of providing a single image of a distributed storage resource. The overarching design goals of such an infrastructure include client performance, global resource utilization, system scalability (providing a single logical view of larger resource and user pools) and application scalability (enabling single applications with large resource requirements). Such a storage infrastructure forms the basis for second generation data-grid efforts underlying massive data handling in high-energy physics, nanosciences, and bioinformatics, among others.

This paper describes some of the foundational technologies underlying such a repository, Plethora, for semi-static peer-to-peer (P2P) networks implemented on a wide-area Internet testbed. In contrast to many current efforts that focus entirely on unstructured dynamic P2P environments, Plethora focuses on semi-static peers with strong network connectivity and a partially persistent network state. In a semi-static P2P network, peers are likely to remain participants in the network over long periods of time (e.g., compute servers), and are capable of providing reasonably high availability and response-time guarantees. The repository integrates novel concepts in locality enhancing overlay networks, transactional semantics for read-write data coupled with hierarchical versioning, and novel erasure codes for robustness. While mentioning approaches taken by Plethora to other problems, this paper focuses on the problem of routing data request to blocks, while integrating caching and locality enhancing overlays into a single framework. We show significant performance improvements resulting from our routing techniques.

1 Plethora: Introduction and Design Principles

The Plethora project at Purdue University aims to build a wide-area read-write storage repository for supporting a single seamless distributed storage resource.

* This research has been supported by the National Science Foundation Grant STI 0334141.

L. Bougé and V.K. Prasanna (Eds.): HiPC 2004, LNCS 3296, pp. 252–261, 2004.
© Springer-Verlag Berlin Heidelberg 2004

In contrast to many current efforts that focus entirely on unstructured dynamic peer-to-peer (P2P) environments, Plethora focuses on semi-static peers with strong network connectivity and a partially persistent network state. In doing so, it alleviates many of the constraints found in conventional P2P networks, focusing instead on mechanisms for supporting expressive storage access and management semantics, exploiting existing Internet infrastructure, and providing performance guarantees in terms of end-user latencies, global resource utilization, and robustness.

At the heart of Plethora's design is a two-level network overlay. A global overlay spans all nodes in the network and is organized much like other P2P systems, which define a collection of peers that cooperatively provide access to resources such as storage. A local overlay, which provides caching mechanisms, lies on top of this global overlay. Nodes belong to both local and global overlays and are organized into local overlays based on proximity information, specifically, the Autonomous Systems (ASs) to which they belong. Algorithms for organizing nodes into local overlays are described in greater detail in Section 3.2. Data accesses are first attempted in the local overlay, and forwarded to the global overlay when no cached copy is found. Specifically, if a Plethora node n requests a globally shared piece of data, the data is routed using a distributed (or consistent) hashing scheme to a node n_l in n's local overlay. If n_l does not have the requested data, a query is sent to the global overlay. This two-tiered search mechanism distinguishes Plethora from other P2P schemes in obvious ways. Most notably, it leverages the physical structure and organization of the Internet to provide efficient routing and lookup of shared data.

Concurrent Access Semantics in Plethora

Plethora is a global object repository upon which classical applications such as file systems, collaborative environments, publishing, etc. can be built. To support such repositories, appropriate concurrent access semantics must be built on top of the Plethora routing core. These semantics must facilitate desirable characteristics of client performance, global resource utilization, system scalability, and application scalability. We start our discussion by presenting the Plethora concurrent access model.

To be a member of a Plethora community, a node n defines two namespaces. The first, called cSpace is a local client namespace in which objects such as files are created and modified by applications that execute on the node. The second, called pSpace is the space used by Plethora to support shared access semantics. Based on this model, we describe the logical operation of the Plethora object sharing mechanism for the following basic tasks: (i) introducing an object into the network, (ii) acquiring an object from the network, and (iii) updating objects in the network. Plethora supports these tasks using local and global object brokers. These brokers are determined by applying appropriate hash functions to the object handle – a local hash function, whose range is the set of nodes in the local overlay and a global hash function, whose range is the set of all nodes in the Plethora community. Note that these ranges can be easily affected by using

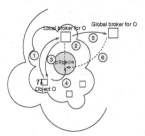

① Node n notifies local broker to initiate Object O.

② Local broker creates reference for O in pSpace.

③ Local broker notifies n of this reference.

④ Node n copies O to this location in pSpace.

⑤ Reference in pSpace sent to global broker.

⑥ Global broker records this reference.

(a) Initiating object O into Plethora.

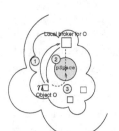

① Node n requests local broker for Object O.

② Local broker sends reference to O in pSpace.

③ Node n creates reference, copies into cSpace.

(b) Searching for object O in Plethora (the case illustrated is one in which object is found in local overlay).

Fig. 1. Protocols for object initiation and lookup. The protocol for updation is similar to initiation augmented by aggregation of commits

appropriate routing table entries in two distinct set of tables without changing the hash function. A local object broker is responsible for keeping track of the shared objects (that hash to it) within the local overlay. A global object broker keeps track of shared objects (that hash to it) in the entire network.

Consider a local object O in node n's cSpace, which needs to be initiated into the network. Node n communicates with a broker for the object in the local overlay. The broker copies the object into node n's pSpace, generates a local reference to this copy and also communicates this reference to the global object broker along with appropriate metadata. At this point, object O is visible to the other nodes. This process is illustrated in Figure 1(a). To acquire object O from the network, a node n first communicates with a local broker for the object. If the object O is cached in the local overlay, a reference to the cached version is returned and a copy subsequently communicated to node d's pSpace. This is illustrated in Figure 1(b). If no cached version exists, a search is initiated by the local broker over the global overlay and a copy is fetched by the broker into node n's pSpace.

To update a shared object O, it must first be copied into a node's cSpace. To make these changes visible to the rest of the network, the node must communicate its intention to have these changes committed to O's local broker. Object O's broker copies this object into node n's pSpace. A new version is recorded at the local broker and the existence of this version is propagated to Object O's global broker. In the event that multiple commits arrive at the local broker for the same object, the broker may choose to either ignore all conflicts among these versions, commit each set of updates as a new version, or attempt more sophisticated conflict resolution and commit only a single resolved version. Yet another option is for the local broker to rely on time stamps to discard all updates other than the most recent set. Plethora provides hooks for each of these policies to be implemented at the local broker. The global broker may apply similar strategies to resolve conflicts between multiple local brokers for the same object.

Several observations can be made about the suitability of Plethora's access semantics for wide-area storage systems. First, updates to objects occur entirely locally once the object is copied into the nodes cSpace. The applications determine when these changes need to be made visible to the rest of the network. Locality of data access in typical applications implies that such a scheme is ideally suited from the point of view of minimizing update traffic. Just as a node aggregates several object updates into a single request to commit to the local broker, the local broker may aggregate several such requests before propagating the version to the global broker. This hierarchical aggregation is critical for scalability to large number of nodes given constraints on available bandwidth.

A novel component of Plethora's software architecture is its use of a versioning system to deal with object conflict resolution and commitment. By supporting versioning semantics as opposed to a copy/ invalidate protocol, Plethora allows concurrent updates to a shared object, and places the burden of proper conflict resolution on the application. In other words, two clients that commit updates to a shared object may see two versions of the object preserved by the object server. They may choose to reconcile these versions by identifying and remedying conflicts in much the same way as version control systems such as CVS or RCS do, or they may choose to simply perform subsequent modifications based on their version or some other version accessible from the local broker. Our primary goal in these design decisions is to provide scalability by taking advantage of the local overlay structure. Our versioning system provides a semantically clean characterization of object modification without requiring global updates on each commit.

2   Plethora and Its Comparison to Related Work

There has been a long history of research in the area of distributed file systems and storage. Plethora's design is inspired by many of these past efforts. Existing systems based on the client-server architecture like AFS [4], NFS, xFS [2], Sprite LFS [10] and Coda [5] do not meet our goals of scalability, availability, and network performance. Farsite [1] is a distributed file system that operates over untrusted environments using randomized replicated storage; the secrecy of file contents is ensured using cryptographic techniques, and Byzantine agreement protocols are used to ensure file and directory integrity. The goals of these systems are only partially aligned with ours. Distributed file systems like Coda or AFS are geared towards a client-server storage model. Decentralized systems like Farsite focus on completely untrusted local-area environments (such as a collection of desktops in a corporate campus). Neither consider routing and location services for geographically-dispersed peers, or consistency semantics when multiple copies of data are cached at distributed sites.

A number of researchers have addressed the problem of providing scalable solutions [13, 14] in a P2P environment. Systems such as Chord [13], Pastry, and Tapestry [14] provide a simple primitive for name resolution: given a file name, return the IP addresses of the nodes that currently have reference to the file. To support this primitive, these systems rely on a *distributed hash table* (DHT) abstraction [13, 14], and provide an upper bound on hop-count of $O(\log n)$, where n is the total number of nodes in the network. This upper bound is achieved using a small amount ($O(\log n)$) of routing information per node. Other systems such as CAN [8] support similar primitives, but have different upper bounds on hop-count subject to varying constraints on per-node routing information. Storage systems that exploit the organization and routing properties of the DHT systems described above have been recently proposed [3, 11, 6, 9, 12].

3   The Plethora Routing Core

Central to the performance of Plethora is an efficient routing core, designed to: (i) optimize locality of data access by efficient replication, (ii) minimize overlay link dilation (stretch) and network congestion, and (iii) support suitable object access semantics and robustness mechanisms. These objectives are achieved in part by organizing the network into two distinct overlays – a local overlay comprising nodes in close network proximity, and a global overlay used to provide universal location services.

3.1   Locality Mechanisms in Plethora

Locality of data access is at the core of scalability of Plethora. This has been recognized by many other projects as well, which use various mechanisms for enhancing locality in overlays. Commonly used methods can be classified as either state-based or static. State-based approaches rely on network state information (latencies, hop-count) for situating an incoming peer in an overlay. These methods include location using landmarks and topological inference. In contrast, static approaches rely on known information from the Internet infrastructure to build local overlays. Plethora bases its design on static information for locality. This is motivated by the availability of standardized information, such as IP domains and Autonomous System (AS) membership, across diverse platforms. Using this information efficiently, however, poses considerable challenges.

With the eventual goal of an AS-based node aggregation in mind, we investigated, in a Gnutella trace, the prefix length of CIDR (Classless Inter Domain Routing) blocks in these systems. The trace contains 3,264 different ASs, which constitute a representative sample of the Internet. The average prefix length of the AS's CIDR blocks was a little over 19 bits. This implies that local overlays based on single ASs constitute very small networks, and are unlikely to yield significant improvements from caching. Plethora therefore aims to build local

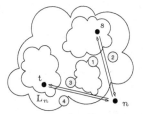

① Node n sends a message to $s = h(n)$.

② Node s sends global routing information to n along with local id. L_n, which n must join (this is determined by Node n's AS).

③ Node n sends a message to $t = h'(n)$ in L_n indicating its desire to join local overlay.

④ Node t sends local routing information to n.

(a) Entering the two-level Plethora routing core.

① Node n sends a message to $t = h'(n)$ requesting data item d in local overlay.

② If Node t has a copy of data item d, steps 2 and 3 are skipped. In step 2, Node t sends a request to Node $s = h(d)$ in global overlay requesting data item d.

③ Node s sends a copy of data item d to Node t.

④ Node t caches a copy of data item d in local overlay L_n and forwards a copy to Node n.

(b) Data access in Plethora.

Fig. 2. Registering peers and locating data objects in Plethora

overlays as aggregates of geographically proximate ASs. For each AS, we determine a ranking sequence of all other ASs close to it (this ranking sequence can be computed using static and/or dynamic state information such as delay or number of hops). This sequence determines the aggregation process. Aggregation is controlled by a root node in the global overlay that is responsible for this AS. The population of a local overlay is controlled by merging and splitting local overlays based on this AS proximity information. The precise algorithms for these operations are described in Section 3.4.

3.2   Building the Plethora Routing Core

The key to building the Plethora routing core is to use the global overlay to maintain state information for local overlay. Specifically, an incoming node n is mapped to a node s in the global overlay. This node s uses n's IP address to determine the AS [7] to which n belongs and maps the incoming node to a local overlay L_n (several ASs might comprise a local overlay depending on peer population). Node s responds to incoming node n's request with the id of the local overlay L_n to which it should join. In addition, node s also sends node n the global routing tables. These correspond to steps 1 and 2 in Figure 2(a). At this point, node n has joined the global overlay and knows that it must join local overlay L_n (note that in this framework, it is possible to use additional state information at node s to determine L_n). Node n then sends a message to a selected node t in local overlay L_n (this node can be selected using a hashcode whose range is in L_n) indicating its desire to join the local overlay. Node t responds to this request with local routing information for overlay L_n. This is illustrated in steps 3 and 4 in Figure 2(a). In this manner, an incoming node joins the global location and local caching network.

3.3   Search in Plethora

The process by which objects are located in Plethora is illustrated in Figure 2(b). An object is first searched in the local overlay using a conventional hash-based query routing scheme. If the object is found, it is fetched from within the local overlay. Otherwise, a query is generated on the global overlay. The object is fetched and cached in the local overlay and forwarded to the requesting node. While current versions of the Plethora core use a simple LRU replacement policy with fixed buffer sizes, the specific caching mechanism is flexible w.r.t. implementation. The performance of this two level scheme is predicated on two critical factors: local overlays must have a high degree of network proximity for meaningful performance gains, and a significant fraction of all accesses must be satisfied from the local overlay (hit ratio). In extensive experiments using access traces as well as synthetic distributions, we have shown that for realistic network sizes, we can achieve hit ratios in excess of 70%. At these hit ratios, we observe over 50% improvement in peer performance with respect to a Pastry peer. With larger buffers and correspondingly larger peer density in local overlays, these savings are expected to grow considerably.

3.4   Plethora Network Maintenance

An important task within the Plethora routing core is to ensure that local overlays do not become too large or too small. In both cases, benefits from localization are diluted, and the performance tends to that of a single global overlay (with some overhead). For this reason, Plethora supports two directives – `pSplit` and `pMerge`. A `pSplit` operation splits a single local overlay into two. This is simply implemented by dropping selected bits (paths) from the routing infrastructure (akin to splitting a single hypercube into two subcubes). It is important that all peers within the same AS drop the same routing paths. A `pMerge` operation conversely merges two local overlays into a single overlay. This is affected by pairing up peers in the two local overlays, having them exchange local routing tables and merging the two with different routing prefixes (akin to merging two subcubes into a hypercube of one larger dimension). Efficient algorithms for `pSplit` and `pMerge` have been developed and their performance characterized within the Plethora core.

3.5   Plethora Routing Performance

To critically examine the performance of the Plethora routing core, we have developed a network simulator that implements the routing schemes of the global and local overlays, implements the algorithms for merging and splitting local overlays, and emulates the underlying network. The topology of the underlying network is generated using the Georgia Tech. transit-stub network topology model (GT-ITM). All experiments are performed on a 32-processor Origin 2000 with 16GB of memory, running IRIX64.

The underlying network topology we use in our experiment contains 10 transit domains, with an average of 10 nodes per transit domain. Each transit node has an average of 10 stub domains attached. There are a total of 1,000 stub domains,

Table 1. Parameters of the simulation experiments

Overlay Nodes	10,000
Network Nodes	110,100
Cache Size per Node	5MB
Distinct Objects	500,254
Mean File Size	5,144 bytes
Max Local Overlay Sizes	200; 300; 400; 500; 1,000; 2,000
Max Delay (D)	30ms; 40ms; 50ms; 100ms; 200ms
α	0.70; 0.75; 0.80; 0.85; 0.90

with an average of 10 nodes per stub domain. A LAN with 10 hosts is connected to each stub node, resulting in a total of 100,000 hosts. The total number of nodes in the underlying network topology is, therefore, 110,100. Each stub domain is considered an autonomous systems. The link delays are selected randomly in the following way: the delay of an edge between two transit domains is in the interval [20-80]ms; the delay of a transit-transit edge in the same transit domain is in the interval [3-23]ms; the delay of a transit-stub edge is in the interval [2-7]ms; the delay of a stub-stub edge is in the interval [1-4]ms; and the delay of an edge connecting a host to a stub node is fixed to 1ms.

The evaluation of a caching scheme requires appropriate dimensioning of the storage available for caching at each node, and a realistic workload. Since there are no publicly available traces that contain file sizes for existing peer-to-peer systems, we use web proxy logs for the distribution of file sizes in the network. The same approach was used to validate PAST in [11]. We use a set of web proxy logs from NLANR[1] corresponding to eight consecutive days in February 2003. The trace contains references to 500,258 unique URLs, with mean file size 5,144 bytes, median file size 1,663 bytes, largest file size 15,002,466 bytes, and smallest file size 17 bytes. The total storage requirement of the files in the trace is 2.4GBytes.

$\alpha = 0.70$					
Size \ Delay	30ms	40ms	50ms	100ms	200ms
200	51.70%	51.38%	51.24%	52.98%	56.64%
300	57.04%	58.40%	60.50%	61.36%	64.33%
400	59.31%	66.03%	66.17%	69.29%	68.88%
500	60.75%	68.26%	68.28%	72.65%	78.17%
1000	60.75%	72.15%	75.64%	86.03%	89.22%
2000	60.75%	72.15%	77.72%	90.01%	94.90%

$\alpha = 0.90$					
Size \ Delay	30ms	40ms	50ms	100ms	200ms
200	71.69%	71.55%	71.50%	72.53%	74.59%
300	74.72%	75.59%	76.82%	77.33%	78.89%
400	75.92%	79.65%	79.83%	81.59%	81.34%
500	76.68%	80.81%	80.94%	83.34%	86.20%
1000	76.68%	82.84%	84.78%	90.38%	92.18%
2000	76.68%	82.84%	86.00%	92.79%	95.88%

Fig. 3. Cache hit ratios for $\alpha = 0.70$ and $\alpha = 0.90$

The main performance measurements that we investigate are the performance gains in response delay and number of packets in the underlying network for

---

[*] http://www.ircache.nlanr.net/

queries in the two-level overlay compared with a single Pastry overlay. The performance gain is defined as: $g = \frac{m_1 - m_2}{m_1}$, where m_1 is the measurement (average delay or lookup messages) in the network without cache, and m_2 is the measurement in the network with cache. The source nodes of the queries are chosen randomly and uniformly, and the objects are accessed according to a Zipf-like distribution, with the ranks of the objects being determined by their position in the original NLANR trace. For the global overlay the Pastry parameters are: $b = 4$, and leaf set size $l = 32$. These parameters are also used in the single Pastry overlay. We measure the impact of the cache hit ratio, the maximum delay used to construct local overlays D, and the maximum number of nodes in a local overlay in the performance gains. The cache hit ratio is the ratio between the number of queries responded in a local overlay by the total number of queries. The parameters of the simulation experiments are summarized in Table 1.

The cache hit ratio is a function of the α value in the Zipf distribution and the maximum number of nodes in a local overlay. Figure 3 illustrates the cache hit ratios obtained with $\alpha = 0.70$ and $\alpha = 0.90$ and the different maximum local overlay sizes. The values in the tables correspond to the minimum and maximum ratios obtained for the parameters, other values of α produce cache hit ratios in those intervals. At these cache hit ratios, the reduction in end-user access latency exceeds 50%.

4   Concluding Remarks

This paper describes the overall design of Plethora – a wide area read-write object repository. It puts Plethora in the context of related efforts, outlines its novel features, describes the Plethora routing core, and demonstrates considerable performance improvements.

Acknowledgements

The first author has been partially funded by CNPq and UFMS, Brazil.

References

1. A. Adya, W. Bolosky, M. Castro, G. Cermak, R. Chaiken, J. Douceur, J. Howell, J. Lorch, M. Theimer, and R. Wattenhofer. FARSITE: Federated, Available, and Reliable Storage for an Incompletely Trusted Environment. In *5th Symposium on Operating Systems Design and Implementation*, Boston, MA, December 2002.
2. T. Anderson, M. Dahlin, J. Neefe, D. Patterson, D. Roselli, and R. Wang. Serverless Network File Systems. In *Proceedings of the 15th Symposium on Operating System Principles. ACM*, pages 109–126, Copper Mountain Resort, Colorado, December 1995.
3. F. Dabek, M. Kaashoek, D. Karger, R. Morris, and I. Stoica. Wide-Area Cooperative Storage with CFS. In *Proceedings of the 18th ACM Symposium on Operating Systems Principles*, Lake Louise, Canada, October 2001.

4. J. Howard, M. Kazar, S. Menees, D. Nichols, M. Satyanarayanan, R. Sidebotham, and M. West. Scale and Performance in a Distributed File System. *ACM Transactions on Computer Systems*, 6(1):51–81, Februrary 1988.
5. J. Kistler and M. Satyanarayanan. Disconnected Operation in the Coda File System. In *13th Symposium on Operating Systems Principles*, Pacific Grove, CA, October 1991.
6. J. Kubiatowicz, D. Bindel, Y. Chen, S. Czerwinski, P. Eaton, D. Geels, R. Gummadi, S. Rhea, H. Weatherspoon, W. Weimer, C. Wells, and B. Zhao. OceanStore: An Architecture for Global-Scale Persistent Storage. In *Proceedings of the Ninth international Conference on Architectural Support for Programming Languages and Operating Systems (ASPLOS 2000)*, Cambridge, MA, November 2000.
7. Z. M. Mao, J. Rexford, J. Wang, and R. H. Katz. Towards an Accurate AS-Level Traceroute Tool. In *Proceedings of the 2003 ACM SIGCOMM Conference on Applications, Technologies, Architectures, and Protocols for Computer Communication*, Karlsruhe, Germany, August 2003.
8. S. Ratnasamy, P. Francis, M. Handley, R. Karp, and S. Shenker. A Scalable Content-Addressable Network. In *Proceedings of the 2001 ACM SIGCOMM Conference on Applications, Technologies, Architectures, and Protocols for Computer Communication*, pages 247–254, San Diego, CA, August 2001.
9. S. Rhea, P. Eaton, D. Geels, H. Weatherspoon, B. Zhao, and J. Kubiatowicz. Pond: The OceanStore Prototype. In *Proceedings of the 2nd USENIX Conference on File and Storage Technologies (FAST '03)*, San Antonio, TX, March 2003.
10. M. Rosenblum and J. Ousterhout. The Design and Implementation of a Log-Structured File System. *ACM Transactions on Computer Systems*, 10(1):26–52, 1992.
11. A. Rowstron and P. Druschel. Storage Management and Caching in PAST, a Large-Scale, Persistent Peer-to-Peer Storage Utility. In *Proceedings of the 18th ACM Symposium on Operating Systems Principles*, Lake Louise, Canada, October 2001.
12. Y. Saito, C. Karamanolis, M. Karlsson, and M. Mahalingam. Taming Aggressive Replication in the Pangaea Wide-Area File System. In *Proceedings of the 5th Symposium on Operating Systems Design and Implementation*, Boston, MA, December 2002.
13. I. Stoica, R. Morris, D. Karger, F. Kaashoek, and H. Balakrishnan. Chord: A scalable Peer-To-Peer lookup service for internet applications. In *Proceedings of the 2001 ACM SIGCOMM Conference on Applications, Technologies, Architectures, and Protocols for Computer Communication*, pages 149–160, San Diego, CA, August 2001.
14. B. Zhao, J. Kubiatowicz, and A. Joseph. Tapestry: An infrastructure for fault-tolerant wide-area location and routing. Technical Report UCB/CSD-0101141, UC Berkeley, Computer Science Division, April 2001.

*i*SAN - An Intelligent Storage Area Network Architecture

Ganesh Narayan and K. Gopinath

Computer Science and Automation,
Indian Institute of Science
{nganesh,gopi}@csa.iisc.ernet.in

Abstract. This paper describes the motivation, architecture and implementation of *i*SAN, an "intelligent" storage area network. The main contributions of this work are: (1) how to architect an intelligent SAN that understands the storage-consumers[1] to serve them better; (2) how to realise this abstract architecture using existing technologies; and (3) to demonstrate the benefits that would accrue from such an intelligent SAN. Our results show that the *i*SAN approach has important benefits when compared with conventional SANs: *i*SAN facilitates storage sharing, is secure and offers better throughput. In this paper, we discuss two case studies as well as discuss how *i*SAN approach has been generic enough to capture a wide range of other requirements of SANs.

1   Introduction

The Internet revolution drives a relentless demand for data to match the accelerating growth in users, digital content and network bandwidth availability. The need to scale data/storage independently has been the primary catalyst for the emergence of a storage tier providing logical/physical separation of storage from the other services in a data center. The result is the advent of an I/O architecture wherein the storage devices and high-speed networks are integrated into forming I/O networks: both the storage and the storage-consumer remain connected to a high speed network and communicate using SCSI commands. These I/O networks, called Storage Area Networks (SANs), provide better scalability and throughput as compared to traditional bus-based storage architectures.

Yet, scalability and throughput requirements are not the only requirements required from such SANs. Multitude of application domains, from content distribution networks to providing storage services , demand a range of properties/services that a successful SAN architecture should support. Unfortunately, SANs, whether based on FC [26] or iSCSI [22], do not export sufficient *functionalities* that are of direct use to storage-consumers. This is because traditionally SANs are seen merely as a replacement for parallel SCSI bus. But as a distributed shared storage system, SAN is more than an extended SCSI bus: SAN based systems demands functionalities which are otherwise not needed in parallel SCSI based systems.

---

[1] A storage-consumer is the software layer that builds storage abstractions from block level storage provided by SANs. This layer typically includes, but not limited to, Volume Managers, File Systems and Data Base Management Systems.

L. Bougé and V.K. Prasanna (Eds.): HiPC 2004, LNCS 3296, pp. 262–273, 2004.
© Springer-Verlag Berlin Heidelberg 2004

For instance, consider storage sharing. Coordinating processes that access shared storage is not a problem in bus-based storage systems as any access to the storage is arbitrated by the storage-consumer to which the storage is physically connected; essentially there is no direct sharing of storage. In a shared storage system like a SAN where storage is directly accessible from multiple storage-consumers, sharing becomes a critical issue.

Besides, contemporary SANs are generally unaware of storage-consumer's exact requirements. For instance, different storage-consumers expect different security guarantees from a SAN depending upon the threat model that they foresee. Hence, it is beneficial to enforce security properties judiciously especially when different security guarantees exhibit significantly differing cost/performance profiles. For example, a storage-consumer that is built assuming a byzantine storage, block level encryption done in SANs may not be of much use. Similarly, with different concurrency control schemes providing varying degrees of consistency guarantees with varying costs[15], one may want to selectively enforce that particular scheme which is economical and best suits the storage-consumer's needs. Hence, SANs should be *intelligent* enough to provide a set of guarantees that serves a storage-consumer best.

In this paper, we propose a novel SAN architecture called *i*SAN. *i*SAN identifies and provides services that are of direct use to the storage-consumers. In providing these services, *i*SAN also "understands" the service semantics sought by the storage-consumers and provides the needed service in a way that best suits the storage-consumer's requirements. The proposed *i*SAN architecture is extensible and generic; it switches the protocol stack based on the storage consumer.

The rest of the paper is organised as follows. Section 2 discusses the requirements that an I/O architecture should satisfy. In Section 3, we describe the architecture of *i*SAN. Section 4 describes our implementation of the proposed architecture which uses Linux kernel and Ensemble [10]. In section 5, we discuss how two important SAN services – concurrency control, security – are realised in *i*SAN. We compare the performance of the suggested solutions with that of existing solutions in Section 6. We conclude the paper in section 8.

2   Design

*i*SAN design is founded on a number of requirements; some of the them, like throughput and scalability, are inherited from basic SAN architecture with little or no enhancements. This section explains each of these requirement and discusses their applicability in state of the art SANs.

Interoperability. In a SAN, a path from a storage-consumer to any storage device may include various combinations and permutations of host bus adaptors, hubs or switches and SCSI peripherals. Not all permutations and combinations are feasible even if all the subsystems have been built using the same network technology, let alone with dissimilar technologies. Thus a critical and essential feature of any SAN architecture is ensuring interoperability of components within the SAN.

Fibre Channel SANs have been suffering from interoperability problems as the higher level FC standards (especially planned at the FC-3 layer) have never been standardised.

Zoning implementation is very much vendor dependent; FSPF protocol which ensures switch-to-switch routing is still proprietary; public and private loop devices still have problems when it comes to fabric logins. While network technologies like TCP/IP and Ethernet have a strong tradition of multi-vendor interoperability, FC devices often do not. Only recently, approaches such as SMIS have shown interoperability at the level of discovery of device information but this does not still guarantee interoperability at the services level. This poses serious problems in the design, procurement, and operation of SANs and interoperability is thus becoming an increasingly important requirement.

Throughput. With high speed network and disk interfaces, SANs are expected to be able to move the data as fast as possible. However, as the carrier bandwidth increases, the transport protocol inefficiencies at the end-points have a negative impact on the effective delay and throughput. For instance, the effective throughput of TCP over a gigabit network, without considerable hardware support from NIC, is only a fraction of the realisable throughput [9]. Even with zero-copy and checksum support from gigabit NICs, TCP has other problems. Extensions have been proposed but not many NIC products properly implement these extensions. In addition, the complexity of TCP has to be taken into consideration. Also, with multiple independent TCP connection(s) between the SCSI end-points, iSCSI handles issues like congestion control less effectively. Importantly, failure detection becomes non-trivial as different TCP connections can break at different time instants, depending upon their respective past activity. Each of these shortcomings can affect the observable throughput in iSCSI SANs.

Even though FC SANs have better throughput than iSCSI SANs in local environments, this advantage is considerably less when it comes to WAN links due to the credit based flow control scheme used in FC SANs [25]. Given these observations, it is appropriate that a SAN leverages a SCSI transport protocol that is fast, efficient and simple, and has better flow control support.

Availability. Today faced with critical need to ensure the availability and continuous operation in spite of isolated failures of disk, switch and links or the catastrophic loss of the computing/communication facilities, SANs need to be highly available. While FC SANs provide subsecond reconfiguration periods in case of a component failure, the traditional Spanning Tree Protocol (STP) employed in Ethernet takes tens of seconds to converge; it is to be noted that during the reconfiguration phase the extended Ethernet LAN is "frozen". STP also has other problems: inefficient bandwidth usage, link blockage and STP is Virtual LAN (VLAN) unaware. A combination of Rapid Spanning Tree protocol (RSTP) and link aggregation would reduce the reconfiguration stalls to even tens of milliseconds. However, RSTP does not use the bandwidth effectively and still has link blockage problems.

Apart from ensuring the availability of SAN infrastructure, a SAN should also provide primitives that help storage-consumers to ensure data availability: multicast is one such primitive that helps in providing availability using replication. However, properties provided by the traditional hardware multicast are not *sufficient* to ensure the mutual consistency of replicated copies; often times the network multicast is combined with protocols providing stronger guarantees like failure atomicity and message ordering to handle replication more effectively. If a SAN is to provide stronger multicast guarantees, it will ease the effort needed in providing transparent, block level replication.

Storage Sharing. An I/O architecture permitting sharing enables seamless fail-over of the storage-consumers that share data sets. Thus storage sharing is crucial to provide uninterrupted service. Shared storage architectures also provide better scalability since storage capacity and processing power could be added dynamically to the pool by adding more storage-consumers and storage. Additionally, data sharing gives high flexibility for dynamic load balancing since the data is uniformly accessible from any storage-consumer. Sharing also facilitates storage consolidation which reduces management and operation costs while increasing system usage.

But to achieve effective data sharing, SANs may need to assure certain properties at the network level. For instance, in order for a storage-consumer to failover correctly, SAN may need to ensure mechanisms such as I/O fencing. In fact, many commercial parallel database clusters expect the underlying cluster transport to provide I/O fencing. Also, concurrent access to the shared storage has to be mediated through certain concurrency control mechanisms. Neither FC nor iSCSI provide any concurrency control primitives; they do not provide I/O fencing also.

Security. The traditional, bus-based based storage-consumers are built assuming that the connected storage is inherently secure. But in a distributed storage system like SAN, such assumptions hold no longer true. In order to bridge the easy migration path for legacy systems, which were built assuming the physical security of storage, SAN should provide means of enforcing the needed security by other means, say, cryptographically. Presentday FC SANs are built around relatively secure fiber transport and are not yet equipped to enforce cryptographic security. On the other hand, iSCSI SANs – which are built around insecure IP networks – enforce the necessary security using cryptographic protocols; to this effect, IPSec/IKE have been chosen as the cryptographic infrastructure for iSCSI SANs.

But, IPSec has many problems that are yet to resolved [8]; so does IKE ([18], [23]). Also, the iSCSI level CRC mechanism and TCP checksum do not co-exist harmoniously owing to strict layering restrictions; with multiple TCP connections and SCSI command ordering in place, handling CRC error induced resynchronizations efficiently could get problematic. Comprehensive IP multicast security is still very much in infancy; of the many suggested key management protocols, only SKIP [2] discusses security in multicast communications explicitly. However, the problem concerning SKIP, and in general, IP multicast is the fact that they are *membership unaware* – a potentially inappropriate design for a restricted environment like SAN.

Intelligence. Traditional storage-consumers interact with the storage using standard storage protocols like SCSI. The storage-consumers are unaware of the underlying storage technology. This enables an easy migration path from a direct access storage system to a SAN. However, the converse is that a SAN is unaware of storage-consumers that could result in poor performance. For instance, in a shared storage system like SAN, the correctness criterion for permissible concurrent interleaving is very much storage-consumer/application dependent.

If a SAN uses strong consistency models like linearizability [16] as default, it may be grossly inefficient since there are many storage-consumers who can manage with much weaker consistency guarantees. Thus a SAN should be intelligent enough to deploy the right concurrency control mechanism that is sufficient for the storage-consumer's requirements; this is especially important as different consistency mechanisms may incur different cost/performance tradeoffs [15]. In general, the data access path of a particular storage-consumer should be efficiently tailored to the exact needs of the storage-consumer and SANs should have provisions for doing so.

Though the "Keep it Simple, Stupid!" [21] approach works well in networks, it may not always be efficient: there are certain problems that do not permit efficient solutions in systems designed strictly using end-to-end arguments. QoS, multicast and VPN are such problems whose efficient solutions demand certain amount of intelligence in the intermediate nodes or *switches*. Thus, SANs should have enough intelligence to handle these critical problems efficiently.

3   Architecture

*i*SAN uses Type I Logical Link Control (LLC) [11] for transport, Group Communication System (GCS) for membership services and VLAN for grouping; VLANs in *i*SAN, pruned based on MAC address, provide efficient application/SCSI level routing. The edge switches that are part of a VLAN form a group with membership respecting strong virtual synchrony (Fig. 1). The group end-point that is housed in an edge switch, called **Sanlet**, acts as a SCSI Target emulator: SCSI commands sent by storage-consumers are received by the Sanlet and are dispatched to the appropriate physical storage device. Thus Sanlet can be seen as a thin Volume Manager residing in the edge switch that manages the virtualized storage. Fig. 2 depicts the flow of data in *i*SAN.

In *i*SAN, service discrimination is done by allowing a group to have a *tailored* protocol stack that matches the storage-consumers. We argue that this approach – called protocol composition, provides enough flexibility so that storage-consumers with very different goals can potentially agree on sharing of a common infrastructure, within which their commonality is captured by layers that they share, and their differences reflected by layers built specifically for their needs. Since the VLAN, by construction, houses storage-consumers with similar requirements, it is natural to use the VLAN tag to choose between various built-in protocol stacks. Presently all *i*SAN stacks share a common set of lower layers – the set formed by set of microprotocols that assure strong virtual synchrony; any further service specialization is done on the foundation of virtual synchrony.

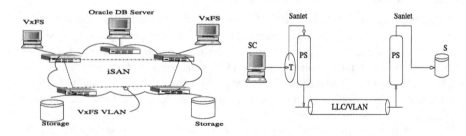

SC–Storage Consumer; S–Storage; PS–Protocol Stack; T–Target

Fig. 1. Usage of VLANs in *i*SAN   **Fig. 2.** Flow of data in *i*SAN

3.1   Design Criteria Revisited

The *i*SAN architecture proposed is interoperable and is likely to provide better through-put and latency guarantees as it is based on standards such as 802.3 LLC and VLAN. By providing virtual synchrony and stronger multicast guarantees (atomicity and ordering), *i*SAN effectively facilitates replication based high availability schemes. However, the high availability of *i*SAN infrastructure itself does not directly follow from the pro-posed architecture. and we do not discuss this aspect of *i*SAN any further in this paper. Based on strong virtual synchrony model, *i*SAN automatically provides basic sharing protection like I/O fencing. Other critical aspects of sharing like concurrency control are discussed in section 5.1. As discussed in section 5.2, *i*SAN deploys Ensemble's *fortress* model of security which is amenable to proofs of correctness. Above all, the protocol composition mechanism presents *i*SAN with an efficient means of distinguishing and servicing different storage-consumers effectively.

4   Implementation

*i*SAN is implemented using Ensemble group communication system [10] and Linux kernel v2.4. The reason for choosing Ensemble are many fold. First, it is a well engineered GCS toolkit with a number of of unique functionalities such as support for composable protocol stacks. Secondly, Ensemble has a powerful, provably correct, multicast aware security infrastructure [19]. Besides, Ensemble is event driven and is considered to be superior to thread based systems like Horus .

Ensemble is implemented using OCaml and is provided as a user level library that could be linked to an Ensemble application. But for our purposes, we needed a version of Ensemble that is written in C so that we could port it to Linux kernel. C-Ensemble serves this purpose precisely. We ported C-Ensemble distribution to Linux kernel with minimal changes: This is achieved by wrapping system calls to provide a libc like interface accessible from inside the kernel and by rewriting the event handling code. We also ported the Ensemble's total, causal ordering protocols to C-Ensemble Linux kernel port because C-Ensemble provides only part of Ensembles' functionality. We added a simple application level credit based flow control protocol to C-Ensemble. We modified both Ensemble and C-Ensemble to include LLC transport provided by Linux native LLC implementation. However, we have not yet ported the security protocols of Ensemble to C-Ensemble. Hence, *i*SAN currently uses the Ensemble/user (v1.33) for the security experiments while the other experiments are done with C-Ensemble/kernel. Sanlet uses a simple implementation of the target emulator software to emulate the SCSI target. This has been written from scratch to be Ensemble friendly. Please consult [17] for further information on implementation and configuring *i*SAN.

5   Case Studies

This section discusses the implementation of two critical storage services – concurrency control and storage security – in *i*SAN and shows how the advantages of the *i*SAN approach. We demonstrate that these two critical services, though seemingly dissimilar, can be implemented efficiently in an iSAN. We believe that this diversity speaks for the generality and expressibility of iSAN.

5.1   Concurrency Control

In a distributed, shared storage system like SAN, the logical view of the storage seen by the storage-consumer can be very much different from the physical view. For example, what the storage-consumers see as a contiguous blocks of storage may not be contiguous at all; worse, may not even be from a single storage device. This is because the underlying storage system may transparently offer functionalities like striping and virtualization. Also, storage systems might impose hidden relationships among the stored data, for example, in the form of shared parity blocks which needs careful interleaving of I/O accesses. Thus, unless proper care is taken to resolve concurrent stripe accesses, a storage-consumer may see inconsistent data irrespective of the fact that it may itself be orchestrating some concurrency control mechanism at higher levels [1].

Concurrency control is the activity of coordinating the actions of processes that operate in parallel, access shared data, and therefore potentially interfere with each other [3]. The unit of a concurrent access, called transaction, consists of several lower level operations which are expected to be executed *atomically*. There are four types of concurrency control schemes that are prevalent in the literature – locking, timestamp ordering(TO)[2], optimistic and hybrid.

Of the four afore mentioned schemes, TO emerges out as the optimal mechanism for shared storage ([1], [24]). However, there are atleast three problems that are associated with TO. First, it requires synchronized clocks. Highspeed networks like SAN may not increase the synchronization accuracy dramatically for atleast two reasons: synchronization accuracy is bound by message delay variance and not by the absolute delay ([14], [4]); also, clock synchronization messages – being few tens of bytes long, may not see significant latency reduction even in gigabit networks. However, highspeed networks will need to handle higher number of active transactions in a given time-slice and hence require *better* clocks. Second, if the transactions are to come in some wildly different order from the original issue order, TO will reject many transactions; [7] shows that probability of such an occurrence could be high. The magnitude of network reordering depends on the existence of redundant links, their configuration and network load; not all of them are completely controllable. Thirdly, the order of transaction executions as governed by the TO scheduler may or may not be conforming to certain expected ordering like causal order [13], depending upon the granularity of clock synchronization and the delay characteristics prevailing. Fig. 3 depicts how causal violation could happen in a master slave clock synchronization setting. Thus, if one needs deterministic ordering of messages, TO cannot be used to enforce such ordering reliably.

iSAN uses message ordering protocols for concurrency control[3]. It is understood that different ordering mechanisms – FIFO, casual, causally constrained total order

---

[2] The timestamp ordering discussed here and elsewhere in the paper is Basic Timestamp Ordering; we also assume strict schedules.

[3] We assume that the storage device commits operations in issue order. This is not a unreasonable assumption as most of the commercial RAID systems guarantee this.

and unconstrained total order[4] – guarantee different consistency semantics with cost and strictness increasing in that order. *i*SAN, being storage-consumer aware, deploys the suitable message ordering that *suffices* the storage-consumer's requirements. To this effect, we have considered five classes of storage-consumers and mapped their requirements to particular message ordering protocol.

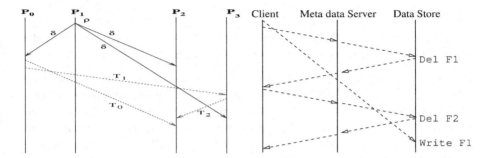

Fig. 3. Master P. broadcasts the recent clock reading δ, which includes the drift ρ, to slaves. T_i represents the clock reading prefixed to the message(m_i) sent. Since these two events, clock correction and message exchange, happen independently, causality violation may occur.

Fig. 4. After sending the *write F1* to the data store, the application deletes the file (*delete F1*) and creates a new file using *create F2*. The metadata server allocates the deallocated blocks from F1 to F2. Thus a delayed *write F1* corrupts the F2's data that was freshly written.

Parallel File Systems. Parallel File Systems (PFS) cater to the I/O requirements of multiprocessor/computer systems. Traditionally, PFS is organised as a set of clients – where the applications run, and servers – which serve the storage. Most PFSs do not favour client side caching and it is to be noted that the PFS clients *can* tolerate minor inconsistencies in the shared state when the conflicts occur. In a PFS VLAN, *i*SAN will not provide any ordering save the FIFO ordering of stripe updates between the client and the IOD. This provides the expected behaviour with little or no additional cost. If one is to deploy TO or strict 2PL, the overhead is very likely to be high as it provides *stricter* consistency guarantee than what is actually needed. Thus, by making use of the semantics of PFS like filesystem, *i*SAN *increases* the amount of concurrency available for shared accesses, leading to better performance.

Hybrid Storage. In many of the SAN based architectures, SAN is hidden behind the fileservers or database servers and the effective throughput seen by the clients is thus still limited at the rate at which these storage-consumers are able to cater to the requests. One way to solve the problem is to let the clients to access storage directly while expecting the storage-consumer to maintain the necessary metadata including the block map information; once the client gets the metadata information from the storage-consumer, say after a file open, it can access the storage directly. Any metadata data update will

---

[4] A total order that respects causal order is referred as causally constrained total order while a total order that may not respect causal order is referred unconstrained total order.

still involve the storage-consumer while the data is accessed directly from the storage. This architecture will be referred here as hybrid storage (often called as out-of-band virtualization).

Given that metadata server and the client may both access the storage simultaneously, one needs to ensure correct interleaving of these operations; Fig. 4 depicts one such problem case that would arise otherwise. A closer examination of Fig. 4 reveals that the problem is indeed due to causal violation: *delete F*1 should have been delivered after *write F*1 message as the latter *causally precedes* the former. So, in hybrid storage, Sanlets will enforce causal ordering of requests. Such mechanisms would improve the asynchrony of the system while adding very little overhead.

Database Systems. Database systems do not favour one serial schedule over the other: all the strict executions are equally correct. Yet, in order to avoid distributed deadlocks, a DB may want to prune a total order out of conflicting transactions as in TO. However, the overhead of synchronizing clocks could be averted if one is to use unconstrained total ordering protocol in place of clocks. The Sanlet connecting DB to storage will thus need to enforce the unconstrained total order and, as a side effect, will solve the problem of mis-ordered transactions. [24] provides a total order based concurrency control protocol that is readily deployable in *i*SAN.

Replication. Replication is an area of interest to both filesystems and databases and hence is of interest to *i*SAN. Replication protocols come in variety of forms and hues, differing in aspects like models, assumptions, mechanisms, guarantees provided, and implementation[28]. Many replication schemes, notably the lazy schemes, will require certain update ordering and in *i*SAN this ordering is effected by using the causally constrained or unconstrained total ordering.

Log Enhanced Filesystems. Log enhanced filesystems like VxFS needs to commit the metadata changes to log before it starts making changes to the on-disk filesystem structure. But in order to ensure that the log writes reach the disk before filesystem updates, the log is usually written *synchronously*. Thus for every metadata change, the filesystem suffers a synchronous log write. But, if the underlying storage layer, .i.e. *i*SAN, is to provide FIFO ordering of commands, the filesystem need not have to write the log records synchronously; it only has to *queue* the log write before the corresponding metadata update. Since the storage layer assures FIFO delivery of commands, by the time the metadata updates reach the disk, the *previously* scheduled log writes would have reached the disk too. Thus, providing FIFO ordering at *i*SAN layer would improve the observed filesystem throughput of a journaling filesystem.

5.2   Storage Security

Organizations increasingly depend on their storage infrastructure for storing critical information. Thus the I/O subsystem should *understand* the sensitivity of the data its serving and should ensure confidentiality, integrity, and availability of the data both in-storage and in-transit. Please consult [17] for further details.

6   Performance

This section describes the experimental setup and results of the conducted experiments. The setup consists of Intel machines ($\geq$ 700MHz) running Linux (v2.4) acting as "edge switches"; these edge switches are connected using a 100Mbps Ethernet. The Target and Initiator are housed in the same switch to reduce the network interruption; it is a priori observed that doing so does not significantly change the performance profile. A FC JBOD, organized as a RAID 5 with three disks, is connected to one of the "edge switches" and is used as the data store. Block level traces for VxFS are generated in a Ultra Sparc machine running Solaris by running 4 benchmarks – ssh, ssl, gcc and postmark [12]. The other traces used are the HP traces [20]. For a detailed description of experimental setup and the related information, please consult [17].

Table 1. CCTRL - Relative cost of different ordering protocols (with FIFO as base)

	fifo	causal	total	total+causal
#3 %	0	7.36	17.43	30.06
#4 %	0	3.64	16.04	20.23

Table 2. CCTRL - Ordered Vs Sync Writes - throughput improvement observed

	ssl	ssh	gcc	postmark
%	11.46	20.90	13.64	15.05

Table 3. CBC Vs ECB – Overhead

	ssl	ssh	gcc	postmark
%	0.97	1.38	0.41	4.73

Table 4. SEC - Throughput improvement for 7% plain data

ssl	ssh	gcc	postmark
28.90	29.67	16.67	11.29

All the above tables depict the percentage of throughput improvement achieved. Table 1 shows that the *relative* overhead of different ordering protocols (with FIFO as the base) is indeed significant; the rows are indexed by the cardinality of the group and the traces used are HP traces. Since we did not have any shared traces, we used the 3 disk streams in HP to emulate shared access. The results signify that the storage consumers indeed benefit from the selective deployment of ordering mechanisms. For instance, causally constrained total order is costlier by 30% compared to FIFO for a group size of 3. However, increasing the group cardinality reduces this performance disparity. This is due to the fact that 100Mbps Ethernet do not have efficient flow control. Lack of proper flow control at the lower layer significantly penalises the low overhead/high throughput FIFO ordering and it explains the reduction in relative overhead. Table 2 compares the performance of ordered log writes compares and synchronous log writes. This supports our argument that ordering of I/O commands helps even in a non distributed setting.

For the security experiments, the throughput difference between ECB-for-all and CBC-for-journal-alone are observed to be with less than 5% range(Table 3). *i*SAN thus achieves increased security at almost negligible cost. The results of experiments wherein the storage-consumer/driver controlling the Sanlet *dynamically* using in-band messages are depicted in table 4. The experiment is conducted by allowing roughly 7% [5] of the block access to be transmitted in plain text; this is because the block level traces generated do not have file names and their attributes. The 7% of block access that are transmitted in plain is uniformly distributed across the total accesses. The results show that the throughput improvement observed is indeed significant.

7 Related Work

Intelligence in iSAN is achieved using dynamic protocol composition which stands comfortably midway between the Turing complete Active Networks [27] and the quasi-static Programmable Networks[6]. Composition, unlike Programmable Networks, provides non trivial specialization, yet, unlike Active Networks, can be very efficient and secure. A work that is very similar in spirit to the proposed architecture is that of Virtual Overlay Networks (VON) [5]. However, our work is novel for many reasons: First, the architecture proposed in [5] is generic while our architecture is tuned to the requirements of SAN. Secondly, [5] is abstract and leaves many engineering issues like selection of the minimal Overlay Network, the vantage point where the code-stubs to be deployed, mode of contacting the code stub etc. open. Our architecture is more concrete, pinning down these crucial design parameters. Thirdly, to our knowledge, ours is the *first* application/realization of [5]. For a detailed comparison of similar works and iSAN, please consult [17].

8 Conclusions

In this paper, we have presented a design and implementation of an intelligent SAN architecture. We have demonstrated how this architecture can be used to efficiently solve some of the critical problems associated with conventional SANs, and evaluated the suitability of solutions. We would like to conclude that iSAN approach of architecting SAN shows great promise as a means of constructing efficient, yet, flexible SANs.

In long run, we would like to extend iSAN design to investigate the following aspects. First, we plan to add a scalable and manageable virtualization architecture to iSAN. The idea is to balance the virtualization overhead with the consumer-awareness of iSAN to arrive at a low cost virtualization scheme. Also, we plan to investigate how increasing asynchrony/concurrency at lower levels would improve performance in the higher layers and to suggest such a scheme as a design principle. Finally, through iSAN research, we are attempting to understand the synergy between virtual synchrony and filesystems.

References

1. K. Amiri, G. Gibson, R. Golding. Highly concurrent shared storage. *ICDCS*, Apr 2000.
2. A. Aziz, T. Markson, H. Prafullchandra. Simple key mgmt for Internet protocols (http://www.skip.org/), 1998.

---

[5] This number we have arrived at after observing the amount of public domain data that we found in our lab machines.

3. Philip A. Bernstein, Vassos Hadzilacos, and Nathan Goodman. *Concurrency Control and Recovery in Database Systems.* 1987.
4. S Biaz and J L. Welch. Closed form bounds for clock synchronization under simple uncertainty assumption. *Info. Processing Letters*, 2001.
5. Ken Birman. Technology requirements for virtual overlay networks. *IEEE Systems, Man and Cybernetics: Special issue on Information Assurance, Vol. 31, No 4*, July 2001.
6. Jit Biswas, et al. Application programming interfaces for networks - IEEE P1520, Jan 1999.
7. C. Bouras and P. Spirakis. Performance modeling of distributed timestamp ordering: Perfect and imperfect clocks. In *Performance Evaluation Journal*, Apr 1996.
8. Niels Ferguson, , and Bruce Schneier. A cryptographic evaluation of IPSec, February 1999.
9. Andrew Gallatin, Jeff Chase, and Ken Yocum. Trapeze/IP: TCP/IP at near-gigabit speeds. In *USENIX Technical Conference*, June 1999.
10. M. Hayden. *The Ensemble System.* PhD thesis, Computer Science Dept, Cornell Univ, 1998.
11. ISO. Logical link control - ISO/IRC 8802-2.
12. J. Katcher. Postmark: A new file system benchmark. TR3022, NetApp, Oct 1997.
13. Leslie Lamport. Time, clocks, and the ordering of events in a distributed system. *Communication of the ACM, vol. 21, no. 7*, July 1978.
14. J Lundelius and N Lynch. An upper and lower bound for clock synchronization. *Information and Control, Vol. 62, Nos. 2/3*, September 1984.
15. David Mosberger. Memory consistency models. *Operating Systems Review*, 1993.
16. Herlihy M.P and Wing J.M. Linearizability: a correctness condition for concurrent objects. *ACM Transactions on Programming Languages and Systems, 12(3)*, October 1990.
17. Ganesh Narayan and K. Gopinath. *i*SAN - an intelligent storage area network architecture. TR-IISc-CSA-2004-6, Computer Science and Automation, Indian Inst of Science, Aug 2004.
18. Radia Perlman and Charlie Kaufman. Analysis of the IPSEC key exchange standard. *IEEE Internet Computing 4(6)*, November 2000.
19. Ohad Rodeh, Ken Birman, and Danny Dolev. The architecture and performance of security protocols in the ensemble group communication system. TR2000-1822, Computer Science Dept, Cornell Univ, Oct 2001.
20. Chris Ruemmler and John Wilkes. Unix disk access patterns. TR, HP Labs, Dec 1992.
21. J. H. Saltzer, D. P. Reed, and D. D. Clark. End-to-end arguments in system design. *ACM TOCS 2(4)*, November 1984.
22. J. Satran, K. Meth, C. Sapuntzakis, M. Chadalapaka, Efri Zeidner. iSCSI, 2001.
23. W A Simpson. IKE/ISAKMP considered dangerous, June 1999.
24. Rashmi Srinivasa. *Network-Aided Concurrency Control in Distributed Databases.* PhD thesis, University of Virginia, January 2002.
25. Nishan Systems. Data storage anywhere, any time - metro and wide area storage networking with Nishan systems IP storage switches, 2000.
26. ANSI NCITS T10/1144D. FC protocol for SCSI, second version (FCP-2), rev 5, Nov 2000.
27. D.L. Tennenhouse, J.M. Smith, W.D. Sincoskie, D.J. Wetherall, G.J. Minden. A survey of active network research. *IEEE Comm. Magazine Vol. 35, No. 1*, Jan 1997.
28. M Wiesmann, F Pedone, A Schiper, B Kemme, and G Alonso. Understanding replication in databases and distributed systems. In *ICDCS*, Apr 2000.

Static Techniques to Improve Power Efficiency of Branch Predictors

Tao Zhang, Weidong Shi, and Santosh Pande

College of computing,
Georgia Institute of Technology, USA
{zhangtao, shiw, santosh}@cc.gatech.edu

Abstract. In this paper, we illustrate the application of two static techniques to reduce the activities of the branch predictor in a processor leading to its significant power reduction. We introduce the use of a static branch target buffer (BTB) that achieves the similar performance to the traditional branch target buffer but which eliminates most of the state updates thus reducing the power consumption of the BTB significantly. We also introduce a correlation-based static prediction scheme into a dynamic branch predictor so that those branches that can be predicted statically or can be correlated to the previous ones will not go through normal prediction algorithm. This reduces the activities and conflicts in the branch history table (BHT). With these optimizations, the activities and conflicts of the BTB and BHT are reduced significantly and we are able to achieve a significant reduction (43.9% on average) in power consumption of the BPU without degradation in the performance.

1 Introduction

Branch prediction has a huge impact on the performance of high end processors which normally have a very deep pipeline. Many studies have been done to improve branch prediction rate using complicated designs, those designs often demand a significant silicon and power budget. As claimed in [5], branch predictor can potentially take up to 10% of the total processor power/energy consumption. With a new metric dimension of power, the focus is how to maintain the same prediction rate as a complex branch predictor but with significantly less power consumption and area. In [5], the authors use two major techniques to reduce branch predictor power consumption, banking and PPD (prediction probe detector). PPD is a specialized hardware that records whether lookups to the BTB or the direction-predictor is necessary, therefore saving power by reducing lookups to BTB and BHT when combined with clock gating.

In this paper, we propose two new optimizations to reduce power consumption of the branch predictor with no degradation in the IPC or prediction accuracy but significant reduction in the branch predictor power consumption. The power consumption of a branch predictor is dominated by the large branch target buffer (BTB) and branch history table (BHT), both of which are introduced to achieve high prediction accuracy in a high-performance processor. To optimize the power

L. Bougé and V.K. Prasanna (Eds.): HiPC 2004, LNCS 3296, pp. 274–285, 2004.
© Springer-Verlag Berlin Heidelberg 2004

consumption of branch target buffer, we introduce static branch target buffer that does not need state updates during runtime except when the program phase changes; thus activities in it are reduced significantly. Using static branch target buffer, we are able to maintain the performance of traditional branch target buffer at the same time eliminate most of the power consumption due to the updates to traditional branch target buffer. To reduce power consumption of branch history table, we combine static branch prediction with hardware dynamic branch prediction. With a hybrid static and dynamic branch prediction, only those branches that are hard to predict statically turn to hardware prediction, reducing branch history table activities and collisions. Such a hybrid predictor can often attain the same prediction rate as a pure hardware predictor with a much smaller branch history table and much less predictor lookups and updates, therefore consumes less power.

2 Static Branch Target Buffer

To achieve a good address hit rate in branch target buffer, modern superscalar processors normally have a very large multi-way branch target buffer. This large buffer leads to high power consumption. Normally the power consumption of the branch target buffer takes at least 50% of the total power consumption of the branch predictor[1].

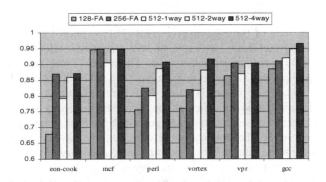

Fig. 1. Address hit rate of different BTBs

An intuitive approach to reduce the power consumption of BTB is to reduce the size of it. BTB is designed to be large because of two possible reasons. First, a large buffer helps reduce conflict misses. Second, a large buffer helps reduce capacity misses. If the main reason for a large BTB is the conflict misses, we can increase the associativity of the BTB. However, according to our experiments, capacity misses are major problems. Some programs have large working sets. To ensure a good address hit rate for those applications, a large BTB must be deployed. Figure 1 shows the address hit rate for six SPEC2000 benchmarks. In this paper, we will mainly study these six benchmarks because they exhibit relatively worse branch prediction

---

[1] In this paper, when we talk about branch predictor, we mean both branch direction predictor and branch target buffer.

performance in our experiments. The configurations are 128-entry fully-associative BTB, 256-entry fully-associative BTB, 512-set 1-way BTB, 512-set 2-way BTB and 512-set 4-way BTB. From Figure 1, for benchmarks like perl, vortex and gcc, even a 256-entry fully-associative BTB cannot achieve a comparable address hit rate to the other configurations that have much less ways but a larger number of entries. Since a fully-associative BTB does not have conflict misses, the above finding shows that some benchmarks have large working sets and complex branch behaviors, requiring a large BTB.

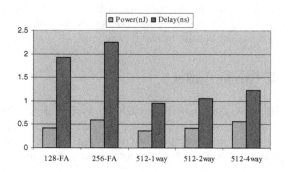

Fig. 2. Per-access Power Consumption and Delay of Different BTB Configurations

The address hit rate of fully-associative BTBs is very good for some benchmarks like mcf. For those benchmarks, we can achieve a comparable address hit rate using a fully-associative BTB with much less entries. However, Figure 2 shows per-access latency (in ns) and per-access power consumption (in nJ) of different BTBs got using CACTI timing and power model (version 3.0). From Figure 2, we can see fully-associative BTBs are not power efficient comparing to multi-way BTBs and the per-access delay of them are several times larger. Such kind of delay is intolerable to high-end processors. So the conclusion here is that we have to maintain a large enough BTB at the same time avoid introducing high associativity into the BTB.

Another way to reduce the power consumption of the BTB is to reduce its activities. With clock gating used extensively in modern processors, the power consumption of a particular function unit is significantly impacted by the activities it undertakes. Unfortunately, to reduce activities of the BTB is also a hard problem. To minimize pipeline stalls, the processor needs to know the target address of a branch as soon as possible if the branch is predicted as taken. The target address is provided by the BTB. In a traditional superscalar processor design, the processor will access BTB during the instruction fetch stage, so that the predicted target address can be fed into next fetch stage without stall. In [5], the authors proposed to reduce the activities due to BTB lookups by only accessing it when necessary. Our work is built upon the optimization proposed in [5]. We assume non-branch instructions have been filtered using a mechanism similar to the one in [5] and they will not lead to BTB lookups. Thus, our scheme can just deal with branch instructions.

We propose static branch target buffer to reduce the activities due to BTB updates. Traditionally, whenever a branch instruction is completed, the BTB is updated to remember its target address. With the optimization in [5] enabled, BTB updates may account for half of the BTB activities. The basic idea is that if we fix the content of the BTB, then no BTB updates are necessary. A naïve implementation could be that the processor preloads BTB content when the program starts and the BTB content is never changed during the program run. This naïve implementation may work for small and simple programs but suffers from a limitation for large and complicated programs, which may run for a long time and exhibit phase change significantly. On the other hand, as shown in [13], although a complicated program does not exhibit a globally stable behavior, its execution can be divided into phases, in each phase the behavior is quite stable. In our static branch target buffer design, program phases are dynamically identified and program phase changes are dynamically detected in the processor. Upon a program phase change, the processor loads the BTB content corresponding to the new phase then the BTB content is fixed until next program phase change.

We use profiling to choose the proper branches to reside in the BTB for each phase. We first categorize branches encountered in the phase according to the set location it will reside in the BTB. For example, a 512-set BTB will have 512 different set locations. Normally, there will be multiple branches belonging to one set location due to collisions. Next, we identify the most frequent ones belonging to a set location through profiling. The number of branches chosen is equal to the number of ways in the BTB. In that way, we choose a subset of branches which just fit into the BTB.

This idea works because program phase changes are infrequent events, otherwise, the power consumed by preloading BTB may eliminate the savings from eliminated BTB updates. We adopted the phase identification scheme from [13], which is very cost effective. Our experiment shows that GCC has the most unstable phase behavior, but on average the phases of GCC still have a length of about 1 million instructions. As pointed out by [13], integer programs tend to have much shorter program phases and much more program phase changes. For floating point programs, the length of program phases is normally tens of million instructions. All of our examined benchmarks are integer programs to stress test our scheme. Our static branch target buffer design will achieve even better results with floating point benchmarks.

There are several pitfalls regarding to static branch target buffer idea. Since we fix the BTB content for each phase now and it is possible that we cannot put all the branches seen in the phase into the BTB, static BTB may degrade address hit rate and runtime performance. However, sacrificing performance means the program will run for a longer time thus the other components in the processor will consume more power! Thus, although static BTB can reduce branch predictor power significantly, if the performance is degraded a lot, we may end up consuming more energy on the whole processor scale. Fortunately, we see near zero performance degradation under our static branch target buffer design. The major reason is that the phase identification scheme works well and captures program phases accurately. Another reason is that lots of BTB misses are actually due to some branches continuously kicking each other out of the BTB, reducing BTB effectiveness. That means fixing the content of BTB may instead help in some cases.

Static branch target buffer also introduces additional overhead to context switches. BTB content now becomes a part of process state. When a process is switched out, the current BTB of the process has to be saved. After a process is switched back, the corresponding BTB needs to be restored. However, context switches can be regarded as rare events in a processor. For example, the default time slice in Linux is 100ms, during which tens of million instructions could have been executed. In our experiments, the power consumption of BTB preloading and phase identification has been modeled and counted in the total power consumption of the processor.

3   Correlation-Based Static Prediction

The power consumption of branch history table is another major source of branch predictor power consumption. Branch history table and branch target buffer normally take more than 95% of total branch predictor power. To reduce the power of BHT, we propose static branch prediction. The basic idea is that those branches that can be statically predicted will not go through the normal BHT lookup algorithm thus will not access BHT. So the accesses to the BHT are reduced, leading to lesser BHT activities thus reduced power consumption. The reduction of BHT accesses also leads to reduction of conflicts in BHT. Thus, a small history table can always achieve the same prediction rate as the traditional predictor with a much larger table. With the same BHT table, static branch prediction may help achieve a better prediction rate due to fewer conflicts in BHT.

The possible strategies of static branch prediction could be always predicting a branch is taken, or always not-taken, or branches with certain op code always taken, or branches with certain op code always not-taken, or always predict backward conditional branches as taken. Another approach for static branch prediction is to rely on compiler program analysis or profiling to come up with a direction hint to a conditional branch. The hint can be encoded with a single bit in the branch instruction. [11] shows that a large percentage of branches are highly biased towards either taken or not-taken. In particular, 26% of conditional branches have a taken-rate of over 95% and another 36% of branches have a taken-rate below 5%. Together, over 62% of conditional branches are highly biased toward one direction. For such conditional branches, their predictions can be hard encoded and this will reduce (1) accesses to the BHT; (2) branch history table entries required for dynamic branch prediction; (3) potential conflicts and aliasing effects in BHT [4, 6]. In our approach, we further extend previous either taken or not-taken static prediction to correlation-based static prediction.

if (x>2)	x>2	y<10	x<y
x = 0;	false	false	true
if (y<10)	false	true	?
y = 0;	true	false	true
if (x<y)	true	true	false

Fig. 3. Example of branch correlation

[12] shows that many conditional branches can be predicted by using only a short global history. This is the manifestation of branch correlation and gives the insight that many conditional branches may have fixed prediction patterns under a short global branch history. Figure 3 gives an example of branch correlation and prediction pattern.

This leads to the design of correlation-based static branch prediction. Correlation-based static branch prediction requires hardware assistance to record the recent global branch history. In many branch predictors, e.g., gshare predictors, such information is already there for dynamic branch prediction. The hardware chooses a prediction based on the current global branch history and the correlation pattern provided by the branch instruction. By using correlation-based static prediction, we can predict more conditional branches statically, which could further reduce area and power consumption of a dynamic branch predictor.

Static correlation pattern is conveyed to the hardware using encoding space in the displacement field of a conditional branch instruction, which is done by the compiler (binary translator). Other instructions are not affected. Conditional branch instructions in most RISC processors have a format such as [op code][disp]. For Alpha 21264 processor, the displacement field has 21 bits. However, most conditional branches are short-range branches. This leaves several bits in the displacement field that can be used for encoding correlation pattern for most conditional branches. In our scheme, we use four bits in the displacement field (bit 4 to bit 1) to encode the prediction pattern based on two most recent branches. There are four possible histories for two branches: 1) [not-taken, not-taken], 2) [not-taken, taken], 3) [taken, not-taken], 4) [taken-taken]. Each bit of the four bits corresponds to the static prediction result for one possibility. Bit 1 corresponds to case 1, and so on. If we statically predict the branch taken, the corresponding bit is set to 1, otherwise, it is set to 0. Using the same example in Figure 3, the encoded correlation pattern for the third branch could be 0101, assume through profiling, we found if the branch history is [not-taken, taken], the third branch has a better chance to be not-taken.

Compiler has the freedom to choose whether correlation-based static prediction should be used. When it becomes unwise to use static prediction, e.g., hard to predict branches, or the displacement value is too large, compiler can always turn back to original branch prediction scheme. In our scheme, we take another bit (bit 0) of displacement field to distinguish extended branches with correlation information from original branches. Thus, we take five encoding bits in total from displacement field for extended branches.

First, we have to decide which conditional branches should use static branch prediction and which should use dynamic branch prediction. A conditional branch is classified as using static prediction if it satisfies the following criteria:

- highly biased towards one direction (taken or not-taken) under at least one correlation path (branch history).
- branch target address's displacement within the range permitted by the shortened conditional branch displacement field.

To evaluate the potential gains offered by correlation-based static prediction, we divide statically predictable conditional branches into three types:

- Non-correlation based type. This corresponds to conditional branches that can be statically predicted using a single-bit direction prediction.
- Correlation-based type. For this type of branches, prediction is biased towards different direction depending on the correlation path.
- Others. This type corresponds to the branches that cannot be categorized to type I or type II.

Table 1 lists the total number of conditional branches for each type, and the total number of dynamic executions made by the branches in each type for six SPEC2000 benchmarks running for 200 million instructions with fast-forwarding 1 billion instructions. As shown in the table, many branches can be predicted statically.

Table 1. Conditional Branch Categorization

	# static branch	# dynamic branch	type 1 static (%)	type 1 dyna. (%)	type 2 static (%)	type 1 + 2 dyna. (%)
eon	4006	14381689	34.95	52.32	37.09	62.18
mcf	678	27299410	60.18	79.92	64.45	80.37
perl	4920	29600039	86.63	68.92	89.59	70.51
vortex	7830	32691578	91.74	84.14	93.74	85.52
vpr	1917	38016432	81.27	54.71	86.18	66.32
gcc	18166	34023907	79.58	49.28	86.03	53.56

One pitfall of encoding correlation information into the branch instruction is access timing. The processor has to know the next PC address for instruction fetch at the end of current instruction fetch stage. Traditionally, during instruction fetch stage, the processor has no access to specific bits of the fetched instruction. To enable correlation-based static prediction, we use a separate extended branch information table (EBIT) with a number of entries exactly corresponding to the level-1 I-cache cache lines. Under our processor model, cache line size is 32B and contains 8 instructions. We impose a limitation that in each cache line, there is at most one extended branch instruction with correlation information, which means that there is at most one extended branch instruction in every 8 instructions. The limitation has minor impact in our scheme. As shown in [5], about 40% of conditional branches have distance greater than 10 instructions. Moreover, only a part of conditional branches will be converted into extended branches. Each entry of EBIT is 8-bit information. Bit 7 indicates whether the cache line has an extended branch instruction. Bit 6 to 4 encodes the position of the extended branch in the cache line. Bit 3 to 0 records the correlation information for the extended branch if there is one. The size of EBIT is 4K bits.

The EBIT is updated with new pre-decoded bits while an I-cache line is refilled after a miss. During each instruction fetch stage, the EBIT is accessed to detect an extended branch instruction and obtain the corresponding correlation information. If the current instruction is an extended branch, further BHT look-up is not necessary. Otherwise, BHT is accessed. Thus, the EBIT access is done before any necessary BHT access. Since the EBIT is small, we assume EBIT access and BHT access together can be done in instruction fetch stage. Note if the current instruction is an original branch, we end up consuming more power since we have to access EBIT too. The validity of correlation-based static prediction relies on the fact that a large percentage of conditional branches can be predicted statically, as shown in Table 1. The power consumption of EBIT is modeled and counted in our experiments.

4   Experiments and Results

We use Simplescalar 3.0 plus wattch version 1.02 for performance simulation and power analysis. We simulated a typical 8-wide 5-stage superscalar processor. We choose BTB and BHT configurations comparable to the ones in [5].

We chose six SPEC2000 integer benchmarks that exhibit relatively worse branch prediction performance. Each benchmark was fast-forwarded 1 billion instructions then simulated for 200 million instructions. The profile information for each benchmark is gathered using standard test inputs. Then standard reference inputs are used to measure performance. Unless explicitly stated, all the branch predictor power consumption results reported are obtained under Wattch non-ideal aggressive clock-gating model (cc3). The power consumption of BTB preloading, phase identification and EBIT is measured using Wattch array structure power model.

Our scheme is built upon the scheme in [5]. We assume only branch instructions will lookup BTB. Thus, BTB updates account for almost half of the BTB activities.

First, we present the results for static branch target buffer. In our study, we assume a 16K PAg direction predictor without static prediction. Three common BTB configurations are examined: 512-set 1-way, 512-set 2-way and 512-set 4-way. Figure 4 shows the normalized address hit rate of static branch target buffer scheme. The address hit rate is normalized to original branch target buffer design. From Figure 4, the impact of our static branch target buffer to address hit rate is minor. From the results, we also observe that the degradation on address hit rate becomes even smaller with the increase of the number of entries in BTB, since now more branches could be preloaded into BTB upon a phase change. *Mcf* benchmark under 512-set 1-way configuration is a corner case in which the address hit rate under static BTB is better than original BTB design. The reason may be destructive aliasing effect. Now that we seldom update BTB, we have much less chance to improperly kick out some branches from BTB.

Figure 5 shows the normalized IPC for static branch target buffer scheme. Note any optimization of power consumption to one component of the processor should not sacrifice the processor performance significantly. Otherwise, the program will run for a longer time and the power savings from the optimized component may be easily killed by more power consumption in other components. From the results, the IPC degradation is very small and become even smaller with a larger BTB. For 512-set 4-way configuration, the degradation is very small thus is not visible in the graph.

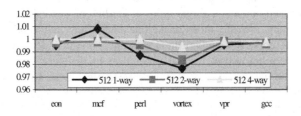

Fig. 4. Normalized Address Hit Rate for Static BTB

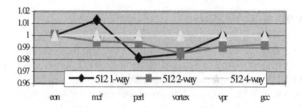

Fig. 5. Normalized IPC for Static BTB

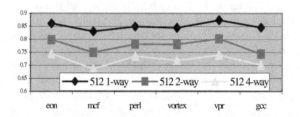

Fig. 6. Normalized Power Consumption for Static BTB

Figure 6 shows the normalized power consumption for the whole branch predictor (BTB + direction predictor) with static BTB design against one with original BTB design. The power saving comes from mostly eliminated BTB updates. The power consumed by BTB preloading and phase tracking reduces the saving but our experiments show they have insignificant impact. After all, our implemented phase tracking hardware is very simple and phase changes are very low-frequency events during execution. For 512-set 1-way configuration, the average power reduction is 14.89%. For 512-set 2-way configuration, the average power reduction is 22.41%. For 512-set 4-way configuration, the average power reduction is 27.88%. For each configuration, static BTB is able to reduce the BTB activities (dynamic power) by almost half. A larger BTB leads to larger power reduction in percentage because the percentage of BTB power is larger in the whole branch predictor power.

Next, we present the results for correlation-based static prediction. Static prediction works along with a dynamic hardware predictor, which handles branches not predicted by static prediction. A large number of hardware branch prediction schemes have been proposed in the literature. It is impossible for us to examine all of them. In this paper, we limited our scope to two types of predictors, i.e., gshare [9] and PAg [1, 2, 10] because they are the most basic ones and they can represent two large categories of predictors. We studied them separately to get a deep understanding of the impacts of correlation-based static prediction to different types of predictors. Further, we can derive the impact of our proposed technique to more complicated predictors like the tournament predictor in Alpha, which is basically composed of two global predictors and one local predictor.

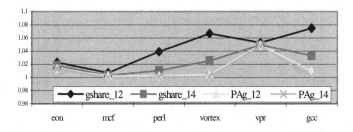

Fig. 7. Normalized Direction Prediction Rate

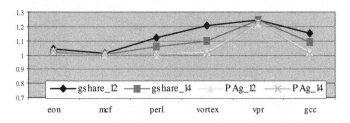

Fig. 8. Normalized IPC

Figure 7 shows branch direction prediction rate under our correlation-based static prediction and dynamic prediction hybrid scheme. The prediction rate is normalized to original pure dynamic direction predictor. In the study of direction predictor, we assume a 512-set 4-way BTB configuration. For gshare-based predictors, we experimented two sizes, 4K and 16K. For PAg based predictors, we also explored two configurations, 4K first level entries –11 bit local history and 16K first level entries – 13 bit local history. From the results, correlation-based static prediction helps prediction rate for gshare-based predictor significantly. For a 4K size gshare predictor, the average improvement in prediction rate is 4.39%. For a 16K size gshare predictor, the average improvement is 2.32%. With the increase of predictor size, the improvement on prediction rate due to correlation-based static prediction becomes smaller. Correlation-based static prediction cannot achieve significant improvement for PAg based predictors, since collision has much smaller impact to the performance of PAg based predictors. For a 4K size PAg predictor, the average improvement is 1.37%. For a 16K size PAg predictor, the average improvement is 1.05%.

Figure 8 shows normalized IPC for the same configurations. For 4K gshare predictor, the average improvement is 13.02%. For 16K gshare predictor, the average improvement is 8.61%. For 4K PAg predictor, the average improvement is 4.97%. For 16K PAg predictor, the average improvement is 4.52%.

Figure 9 shows the normalized whole branch predictor power consumption under correlation-based static prediction and dynamic prediction hybrid scheme. For 4K gshare predictor, the average power reduction is 6.2%. For 16K gshare predictor, the average power reduction is 7.2%. For 4K PAg predictor, the average reduction is 14.8%. For 16K PAg predictor, the average reduction is 22.2%. The reduction for gshare predictors is much smaller since there is no local history table and the power consumed by the EBIT is relatively large.

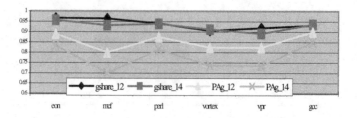

Fig. 9. Normalized Predictor Power Consumption

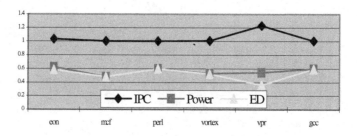

Fig. 10. Results for Combined Architecture

Our static BTB and correlation-based static prediction could be integrated together into the processor, so that we can achieve significant branch predictor power reduction at the same time with at least no performance degradation in branch predictor. For some benchmarks, the performance of branch predictor is actually improved. Figure 10 shows the results after integration. All the results are normalized against original branch predictor design. We assume a 512-set 4-way BTB configuration and a 16K PAg direction predictor. From the results, for all benchmarks, our optimizations never degrade IPC, i.e., the processor performance. For benchmark *vpr*, the performance is improved significantly instead. The average power consumption reduction is 43.86%. We also show the energy-delay product results, which is a metric to show the trade off between power consumption and performance. The average reduction in energy-delay (ED) product is even better, which is 47.73%.

5   Conclusion

In this paper, we proposed two optimizations to reduce the power consumption of the branch predictor in a high-performance superscalar processor. We raised the idea of static branch target buffer to eliminate most of the update power of BTB. We also pointed out that a hybrid branch predictor combining static prediction and dynamic prediction not only can reduce branch predictor size and destructive aliasing but branch prediction power consumption as well. We extended the conventional static prediction by putting branch correlation pattern into spare bits in the conditional branches when they are not used. Such correlation-based static prediction can further improve static branch prediction rate comparing to single direction static prediction. We explored the effects of the two proposed optimizations on the trade off between performance and power consumption. Our simulation results show that the integration of the proposed optimizations can reduce the branch predictor power consumption by 44% and branch predictor energy-delay product by 48% while never degrading overall processor performance.

References

[1] T. Y. Yeh, and Y. N. Patt. "Two Level Adaptive Branch prediction". 24th ACM/IEEE International Symposium on Microarchitecture, Nov. 1991.

[2] T. Y. Yeh, and Y. N. Patt. "A Comparison of Dynamic Branch Predictors that Use Two levels of Branch History". 20th Annual International Symposium on Computer Architecture, May 1996.

[3] Cliff Young, Nicolas Gloy, and Michael D. Smith. "A Comparative Analysis of Schemes For Correlated Branch Prediction". ACM SIGARCH Computer Architecture News , Proceedings of the 22nd annual International Symposium on Computer Architecture May 1995, Volume 23 Issue 2.

[4] S. Sechrest, C. C. Lee, and Trevor Mudge. "Correlation and Aliasing in Dynamic Branch Predictors". ACM SIGARCH Computer Architecture News , Proceedings of the 23rd annual international symposium on Computer architecture May 1996, Volume 24 Issue 2.

[5] D. Parikh, K. Skadron, Y. Zhang, M. Barcella, and M. Stan. "Power Issues Related to Branch Prediction". In Proc. of the 2002 International Symposium on High-Performance Computer Architecture, February, 2002, Cambridge, MA.

[6] Harish Patil and Joel Emer. "Combining static and dynamic branch prediction to reduce destructive aliasing". Proceedings of the 6th Intl. Conference on High Performance Computer Architecture, pages 251-262, January 2000.

[7] Cliff Young, Michael D. Smith. "Improving the Accuracy of Static Branch Prediction Using Branch Correlation". ASPLOS 1994: 232-241.

[8] Cliff Young, Michael D. Smith. "Static correlated branch prediction". TOPLAS 21(5): 1028-1075. 1999.

[9] S. McFarling. "Combining branch predictors". Tech. Note TN-36, DEC WRL, June 1993.

[10] Shien-Tai Pan, Kimming So, Joseph T. Rahmeh, "Improving the Accuracy of Dynamic Branch Prediction Using Branch Correlation", ASPLOS 1992: 76-84.

[11] Michael Haungs, Phil Sallee, Matthew K. Farrens. "Branch Transition Rate: A New Metric for Improved Branch Classification Analysis". HPCA 2000: 241-250.

[12] D. Grunwald, D. Lindsay, and B. Zorn. "Static methods in hybrid branch prediction". In Proc. Of the International Conference on Parallel Architectures and Compilation Techniques (PACT), Oct. 1998. Pages:222 – 229.

[13] A. S. Dhodapkar and J. E. Smith, "Managing Multi-Configuration Hardware via Dynamic Working Set Analysis," Proc. of the 29 Intl. Sym. on Computer Architecture, May 2002, pp. 233 –244.

Realistic Workload Scheduling Policies for Taming the Memory Bandwidth Bottleneck of SMPs

Christos D. Antonopoulos[1,*], Dimitrios S. Nikolopoulos[1], and Theodore S. Papatheodorou[2]

Department of Computer Science, The College of William & Mary,
118 McGlothlin-Street Hall, Williamsburg, VA 23187-8795, U.S.A.
{cda, dsn}@cs.wm.edu,
High Performance Information Systems Lab,
Computer Engineering & Informatics, Department,
University of Patras, 26500 Patras, Greece
tsp@hpclab.ceid.upatras.gr

Abstract. In this paper we reformulate the thread scheduling problem on multiprogrammed SMPs. Scheduling algorithms usually attempt to maximize performance of memory intensive applications by optimally exploiting the cache hierarchy. We present experimental results indicating that - contrary to the common belief - the extent of performance loss of memory-intensive, multiprogrammed workloads is disproportionate to the deterioration of cache performance caused by interference between threads. In previous work [1] we found that memory bandwidth saturation is often the actual bottleneck that determines the performance of multiprogrammed workloads. Therefore, we present and evaluate two realistic scheduling policies which treat memory bandwidth as a first-class resource. Their design methodology is general enough and can be applied to introduce bus bandwidth-awareness to conventional scheduling policies. Experimental results substantiate the advantages of our approach.

1 Introduction

Conventional schedulers for shared-memory multiprocessors are practically organized around the well-known UNIX multilevel priority queue mechanism, with limited extensions for support of multiprocessor execution. These schedulers try to achieve a balanced allocation of threads to processors. They also favor *cache affinity*, by preserving a long-term association between threads and the processors they are executed on.

---

* This work has been partially carried out while the first author was with the High Performance Information Systems Lab, University of Patras, Greece.

L. Bougé and V.K. Prasanna (Eds.): HiPC 2004, LNCS 3296, pp. 286–296, 2004.
© Springer-Verlag Berlin Heidelberg 2004

This paper argues that for memory-intensive multiprogrammed workloads, it is often the consumed memory bandwidth and not necessarily the cache affinity that should be considered as a first-class citizen in the development of an effective multiprocessor kernel scheduler. Our recent work [1] indicated that bus saturation can be harmful enough to nullify the benefits of parallelism. This paper shows that in certain memory-intensive workloads, memory bandwidth saturation is more harmful for performance that the loss of cache affinity.

We introduce two realistic workload schedulers, Bus Bandwidth-Aware Round Robin (B^2ARR) and Bus Bandwidth-Aware Dynamic Space Sharing (B^2ADSS), that effectively control bandwidth consumption. The methodologies used for their design are general and can be applied to introduce bus bandwidth consciousness to conventional, bandwidth-oblivious policies. The effectiveness of the new scheduling policies is evaluated using multiprogrammed workloads which consist of instances of NAS benchmarks [2].

Cache affinity scheduling is well studied in previous work [3, 4, 5]. Depending on the workload, it may provide substantial impact over a naive scheduling algorithm. This paper shows that for memory-intensive workloads, bandwidth consumption generally has a more significant impact than cache affinity, therefore it should be integrated as a criterion in multiprocessor schedulers. However, our policies do not prevent the use of cache affinity heuristics, since they are in general orthogonal to the criteria used for cache affinity preservation. Cache affinity heuristics usually control the association between processor and threads, whereas our policies focus on the optimal selection of coscheduled threads.

Symbiotic job scheduling [6, 7] proposes the use of event monitoring hardware to infer the interference between threads on shared execution resources and make informed online scheduling decisions based on this interference. It targets simultaneous multithreaded processors. Out policies target more conventional SMPs, however they could benefit multithreaded processors as well. They use a single additional metric (bus bandwidth consumption) as opposed to a complete array of microarchitectural events, in the case of symbiotic job scheduling. They are implemented with the efficiency of on-line monitoring as a principle, as opposed to symbiotic co-scheduling, which is a simulation-driven study.

Scheduling with runtime metrics such as the runtime speedup of parallel applications in multiprogrammed workloads has been investigated in related work [8, 9]. Our algorithms also use runtime metrics to improve scheduling decisions, however they focus on memory bandwidth consumption, a specific aspect of system performance, with a farther goal of maximizing system throughput.

This paper is organized as follows: Section 2 outlines the software and hardware configuration of our experimental platform. In Section 3 we present experimental results which indicate that cache affinity is not as important as bus bandwidth consumption for the performance of multiprogrammed SMPs. Section 4 introduces B^2ARR and B^2ADSS, two bus bandwidth-aware scheduling algorithms. In Section 5 we provide experimental evidence on the efficiency of the new algorithms. Finally, Section 6 concludes the paper.

2   Experimental Platform Configuration

For the purposes of our work we have used the NANOS compilation and execution environment on a system running Linux 2.4.25. The environment consists of an OpenMP compiler, a run-time threads package and a CPU manager. The front-end of the environment is NanosCompiler [10], an OpenMP Fortran77 compiler which creates executables that can dynamically adapt the degree of their parallelism to the available processors.

We have developed a customized user-level CPU manager for testing kernel scheduling policies, without actual kernel hacking. Our CPU manager borrows several ideas from the NANOS CPU manager which we co-developed for cache-coherent NUMA multiprocessors [11], but uses a simplified internal structure and interface. The CPU manager communicates its scheduling decisions to the applications to allow them adapt to their execution environment. Moreover, it allows them to recover from inopportune thread preemptions by resuming preempted threads at the expense of executing threads of the same application.

The policies we introduce exploit performance related information available by all modern processors through performance monitoring counters. We have experimented on a dedicated, bus-based SMP system. The system is equipped with 4 hyperthreaded Intel Xeon MP processors, running at 1.4 GHz, with 256KB L2 cache each. We had to disable hyperthreading due to limitations in the concurrent performance monitoring of threads executing on the same processor. The system is also equipped with 1GB main memory. The practically attainable bandwidth of the bus which connects processors to main memory has been experimentally evaluated to be 1797 MB/sec.

Throughout our experiments we have used class W, OpenMP versions of benchmarks from the NAS 2.3 suite [2]. We did not use a higher class of NAS, such as class A or B, because the memory footprints of most benchmarks are large with respect to the physical memory of our system. In any case, the computational weight of applications is not as important as their computation / memory transfers ratio. We have also used three synthetic microbenchmarks: FLUSH, BBMA and nBBMA. Each time FLUSH is executed on a processor, it completely flushes the cache and then reuses data from it until getting suspended by the scheduler. BBMA is similar to FLUSH. The sole difference is that, beyond flushing the cache, BBMA continuously performs back-to-back accesses to the main memory. A single instance of BBMA can bring the system bus close to the limit of saturation. nBBMA, in turn, causes practically negligible interference to both the cache of the processors that execute it and the system bus.

3   Sensitivity of Workload Performance to Cache Affinity and Bus Saturation

In this section, we present experiments that quantify the impact of cache affinity and memory bandwidth-saturation on the performance of multiprogrammed

workloads. The experiments have been executed using the CPU manager with a
round-robin scheduling policy. The use of the CPU manager allows applications
to adapt to the available processors and to minimize the adverse effects due to
the non-coscheduled execution of their threads.

We executed 2 sets of experiments in order to evaluate the effect of multiprogramming on cache performance and the effect of cache affinity on workload performance. In the first set, each application is executed alone, using 4
threads. In the second set, we execute each application, which again requests 4
threads, together with 1, 2, 4 and 8 instances of the FLUSH microbenchmark.
The slowdowns and the normalized L2 cache miss rates (CMR) with respect to
the standalone execution are depicted in Figure 1.

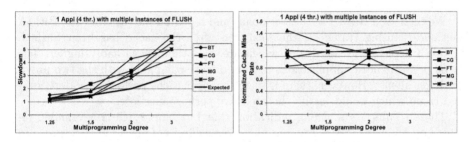

Fig. 1. Slowdowns (left) and normalized cache miss rates (right) from the multiprogrammed execution of NAS applications with instances of FLUSH

Despite the close cooperation between applications and the CPU manager,
applications suffer a performance penalty higher than expected, i.e. equal to the
multiprogramming degree. Following the common belief, one would attribute
the excessive performance penalty to the cache pollution introduced by FLUSH.
However, the normalized CMR diagram reveals that, in most cases, the cache
performance of this class of applications does not degrade severely due to FLUSH
interference. Some workloads even experience a cache performance improvement
in the presence of multiprogramming. As the CPU manager reduces the number
of processors allocated to each application, applications react by reducing the
number of threads they create and use. As a consequence, the effects of true- and
false-data sharing are minimized. Moreover, in case two or more threads which
share data happen to time-share the same processor, each one may benefit from
the data fetched to the L2 cache by the others.

As a next step, we repeated the same experiments using nBBMA instead
of FLUSH. nBBMA does not interfere with processor caches. If cache affinity
was a determinative factor for workload performance, the slowdowns suffered by
applications should this time be lower. The results are summarized in Figure 2.
The comparison with Figure 1 reveals similar performance deterioration in both
cases. The cache performance of applications is also similar (the diagram is not
reported due to space limitations).

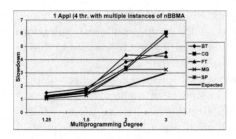
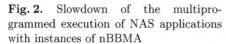
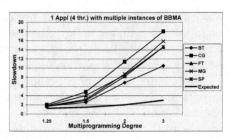

Fig. 2. Slowdown of the multiprogrammed execution of NAS applications with instances of nBBMA

Fig. 3. Slowdown of the multiprogrammed execution of NAS applications with instances of BBMA

We then used BBMA in the workloads and repeated the experiments. BBMA continues causing traffic on the bus even after the cache has been flushed. Figure 3 depicts the slowdowns which are, this time, remarkably higher. The cache performance of applications is more diverse compared with the two previous experiments. However, there is again no clear cache performance deterioration trend which would justify the excessive performance loss.

These results are a strong indication that bus bandwidth is a valuable resource on bus-based SMP systems. The diverse effects of bus saturation prove to be more harmful for performance than the loss of locality on multiprogrammed, multiprocessor systems.

4   Bus Bandwidth Conscious Scheduling Policies

We introduce B^2ARR and B^2ADSS, two realistic scheduling policies which target bus bandwidth as a scheduling resource of primary importance. The policies are based on typical round-robin (RR) and on a variant of dynamic space sharing (DSS) presented in [12]. The scheduling quantum is fixed to 100 msec, equal to the scheduling quantum of the standard Linux scheduler. Ready to execute threads are conceptually organized as a linked list.

At the end of each scheduling quantum B^2ARR deallocates all executing threads are enqueues them to the tail of threads queue. It also updates the performance related statistics for all threads that executed during the latest quantum and calculates the Bus Transactions Rate (BTR) of each thread. BTR, measured as transactions/μsec, is a metric of the bandwidth consumption of the thread during its last execution and is used as an estimation of its future requirements. The Available Bus Transactions Rate (ABTR) is then initialized to be equal to the Systems Bus Transactions Rate (SBTR), a constant value which characterizes the system bus throughput. ABTR represents the available bus transactions rate for allocation to the remaining processors. It is calculated by subtracting the requirements of already allocated threads from SBTR.

Scheduling is divided in 2 phases. During the first phase, a portion of the system processors are allocated in a round robin way to the threads that reside at the head of the threads queue. Each time a processor is allocated to a thread, the BTR of that thread is subtracted from ABTR and the thread is dequeued from the threads queue. The remaining system processors are allocated to threads during the second scheduling phase, following bus bandwidth consumption criteria. The processors are allocated in rounds, one processor at a time. At the beginning of each round the policy calculates the Average Bus Transactions Rate per Unallocated Processor ($ABTR_{proc}$). $ABTR_{proc}$ corresponds to the bus transactions requirements of the ideal candidate for scheduling in this round. All threads in the queue are then scanned in order to locate the fittest thread for allocation in that round. Formula 1 estimates thread fitness:

$$Fitness = \frac{1000}{1 + |ABTR_{proc} - BTR|} \tag{1}$$

The fitness value quantifies the distance between the estimated (BTR) and the ideal ($ABTR_{proc}$) bus bandwidth consumption. At the end of each round a processor is allocated to the fittest thread. Formula 1 favors an optimal bus bandwidth exploitation. If, for example, threads with high bandwidth requirements have already been allocated, $ABTR_{proc}$ is low and threads with low requirements are preferred for allocation. The formula works well even in cases bus saturation can not be avoided. If the bus gets overcommitted, $ABTR_{proc}$ turns negative. As a result, the thread with the lowest bus bandwidth requirements is the fittest.

Bookkeeping and statistics collection are organized in B^2ADSS similarly to B^2ARR, however the scheduler also calculates the Average System Bus Transactions Rate ($SBTR_{avg}$), i.e. the average bus transactions rate of all active threads in the system. During the first phase of B^2ADSS , the typical DSS algorithm is applied to allocate processors to applications. However, it allocates a multiple *mult* of system processors (*mult * System Processors*). At the second phase, the scheduler forms *mult* chunks of threads to execute during the next *mult* quanta using bus bandwidth optimization criteria. Only threads that belong to applications which have been allocated processors during the first phase are candidates for selection. Formula 1 is used again for the fitness characterization of threads, however the target this time is to achieve a bus bandwidth utilization from each chunk of threads as close as possible to $SBTR_{avg}$.

The design of the new scheduling policies implicitly provides two general methodologies which can be applied to introduce bus bandwidth-consciousness to conventional, bandwidth-oblivious policies. The first method is to allocate a subset of system processors with the conventional policy. The remaining processors are allocated to threads with the goal of optimizing bus bandwidth consumption. The decision on the percentage of processors that are allocated by the conventional policy introduces an interesting tradeoff. If that percentage is low, threads with bus bandwidth requirements 'incompatible' with those of the other threads in the workload may experience large delays between two consecutive activations by the scheduler. On the other hand, allowing the conventional policy to allocate too many threads minimizes the opportunities for optimally co-scheduling threads that optimize bus bandwidth usage. In B^2ARR we have heuristically chosen to allocate 50% of the system processors using RR.

The second method applies the conventional policy to allocate processors to applications for a scheduling epoch, namely a number of scheduling quanta. The chunks of specific threads that execute during each quantum are then formed with the objective of equilibrating bus bandwidth consumption among quanta. A similar tradeoff applies to this methodology as well. Using an epoch of few quanta may not allow an optimal exploitation of bus bandwidth. On the other hand, a wide epoch would introduce a significant delay between the first phase of scheduling and the actual execution of the threads in the last quanta of the epoch, increasing the risk of applying outdated scheduling decisions. For B^2ADSS the epoch has been heuristically chosen to be 2 quanta long.

The heuristic choice of the values of 50% and 2 for the percentage of processors allocated by the conventional policy and the epoch length respectively, has been experimentally driven. However, we intend to evaluate other heuristics as well. For example, it might be beneficial to allocate a percentage of the available bus bandwidth using the conventional policy, instead of a percentage of system processors.

In previous work [1] we have presented two variants of a scheduling policy which also schedules applications taking into account their bus bandwidth requirements. The two variants, namely Latest Quantum Gang (LQG) and Quanta Window Gang (QWG) are gang-like and target each application as a single entity. The fitness metric used to select and schedule applications is quite similar to Equation 1. LQG uses performance data collected only during the latest execution of each application, whereas QWG uses the average over a moving window which spans several previous quanta.

LQG and QWG, as typical gang-like policies, have the disadvantage of often resulting to suboptimal utilization of system processors. The rigid requirement of co-executing all application threads may leave processors idle during one or more scheduling quanta. B^2ARR and B^2ADSS alleviate this disadvantage by allowing an arbitrary number of threads of each application to execute simultaneously. On the other hand, the concurrent execution of all application threads minimizes synchronization and unbalancing problems which may appear due to inopportune preemptions of threads by the OS scheduler. However, the adaptability of applications created by the NanosCompiler, combined with the information and mechanisms offered by the CPU manager, allow threads scheduled with B^2ARR and B^2ADSS to minimize the adverse effects of such situations.

The new policies do not require any *a-priori* knowledge on the requirements of threads and their interaction with the hardware. Instead, they exploit performance data monitored in the past to predict thread behavior in the near future. This property, combined with the aforementioned characteristics make the new policies flexible and realistic. As a result, they are good candidates for adoption in a real-world system.

5 Experimental Evaluation

In order to evaluate the effectiveness of the new policies we have executed a set of workloads using the CPU manager with the new scheduling policies, the corresponding bandwidth-oblivious policies and LQG. Moreover, we have scheduled the same workloads using the native Linux scheduler, without the intervention of the CPU manager. We did not experiment with QWG, since NAS applications have regular, smooth transaction patterns and are generally insensitive to external noise. For such applications, LQG performs better than QWG [1].

Table 1. Workload Composition

ID	Description	Max. Multipr. Degree	ID	Description	Max. Multipr. Degree
A.1	2BT(3)+2CG(1)	2	B.3	2BT(4)+2SP(4)	4
A.2	4BT(1)+4CG(1)	2	C.1	(FT(1);FT(3))+(MG(3);MG(1))+	3
A.3	2BT(4)+2CG(4)	4		+(FT(3);FT(1))+(MG(1);MG(3))	
B.1	2BT(1)+2SP(3)	2	C.2	4FT(1)+4MG(1)	2
B.2	4BT(1)+4SP(1)	2	C.3	2FT(4)+2MG(4)	4

Table 1 describes the workloads we have used. A(n) means that application A is executed with n threads. A+B represents the concurrent execution of applications A and B. Similarly, mA represents the concurrent execution of m instances of application A. Finally, A;B means that application B starts right after the termination of application A. The rightmost column of Table 1 reports the maximum multiprogramming degree exposed by each workload. However, the actual multiprogramming degree may vary during execution.

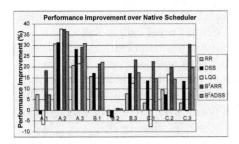

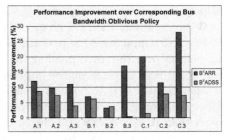

Fig. 4. Average workload turnaround time improvement, with respect to the execution with the native Linux scheduler

Fig. 5. Average workload turnaround time improvement when B˙ARR and B˙ADSS are applied, compared with the execution with RR and DSS respectively

Figure 4 depicts the performance improvement attained by the new policies, executed in the context of the CPU manager, over the native Linux scheduler. A first important observation is that our CPU manager, even with bandwidth-oblivious policies such as RR or DSS, outperforms the native Linux scheduler by 11% and 14% respectively. This performance improvement can be attributed to the information and mechanisms provided to applications by the CPU manager, in order to assist them adapt to the available processors and make progress on the critical path of their computation. Only in few cases (workloads A.1, B.2 and C.3) RR and DSS perform slightly worse (up to 3.2%) than the native scheduler. The average performance improvements attained by bus bandwidth-conscious policies, namely LQG, B^2ARR and B^2ADSS, are 9%, 23% and 18% respectively. In 2 workloads (A.1 and C.1) LQG performs worse than Linux scheduler. The reason is explained in detail in the next paragraph.

We then compare the performance of bandwidth-conscious policies with that of RR. The comparison isolates the performance gains due to application adaptability to the available processors and focuses on the impact of the policies themselves. LQG is in average 1% worse than RR. Although LQG performs in most cases better than RR, 3 workloads experience severe performance degradation. These workloads reveal the fundamental weakness of gang-like policies: the rigid rule of scheduling all threads of each application together results to low utilization of system processors. If the three problematic workloads are excluded, LQG is 4% more efficient than RR. In workload C.3 LQG also performs 4% worse than RR. In this case, RR forces the application to reduce the degree of parallelism in the presence of multiprogramming instead of time-sharing applications on the same processors. This choice proves to be beneficial for performance. B^2ARR and B^2ADSS, on the other hand, do not suffer from the same problems as LQG. They perform, in average, 28% and 17% better than RR.

As a next step, we quantify the performance improvement attained by B^2ARR and B^2ADSS over LQG. Although not co-scheduling application threads may introduce overheads, B^2ARR and B^2ADSS perform in average 13% and 9% better than LQG. Workloads A.2, B.2 and C.2 are the only exceptions. In these cases B^2ADSS performs up to 3% worse than LQG. Since all applications that participate in these workloads are single-threaded, they contribute equally to the total workload and the DSS phase of B^2ADSS has practically no effect.

Figure 5 summarizes the performance gains of bus bandwidth-aware scheduling policies over the corresponding bus bandwidth-oblivious ones. This comparison quantifies the performance improvement due to the optimal exploitation of bus bandwidth. It is important to notice that in all cases the new scheduling policies perform better than conventional policies. B^2ARR is in average 13% faster than RR. B^2ADSS also outperforms DSS, this time by 5%.

We expect our policies to perform even better on a system where more than 4 processors share a bus, since the bus saturation problem our policies cope with will be more acute on such a machine. However, commercial, bus-based SMPs are usually limited to 8 processors due to bus scalability issues. Multiple buses are used to integrate even 8 processors on a bus-based system. In such architectures, a slightly modified version of our policies would have additional choices for optimal bandwidth exploitation. It would be possible to even move threads among buses in order to optimize the bus bandwidth usage on each bus.

Significant effort has been paid to enhance the scalability of the Linux scheduler on large-scale multiprocessors. A new O(1) scheduler, present in 2.6.x kernels, allows constant overhead scheduling, independently of the number of tasks (N) and processors (P) in the system. The overhead related to the preservation of load-balancing between processors grows linearly with the number of processors. The current implementation of our scheduling policies has an $O(N^2)$ overhead, which can be reduced to $O(N\log N)$ if tasks are organized in priority queues, according to their BTR. The overhead is in any case higher than that of the standard Linux scheduler. However, as aforementioned, the policies target bus-based SMPs, which are limited to small- or medium-scales. For such systems the overhead of our policies has practically proven to be negligible.

6   Conclusions

In this paper we first presented experimental results which indicate that, for the class of memory-intensive, numerical applications, executed on multipro-grammed SMP systems, it is often bus bandwidth consumption and not cache affinity that determines application performance. Driven by this observation, we introduced B^2ARR and B^2ADSS, two realistic scheduling policies that target bus bandwidth as a top-importance scheduling resource. The new policies measure the bus-bandwidth requirements of threads during their execution and use the collected performance data to estimate thread behavior in the close future. They select threads to be co-scheduled during each quantum with the goal of neither wasting bus bandwidth, nor saturating the bus. The scheduling policies have been implemented in the context of a CPU manager, a user-level process that applies scheduling policies and precisely controls thread execution. The CPU manager is part of a compilation and execution environment which allows multithreaded applications to minimize the adverse effects of multiprogramming.

We evaluated the effectiveness of B^2ARR and B^2ADSS using workloads consisting of instances of applications from the NAS benchmarks suite. Our algorithms have been compared with the native Linux scheduler, the correspond-ing bus bandwidth-oblivious policies RR and DSS, and LQG, a gang-like, bus bandwidth-aware scheduling policy presented in our earlier work. B^2ARR and B^2ADSS attained significant performance gains over the Linux scheduler. More-over, they turned out to be an important improvement over RR, DSS and LQG.

We plan to investigate the impact of sharing resources other than bus band-width on job scheduling. Such an investigation would be of particular interest for emerging architectures such as SMTs or HyperThreaded (HT) processors, where various execution units and levels of the on-chip memory hierarchy are shared among threads. We also intend to apply the idea of optimally using the available memory bandwidth to other levels of the memory hierarchy, beyond the front-side bus. Possible targets are the bandwidth of cache ports in SMTs, HT, and multi-core processors, or the bandwidth of network links in clusters of SMPs.

Acknowledgements

The first author is supported by a grant from 'Alexander S. Onassis' public benefit foundation and the European Commission through IST grant No. 2001-

33071. The first two authors are supported by NSF awards ITR/ACI-0312980 and CAREER/CCF-0346867.

References

1. Antonopoulos, C.D., Nikolopoulos, D.S., Papatheodorou, T.S.: Scheduling Algorithms with Bus Bandwidth Considerations for SMPs. In: Proceedings of the 33d International Conference on Parallel Processing (ICPP '03), Kaohsiung, Taiwan, ROC (2003) 547–554
2. Jin, H., Frumkin, M., Yan, J.: The OpenMP Implementation of NAS Parallel Benchmarks and its Performance. Technical Report NAS-99-011, NASA Ames Research Center (1999)
3. Squillante, M., Lazowska, E.: Using Processor-Cache Affinity Information in Shared-Memory Multiprocessor Scheduling. IEEE Transactions on Parallel and Distributed Systems **4** (1993) 131–143
4. Torrellas, J., Tucker, A., Gupta, A.: Evaluating the Performance of Cache-Affinity Scheduling in Shared-Memory Multiprocessors. Journal of Parallel and Distributed Computing **24** (1995) 139–151
5. Vaswani, R., Zahorjan, J.: The Implications of Cache Affinity on Processor Scheduling for Multiprogrammed Shared Memory Multiprocessors. In: Proc. of the 13th ACM Symposium on Operating System Principles (SOSP'91), Pacific Grove, California (1991) 26–40
6. Snavely, A., Tullsen, D.: Symbiotic Job Scheduling for a Simultaneous Multithreading Processor. In: Proc. of the 9th International Conference on Architectural Support for Programming Languages and Operating Systems (ASPLOS'IX), Cambridge, Massachusetts (2000) 234–244
7. Snavely, A., Tullsen, D., Voelker, G.: Symbiotic Jobscheduling with Priorities for a Simultaneous Multithreading Processor. In: Proc. of the ACM 2002 Joint International Conference on Measurement and Modeling of Computer Systems (SIGMETRICS'2002), Marina Del Rey, CA (2002) 66–76
8. Corbalan, J., Martorell, X., Labarta, J.: Performance Driven Processor Allocation. In: Proc. of the 4th USENIX Symposium on Operating System Design and Implementation (OSDI'2000), San Diego, California (2000)
9. Nguyen, T., Vaswani, R., Zahorjan, J.: Maximizing Speedup through Self-Tuning Processor Allocation. In: Proc. of the 10th IEEE International Parallel Processing Symposium (IPPS'96), Honolulu, Hawaii (1996) 463–468
10. Ayguadé, E., Gonzàlez, M., Labarta, J., Martorell, X., Navarro, N., Oliver, J.: NanosCompiler: A Research Platform for OpenMP Extensions. Technical Report UPC-DAC-1999-39, Dept. D' Arquitectura de Computadors - Universitat Politècnica de Catalunya (1999)
11. Martorell, X., Corbalan, J., Nikolopoulos, D.S., Navarro, N., Polychronopoulos, E.D., Papatheodorou, T.S., Labarta, J.: A Tool to Schedule Parallel Applications on Multiprocessors. The NANOS CPU Manager. In: Proceedings of the 6th IEEE Workshop on Job Scheduling Strategies for Parallel Processing (JSSPP'2000). Volume 1911., LNCS (2000) 87–112
12. Polychronopoulos, E.D., Martorell, X., Nikolopoulos, D.S., Labarta, J., Papatheodorou, T.S., Navarro, N.: Kernel-Level Scheduling for the Nano-Threads Programming Model. In: Proceedings of the 12th ACM International Conference on Supercomputing (ICS'98), Melbourne, Australia, ACM Press (1998) 337–344

A Parallel State Assignment Algorithm for Finite State Machines

David A. Bader[1,*] and Kamesh Madduri[2,**]

* Electrical and Computer Engineering Department,
University of New Mexico, Albuquerque, NM 87131
* Department of Electrical Engineering,
Indian Institute of Technology – Madras, India 600036
{dbader, kamesh}@ece.unm.edu

Abstract. This paper summarizes the design and implementation of a parallel algorithm for state assignment of large Finite State Machines (FSMs). High performance CAD tools are necessary to overcome the computational complexity involved in the optimization of large sequential circuits. FSMs constitute an important class of logic circuits, and state assignment is one of the key steps in combinational logic optimization. The SMP-based parallel algorithm – based on the sequential program JEDI targeting multilevel logic implementation – scales nearly linearly with the number of processors for FSMs of varying problem sizes chosen from standard benchmark suites while attaining quality of results comparable to the best sequential algorithms.

1 Introduction

Parallel architectures have promised high performance computing, but their use remains largely restricted to well structured numeric applications. Exploiting parallelism at the level of large distributed memory systems is hindered by the cost of message passing. However, Symmetric Multiprocessors (SMPs) with modest shared memory have emerged as an alternative platform for the design of scientific and engineering applications. SMP clusters are now ubiquitous for high-performance computer, consisting of clusters of multiprocessors nodes (e.g., IBM Regatta, Sun Fire, HP AlphaServer, and SGI Origin) interconnected with high-speed networks (e.g., vendor-supplied, or third party such as Myricom, Quadrics, and InfiniBand). Current research has shown that it is possible to design algorithms for irregular and discrete computations [1–4] that provide efficient and scalable performance on SMPs.

With the rapid strides in VLSI technology, circuit design and analysis are becoming increasingly complex. There is a growing need for sophisticated CAD

---

* This work was supported in part by NSF Grants CAREER ACI-00-93039, ITR ACI-00-81404, ITR EIA-01-21377, Biocomplexity DEB-01-20709, and ITR EF/BIO 03-31654; and DARPA contract NBCH30390004.
** Supported in part by an NSF Research Experience for Undergraduates (REU) grant.

© Springer-Verlag Berlin Heidelberg 2004

tools that can handle large problem sizes and quicken the synthesis, analysis, and verification steps in VLSI design. CAD applications are inherently unstructured and non-numerical, making them difficult to parallelize effectively. The state assignment problem for Finite State Machines is one such application. It can be formulated as an optimization problem that is NP-complete. Sequential heuristics that try to solve this task are computationally intensive and fail for large problem instances. The parallel implementation discussed in the paper, which is based on the sequential algorithm JEDI [5], overcomes this limitation and attains better results, i.e., designs with fewer *literals* (a Boolean variable or its negation) and hence, faster circuits with reduced size and power consumption, as well as faster execution times for the design and analysis.

The Finite State Machine (FSM) is a device that allows simple and accurate design of sequential logic and control functions. Any large sequential circuit can be represented as an FSM for easier analysis. For example, the control units of various microprocessor chips can be modeled as FSMs. FSM concepts are also applied in the areas of pattern recognition, artificial intelligence, language and behavioral psychology.

An FSM can be optimized for area, performance, power consumption, or testability. The various steps involved in optimizing an FSM are state minimization, state assignment, logic synthesis, and logic optimization. SIS [6] is a popular tool for synthesis and optimization of sequential circuits. Many different programs and algorithms have been integrated into SIS, allowing the user a range of choices at each step of the optimization process. The first step of optimization, state minimization, is performed using STAMINA [7]. Two state assignment programs, NOVA [8] and JEDI, are distributed with SIS. After state assignment, the resulting logic for the output can be minimized by the logic minimizer ESPRESSO [9] which tries to find a logic representation with the minimum literals while preserving the functionality of the FSM.

2   Problem Overview

The State assignment problem deals with assignment of unique binary codes to all of the states of the FSM so that the resulting binary next-state and output

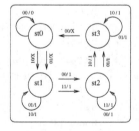

Fig. 1. *train4* FSM

functions can be implemented efficiently. Most of the algorithms optimize for the area of logic implementation. The number of literals in the factored form of logic is the accepted and standard estimate [10] for area requirements of a logic implementation. The parallel algorithm discussed here minimizes this measure and thus optimizes for the area.

To illustrate the significance of state assignment, consider the following example. The FSM *train4* (Fig. 1) consists of four symbolically encoded states *st0*, *st1*, *st2* and *st3*. These states can be assigned unique states by using a minimum of two bits, say y_1 and y_2. The input can be represented by two bits x_1 and x_2 and the the single bit output by z. y_{1ns} and y_{2ns} represent the next state bits of y_1 and y_2, respectively. Suppose the following states are assigned:

$$st0 \leftarrow 00, st1 \leftarrow 01, st2 \leftarrow 11, st3 \leftarrow 10.$$

The resulting logic equations after *multilevel logic optimization* would be

$$y_{1ns} = z(x_1 y_1 + x_2 y'_{2ns} + k) + x'_1 x'_2 y_{2ns}$$
$$y_{2ns} = y'_1 k'(x_1 + x_2) + y_2(x'_1 x'_2 + k)$$
$$z = y'_2(y'_1 + k)$$
$$k = x_1 x_2$$

This implementation results in a literal count of 22 after logic optimization and requires 15 gates to realize assuming the complements of inputs are present. However, if the state assignments were

$$st0 \leftarrow 00, st1 \leftarrow 10, st2 \leftarrow 01, st3 \leftarrow 11,$$

the logic equations after optimization would be

$$y_{1ns} = z(y_1 + y'_{2ns})$$
$$y_{2ns} = x_1 x'_2 + x'_1 x_2$$
$$z = y_1 y_2 y'_{2ns} + y'_1 y'_2$$

The literal count is only 12 in this case and 8 gates are sufficient to implement the logic. Optimal State assignment becomes computationally complex as well as crucial for larger FSMs.

2.1   Previous Research

There is significant prior research in the area of state assignment algorithms. KISS is one of the first algorithms proposed targeting a PLA-based implementation. NOVA improves on KISS and is based on a graph embedding algorithm. Genetic algorithms [11] and algorithms based on FSM decomposition [12] are also proposed.

The sequential logic optimization tool SIS uses algebraic techniques to factor the logic equations. It minimizes the logic by identifying common subexpressions.

The algorithms MUSTANG [13] and JEDI use this fact and try to maximize the size as well as the number of common subexpressions. These algorithms target a multilevel logic implementation and the number of literals in the combinational logic network after logic optimization is taken as the measure of the quality of the solution.

The ProperCAD project [14] aims to develop portable parallel algorithms for VLSI CAD applications. Parallel algorithms for state assignment [15] based on MUSTANG and JEDI have also been proposed as part of this project. However, the speedups in case of shared memory implementations were not significant. The shared memory algorithm discussed here improves on this work as it attains better speedups without compromising the quality of results.

3   JEDI

JEDI is a state assignment algorithm targeting a multi-level logic implementation. The program involves two main stages, the weight assignment stage and the encoding stage.

Weight Assignment: In this stage, JEDI assigns weights to all possible pairs of states, which is an estimate of the affinity of the states to each other. It chooses the best of four heuristics for this purpose: input dominant, output dominant, coupled, and variation.

The input dominant algorithm assigns higher weights to pairs of present states, which assert similar outputs and produce similar sets of next states. It has the effect of maximizing the size of common cubes in the implemented logic function. The output dominant algorithm assigns higher weights to pairs of next states, which are generated by similar input combinations and similar sets of present states. It has the effect of maximizing the number of common cubes in the logic function. The coupled algorithm adds up the weights generated by both the input and output dominant algorithms and the variation algorithm takes into consideration the number of input and output bits also in the weight computation.

In the input dominant heuristic, an *Input State Assignment matrix* M_I is computed. This matrix is an $N_s \times N_s$ symmetric matrix with each element m_{ij} corresponding to the weight of the edge (s_i, s_j). In general, m_{ij} is defined by $m_{ij} = \sum_{a=1}^{O_i} \sum_{b=1}^{O_j} P(o_{ia}, o_{jb})$ where O_i and O_j are the number of transitions out of states s_i and s_j of the STG, o_{ia} is the set of binary outputs produced by transition a out of state s_i, and $P(o_{ia}, o_{jb})$ are the corresponding product terms.

To illustrate these weight assignment algorithms, consider again the example of the FSM *train4* (Fig. 1). It has four states: *st0*, *st1*, *st2*, and *st3*. The input is two bits and the output is a single bit. Let us determine the weight of the edge (*st1*, *st2*).

Firstly, for the states *st1* and *st2*, the states they fan out to and their corresponding frequency are determined. The FSM, represented in the State Transition Table (STT) form is input to JEDI. A transition corresponds to an edge in the State Transition Graph (STG) or a row in the STT.

$$st1 \rightarrow (st1^2, st2^2), st2 \rightarrow (st2^2, st3^2)$$

Next, all the outputs produced by these transitions are inspected. In this case, there is a single output bit and both *st1* and *st2* produce an output of 1 for all of the transitions. Thus, the value of $m_{12} = m_{21} = 4 * 4 = 16$ in this case.

For the output dominant case, an *Output State Assignment matrix* is defined and the weights are calculated similarly.

Encoding Phase – Simulated Annealing: JEDI tries to encode states with high edge weights closely. This step can be formulated as minimization of the sum $\sum_{i=1}^{N_s} \sum_{j=1}^{N_s} m_{ij} * dist(s_i, s_j)$ where m_{ij} is the edge weight from the input assignment matrix and $dist(s_i, s_j)$ denotes the Hamming distance between two codes s_i and s_j when they are encoded using the minimum number of bits. The Hamming distance between two binary codes is the number of bits in which the codes differ. For example, the codes 1000 and 1110 have a Hamming distance of 2.

Encoding is done using Simulated Annealing, a probabilistic hill climbing heuristic. The general algorithm is as follows:

1. Start with an initial temperature T and a random configuration of states.
2. For a given temperature, pick two states at random and assign new encodings or exchange their encodings.
3. Calculate the change in cost function.
4. Accept the exchange for a decrease in the cost function. Allow some encodings to be accepted even if it leads to an increase in the cost function in order to avoid local minima.
5. Repeat steps 2–4 until a certain number of moves are made. Then lower the temperature and continue with the process.
6. Stop when the temperature reaches the minimum temperature.

After the encoding and simulated annealing is done, the output can be written in a BLIF (Berkeley Logic Interchangeable Format) file. This file is then passed into the sequential logic synthesizer, and logic optimization is carried out using ESPRESSO. The required parameters such as the literal count and final output logic then can be retrieved for further analysis.

4   Parallel Implementation

We parallelize JEDI using the SMP Node primitives of SIMPLE [16], a methodology for developing high performance programs on clusters of SMPs. Both the weight computation and the encoding stages (the computationally expensive steps) have been parallelized. Our source code for parallel JEDI is freely-available from our web site, http://hpc.ece.unm.edu/.

The programming environment for SMPs is based upon the SMP node library component of SIMPLE, that provides a portable framework for developing SMP algorithms using the single-program multiple-data (SPMD) programming style.

This framework is a software layer built from POSIX threads that allows the user to use either the already developed SMP primitives or the direct thread primitives. The SMP Node library contains a number of SMP node algorithms for barrier synchronization, broadcasting the location of a shared buffer, replication of a data buffer and memory management. In addition to these functions, there is a *parallel do* that schedules n independent work statements implicitly to p processors as evenly as possible.

Weight Computation Stage: The parallel algorithm for weight computation is detailed in Algorithm 1. To calculate the weight matrix, all of the state pairs are inspected first. For state i, the edge weights of state pairs $(1, i)$ to $(i - 1, i)$ are checked. The $i - 1$ states are distributed among the processors using the *pardo* primitive. Thus no two processors get the same edge pair and so there are no conflicts. Each processor updates the weight matrix independently. Since it is a shared memory implementation, no merging of local weight matrices is required.

Result : Weight computation of the states in parallel
compute the weight assignment matrix;
(the inner loop is executed in parallel:)
for $i = 1$ *to* N_s **do**
  for $j = 1$ *to* $i - 1$ **in parallel do**
    calculate the edge weight (s_i, s_j);
  end
end
synchronize;

Algorithm 1: Weight computation stage after the KISS format file is read into the appropriate data structure

Encoding Stage – Simulated Annealing: The encoding stage involves assignment of unique binary codes to each state such that the literal count of the combinational logic is minimized. This is done using simulated annealing, which is a computationally intensive process. A lot of research has been done in parallelizing it for the placement problem in VLSI CAD applications [17].

Our implementation implements simulated annealing using the *divide and conquer* strategy as well as the *parallel moves* technique. Previous research shows that these techniques of parallel simulation are well-suited for shared memory multiprocessors [18].

A unique global configuration is maintained in *divide and conquer* and *parallel moves*, which simplifies the implementation for shared memory. The principle of parallel moves applies multiple moves to a single configuration simultaneously. On the other hand, the divide and conquer method lets processors make simultaneous moves within preset boundaries of the configuration.

We use the same default Initial and Stopping temperatures of JEDI in the parallel implementation. At higher temperatures, some encodings are accepted even if they do not reduce the cost function in order to avoid local minima. But for low temperature values, the acceptance rate decreases. Hence, the number of attempted moves increases as the temperature increases. The number of moves are also varied according to the problem size.

Algorithms 2 and 3 detail the Simulated Annealing strategies.

Result : Simulated Annealing by 'Divide and Conquer'
(initialization:)
1. Set the initial and stopping temperatures;
2. Set the cooling parameter;
3. Calculate the initial no. of moves per stage;
4. Divide the state space into P partitions;
(annealing in parallel:)
while *Current temp > Stopping Temp* **do**

    for $i = 1$ *to maxgen* **do**

        (*maxgen* is the max. no. of moves for a temp. value);
        compute cost function in the local encoding space;
        generate two random codes;
        check whether they are already assigned;
        if *states are assigned* **then**
            exchange the codes of the two states;
        else
            assign new codes to states;
        end
        calculate the additional cost in the local space;
        accept or reject the exchange;
        update the local data structure;

    end
    synchronize;
    update the temperature value;
    calculate the cost function of the entire state space;
end

Algorithm 2: Simulated Annealing by 'divide and conquer'

Error and Quality Control: The parallel simulated annealing stage has been implemented such that there are no conflicts. In the parallel moves method of annealing, each processor can choose moves independently from the entire set of available moves. When moves are evaluated in parallel, it is important to control how the moves are to be accepted. Firstly, it must be ensured that moves in parallel are not contradictory. For example, two different codes must not be assigned to the same state in parallel. This case does not arise due to the global data structure and shared memory configuration.

In case of sequential simulated annealing, the cost function needs to be determined only once for a particular temperature, and then, only the change in the cost is evaluated when deciding whether to accept or reject a state. But in parallel, the cost is evaluated every time a decision has to be made, since the global

Result : Simulated Annealing by 'Parallel moves'
(initialization:)
1. Set the initial and stopping temperatures;
2. Set the cooling parameter;
3. Calculate the initial no. of moves per stage;
(annealing in parallel:)
while *Current temp > Stopping Temp* **do**
    for $i = 1$ *to maxgen* **do**
        (*maxgen* is the max. no. of moves for a temp. value);
        (the *maxgen* moves are divided among p processors);
        compute cost function;
        generate two random codes;
        check whether they are already assigned;
        if *codes are assigned* **then**
            exchange the codes of the two states;
        else
            assign new codes to two states;
        end
        calculate the additional cost;
        accept or reject the exchange;
        update the data structure;
    end
    synchronize;
    update the temperature value;
end

Algorithm 3: Simulated Annealing by 'parallel moves'

data structure can be changed by any processor. This is an additional overhead in the case of the parallel moves implementation. But this does not affect the quality of the annealing. However, it is possible that there is a repetition of the same moves by different processors leading to redundant moves.

In case of the divide and conquer strategy, a serious issue needs to be considered. If the local state space partitions are static, then the number of potential moves is reduced as the number of processors increase. This would lead to a degradation of quality, and the algorithm may not even converge. To avoid this, the states need to be shuffled and distributed in such a manner that it is possible to realize all possible moves in the global state space. This is ensured by changing the partitions for a change in temperature so that all the possible moves are probable. However, the probability that a move is realized decreases when compared to the sequential algorithm. This leads to a degradation in the quality for some runs.

To illustrate how the repartitioning is done, consider the following example. Suppose an FSM has 28 states. The minimum number of bits needed to encode all states is 5, and this gives a state space of size 32. Consider a four processor parallel annealing algorithm carried out in this space. Each processor is assigned

8 codes initially, processor 1 getting codes 0–7, processor 2 getting codes 8–15, and so on. If no repartitioning is done, then the exchange of states 1 & 8, 2 & 9, etc., is not possible. However, for the next temperature change, if the partitions are modified such that processor i gets codes $4k + i$, for $0 \leq k < 8$, (processor 0 getting $0, 4, 8, 12, \ldots, 28$, and so on), then all exchanges are theoretically possible. This partitioning scheme can be be extended for a general case, an N-bit encoding space and $P = 2^k$ processors.

5   Experimental Results

We present an experimental study to compare our parallelized state assignment algorithm to the state-of-the-art sequential approach, both in terms of quality and running time. As we will show, our new approach scales nearly linearly with the number of processors, while maintaining a similar quality of solution. We compare our parallel JEDI approach with a simplified version of the sequential algorithm distributed with SIS. These algorithms converge to the same solutions and preserve most of the features of the original algorithm. The FSMs used for testing are from the MCNC benchmark suite. After state assignment, the output in PLA file format is passed through MV-SIS [19], a multilevel multi-value logic synthesis tool from Berkeley. The quality of the solution is given by the literal count after multilevel logic optimization. Lower literal count implies less area of implementation. We use the SMP programming environment previously de-

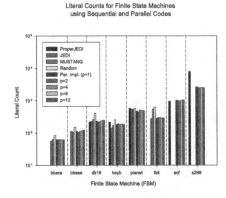

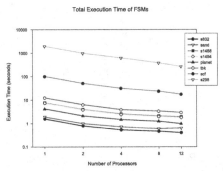

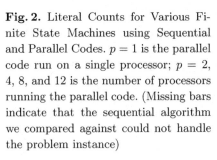

Fig. 2. Literal Counts for Various Finite State Machines using Sequential and Parallel Codes. $p = 1$ is the parallel code run on a single processor; $p = 2$, 4, 8, and 12 is the number of processors running the parallel code. (Missing bars indicate that the sequential algorithm we compared against could not handle the problem instance)

Fig. 3. Total Execution time (in seconds) for a suite of Finite State Machines. Note that this is a log-log plot

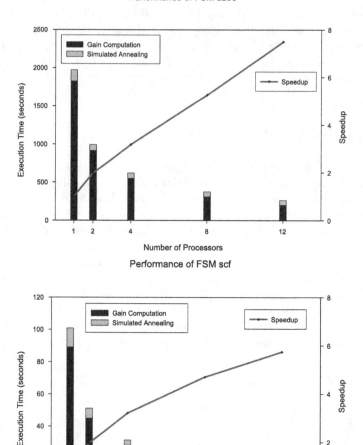

Fig. 4. Performance for the weight computation and simulated annealing steps for two different Finite State Machines, s298 (top) and scf (bottom)

scribed and discuss next the multiprocessor environment used in this empirical study.

We test the shared-memory implementation tested on a Sun E4500, a uniform-memory-access (UMA) shared memory parallel machine with 14 UltraSPARC II 400MHz processors and 14 GB of memory. Each processor has 16 Kbytes of direct-mapped data (L1) cache and 4 Mbytes of external (L2) cache.

We measure the quality of the solution by the literal count after multilevel logic optimization (see Fig. 2) for FSMs of varied problem sizes, when the parallel algorithm is run on one processor. (See [20] for corresponding tables.) The count reported is the best value obtained among the four weight generation heuristics. The literal count reported by JEDI in [5] and the uniprocessor results obtained by the parallel implementation ProperJEDI [15] are also listed for comparison.

The quality of the solution for multiprocessor runs is reported in detail in [20]. Simulated annealing on multiple processors leads to slightly varying results for successive runs of the same problem, so we report the average literal count obtained over five test runs. The results listed are the ones obtained from the parallel moves simulated annealing technique as they give better results than the divide and conquer technique.

Figure 3 shows the total execution time for different FSMs on multiprocessor runs. (See [20] for detailed running times.) The reported speedups are good for all problem sizes (number of states) but are higher for larger problem sizes. Figure 4 shows the execution time and corresponding speedup for the weight computation and simulated annealing steps separately for two different FSMs. It is observed that the execution time for the weight computation phase scales nearly linearly with the number of processors for all problem sizes.

6   Conclusions

Our new parallel implementation of the popular state assignment algorithm JEDI has been developed, specifically targeting shared memory multiprocessors. The present sequential algorithms fail to give good results for large Finite State Machines. However with the parallel implementation, significant reduction in implementation time can be achieved without compromising the quality of the solution. The algorithm can also be re-structured in order to run on distributed memory systems. In the gain computation step, the local data structures need to be merged to generate the weight assignment matrix and in the simulated annealing step, periodic global update of the data structure has to be done to avoid conflicting state assignments. For future work, better results can be obtained if parallel logic optimization tools are also developed.

References

1. Bader, D., Illendula, A., Moret, B.M., Weisse-Bernstein, N.: Using PRAM algorithms on a uniform-memory-access shared-memory architecture. In Brodal, G., Frigioni, D., Marchetti-Spaccamela, A., eds.: Proc. 5th Int'l Workshop on Algorithm Engineering (WAE 2001). Volume 2141 of Lecture Notes in Computer Science., Århus, Denmark, Springer-Verlag (2001) 129–144

2. Bader, D., Sreshta, S., Weisse-Bernstein, N.: Evaluating arithmetic expressions using tree contraction: A fast and scalable parallel implementation for symmetric multiprocessors (SMPs). In Sahni, S., Prasanna, V., Shukla, U., eds.: Proc. 9th Int'l Conf. on High Performance Computing (HiPC 2002). Volume 2552 of Lecture Notes in Computer Science., Bangalore, India, Springer-Verlag (2002) 63–75

3. Bader, D.A., Cong, G.: A fast, parallel spanning tree algorithm for symmetric multiprocessors (SMPs). In: Proc. Int'l Parallel and Distributed Processing Symp. (IPDPS 2004), Santa Fe, NM (2004)

4. Bader, D.A., Cong, G.: Fast shared-memory algorithms for computing the minimum spanning forest of sparse graphs. In: Proc. Int'l Parallel and Distributed Processing Symp. (IPDPS 2004), Santa Fe, NM (2004)

5. Lin, B., Newton, A.: Synthesis of multiple level logic from symbolic high-level description language. In: Proc. of the IFIP TC 10/WG 10.5 Int'l Conf. on Very Large Scale Integration, Germany (1989) 414–417

6. Sentovich, E., Singh, K., Lavagno, L., Moon, C., Murgai, R., Saldanha, A., Savoj, H., Stephan, P., Brayton, R., Sangiovanni-Vincentelli, A.: SIS: A system for sequential circuit synthesis. Electronics Research Laboratory, University of California, Berkeley. Ucb/erl m92/41 edn. (1992)

7. Rho, J.K., Hachtel, G., Somenzi, F., Jacoby, R.: Exact and heuristic algorithms for the minimization of incompletely specified state machines. IEEE Trans. Computer-Aided Design **13** (1994) 167–177

8. Villa, T., Sangiovanni-Vincentelli, A.: NOVA: State assignment of finite state machines for optimal two-level logic implementation. IEEE Trans. Computer-Aided Design **9** (1990) 905–924

9. Theobald, M., Nowick, S., Wu, T.: Espresso–HF: A heuristic hazard-free minimizer for two-level logic. In: Proc. 33rd ACM Design Automation Conf., Las Vegas, NV (1996) 71–76

10. Brayton, R., McMullen, C.: Synthesis and optimization of multistage logic. In: Proc. IEEE Int'l Conf. Computer Design (ICCD), Portchester, NY (1984) 23–30

11. Almaini, A., Miller, J., Thomson, P., Billina, S.: State assignment of finite state machines using a genetic algorithm. IEEE Proc. Computers and Digital Techniques **142** (1995) 279–286

12. Ashar, P., Devadas, S., Newton, A.: A unified approach to the decomposition and re-decomposition of sequential machines. In: Proc. 27th ACM/IEEE Design Automation Conf., Orlando, FL (1990) 601–606

13. Devadas, S., Ma, H.K., Newton, A., Sangiovanni-Vincentelli, A.: MUSTANG: State assignment of finite state machines for optimal multi-level logic implementations. IEEE Trans. Computer-Aided Design **7** (1988) 1290–1300

14. Ramkumar, B., Banerjee, P.: ProperCAD: A portable object-oriented parallel environment for VLSI CAD. IEEE Trans. Computer-Aided Design **13** (1994) 829–842

15. Hasteer, G., Banerjee, P.: A parallel algorithm for state assignment in finite state machines. IEEE Transactions on Computers **47** (1998) 242–246

16. Bader, D.A., JáJá, J.: SIMPLE: A methodology for programming high performance algorithms on clusters of symmetric multiprocessors (SMPs). Journal of Parallel and Distributed Computing **58** (1999) 92–108

17. Kim, S., Chandy, J., Parkes, S., Ramkumar, B., Banerjee, P.: ProperPLACE: A portable parallel algorithm for standard cell placement. In: Proc. 8th Int'l Parallel Processing Symp. (IPPS'94), Cancún, Mexico (1994) 932–941

18. Kravitz, S., Rutenbar, R.: Placement by simulated annealing on a multiprocessor. IEEE Trans. Computer-Aided Design **6** (1987) 534–549

19. Gao, M., Jiang, J.H., Jiang, Y., Li, Y., Sinha, S., Brayton, R.: MVSIS. In: Proc. Int'l Workshop on Logic Synthesis, Tahoe City, CA (2001) 138–144

20. Bader, D.A., Madduri, K.: A parallel state assignment algorithm for finite state machines. Technical report, Electrical and Computer Engineering Department, The University of New Mexico, Albuquerque, NM (2003)

A Novel Scheme to Reduce Burst-Loss and Provide QoS in Optical Burst Switching Networks

Ashok K. Turuk and Rajeev Kumar

Department of Computer Science and Engineering,
Indian Institute of Technology Kharagpur,
Kharagpur, WB 721 302, India
{akturuk, rkumar}@cse.iitkgp.ernet.in

Abstract. Burst loss due to contention is a major issue in optical burst switching networks. In this paper, we propose a contention resolution scheme that uses a offset time different from that of conventional optical burst switching (OBS) to reduce burst loss and to provide QoS in optical burst switching networks. The proposed scheme can be tuned to both prioritized traffic and delay constraint traffic by changing the offset time. For selecting a data-channel, we propose three channel selection algorithms, namely Least Recently Used (LRU), First Fit (FF), and Priority Set (PS) algorithms. We simulate and compare proposed scheme with the preemptive priority just-enough-time (PPJET) contention resolution scheme. We consider bursty traffic in our simulation. It is found that our scheme outperforms PPJET in burst-loss.

1 Introduction

There has been a phenomenal increase in the number of Internet users and the variety of Internet applications in recent years. This has resulted in exponential growth of Internet traffic, demanding a huge bandwidth at the backbone network. To meet this growing demand for bandwidth, wavelength division multiplexing (WDM) network has become the de-facto choice for the backbone network. IP over WDM networks have drawn much attention among researchers, and many integration schemes between IP and WDM layers have been proposed [1].

To carry IP traffic over WDM networks three switching technologies have been studied: optical circuit switching, packet switching and burst switching. Optical circuit switching and packet switching have their own limitations when applied to WDM networks. Circuit switching is not bandwidth efficient unless the duration of transmission is greater than the circuit establishment period. It is shown that establishment of circuits (lightpaths) in optical networks is an NP-hard problem [2]. Many heuristics and approximation algorithms exist for establishing lightpaths in optical networks e.g., see [3] and the references therein. Packet switching is hop-by-hop store and forward scheme and, needs buffering and processing at each intermediate node. It is flexible and bandwidth efficient. However, technology for buffering and processing in optical domain is yet to mature for this scheme to commercialize. Fiber delay lines proposed in literature provide limited buffer capability and are suitable for delays of fixed duration only.

L. Bougé and V.K. Prasanna (Eds.): HiPC 2004, LNCS 3296, pp. 309–318, 2004.
© Springer-Verlag Berlin Heidelberg 2004

In this context optical burst switching (OBS) is emerging as the new switching paradigm for the next generation optical networks [4, 5]. It combines features of both circuit and packet switching. As such there exists no formal definition of OBS, the features defined by Yoo and Qiao [4] for OBS have become de-facto standards. The burst-size granularity (which lies between circuit and packet switching), separation of control and data bursts, one-way (for most cases) or two-way reservation scheme, and no optical buffering are important characteristics of the OBS paradigm.

Some major issues in optical burst switching networks are: (i) contention resolution, (ii) burst assembly, and (iii) quality-of-service (QoS) support. In a buffer-less OBS network contending burst is lost. Therefore, burst-loss should be minimized in OBS networks, is the key design parameter. A few approaches to contention resolution used in OBS are: buffering [6], deflection routing [7], burst segmentation [8, 9] and window based technique [10]. Burst assembly is the process of aggregating and assembling IP packets into bursts[11]. With increase in variety of Internet applications, different applications such as voice-over-IP (VoIP), video-on-demand, video conferencing etc. demand different QoS requirements. To meet the QoS requirements of different applications, IETF proposed IntServ and DiffServ schemes. However, such conventional priority schemes are defined for electronic domain which trivially mandates the use of buffers at intermediate nodes. Such schemes cannot be used directly to support QoS in buffer-less OBS networks. Thus, any scheme to support *differential* QoS requirements in OBS networks should not mandate the use of buffers at intermediate nodes.

Many schemes, in recent years, have been suggested to support priority based QoS in OBS networks. All the proposed schemes have tried to reduce the burst loss. It is not the burst-loss only but also the number of packet-loss that matters. For example, consider three bursts b_1, b_2 and b_3 of size 10, 20 and 50 number of packets each. A loss of any of the bursts indicates 33% of the burst loss. However, if we consider loss in terms of packets, the packet-loss comes out to be 14%, 28% and 64% respectively. Therefore, it is desirable that a contention resolution scheme should take care of the losses calculated in terms of packets. Consider Fig. 1(a) and assume both requests r_1 and r_2 have the same priority and arrive at a node at the same time. In OBS, the burst b_2 is always dropped. However, if the burst-size is taken into account the larger burst b_2 could have succeeded and the smaller burst b_1 is dropped. This will result into larger number of packets transmitted and higher resource utilization.

Next, consider Fig. 1(b) having two bursts of the same priority. In OBS, both the bursts are dropped. However, taking the burst-size or the number of hops traversed into contention resolution scheme one of the bursts succeeds. This gives rise to lower burst loss and larger number of packets transmitted. Thus, if we consider two more parameters – burst-size and number of hops traversed – in resolving contention, this will guarantee that at least one of the bursts succeeds and larger number of bits transmitted.

End-to-end delay is another key parameter for QoS provisioning. All delay sensitive applications demand that end-to-end delay is bounded by the delay constraints imposed by the respective application. Contention should be resolved by considering the delay factor too.

In this paper, we present a flexible algorithm for contention resolution to support a larger set of QoS parameters in OBS networks. We consider packet loss and number of

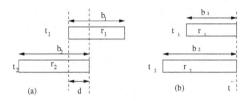

Fig. 1. Illustrations for Contention Resolution : (a) two requests are partially overlapped for a period d, (b) two requests have the same reservation instance

hops traversed, in addition to priority, for resolving contention. Our scheme is generic and can easily be adapted to satisfy delay-constraints. The aim is to reduce blocking probability of the bursts arising due to resource contention at intermediate nodes as well as to meet the delay constraints of the delay sensitive traffic. The proposed scheme guarantees that at least one of the bursts succeeds when contention occurs; the contention should be resolved in accordance with satisfaction of QoS parameters. To select data-channel, we propose three channel selection algorithms – (i) *Least Recently Used* (LRU), (ii) *First Fit* (FF), and (iii) *Priority Set* (PS). Channel selection algorithms run at the ingress routers to select a data-channel for reservation and for subsequent transmission. We evaluate the proposed scheme with the above channel selection algorithms.

The rest of the paper is organized as follows. Section 2 explains the contention resolution technique; few assumptions and notations used are described in Sub-section 2.1. Channel selection algorithms are explained in Section 3. Simulation results are presented in Section 4 and compared with PPJET. Finally, some conclusions are drawn in Section 5.

2   Proposed Contention Resolution Scheme

2.1   Assumptions and Notations

We model an optical network by means of a undirected graph $G(V, E)$ where V is the set of vertices (nodes) and E represents the set of links/edges in the network. Two types of nodes (here after, we use the terms node and router interchangeably) are identified: edge routers and core routers. Every edge router has $(n_e - 1) \times P$ electronic buffers where n_e is the number of edge routers, P is the number of priority classes supported in the system. Each buffer belongs to a specific pair of priority class and an egress router. The core router has no buffer; this is a desirable feature of the optical burst switching networks. A core router acts as a transit router for data-traffic. Thus, the data-traffic remains in optical domain from ingress to egress router. Propagation delay, t_p, between every pair of adjacent vertices in graph G is assumed to be the same. Processing delay of the control packet at each router is assumed to be δ. We use the following notations in rest of the paper:

$H_t^{sd}(r)$: Number of hops for the request r between source - destination pair (s, d),
$H_i^{sd}(r)$: Remaining number of hops for the request r between source - destination pair (s, d) at node i.

Original burst: A burst for which resources are already reserved at the core router,
Contending burst: A burst whose reservation request has resulted in resource contention at the core router.

We define the following three situations that can occur when an intermediate router receives a reservation request:

- *No contention* (NC): When no contention for resources occurs at the intermediate core router.
- *Contention resolved* (CR): When a contention occurs at an intermediate core router i and for at least one of the requests $H_i^{sd}(r) > H_t^{sd}(r)/2$.
- *Contention-not-resolved* (CNR): When contention occurs at intermediate core router(s) and for none of the request $H_i^{sd}(r) > H_t^{sd}(r)/2$.

2.2   Proposed Scheme

OBS is based on either *one way* or *two way* reservation protocol. The minimum latency in one way reservation protocol is $P + \delta \cdot H$ where the minimum latency in two way reservation protocol is $2P + \delta \cdot H$. Our proposed scheme is a one way reservation protocol however it differs from other OBS schemes in two aspects - one, the offset time, and second, the methods adopted for contention resolution. In other OBS schemes, the offset time is $\delta \cdot H$ where δ is the processing delay of control packet at each node and, H is the number of hops between source-destination pair. In our scheme, we take the offset time to be $P + \delta \cdot H$ where P is the propagation delay between source-destination pair. The need for the additional P units of time is explained subsequently. The minimum latency of burst in other OBS schemes, is $P + \delta \cdot H$ which is same if a burst is sent along with control packet in optical packet switching. The minimum latency in optical circuit switching is $3P + \delta \cdot H$. In the proposed scheme, the minimum latency of a burst is $2P + \delta \cdot H$. Thus, we can say the minimum latency in our scheme is identical to OBS with *two way* reservation protocol. However, the proposed scheme is a *one way* reservation protocol where each burst experiences an additional delay of P units. The scheme is also tunable to delay sensitive traffic. For delay sensitive traffic the offset time in the proposed scheme is taken to be $\delta \cdot H$ which is the same as that in OBS. However, this offset can be made adaptive to the application needs. In the scheme, if a contention occurs and the situation is a *CR* one as mentioned in Section 2.1 then a burst is further delayed for the contention period. However, this delaying technique of our scheme is not applicable in case of delay sensitive traffic. For delay sensitive traffic if the required resources are not available within that amount of time, the burst is dropped.

Secondly, the proposed scheme differs from OBS in the method adopted for contention resolution. In OBS, the resource conflict is resolved on the basis of request priority and the time instance for which request is made. In addition to the above two parameters, we take burst-size and the number of hops traversed to resolve contention. A higher priority request is given a priority. However, for the same priority requests, the one that has traversed the maximum number of hops, is accepted for better resource utilization. The request that has traversed the maximum number of hops have more resources reserved on the path. Accepting this request will give rise to higher resource utilization. For same priority and the equal number of hops traversed the burst that has larger burst-size is accepted. For same priority, equal number of hops traversed and the same burst-size their instance of reservation is taken for resolving the conflict. Thus, the tie in contention resolution is resolved in order of priority, number of hops traversed, burst-size and delay.

Next, we explain the basis of having P additional units of delay in offset time with the help of timing diagrams illustrated in Figures 2 and 3. The total delay encountered by a control packet for source-destination pair (s, d) is no greater than $\Delta = \delta \times H_t^{sd}(r)$. The offset-time, T, in OBS is taken to be at least Δ. In Fig. 2 the number of hops between source - destination pair (s, d) is 4. Therefore, the offset-time T in OBS is 4δ. In OBS, if a contention occurs say at node A or at node C then the burst is dropped at A or at B as shown in Figures 2(b) and 2(c) respectively. With this offset time a contending burst cannot be further delayed.

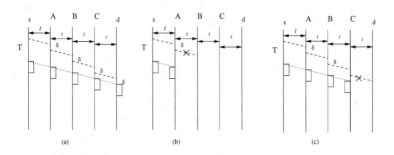

Fig. 2. Timing Diagram of burst switching network: (a) no contention occurs at intermediate nodes, (b) contention occurs at node A and (c) contention occurs at node C

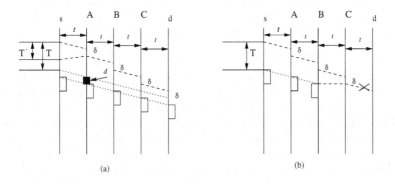

Fig. 3. Timing Diagram for the proposed scheme: (a) contention at node A is resolved, and (b) contention at node C but the burst is dropped at node B

In our proposed scheme, the offset, T, between source-destination pair (s, d) is taken to be $(t + \delta)H_t^{sd}(r)$. For the above example the offset time between source-destination pair (s, d) is $4(t + \delta)$. Let us consider Fig. 2(b) where contention has occurred at node A and, d be the duration of the contention period. The control packet has taken *one* hop to reach the node A from the source s. If a message is sent from node A to the source s to delay the transmission of burst for the contention period d it will reach s at $T' = 2(t + \delta)$ after the source s has sent the control packet (Fig. 3(a)). The offset-time $T > T'$ i.e., source s will receive the message to delay the transmission before the expiry of offset-time. Hence, the transmission of the burst is delayed and is not dropped at node A as shown in Fig. 3(a).

Let us consider Fig. 2(c) where contention has occurred at node C and, d be the duration of the contention period. If a message is sent from node C to the source s to delay the transmission of burst for the contention period d it will reach s at $T' = 6(t + \delta)$. The offset-time $T < T'$, i.e., source s will receive the delay message after it has transmitted the burst and the burst is dropped at node C. Therefore, instead of sending a delay message if a resource-release message is sent from node C, the message will release the resources reserved at node B before the burst arrives at node B and is dropped at node B rather than at node C. This gives rise to better utilization of the resources on link BC which was earlier occupied by the request.

Thus, in the proposed scheme, a contention occurs at node i and $H_i^{sd}(r) > H_t^{sd}(r)/2$ (this is the CR situation as described in Section 2.1), a message is sent to delay the transmission of the burst for duration of the contention period. For $H_i^{sd}(r) <= H_t^{sd}(r)/2$ (this is the CNR situation as described in Section 2.1), a message is sent to release the reserved resource.

We illustrate below possible cases of contention and the way contention is resolved in the scheme. For all the cases we refer to Fig. 1(a). In Fig. 1(a), the value of t_1 and t_2 indicates the time of arrival of requests r_1 and r_2 respectively, at a core router i. Burst-size of the requests r_1 and r_2 is indicated by b_1 and b_2 respectively. Below, we give interpretation for Case 1; the rest of the cases are interpreted in the same way. In Case 1 the contention has occurred due to the arrival of requests r_1 and r_2 at the core router i at the same time ($t_1 = t_2$). The remaining number of hops to be traversed for request r_1 is $H_i^{sd}(r_1) > H_t^{sd}(r_1)/2$ and, for r_2 is $H_i^{mn}(r_2) > H_t^{mn}(r_2)/2$. We have assumed that the contention has occurred at the core router i.

Case 1: $t_1 = t_2$, $H_i^{sd}(r_1) > H_t^{sd}(r_1)/2$, $H_i^{mn}(r_2) > H_t^{mn}(r_2)/2$.
Accept the high priority request and send a message to the ingress router of low priority request to delay the transmission for the contention period d. For same priority of both the requests, accept the request that has traversed the maximum number of hops and send a message to the ingress router of the other request to delay the transmission for the contention period d. For same priority and the equal number of hops traversed, accept the request that has larger burst-size and send a message to the ingress router of other request to delay the transmission for the contention period d.

Case 2: $t_1 = t_2$, $H_i^{sd}(r_1) \leq H_t^{sd}(r_1)/2$, $H_i^{mn}(r_2) > H_t^{mn}(r_2)/2$.
Accept the high priority request. If the low priority request is r_1 then it is dropped else a message is sent to the ingress router of r_2 to delay the transmission for the contention period d. For same priority of both the requests, accept the one that has traversed the maximum number of hops. Other request is processed as explained. For same priority

and equal number of hops traversed, accept the one with higher burst-size. Other request is processed as explained earlier.

Case 3: $t_1 = t_2$, $H_i^{sd}(r_1) > H_t^{sd}(r_1)/2$, $H_i^{mn}(r_2) \leq H_t^{mn}(r_2)/2$.
Requests are processed as explained in Case 2. Here, the request that is to be dropped is r_2.

Case 4: $t_1 = t_2$, $H_i^{sd}(r_1) \leq H_t^{sd}(r_1)/2$, $H_i^{mn}(r_2) \leq H_t^{mn}(r_2)/2$.
Requests are processed as in Case 1. Here the request that is not accepted is dropped.

Case 5: $t_1 < t_2$, $H_i^{mn}(r_2) > H_t^{mn}(r_2)/2$.
In this case the request r_1 has arrived before r_2 and resources are already reserved for the request r_1. For request r_2 a message is sent to the ingress router to further delay the transmission of burst for the contention period d.

Case 6: $t_1 < t_2$, $H_i^{mn}(r_2) \leq H_t^{mn}(r_2)/2$.
As in Case 5 resources are already reserved for the request r_1. Request r_2 is dropped.

Case 7: $t_1 > t_2$, $H_i^{mn}(r_2) > H_t^{mn}(r_2)/2$.
In this case request r_1 has arrived at a later point of time than r_2 and is contending with request r_2. Requests are processed similar to Case 5.

Case 8: $t_1 > t_2$, $H_i^{sd}(r_1) \leq H_t^{sd}(r_1)/2$.
As in Case 7, request r_1 has arrived at a later point of time than r_2 and is contending with request r_2. Requests are processed similar to Case 6.

In the above cases, cases 4, 6 and 8 are *CNR* situations and the rest are *CR* situations as defined in Section 2.1.

3   Channel Selection Algorithms

In this section, we describe three channel selection algorithms called (i) *Least Recently Used* (LRU), (ii) *First Fit* (FF), and (iii) *Priority Set* (PS) algorithms used in channel selection for our proposed contention resolution scheme (Section 2.2). The channel selection algorithms are run only at the edge routers to find the data-channels for which reservation request is to be made and subsequently transmit the data-burst. In LRU, a data-channel which is idle for the maximum duration is selected. In FF, data-channels are searched from the lowest index and the one that is available first, is selected. For example, consider Fig. 4, LRU channel selection algorithm selects the data-channel 2 as it is idle for the maximum duration where as FF channel selection algorithm selects the data-channel 0.

In PS approach, we decompose the set of data-channels, S, into P sub-sets, S_i, of data-channel. P is the number of priority classes supported. $S = S_0 \cup S_1 \cup \cdots \cup S_{P-1}$. A priority class i selects the data-channel from the set S_i. If no data-channel is available in the set S_i then it selects from the set S_{i-1} and if not available then from the set S_{i-2}. This process is iterated till the lowest priority set S_0 is searched. If no data-channel is available in the set S_0 then the burst is dropped at the ingress router. The number of data channels in the set S_i is in proportion to the traffic of priority class i.

For the priority class 0, if no data-channel is available in the set S_0 then the burst is dropped at the ingress router. To illustrate the working of Priority Set approach, we consider two priority classes 0 and 1; class 1 has higher priority than class 0. We divide the available data-channel as shown in Fig. 5 in two sets $S_0 = \{0, 1\}$ and $S_1 = \{2, 3\}$. Let a burst of class 1 arrive at t_a and it is to be transmitted at t_s after the base-offset time t_{offset}. Since all the data-channels in the set S_1 are busy at t_s, channel 0 from the set S_0 is selected.

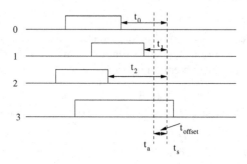

Fig. 4. Illustration for selection of data-channel in LRU and FF algorithms

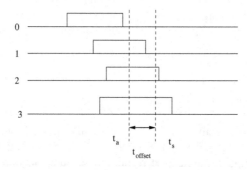

Fig. 5. Illustration for selection of data-channel in PS algorithm

4   Simulation Results

We assume the following time-units for different tasks to carry out the simulation. The propagation delay, t_p, between any two adjacent nodes in the burst switching network is assumed to be $1ms$. The processing time of control packet at the router is assumed to be $2\mu s$. We assume there is no wavelength conversion and there exists no optical buffer in the switch. For simplicity and without loss of generality, we consider two classes of traffic: class 0 (low priority) and class 1 (high priority). We generate high priority traffic with a probability of 0.4. Traffic is generated only at the edge router and, the load is measured in Erlang.

We compare the simulation results of our proposed scheme with PPJET [12]. We consider burst blocking probability as the performance metric for comparison. We have taken number of wavelengths available on each link to be *seven*. Traffic in the Internet is reported to be bursty in nature [13]. We consider bursty traffic with Pareto ($\alpha = 1.1$) distributed burst length and Pareto ($\alpha = 1.1$) distributed inter-arrival time.

We include the overall burst loss for the proposed scheme with three channel selection algorithms, in Fig. 6 and compare with PPJET. It is observed from Fig. 6 that the overall burst loss in our scheme is lower than that in PPJET. Of the proposed channel selection algorithms, LRU algorithm gives lower overall burst loss, and PS gives higher. The higher overall burst loss in PS is due to the higher low priority burst loss. We generated many more results through simulation by varying various parameters; the detailed results will be presented during the conference.

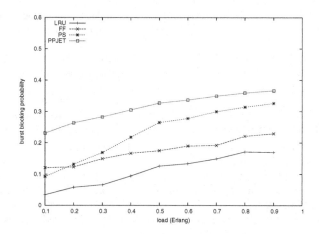

Fig. 6. Overall burst loss in the proposed scheme with different channel selection algorithms and PPJET. Pareto distributed burst-size and Pareto distributed inter-arrival of burst is considered

5   Conclusions

In this paper, we proposed a contention resolution scheme for OBS networks. The scheme takes the following three parameters – priority, number of hops traversed and burst-size - into account to resolve contention. The proposed scheme is adaptable to both prioritized and delay constraint traffic. We also proposed three channel selection algorithms called, LRU, FF and PS algorithms to select data-channel at the ingress router for the proposed scheme. We simulate our scheme with each of the channel selection algorithms and compare the results with PPJET. We consider bursty traffic in our simulation. Simulations were carried out for prioritized traffic. We observed that LRU channel selection algorithm gives lower overall burst loss. In addition, Priority Set channel selection algorithm gives the lowest high priority burst loss.

We compared our scheme with another contention resolution scheme called PPJET. We found lower overall blocking probability in our proposed scheme using LRU than

PPJET for all load. The proposed scheme using PS channel selection algorithm gives the lower blocking for high priority traffic than PPJET. Thus we can conclude that if a lower overall burst loss is required then our scheme with LRU selection algorithm can be used. If a low blocking of high priority traffic is desired then the proposed scheme with PS algorithm may be the choice.

The lower blocking in our scheme comes with an additional delay. In PPJET an incoming burst is delayed for an amount of time which is equal to the total processing time of the control token at each node. However, in our scheme an additional delay which is equal to the propagation time between source to destination, is involved.

References

1. Yao, S., Yoo, S.B., Mukherjee, B., Dixit, S.: All-Optical Packet Switching for Metropolitan Area Networks: Opportunities and Challenges. IEEE Communications Magazine **39** (2001) 142 – 148
2. Chlamtac, I., Ganz, A., Karmi, G.: Lightpath Communications: An Approach to High Bandwidth Optical WANs. IEEE Transactions on Communications **40** (1992) 1171 – 1182
3. Dutta, R., Rouskas, G.: A Survey of Virtual Topology Design Algorithm for Wavelength Routed Optical Networks. Optical Network Magazine **1** (2000) 73–89
4. Yoo, M., Qiao, C.: Optical Burst Switching (OBS) - A New Paradigm for an Optical Internet. Journal of High Speed Network **8** (1999) 69 – 84
5. Qiao, C., Yoo, M.: Choices, Features and Issues in Optical Burst Switching. Optical Network Magazine **1** (2000) 36 – 44
6. Yoo, M., Qiao, C., Dixit, S.: QoS Performance in IP over WDM Networks. IEEE Journal on Selected Areas in Communications, Special Issues on Protocols for Next Generation Optical Internet **18** (2000) 2062 – 2071
7. Kim, H., Lee, S., Song, J.: Optical Burst Switching with Limited Deflection Routing Rules. IEICE Trans. Commun. **E86-B** (2003)
8. Vokkarane, V.M., Zhang, Q., Jue, J.P., Chen, B.: Generalized Burst Assembly and Scheduling Techniques for QoS Support in Optical Burst-Switched Networks. In: Global Telecommunications Conference, 2002, GlOBECOM'02. Volume 3. (2002) 2747 – 2751
9. Zhang, Q., Vokkarane, V.M., Chen, B., Jue, J.P.: Early Drop Scheme for Providing Absolute QoS Differentiation in Optical Burst-Switched Networks. In: Workshop on High Performance Switching and Routing, 2003, HPSR. (2003) 153 – 157
10. Farahmand, F., Jue, J.P.: Look-ahead Window Contention Resolution in Optical Burst Switched Networks. In: Workshop on High Performance Switching and Routing, 2003, HPSR. (2003) 147 – 151
11. Düser, M., Bayvel, P.: Analysis of a Dynamically Wavelength-Routed Optical Burst Switched Network. Journal of LightWave Technology **20** (2002) 574 – 585
12. Kaheel, A., Alnuweiri, H.: A Strict Priority Scheme for Quality-of-Service Provisioning in Optical Burst Switching Networks. In: Proceedings of the Eight IEEE International Symposium on Computers and Communication (ISCC'03). (2003) 16 – 21
13. Paxson, V., Floyd, S.: Wide Area Traffic: The Failure of Poisson Modeling. IEEE/ACM Transaction on Networking **3** (1995) 226 – 244

Single FU Bypass Networks for High Clock Rate Superscalar Processors

Aneesh Aggarwal

Department of Electrical and Computer Engineering,
Binghamton University, Binghamton, NY 13902
aneesh@binghamton.edu

Abstract. Microprocessors depend heavily on broadcast-based bypass networks, to eliminate pipeline hazards arising due to data dependencies. However, even though bypassing is logically simple, increasing clock speeds make broadcasting slower and difficult to implement, especially for wide issue and deeply pipelined processors. The problem is exacerbated by shrinking feature size, as wire delays become more important than the gate delays.

In this paper, we propose Single FU bypass networks for high clock rate superscalar processors where, instead of a fully connected broadcast-based bypass network, results from an FU are forwarded only to itself. The new bypass network design is based on the observations that a result produced by an instruction is mostly required by just one other instruction and that the operands of many instructions come from a single other instruction. The new bypass network results in significant reduction in the data forwarding latency, while incurring only a small impact (about 2% for most of the SPEC2K benchmarks) on the instructions per cycle (IPC) count. However, reduced bypass latency has a high potential for increased clock speeds. Single FU bypass networks are also much more scalable than the broadcast-based bypass networks, for more wide and more deeply pipelined future microprocessors.

1   Introduction

The bypass network lies in the most critical loop in pipelined processors that enables data dependent instructions to execute in consecutive cycles [5]. Prior studies [4][9][10] have shown that an increase of a single cycle in this critical loop reduces the instruction throughput dramatically. Most modern processors use a broadcast-based bypass network, where a result produced by a functional unit (FU) is made available at the inputs of all the other FUs. With a broadcast-based bypass network, bypassing can take significant amounts of wiring area on the chip [2], especially for wide-issue and deeply pipelined processors. In addition to a large wiring area, studies [2][6] show that with broadcast-based bypass networks, the wire complexity grows proportional to the square of the issue width and the pipeline depth. This leads to a significant increase in the wire path delay from the source to the destination. This problem is further

L. Bougé and V.K. Prasanna (Eds.): HiPC 2004, LNCS 3296, pp. 319–332, 2004.
© Springer-Verlag Berlin Heidelberg 2004

exacerbated with the growing wire delays in the sub-micron technology era [1][11]. In addition, large number of bypass paths increase the fan-out delay at the source and the fan-in delay at the destination, by increasing both the capacitive load within the network and the multiplexor complexity at each destination. The fan-out at each source and the fan-in at each destination increases roughly with the product of the pipeline depth and the pipeline width [8]. In fact, bypass network latency is expected to set the cycle time of future microprocessors [6][8].

The overall impact of the broadcast-based bypass network complexity is that multiple cycles may be required to forward the values [8]. With a multi-cycle bypass network, dependent instructions are not able to execute in consecutive cycles, resulting in a decrease in the instruction throughput and a corresponding decrease in the overall performance. Increased bypass widths not only impact performance, but also increase the power consumption (which has become a primary design issue [22][23]) due to the wide multiplexors at each destination and the increased number of long wires, and reduce reliability by increasing the cross-talk between the wires and by reducing the signal strength.

In this paper, we propose a Single FU bypass network for high clock rate superscalar processors. In this bypass network, instead of a fully connected broadcast-based forwarding, an FU's output is only forwarded to its own inputs. Single FU bypass network facilitates low latency and energy-efficient data-forwarding because of a reduction in the fan-in at the inputs of the FUs, a reduction in the fan-out at the outputs of the FUs, and a reduction in the lengths and the number of bypass paths. We found that with a Single FU bypass network, the bypass latency reduces significantly, while the IPC is only about 2% less than that of a broadcast-based bypass network. We also discuss further reducing the number of bypass paths using a Single Input Single FU bypass network.

The rest of the paper is organized as follows. Section 2 discusses the motivation behind Single FU bypass networks. Section 3 presents a description of the Single FU bypass network. Section 4 presents IPC results and their analysis, and techniques to reduce the IPC impact of Single FU bypass network. Section 5 discusses further reduction of bypass paths using Single Input Single FU bypass network. Section 6 concludes the paper.

2   Motivation and Background

2.1   Impact of Multi-cycle Forwarding

First, we measure the impact of increased forwarding latency on IPC. For this, we use the parameters given in Table 1 (on page 15) for a superscalar pipeline shown in Figure 1[1]. The forwarding latency is increased from 0 cycles to 2 cycles. Dependent instructions can execute in consecutive cycles only with a

---

[*] Our pipeline resembles closely to that of Intel Pentium 4 Netburst architecture [7].

0-cycle forwarding latency. Figure 2 shows the IPC (along the Y-axis), with varying forwarding latencies, for many of the SPEC2000 Integer and Floating Point benchmarks.

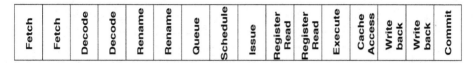

Fig. 1. Base Pipeline

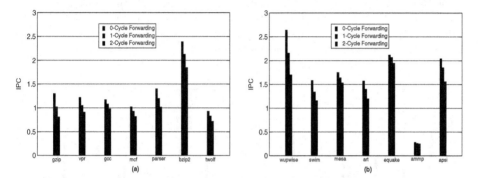

Fig. 2. Impact of Multi-cycle Bypass Networks (a)Integer Benchmarks (b)Floating Point Benchmarks

As can be seen in Figure 2, there is a significant reduction in the IPC counts of the programs as the forwarding latency is increased. For instance, as compared to a 0-cycle forwarding latency, IPC reduces by about 15% for a 1-cycle forwarding latency for both integer and floating point benchmarks. The impact of increased forwarding latency is relatively higher for higher IPC benchmarks. In addition, the IPC impact is very similar for all the integer benchmarks (except for gcc, which has a relatively lower impact). The IPC impact for the FP benchmarks, on the other hand, is much more varied.

2.2   Data Dependence Characteristics

Next, we look at the typical data dependence characteristics in the programs. For this, we measure the type of instruction producing a value and the type and number of instructions using that value. The type of an instruction is defined by type of the functional unit it uses to execute. We define 6 types of instructions: IALU (simple integer instructions using an ALU), IMULT (complex integer instructions using a multiplier), LOAD (load instructions), STORE (store instructions), FPALU (simple floating point instructions using an ALU), and FPMULT (complex floating point instructions using a multiplier). These measurements depict the actual data dependence present in the dynamic trace

of instructions along the correct execution path of the programs. However, these
statistics may vary depending on the compiler optimizations performed, and on
the Instruction Set Architecture used. We work with the PISA ISA, which is a
typical RISC ISA. Figure 3 presents these measurements in the form of a stacked
bar graph. For each integer benchmark, there are 2 sets of stacked bars, where
each stacked bar represents the type of instruction producing the register value.
Similarly, for each FP benchmark, there are 4 sets of stacked bars. The value
on top of each stacked bar represents the percentage of results (out of the total)
produced by instructions of that particular type and used by other instructions
(depicted by the stacks). The total of all the stacked bars is less than 100%
because some results are not used at all and some results are produced by instructions of other types. For instance, for `gzip`, about 35% of the values (out of
the total produced) are produced by an IALU instruction and used by just one
other IALU instruction. `Equake` has 0 FP instructions, because FP instructions
in `equake` are encountered beyond the range of instructions simulated.

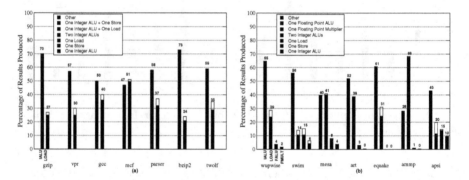

Fig. 3. Result Usage Characteristics for SPEC2K (a) Int and (b) FP Benchmarks

As can be seen in Figure 3, most of the values (about 70%) produced are only
used by just one instruction. Figure 3 also explains the different reduction in IPC
observed for different benchmarks, in Figure 2. In general, for benchmarks where
more results are used by just one other instruction and fewer results are used
by store instructions, increased forwarding latency has a larger impact on IPC.
This is because, when a result is used by more instructions, after the forwarding
latency delay more instructions get ready at the same time, resulting in less
effective average forwarding latency per instruction. The effect is exactly opposite
in case a result is mostly used by just one instruction. Similarly, if the number of
results used by store instructions is higher, the impact of increased forwarding
latency is lower, because a delayed store instruction does not have a significant
impact on the execution of other instructions (store instructions do not produce
any result). This is the reason for high IPC impact for `wupwise`, `swim`, `art`, and
`apsi`, and low IPC impact for `gcc`. `Ammp` also has a high percentage of instructions
with just one consumer, but its extremely low IPC results in a low IPC impact in

its case. Other factors affecting the IPC impact of higher bypass latency are the percentage of results (out of the total produced) that are actually used, branch prediction accuracy, and load miss rate.

Next we look at the data dependence characteristics of programs from the perspective of the consumer. Figure 4 presents these statistics[2] in a manner similar to Figure 3, except that each stack now represents operand-producing instructions rather than result-using instructions. For instance, for gzip, about 68% of the instructions (out of the total executed) are IALU instructions, and about 30% of instructions (out of the total executed) are IALU instructions whose operands are produced by just one other IALU instruction. An important observation that can be made from the Figure 4 is that a significant percentage (about 70%) of the total instructions executed have their operands produced either by just one other instruction or by no instructions. The load instructions that do not have any producer instructions for their operands are the ones which use the same register operand and this register operand is produced before the start of the collection of the statistics (we collect the statistics after skipping the first 500 million instructions). The integer ALU instructions that do not have any producer instructions for their operands are mostly the ones which load an immediate value, or the ones that use register r0 to set a register to an immediate value.

Overall, it is observed that for most of the instructions, their results are used by just one other instruction and their operands are produced by just one other instruction. This motivates us to investigate *Single FU* bypass networks, where the results produced in a FU are only forwarded to its own inputs .

2.3   Related Work

Bypassing is an old idea and was first described in 1959 by Bloch [3]. Since then, the issue width of the processors and the number of functional units in the processors have increased considerably. Unfortunately, not enough work has been done for efficient data forwarding. Here, we classify the proposed efficient bypass networks into 2 broad categories: *limited bypassing*, and *partitioned bypassing*.

In *limited bypassing*, certain paths are missing from the bypass network. Ahuja et al [2] study bypass networks where the results from the FUs are forwarded to only one of the inputs of all the FUs. They propose simple code transformations such as interchange of operands and instruction scheduling to avoid the stalls generated due to missing bypasses. For efficient bypassing, the Pentium 4 processor [7], limits the number of bypass inputs into each FU as well as the number of bypass outputs from each FU. For this, [7] uses a complex multi-stage bypass network that stores and selectively forwards the results to be bypassed from the latter stages of the pipeline.

[.] We only show Integer instructions even for the FP benchmarks because of the very low percentage of FP instructions in the benchmarks. However, the FP instructions showed very similar characteristics.

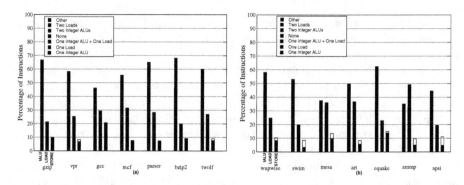

Fig. 4. Operand Production Characteristics for SPEC2K (a) Int and (b) FP Benchmarks

Partitioned bypassing is used in clustered processor architectures. In these architectures, there is typically a broadcast-based bypassing within a cluster, and either broadcast-based inter-cluster bypassing [16][15][13][6], or point-to-point inter-cluster bypassing [14][17][18][19][20]. Intra-cluster bypassing is fast because of reduction in the number of the bypass paths, whereas inter-cluster bypassing may take additional cycles because of longer wires and/or multiple hops required. However, clustered architectures do not achieve the single-threaded IPCs obtained with a centralized superscalar [21].

3   Single FU (SFU) Bypass Network

3.1   Basic Idea

Based on the observations in Section 2, we propose *Single FU (SFU)* bypass networks, where the results produced in a FU are forwarded only to itself, thus reducing the bypass network latency and facilitating high clock rates.

Figure 5(a) shows a Pentium 4 [7] style broadcast-based bypass network for the integer units. A similar bypass network could be implemented for the floating point units. This design reduces the fan-in and the fan-out for the FUs, as well as the number and lengths of the bypass paths. The multi-stage bypass network is responsible for forwarding the correct values from the other FUs and the latter stage of the pipeline. Figure 5(b) shows one configuration of the SFU bypass network for the same set of integer FUs. In this configuration, the output of an ALU is immediately forwarded to only its own inputs. However, the values loaded are typically required in IALU instructions. Hence, instead of forwarding the output of the load unit to itself, it is forwarded to one of the ALUs. In addition, load units typically read the base address from the register file or are, in some cases, forwarded the value from the ALUs. Hence, the output of one of the ALUs is also forwarded to the load unit. Without loss of generality, in Figure 5(b), the output of ALU2 is also forwarded to the load unit and that of load unit is forwarded to ALU2. The rest of the bypass network remains the same. In this configuration, the results from an FU are immediately available to the ones it is directly connected to, and are available to all the FUs after an additional cycle. We call this bypass network as *Limited SFU (LSFU)* bypass network. Figure 5(c) illustrates another configuration of the SFU bypass network. In this configuration, the multiplier and store units are completely isolated, *i.e.* their results are available only from the register file and they read their operands only from the register file. In addition, the results from an FU are only available to the one it is directly connected to, even for bypasses from the latter stages of the pipeline. We call this bypass network as *Extreme SFU (ESFU)* bypass network. For all the configurations, a similar bypass network can be assumed for the floating point units. The primary advantages of a SFU bypass network design include shorter and fewer bypass paths, reduced fan-in and fan-out for each FU, and narrower multiplexors. This not only reduces the bypass latency, but also significantly reduces the bypass power consumption.

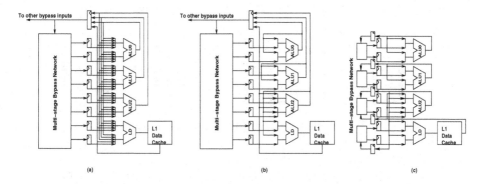

Fig. 5. (a) Conventional Bypass Network; (b) Limited SFU; (c) Extreme SFU

3.2   FU Assignment

The performance of the new SFU bypass network design relies heavily on the ability to assign instructions to the FUs where their operands are available through the bypass network. We propose a post-schedule FU assignment scheme. In this scheme, the FUs assigned to the instructions by the select logic are selectively discarded[3].

Once an instruction is scheduled for execution, it is assigned an FU based on where its operands will be available. Figure 6 shows the new pipeline for this FU assignment scheme. All the scheduled instructions access an *FU table* (in the *FU assign/arbiter* stage), for each valid operand, to get an FU assigned to them. From the *FU table*, the following information is obtained for each valid operand: (i) whether it is available from the bypass network, and if it is, then

---

[*] However, the conventional select logic is still needed to select the right instructions (based on the priority scheme used, which could be the "oldest" first) to be scheduled.

in which FU, or (ii) whether it is available from the register file. Based on the information obtained regarding an instruction's operands, an FU is assigned for the instruction as follows:

- If an instruction has only one operand with a valid FU where the operand is available from the bypass network, and the other operand is either not present or is available from the register file or is available in any FU from the bypass network (for the LSFU network), then it is assigned the valid FU.
- If an instruction has multiple operands with valid FUs or an operand cannot be obtained from the bypass network, then that instruction is marked "unscheduled" and it remains in the issue queue.
- If an instruction does not have any register operands or all its operands are available from the register file or all its operands are available in all the FUs from the bypass network (for the LSFU network), then the FU assigned by the select logic is used.

Once the FU for an instruction is decided, the *FU table* is updated.

Fetch	Fetch	Decode	Decode	Rename	Rename	Queue	Schedule	FU assign/ arbiter	Issue	Register Read	Register Read	Execute	Cache Access	Write back	Write back	Commit

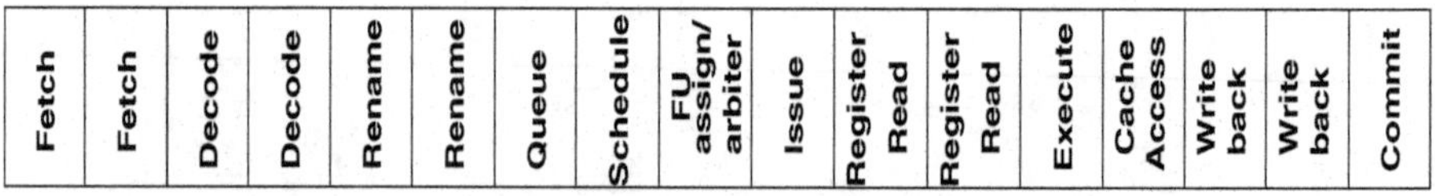

Fig. 6. Post-schedule FU Assignment Pipeline

In this scheme, one apparent issue that needs to be addressed is what happens if multiple scheduled instructions are assigned the same FU? This issue can be resolved by using *FU arbitration*, once FUs are assigned. All the scheduled instructions (that get an FU assigned) send a request for the assigned FU. FU arbiters grant the requests of the scheduled instructions, based on priorities which could be the same as that used by the scheduler. In case an instruction cannot acquire the assigned FU, it is "unscheduled" and is again scheduled in the following cycles.

Figure 7 shows the operation of the *FU assign/arbiter* stage for one instruction. The register operands are used to index into the *FU table* (storing the FUs where the registers are produced) and the *valid bit table* (indicating whether the *FU table* entry is valid). Based on the valid bits and the FU mappings read from the table, an FU is assigned to the instruction. Based on the assigned FU, the instruction sends a request to that particular FU's arbiter, and the *FU table* is updated simultaneously.

To perform FU assignment and arbitration in a single pipeline stage, these operations need to be fast. *FU arbitration* is very similar to the select logic, but of significantly lower complexity than the select logic (because of the maximum number of requests that can be generated). Our investigations (using the calculations in [6]) suggest that the latency of *FU arbitration* is about 60% less than

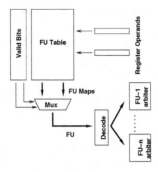

Fig. 7. Schematic FU assign/arbiter stage

that of the select logic, for the parameters in Table 1. For a faster FU table access (for fast FU assignment), we propose a FU table organization in which the FUs for multiple registers are stacked together in a single FU table entry. The higher end bits of a register tag are used to index into the table and the lower end bits give the offset of the FU for a particular register tag. With this design, the access latency of a FU table is about 90% less than that of a 128-entry physical register file used for a 6-way issue processor, based on the calculations in [6]. The FU table is read during *FU assignment* and updated during *FU arbitration*. With such small FU assignment and arbitration latencies, we assume a single *FU assign/arbiter* stage.

To execute the dependent instructions in consecutive cycles, a scheduled instruction immediately wakes up the dependent instructions. In this case, "unscheduling" a scheduled instruction may lead to the consumer instruction getting executed before the producer instruction. This situation is avoided by keeping a bit-vector (of size equal to the number of physical registers) to indicate whether instruction producing a particular register has been dispatched to the FUs. The instructions check this bit-vector in parallel to FU arbitration. If the producer of an instruction's operand has not been dispatched to the FU, it is also "unscheduled".

4   Results

4.1   Experimental Setup

The processor parameters used in our experiments are given in Table 1. We use a modified version of the Simplescalar simulator [24] for our experiments. For benchmarks, we use benchmarks from the 7 Integer (gzip, vpr, gcc, mcf, parser, bzip2, and twolf) and 7 FP (wupwise, art, swim, ammp, equake, apsi, and mesa) benchmarks from the SPEC2K benchmark suite compiled with the options provided with the suite. Latency calculations are performed for a 0.18μm feature size.

Table 1. Default Parameters for the Experimental Evaluation

Parameter	Value	Parameter	Value
Fetch/Decode Width	8 instructions	Instr. Window Size	128 instructions
Phy. Register File	128 Int/ 128 FP	Int. FUs	3 ALU, 1 Mul/Div, 2 ld/2 st
Issue/Commit Width	6 instructions	FP FUs	3 ALU, 1 Mul/Div
Branch Predictor	bi-modal 4K entries	BTB Size	2048 entries, 2-way assoc.
L1 - I-cache	32K, direct-map, 2 cycle latency	L1 - D-cache	32K, 4-way assoc., 2 cycle latency
Memory Latency	40 cycles first chunk 1 cycles/inter-chunk	L2 - cache	unified 512K, 8-way assoc., 6 cycles

In our experiments, we experiment with 4 different bypass network configurations — a fully connected 0-cycle latency bypass network (*FUL0*), a fully connected 1-cycle latency bypass network (*FUL1*), *LSFU*, and *ELSFU*. The forwarding latencies for the *LSFU* and *ELSFU* bypass networks is 0 cycles for direct connections.

4.2   IPC Results and Analysis

We use the calculations in [6] to compare the forwarding latencies of the *FUL0* bypass network (Figure 5(a)) and the *SFU* bypass network (Figure 5(b)), for the FUs given in Table 1. Note that, we assume that only the ALUs and the load units are directly connected to each other in Figure 5(a) and values from the other FUs are forwarded using the multi-stage bypass network. We found that the forwarding latency of the *SFU* bypass network is about 70% less than the *FUL0* bypass network.

Figure 8 gives the IPCs for the SPEC 2000 INT and PF benchmarks. As seen in Figure 8, the IPC from *LSFU* bypass network is very close to that obtained from the *FUL0* network, and significantly higher than that from the *FUL1* network. However, the *ELSFU* network with minimal bypass hardware incurs more IPC impact than the *LSFU* network because of more instructions getting delayed due to extremely limited bypassing. However, for many of the benchmarks, *ELSFU* configuration is quite close to or even higher than the *FUL1* network. *LSFU* network performs better than the *FUL1* network, because in case of the *FUL1* network, all the instructions incur a 1 cycle forwarding latency. However, in the *LSFU* network, most of the instructions do not incur any delays due to forwarding, a few instructions suffer a 2-cycle delay because of getting re-scheduled and a negligible number of instructions incur a delay of more than 2 cycles (if they do not get any FU during the second try as well). The fact that most of the instructions do not incur any forwarding latency delays is also the reason that the IPC of the *ELSFU* network does not reduce significantly. However, in case of *ELSFU* network, if an instruction is not able to get its operands from the bypass network, then it has to wait at least 4 cycles so that it can read the operand values from the register file (it takes 2 cycles to write the values into the register file).

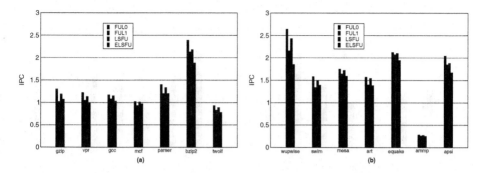

Fig. 8. IPCs for SPEC 2000 (a) INT and (b) FP Benchmarks

4.3 Lower Priority to Branch Instructions

One of the main reasons for the performance impact when using single FU bypass network is the delayed execution of some of the instructions. To recover the
performance loss, we investigate a technique that gives lower priority to branch
instructions. Not all the branch instructions affect performance. Only the mispredicted branches affect performance, and delayed execution of correctly predicted
branches does not impact performance. However, branches with low prediction
accuracy need to be executed at the earliest to know their outcome. Branch instructions, on the other hand, compete with other instructions for the valuable
forwarding paths. For instance, if a branch instruction and a result-producing
instruction are both dependent on the same instruction, then both the instructions will be assigned the FU used by the producer instruction. In this case, if the
branch instruction gets the FU, and the result-producing instruction is delayed,
the performance will be hit. However, if the branch instruction is delayed, then
the performance is hit only if the branch instruction is mispredicted. Hence, to
improve performance with a SFU bypass network, branch instructions are given
a lower priority during *FU arbitration*. For this, each instruction is assigned a
bit (called the "type bit"), which indicates whether the instruction is a branch
instruction or not, and in case of a collision for the same FU, lower priority
branch instruction gets "un-scheduled".

Figure 9 shows the IPC results when this technique is employed for the SFU
bypass network. As can be seen in Figure 9, a significant performance improvement is observed for many benchmarks, bringing the IPC of the *LSFU* bypass
network almost equal to that of the *FUL0* network.

5   Single Input Single FU Bypass Network

The bypass network complexity can be further reduced by half by forwarding the
values to only one of the inputs of the FUs [2]. Since, in the SFU bypass network
design, only the instructions that have at most one operand forwarded from
the bypass network are scheduled, having a single input forwarding is a natural
extension of this design. This would require a switch in the operand locations in

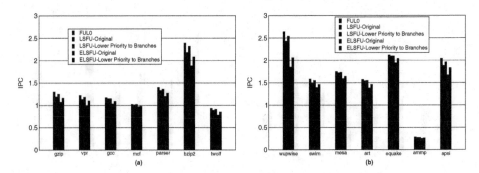

Fig. 9. IPCs for SPEC 2000 (a) INT and (b) FP Benchmarks

the instructions, so that the correct operand is bypassed the correct value. Since the values are forwarded to the inputs of the same FU in the SFU design, the switch can be performed once the operand that will be forwarded is known. No additional performance loss is observed for a Single Input SFU bypass network, because as explained earlier, only the instructions that need at most one operand from the bypass network are scheduled for execution. However, with single input SFU bypass network, the forwarding latency further reduces to about 85% less than the broadcast-based bypass network. This kind of data forwarding has one of the minimum bypass network hardware, apart from the case when there is no data forwarding.

6   Conclusions

Microprocessors usually use broadcast-based bypass networks to execute dependent instructions in consecutive cycles, to better performance. However, for a wide-issue and deeply pipelined processor, broadcasting the results can take multiple cycles, especially with the wire delays increasing in the sub-micron technology era, reducing performance significantly.

In this paper, we observed that the results of most of instructions are used by just one other instruction and the operands of many instructions come from a single other instruction. Based on this observation, we proposed a Single FU bypass network, where the results for an FU are only bypassed to its own inputs, thus reducing the bypass network complexity significantly, and facilitating fast forwarding. Our studies showed that the forwarding latency can be reduced by more than 70%, while incurring a small IPC impact of about 2% for most of the benchmarks. Since the bypass network is a big factor in determining the clock speed, Single FU bypass networks have a high potential of increasing the clock speed. Single FU bypass networks are also much more scalable than the broadcast-based bypass networks, as the future microprocessors become more wide and more deeply pipelined.

References

1. V. Agarwal, M. S. Hrishikesh, S. W. Keckler and D. Burger, "lock rate versus IPC: the end of the road for conventional microarchitectures," *Proc. ISCA-27*, 2000.
2. P. Ahuja, D. Clark, and A. Rogers, "The performance impact of incomplete bypassing in processor pipelines," *Proc. Micro-28*, 1995.
3. E. Bloch, "The Engineering Design of the Stretch Computer," *Proc. Eastern Joint Computer Conference*, 1959.
4. M. Brown, J. Stark and Y. Patt, "Select-free Instruction Scheduling Logic," *Proc. Micro-34*, 2001.
5. J. Hennessy and D. Patterson, "Computer Architecture: A Quantitative Approach," Morgan Kaufmann Publishers, 2002.
6. S. Palacharla, N. P. Jouppi and J. E. Smith, "Complexity-Effective Superscalar Processors," *Proc. ISCA*, 1997.
7. G. Hinton, et al, "A 0.18-um CMOS IA-32 Processor With a 4-GHz Integer Execution Unit," *IEEE Journal of Solid-State Circuits*, Vol. 36, No. 11, Nov. 2001.
8. K. Sankaralingam, V. Singh, S. Keckler and D. Burger, "Routed Inter-ALU Networks for ILP Scalability and Performance," *Proc. ICCD*, 2003.
9. E. Sprangle and D. Carmean, "Increasing Processor Performance by Implementing Deeper Pipelines," *Proc. ISCA-29*, 2002.
10. J. Stark, M. Brown and Y. Patt, "On Pipelining Dynamic Instruction Scheduling Logic," *Proc. Micro-33*, 2000.
11. The National Technology Roadmap for Semiconductors, Semiconductor Industry Association, 2001.
12. E. Rotenberg, Q. Jacobson, Y. Sazeides, and J. Smith, "Trace Processors," *Proc. 30th International Symposium on Microarchitecture*, 1997.
13. D. Leibholz and R. Razdan, "The Alpha 21264: A 500 MHz Out-of-Order Execution Microprocessor," *Proc. Compcon*, pp. 28-36, 1997.
14. K. Farkas, P. Chow, N. Jouppi, and Z. Vranesic, "The Multicluster Architecture: Reducing Cycle Time Through Partitioning," *Proc. 30th International Symposium on Microarchitecture*, 1997.
15. R. Canal, J. M. Parcerisa, and A. Gonz?lez, "Dynamic Cluster Assignment Mechanisms," *Proc. Int. Symp. on High-Performance Computer Architecture (HPCA-6)*, 2000.
16. A. Baniasadi and A. Moshovos, "Instruction Distribution Heuristics for Quad-Cluster, Dynamically-Scheduled, Superscalar Processors," *Proc. International Symp. on Microarchitecture (MICRO-33)*, 2000.
17. J. M. Parcerisa, J. Sahuquillo, A. Gonzalez and J. Duato, "Efficient Interconnects for Clustered Microarchitectures," *Proc. PACT-11*, 2002.
18. R. Nagarajan, et al, "A design space evaluation of grid processor architectures," *Proc. Micro-34*, 2001.
19. E. Waingold, et al, "Baring it all to software: RAW machines," *IEEE Computer*, 30(9):86-93, September 1997.
20. M. Fillo, et al, "The M-Machine Multicomputer," *Proc. Micro-28*, 1995.
21. A. Aggarwal and M. Franklin, "Instruction Replication: Reducing Delays due to Inter-Communication Latency," *Proc. PACT*, 2003.

22. M. K. gowan, et. al., "Power Considerations in the Design of the Alpha 21264 Microprocessor," *Proc. DAC*, 1998.
23. V. Tiwari, et al., "Reducing Power in High-performance Microprocessors," *Proc. DAC*, 1998.
24. D. Burger and T. Austin, "The Simplescalar Tool Set," *Technical Report*, Computer Sciences Department, University of Wisconsin, June 1997.

DSP Implementation of Real-Time JPEG2000 Encoder Using Overlapped Block Transferring and Pipelined Processing

Byeong-Doo Choi[1], Min-Cheol Hwang[1], Ju-Hun Nam[1],
Kyung-Hoon Lee[2], and Sung-Jea Ko[1]

[1] Department of Electronics Engineering,
Korea University, 5-1 Anam-Dong,
Sungbuk-ku, Seoul 136-701, Korea
[2] Electronics and Telecommunications Research Institute
sjko@dali.korea.ac.kr

Abstract. This paper presents a DSP implementation of real-time JPEG 2000 encoder system. Among several modules in JPEG 2000 encoder, the lifting algorithm for discrete wavelet transform (DWT) and the embedded block coding with optimized truncation (EBCOT) comprise more than 85% of the encoding complexity. Thus, it is very important to design and optimize these two modules in order to increase the encoding performance. First, we propose a overlapped block transferring (OBT) method that can significantly improve the performance of the lifting algorithm for DWT by increasing the cache hit rate. Next, we introduce a pipelined processing of passes (PPP) method for fast implementation of EBCOT Tier-1. This method reduces the processing time of EBCOT Tier-1 by processing the three coding passes of the same bit-plane like pipeline. Moreover, we propose a computationally efficient method of EBCOT Tier-2 to predict the truncation point by using the temporal redundancy in the image sequence. Experimental results show that our developed Motion-JPEG 2000 DSP system meets the common requirement of the real-time video coding [30 frames/s (fps)] and is proven to be a practical and efficient DSP solution.

1 Introduction

JPEG2000 compression standard has been created to provide high compression efficiency compared to JPEG [1]. It includes a rich set of features such as improved compression efficiency, lossy to lossless compression, multiple resolution representation, embedded bit-stream, region-of-interest (ROI) coding, and error resilience [2, 3]. Motion-JPEG2000 (MJP2) is intended to create a new coding system required by video communication market and applications based on JPEG2000. It is notable that Motion-JPEG2000 provides high compression performance, strong error resilience, and good perceptual image quality [4].

L. Bougé and V.K. Prasanna (Eds.): HiPC 2004, LNCS 3296, pp. 333–341, 2004.
© Springer-Verlag Berlin Heidelberg 2004

Among several modules in JPEG2000 encoder, the lifting algorithm for discrete wavelet transform (DWT) and the embedded block coding with optimized truncation (EBCOT) comprise more than 85% of the encoding complexity. Thus, it is very important to design and optimize these two modules in order to increase the performance. The latest DSP chip can enable the real-time implementation of the DWT and adaptive binary arithmetic coding. Utilizing the hardware features of the DSP chip, we optimize wavelet filtering and the EBCOT algorithm.

In this paper, we propose the overlapped block transferring (OBT) method, based on the cache performance to improve DWT. Instead of the line-based lifting scheme, an image is divided into overlapped subblocks and then each overlapped subblock is processed by a 2-D lifting algorithm to increase the cache hit rate. We show that the OBT-based lifting scheme can increase the performance of the DWT drastically. Next, we propose a pipelined processing of passes (PPP) for fast implementation of EBCOT Tier-1. This method reduces the processing time of EBCOT Tier-1 by processing the three coding passes of the same bitplane in parallel. Moreover, we propose a scheme to reduce the computation of EBCOT Tier-2 by predicting the truncation point using the temporal redundancy in the image sequence instead of calculating the optimal bit-rate of each frame independently.

The paper is organized as follows: The OBT-based lifting scheme is proposed in Section 2, and pipeline processing of passes for EBCOT is proposed in Section 3. We propose a computational reduction method of EBCOT Tier-2 in Section 4. In Section 5, the performance of the proposed system is discussed and conclusions are given in Section 6.

2   OBT-Based Lifting Scheme for Efficient Cache Utilization

The lifting algorithm is a fast computing technique of the DWT. However, in a point of view of memory management, it still has severe cache-miss problems during the execution of the vertical wavelet filtering. A number of cache-misses make the processing time increase critically. Thus, even though the lifting algorithm requires few execution of CPU, the processing time of DWT cannot be reduced remarkably without the memory management to reduce cache-misses.

This problem can be improved by partitioning an entire image to blocks. Conventionally, in order to perform a lifting scheme, the image rows are filtered in the horizontal direction, and the image columns are filtered in the vertical direction. However, our approach partitions an entire image into blocks to fit into the cache size and reorders the processing sequence to be processed block by block. It reduces the cache miss rate because data, which is fetched in horizontal filtering, is remained on cache until the vertical filtering of current block is completed. Thus, a whole image data can not be loaded on data cache.

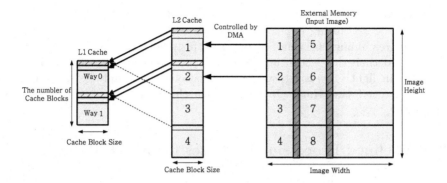

Fig. 1. Memory manipulation of proposed OBT

This method is seemed to remove perfectly a cache-miss problem. However, two problems exist in the proposed method. One is that coefficients of edge cannot be filtered independently without coefficients of adjacent blocks. The other is that the data, which is not aligned by cache block size, must be fetched two times. This means that the redundancy of re-fetch exists yet.

In order to solve these two problems, we propose an overlapped block transferring (OBT) method. This method is based on hierarchical memory architecture. The memory architecture of JPEG2000 encoding system is composed of two layers of data caches (L1D, L2D) and an external memory.

The principle of the OBT method is that DMA operating independently without CPU execution transfers the image data to L2D from the external memory by block size equal to the cache size. Unlike the external memory, the address of the data memory on L2D is aligned by cache block size. Since the L2D can not hold a whole image of large size, DMA transfers data blocks from the external memory to L2D and from L2D to the external memory repeatedly like double-buffering. Fig. 1 shows this mechanism. The data of the first column of blocks is transferred to L2D. After the first column of blocks is processed, this data is moved to the external memory and the next column of blocks is transferred to same location of L2D by DMA.

Fig. 2 shows that the adjacent blocks are overlapped with each other along the horizontal direction. Area 1 in light gray is completely wavelet processed,

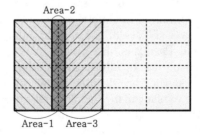

Fig. 2. Overlapped block configuration

whereas Area 2 in dark gray contains data lifted partially. Thus, the next block for the 2-D lifting is placed to include Area 2 as well as Area 3 in light gray. The remaining horizontal lifting steps for pixel values in Area 2 are completed, and then the 2-D lifting scheme is processed for Area 3. As a result, the data in light gray is fetched onto cache one time, and the data in dark gray is fetched two times. It means that the cache miss rate is reduced drastically.

3   Pipelined Processing of Passes for EBCOT Tier-1

Embedded block coding with optimal truncation (EBCOT) is the most complicated part in JPEG2000. The context of a sample coefficient is formed according to the significant state of the sample and its eight neighbors within 3x3 context window. Then the context data is processed by the arithmetic coder. Each bit-plane is encoded through three coding passes, called significant propagation pass (Pass 1), magnitude refinement pass (Pass 2) and clean up pass (Pass 3). In conventional EBCOT method, each pass is processed independently, although the processing of these three passes is very similar. Thus, if the redundancy between three passes, e.g. extraction of bit-plane data and context information from image data (16bits), is removed, it is possible to reduce remarkably the processing time of EBCOT.

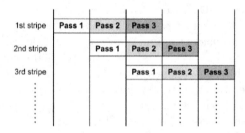

Fig. 3. Pipeline processing of three passes

Parallel processing can be utilized to remove the redundancy. However, due to the dependency of passes, it is difficult to process three passes of one stripe simultaneously. In detail, in order to process Pass 2 of current stripe, the context information of current and adjacent stripes, which is updated by processing Pass 1, is required. In case of Pass 3, the context information to be acquired by processing Pass 1 and Pass 2 is also needed. We use a pipelined processing of passes (PPP) scheme as an alternative method of parallel processing. The strategy is to process the three coding passes of the same bit-plane using pipeline architecture as shown in Fig. 3. First, the bit-plane data of the first stripe is calculated for Pass 1. Second, the bit-plane data of the first and second stripes is calculated for Pass 2 and Pass 1 respectively. Third, the bit-plane data of the first, second and third stripes is processed for Pass 3, Pass

2 and Pass 1 respectively. Third step is iterated until the coding block is completely processed. As a consequence, all the three passes are processed in one scan.

4   Fast Rate Distortion Optimization Method for EBCOT Tier-2

Post-Compression Rate-Distortion Optimization (PCRD Opt.) can find the truncation point which attains the best image quality. However, this scheme requests Tier-1 coding of all the bit-plane data, which is the most computationally intensive process in JPEG2000 coding.

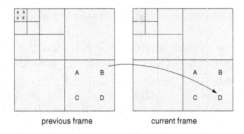

previous frame   current frame

Fig. 4. Example of the adaptive prediction scheme

Our approach predicts the truncation point by using the temporal redundancy in the image sequence, which is viable for real-time processing. Because two adjacent frames in the image sequence tend to have temporal redundancy, truncation points of code-blocks in the current frame can be very close to ones in the previous frame. Therefore, using the number of coding passes of code-blocks in the previous frame, we can predict the number of coding passes in the current frame.

In the proposed method, the first frame is encoded in the full mode due to the lack of the reference frame. Then the currently encoded frame becomes the reference frame for the next frame. The truncation point of each code-block in the next successive frames is predicted using that of the previous frame. Each sub-band is further divided into several blocks. A best matching block is selected among the blocks in the same sub-band of the previous frame using the Sum of Absolute Difference (SAD). The truncation point of the best matching block becomes the predicted truncation point of the block at the same position of the current frame. Since the LL sub-band is a downscaled image of the whole image, the motion activities for the sub-bands are highly correlated. Hence, the procedure to find the best matching block is performed only for the LL sub-band and the location information of the best matching block obtained in the LL sub-band is utilized for the other sub-bands (LH, HL, and HH).

5   Experimental Results

The proposed OBT-based lifting scheme, the PPP method and fast PCRD Opt. method for EBCOT are demonstrated in this section. To prove the effectiveness of the proposed method, simulations were conducted using on a TMS320C6416 (600Mhz, 4800MIPS).

Table 1 shows the number of cache misses produced by the proposed OBT method. In the horizontal filtering, the OBT method produces more cache misses than the conventional method, because the data of overlapped area are fetched twice. However, in the vertical filtering, the OBT method completely removes the cache misses. Consequently, the OBT method reduces the cache-miss rate by 98%.

Table 2 shows the processing time of the DWT using the two existing method, which are Meerwald's method (Row extension, Aggregation) [7] and Chatterjee's method (Strip-mining, Data layout) [8], and the proposed OBT method. For all image sizes, there is no improvement in the horizontal filtering. But all three methods are effective in vertical filtering. Row extension, aggregation and the combination of both methods reduce the processing time by 78, 88 and 90%, respectively, in the vertical filtering. Strip mining, recursive data layout and the combination of both methods reduce the processing time by 73, 66 and 82%, respectively, in the vertical filtering. Our method reduces the processing time by 98% in the vertical filtering. Note that the speed of horizontal filtering is almost identical to that of vertical filtering. It means that the proposed method eliminates most cache misses in vertical filtering, as we expected.

Table 3 shows the performance improvement by using the proposed PPP algorithm for EBCOT Tier-1. As shown in Table 3, for Pass 1, the proposed method does not affect the execution time because there is no difference between the proposed method and the conventional method. However, for Pass 2 and 3, the proposed method reduces the calculation time up to 41% (Pass 2) and 32% (Pass 3). This result indicates that the proposed method significantly reduces the

Table 1. Comparison of number of cache misses

Image size	Lifting direction	Number of cache misses	
		Conventional lifting	Overlapped Block Transferring
256x256	Horizontal	1024	1280
	Vertical	65,536	0
512x512	Horizontal	4096	4608
	Vertical	262,144	0

Table 2. Comparison of processing time of wavelet lifting scheme

Different method		Execution time of DWT			
		Horizontal(ms)	Vertical(ms)	Total(ms)	Speed-up
Image size: 256x256					
Original wavelet-lifting		2.65	111.63	120.28	1
Meerwald's method	Row extension	2.85	24.66	27.51	4.38
	Aggregation	2.95	14.14	17.09	7.04
	Combination	2.88	10.88	13.76	8.74
Chatterjee's method	Strip-mining	2.71	32.27	33.98	3.54
	Data layout	2.87	41.12	43.99	2.76
	Combination	2.77	20.26	23.03	5.22
Overlapped block transferring		3.81	3.22	7.03	17.18
Image size: 512x512					
Original wavelet-lifting		12.74	659.35	672.09	1
Meerwald's method	Row extension	12.98	143.77	156.75	4.28
	Aggregation	12.85	77.15	89.10	7.54
	Combination	13.02	61.27	74.29	9.04
Chatterjee's method	Strip-mining	12.89	175.90	188.79	3.56
	Data layout	12.95	225.71	238.66	2.82
	Combination	12.87	115.40	128.27	5.24
Overlapped block transferring		17.94	17.35	35.29	19.04

Table 3. Comparison of consuming time of EBCOT between conventional method and pipelined processing of passes

Image		Lena	Baboon	Peppers
Conventional method (ms)	Pass 1	297.8	277.9	269.7
	Pass 2	140.3	156.8	157.2
	Pass 3	522.8	531.7	533.9
PPP method (ms)	Pass 1	298.6	281.6	272.8
	Pass 2	88.5	96.3	92.6
	Pass 3	357.4	369.5	378.3
Improvement		23%	25%	24%

processing time for scanning and masking in case of Pass 2 and 3 by reusing the parameter and data used in Pass 1. Generally, the computation complexity of the whole EBCOT Tier-1 can be reduced by 24% as compared with the conventional architecture.

Fig. 5 shows the comparison of the normalized processing time of EBCOT Tier-2 at various bit rates. As shown in Fig. 5, in case of the full mode, the processing time of EBCOT Tier-2 is independent of the bit rate. Note that

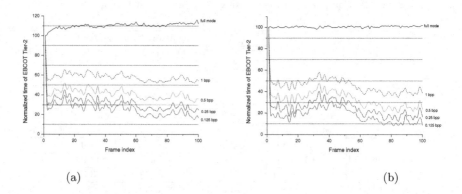

(a)   (b)

Fig. 5. Comparison of the normalized processing time of EBCOT Tier-2. (a) "Football" sequence. (b) "Flower Garden" sequence.(Grayscale, 512x512 size, 1 tile, 1 layer, 5 decomposition levels, 5/3 filter)

the proposed method reduces considerable computation since a large part of unnecessary processes is skipped. The proposed method reduces the processing time of EBCOT Tier-2 down to about 40% at 0.5bpp(1/16) compression.

6   Conclusions

In this paper, we have presented a real-time embedded Motion-JPEG 2000 encoding system using a fixed-point DSP chip. To improve the performance of the system, we have proposed OBT-based lifting scheme to increase the cache hit rate. The OBT-based lifting scheme is over five times faster than the line-based lifting scheme. In addition, we showed that the proposed PPP algorithm and fast rate distortion optimization method can significantly reduce the execution time of EBCOT. Consequently, the Motion-JPEG2000 implementation on a DSP meets common requirement of real-time video coding [30 frames/s (fps)] and is proven to be a practical and efficient DSP solution.

References

1. Rabbani, M., Joshi, R.: An overview of the JPEG2000 still image compression standard. Signal Processing and Image Communication. 17 (2002) 3-48
2. Taubman, D.S., Marellin, M.W.: JPEG2000: Image compression fundamentals, standards and practice. Kluwer Academic Publishers. (2002)
3. Information Technology - JPEG2000 Image coding system:Part 1. ISO/IEC International Standard. 15444-1 (2000)
4. Yu, W., Qiu, R., Fritts, J.: Advantages of motion-jpeg2000 in video processing. in Proceedings of the SPIE, Visual Commmunications and Image Processing. 4671 (2002) 635-645

5. Daubechies, I., Sweldens, W.: Factoring wavelet transforms into lifting scheme. The J. of Fourier Analsys and Applications. 4 (1998) 247-269
6. Taubman, D.S.: High performance scalable image compressin with EBCOT. IEEE Trans. Image Processing 9 (2000) 1158-1170
7. Meerwald, P., Norecn, R., Uhl, A.: Cache issues with JPEG2000 wavelet lifting, Proc. SPIE, Electron. Imaging, Vis. Commun. Image Process. 4671 (2002) 626-634
8. Chatterjee, S., Brooks, C.D.: Cache-efficient wavelet lifting in JPEG2000. IEEE Int. Conf. on Multimedia and Expo. 1 (2002) 797-800

Dynamic Load Balancing for a Grid Application

Menno Dobber, Ger Koole, and Rob van der Mei

Vrije Universiteit, De Boelelaan 1081a, 1081 HV Amsterdam,
The Netherlands
{amdobber, koole, mei}@few.vu.nl
http://www.cs.vu.nl/~amdobber

Abstract. Grids functionally combine globally distributed computers and information systems for creating a universal source of computing power and information. A key characteristic of grids is that resources (e.g., CPU cycles and network capacities) are shared among numerous applications, and therefore, the amount of resources available to any given application highly fluctuates over time. In this paper we analyze the impact of the fluctuations in the processing speed on the performance of grid applications. Extensive lab experiments show that the burstiness in processing speeds has a dramatic impact on the running times, which heightens the need for dynamic load balancing schemes to realize good performance. Our results demonstrate that a simple dynamic load balancing scheme based on forecasts via exponential smoothing is highly effective in reacting to the burstiness in processing speeds.

1   Introduction

Often, grid environments are seen as the successors of distributed computing environments (DCEs). Nevertheless, these two environments are fundamentally different. A DCE environment is rather predictable: the nodes are usually homogeneous, the availability of resources is based on reservation, the processing speeds are static and known beforehand. A grid environment, however, is highly unpredictable in many respects: resources have different and usually unknown capacities, they can be added and removed at any time, and the processing speeds fluctuate over time. In this context, it is challenging to realize good performance of parallel applications running in a grid environment. In this paper we focus on the fluctuations in processing speeds. First, we conducted elaborate experiments in the Planetlab [1] grid environment, in order to investigate over which time scales the processing speed fluctuates. Experimental results show fluctuations over different time scales, ranging from several seconds to minutes. Second, we analyze the potential speedup of the application by properly reacting to those fluctuations. We show that dynamic load balancing based on forecasts obtained via exponential smoothing can lead to a significant reduction of the running times.

Fluctuations in processing speeds are known to have an impact on the running times of parallel applications, and several studies on analyzing the impact of the fluctuations on the running time have been conducted. However, these

L. Bougé and V.K. Prasanna (Eds.): HiPC 2004, LNCS 3296, pp. 342–352, 2004.
© Springer-Verlag Berlin Heidelberg 2004

fluctuations are typically artificially created, and hence controllable (see for example [2]), whereas the fluctuations in grid environments are not controllable. In the research community several groups focus on performance aspects of grid applications. A mathematicians' approach is to develop stochastic models for processing speeds, latencies and dependencies involved in running parallel applications, to propose algorithms for reduction of the running time of an application, and to provide a mathematical analysis of the algorithm [3, 4, 5]. Such a mathematical approach may be effective in some cases; however, usually unrealistic assumptions have to be made to provide a mathematical analysis, which limits the applicability of the results. On the other hand, computational grid experts develop well-performing strategies for computational grids, i.e., connected clusters consisting of computational nodes. However, due to the difference in fluctuations between general grid environments and computational grids, the effectiveness of these strategies in a grid environment is questionable [6]. A third group of researchers focuses on large-scale applications with parallel loops (i.e., loops with no dependencies among their iterations) [7, 8], combining the development of strategies based on a probabilistic analysis with experiments on computational grids with regulated load. However, due to the absence of dependencies among the iterations of those applications, these strategies are not applicable to parallel applications with those dependencies. These observations heighten the need for an integrated analysis of grid applications (including dependencies among their iterations), combining a data approach with extensive experimentation in a grid environment.

The increasing popularity of parallel applications in a grid environment creates many new challenges regarding the performance of grid applications, e.g., in terms of running times. To this end, it is essential to reach a better understanding of (1) the nature of fluctuations in processing speeds and the relevant time scale of these fluctuations, (2) the impact of the fluctuations on the running times of grid applications, and (3) effective means to cope with the fluctuations. The goal of this paper is to address these questions by combining results from lab experiments with mathematical analysis. To address these questions we have performed extensive test-lab experiments in a grid environment called Planetlab [1] with the classical Successive Over Relaxation (SOR) application. First, we provide a data analysis showing how processing speeds change over time. The results show fluctuations over multiple time scales, ranging from seconds to minutes. Then, we focus on the impact of the fluctuations on the running times for SOR at different time scales. The results show a dramatic influence of fluctuating processing speeds on running times of parallel applications. Subsequently, we focus on a dynamic load balancing scheme to cope with the fluctuations in processing speeds. We show that significant reductions in running times can be realized by performing load balancing based on predictions via the classical exponential smoothing technique.

This paper is organized as follows. In Section 2 we will describe the Planetlab testbed environment and the SOR application used in our experiments. Section 3 will show the data collection procedure and results about the different time scales of the fluctuations in processing speeds. In Section 4 different load balancing strategies will be presented. Finally, in Section 5 the results and in Section 6 the conclusions will be addressed.

2   Experimental Setup

Experiments were performed with a parallel application on a grid test bed. A main requirement for the test bed is that it needs to use a network with intrinsic properties of a grid environment: resources with different capacities and many fluctuations in load and performance of geographically distributed nodes. We have performed our experiments on the Planetlab test bed [1], which meets these requirements. PlanetLab is an open, globally distributed processor-shared network for developing and deploying planetary-scale network services.

The application has also been carefully chosen so as to meet several requirements. On the one hand, the application must have dependencies between its iterations, because most of the parallel applications have that property, while on the other hand the structure of the dependencies should be simple. A suitable application is the Successive Over Relaxation (SOR) application. SOR is an iterative method that has proven to be useful in solving Laplace equations [9]. Our implementation of SOR deals with a 2-dimensional discrete state space $M \times N$, a grid. Each point in the grid has 4 neighbors, or less when the point is on the border of the grid. Mathematically this amounts to taking a weighted average of the values of the neighbors and its own value. The parallel implementation of SOR is based on the Red/Black SOR algorithm [10]. The grid is treated as a checkerboard and each iteration is split into phases, red and black. During the red phase only the red points of the grid are updated. Red points only have black neighbors, and no black points are changed during the red phase. During the black phase, the black points are updated in a similar way. Using the Red/Black SOR algorithm, the grid can be partitioned among the available processors. All processors can update different points of the same color in parallel. Before a processor starts the update of a certain color, it exchanges the border points of the opposite color with its neighbors. Figure 1 illustrates the use of SOR over different processors.

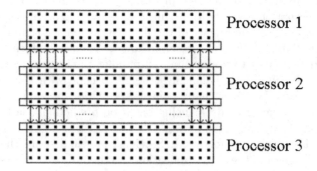

Fig. 1. Successive Over Relaxation

3   Analysis of Fluctuations in Processing Speeds

To characterize the fluctuations in processor speeds in a grid environment, we collected data about the processor speeds and the communication times by doing 10 runs of Red-Black SOR with a grid size of 5000×1000. Interrupted runs were omitted. One run consists of 1000 iterations, from which there are 3 warming up, 994 regular, and 3 cooling down iterations. Every iteration has two phases (see Section 2), which leads to 1988 data lines per run. To increase the running times such that parallellisation improves performance we repeated each iteration 50 times. This corresponds to a grid size of $25 \cdot 10^4 \times 10^3$. Table 1 lists the sites and node names we used during the experiments. For every run we used 4 independently chosen sites from that table. We collected data about calculation times and receive times of each node, and wait times and send times between all nodes. The calculation time is the time a node uses to compute one calculation of one iteration, the wait time is the time a node has to wait for data of its neighbors before it can do a new step, the send time is the time a node uses to send all the relevant data to its neighbors and receive the acknowledgement, and the receive time is the time a node uses to load the relevant data of its neighbors from the received-data table. We do not run other applications on the same nodes during our runs to create changing load on the processors.

Figures 2 and 3 show the calculation times as a function of the iteration number for a set of 250 iterations for different sites. Figures 4 and 5 show the results for the send times. The receive times were found to be negligible (mostly less than 0.5 ms).

The results presented in Figures 2 to 5 reveal a number of interesting phenomena. First, we observe that fluctuations in the calculation times and the send times are considerable. We also observe fluctuations on multiple time scales. On the one hand there are short-term fluctuations in both the calculation times and the send times, on the order of seconds. On the other hand, we observe long-term fluctuations, as can be seen from Figure 2. These fluctuations are presumably caused by a changing load at the processor. The long-term fluctuations in calculation times suggest that reduction of the running times can be realized by dynamically allocating more tasks to relatively fast processors. Second, Figures 2–5 show that the burstiness in the send times is larger than in the calculation times. We observe that the send times do not have a long-term effect, whereas the calculation times often show huge long-term fluctua-

Table 1. Used nodes in our experiments

Site	Abbreviation	Nodename
University of Utah	utah1	planetlab1.flux.utah.edu
University of Washington	wash1	planetlab01.cs.washington.edu
University of Arizona	arizona1	planetLab1.arizona-gigapop.net
California Institute of Technology	caltech1	planlab1.cs.caltech.edu
University of California, San Diego	ucsd1	planetlab1.ucsd.edu
Boston University	boston1	planetlab-01.bu.edu

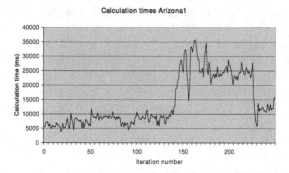

Fig. 2. Calculation times of 250 iterations in Arizona1

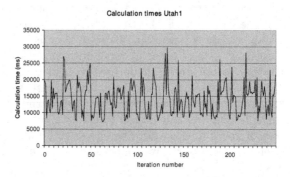

Fig. 3. Calculation times of 250 iterations in Utah1

tions. This observation also suggests that there is a great potential reduction in calculation times, which can be achieved by adapting the load according to the current speeds of the processors. Note that those findings correspond with the results about fluctuations of CPU availability in time-shared unix systems [11]. In this paper we do not investigate the causes of those fluctuations, but we are interested in how to deal with them. That corresponds with the idea that it will be hard to retrieve causes of fluctuations in the future grid.

4   Load Balancing Strategies

Load balancing is an effective means to reduce running times in a heterogeneous environment with fluctuating processing speeds. In this section we quantify the feasible reduction in running time by using different load balancing strategies. We consider two types of load balancing strategies: static and dynamic.

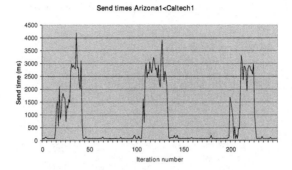

Fig. 4. Send times of 250 iterations from Caltech1 to Arizona1

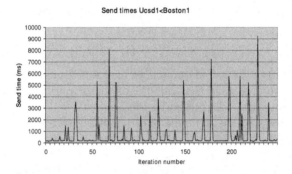

Fig. 5. Send times of 250 iterations from Boston1 to Ucsd1

4.1   Definitions

Static Load Balancing (SLB) strategies use a number of "cold iterations" to estimate the average processor speeds, in order to balance the load. Define

$S(n) :=$ total running time with SLB using the average calculation times
of the first n iterations.

Note that the special case $S(0)$ corresponds to the running time of a run without load balancing, that is, with equal loads.

Berman et al. [12] show that forecasting the performance of the network is useful for Dynamic Load Balancing (DLB). Several prediction methods have been developed for network performance and CPU availability [13]. We use the method of Exponential Smoothing (ES) to predict calculation times (see also [14]). ES appears to be a very simple and effective method to reduce running times. ES is a forecasting method that on the one hand filters out outliers in the data, and on the other hand reacts quickly to long-term changes. Denote by y_n the realization of the n-th iteration step, and let $\hat{y}_n$ denote the prediction of y_n. Then ES is based on the following recursive scheme:

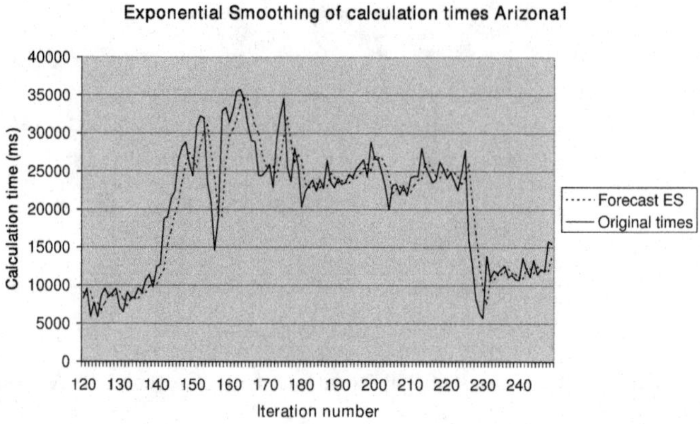

Exponential Smoothing of calculation times Arizona1

Fig. 6. Exponential Smoothing of calculation times of Arizona1

$$\hat{y}_n = \alpha y_{n-1} + (1 - \alpha)\hat{y}_{n-1} \ . \tag{1}$$

Figure 6 shows the calculation times as a function of the iteration sequence number. The results show that our ES predictor performs very well: even the high fluctuations are well tracked by the forecasts.

In the context of the dynamic load balancing strategies the ES-based prediction $\hat{y}_n$ represents the predicted calculation time in the n-th iteration, $n = 1, 2, \ldots$.

If we want to change the load of the processors in the case of Dynamic Load Balancing we have to move rows in the grid from one processor to the other. To avoid excessive communication we introduce a parameter T indicating how often we move rows around. Define for the ES-based Dynamic Load Balancing strategies:

$$D(\alpha, T) := \text{running time with DLB using ES with parameter } \alpha,$$

and load balancing every T iterations.

4.2   Calculation of Running Times

As described in Section 3 we collected data about calculation and send times by doing 10 runs of the Red-Black SOR. In the subsection before we defined several Static and Dynamic Load Balancing strategies. In this subsection we will describe the calculation methods we used to generate estimates of the running times of runs using the optimal static and dynamic strategies from the datasets of the original runs.

We assume a linear relation between the number of tasks (in SOR: the number of rows) and the calculation times of those tasks together, and also a linear relation between the amount of data (in SOR: the number of rows) sent by the application, and the total send time. We also assume that the overhead involved in load balancing is negligible in the long time scale (of minutes) considered here, because calculation times are significantly higher than the overhead.

To start, we explain how we compute an estimation of the lowest possible running time under a DLB strategy, denoted by D^*. To this end, we use the measured calculation times in the original run to estimate the lowest possible calculation time with the optimal DLB strategy. Let $nolb\_rows_j$ be the number of rows assigned to processor $j = 1, \ldots, P$ in the original run, and let $nolb\_calc_{i,j}$ be the measured calculation time on processor j for the ith iteration, $i = 1, \ldots, I$. Then, $nolb\_rows_j/nolb\_calc_{i,j}$ is an approximation of the number of rows that can be executed during iteration i on processor j per time unit. The total processing rate for iteration i is therefore $\sum_j nolb\_rows_j/nolb\_calc_{i,j}$, and the time it takes under a perfect DLB strategy to do iteration i is therefore

$$D^*\_calc_i = \frac{\sum_{j=1}^P nolb\_rows_j}{\sum_{j=1}^P nolb\_rows_j/nolb\_calc_{i,j}} = \frac{(\sum_{j=1}^P nolb\_rows_j)(\prod_{j=1}^P nolb\_calc_{i,j})}{\sum_{j=1}^P (nolb\_rows_j \prod_{k \neq j} nolb\_calc_{i,k})} . \tag{2}$$

Note that $D^*\_calc_i$ is the estimated calculation time for iteration i with the optimal dynamic load balancing strategy, assuming all processor speeds are known in advance, and that $D^*\_calc_i$ is the same for all processors. In this calculation we assumed that the overhead in realizing the dynamic load balancing is negligible.

Now we focus on the calculation of an estimation of the lowest possible running time under SLB, denoted by S^*. With respect to the dynamic situation, we compute the average processing rate over the whole run, and not the rate per iteration. This rate is given by

$$\frac{nolb\_rows_j}{\frac{1}{I}\sum_{i=1}^I nolb\_calc_{i,j}} .$$

Thus the number of rows $S^*\_rows_j$ that have to be assigned to processor k under the SLB strategy is equal to

$$S^*\_rows_j = \sum_{k=1}^P nolb\_rows_k \frac{\frac{nolb\_rows_j}{\frac{1}{I}\sum_{i=1}^I nolb\_calc_{i,j}}}{\sum_{k=1}^P \frac{nolb\_rows_k}{\frac{1}{I}\sum_{i=1}^I nolb\_calc_{i,k}}} , \tag{3}$$

and we estimate that iteration i on processor j takes

$$S^*\_calc_{i,j} = \frac{S^*\_rows_j}{nolb\_rows_j} nolb\_calc_{i,j} . \tag{4}$$

time units.

To calculate the running times of the Static and Dynamic Load Balancing strategies we first derive the number of rows, $S(n)\_rows_j$ and $D(\alpha, T)\_rows_{i,j}$ respectively, each processor j receives from the strategy in each iteration i. For this step we used the methods described in the previous subsection: for calculating $S(n)\_rows_j$ we used the first n iterations and for $D(\alpha, T)\_rows_{i,j}$ Exponential Smoothing. Next, with the following formulas we compute the new calculation times of the strategies for each processor j in iteration i:

$$S(n)\_calc_{i,j} = \frac{S(n)\_rows_j}{nolb\_rows_j} nolb\_calc_{i,j} \;, \tag{5}$$

$$D(\alpha, T)\_calc_{i,j} = \frac{D(\alpha, T)\_rows_{i,j}}{nolb\_rows_j} nolb\_calc_{i,j} \;. \tag{6}$$

Above, we explained how we calculated the new calculation times for each strategy in each iteration. Finally, we put those new calculation times and the send times of the original run in a plain model to derive the new wait times and the estimated running times of the different strategies.

5   Performance Comparison: Experimental Results

To compare the performance under different load balancing strategies, we have estimated the running times under a variety of static and dynamic load balancing schemes. We define the speedups of $S(n)$, $D(\alpha, T)$, S^* and D^* as the number of times those strategies are faster than the run without load balancing:

$$\text{speedup } S(n) := \frac{S(0)}{S(n)} \;, \tag{7}$$

$$\text{speedup } D(\alpha, T) := \frac{S(0)}{D(\alpha, T)} \;, \tag{8}$$

$$\text{speedup } S^* := \frac{S(0)}{S^*} \;, \tag{9}$$

$$\text{speedup } D^* := \frac{S(0)}{D^*} \;. \tag{10}$$

Table 2 shows the speedups that can be made by load balancing on the basis of ES predictions, compared to the case with no load balancing, for a variety of load balancing strategies. Based on extensive experimentation with the value of α, we found that a suitable value of α is 0.5.

The results shown in Table 2 lead to a number of interesting observations. First, we observe that there is a high potential speedup by properly reacting to fluctuations of processing speeds by dynamic load balancing. The potential speedup is shown by the speedup of D^* in Table 2; in the optimal dynamic load balancing case it is possible to obtain a speedup of 2.5 minus the overhead for the running times, and in 20% of the iterations even more than 3.6. Second, we observe that despite the inaccuracy in the predictions of the calculation times the speedup factor by applying dynamic load balancing is still close to the "theoretical" optimum. Even load balancing every 200 iterations, which relatively causes almost no extra overhead compared to the total running time, leads to an average speedup of 2.0 compared to the case of no load balancing. Third, we also observe that even if a better static load balancing scheme is used as a benchmark, the speedup factor realized by implementing a dynamic load balancing scheme is still significant.

Table 2. Relative improvements compared of different load balancing strategies (compared to no load balancing)

LB strategy	Mean speedup of 10 runs
$S(0)$	1.0
$S(1)$	1.2
$S(10)$	1.2
$S(20)$	1.3
S^*	1.9
$D(0.5, 1)$	2.5
$D(0.5, 2)$	2.5
$D(0.5, 3)$	2.4
$D(0.5, 4)$	2.4
$D(0.5, 5)$	2.4
$D(0.5, 10)$	2.3
$D(0.5, 20)$	2.3
$D(0.5, 30)$	2.3
$D(0.5, 40)$	2.3
$D(0.5, 50)$	2.2
$D(0.5, 100)$	2.2
$D(0.5, 200)$	2.0
$D(0.5, 300)$	2.0
$D(0.5, 400)$	1.8
$D(0.5, 500)$	1.7
D^*	2.5

6   Conclusions and Further Research

The results presented in this paper raise a number of challenges for further research. First, the results demonstrate the importance of effectively reacting to randomness in a grid environment. The development of robust grid applications is a challenging topic for further research. Second, in the results presented here we have focused on the fluctuations in processing speeds. However, in data-intensive grid applications the fluctuations in the available amount of network capacity may be even more important than fluctuations in processor speed. To this end, extensive experiments need to be performed to control changing network capacities. Third, more research has to be done on the aspect of selecting the best predicting methods for processor speeds. With those methods general dynamic load balancing algorithms for regularly used parallel applications have to be developed. Finally, in this paper we focus on the SOR application, which has a relatively simple linear structure (see Figure 1). One may suspect that even larger improvements of the running times may be obtained for more complex computation-intensive applications with more complex structures, which is an interesting topic for further research.

Acknowledgements

We are indebted to Mathijs den Burger, Thilo Kielmann, Henri Bal, Jason Maassen and Rob van Nieuwpoort for their useful comments.

References

1. ⟨http://www.planet-lab.org⟩
2. Banicescu, I., Velusamy, V.: Load balancing highly irregular computations with the adaptive factoring. In: Proceedings of the International Parallel and Distributed Processing Symposium (IPDPS). (2002)
3. Attiya, H.: Two phase algorithm for load balancing in heterogeneous distributed systems. In: Proceeding of the 12th Euromicro conference on parallel, distributed and network-based processing. (2004)
4. Shirazi, B.A., Hurson, A.R., Kavi, K.M.: Scheduling and Load Balancing in Parallel and Distributed Systems. IEEE CS Press (1995)
5. Zaki, M.J., Li, W., Parthasarathy, S.: Customized dynamic load balancing for a network of workstations. Journal of Parallel and Distributed Computing **43** (1997) 156–162
6. Nemeth, Z., Gombas, G., Balaton, Z.: Performance evaluation on grids: Directions, issues and open problems. In: Proceedings of the 12th Euromicro Conference on Parallel, Distributed and Network-based Processing. (2004)
7. Banicescu, I., Liu, Z.: Adaptive factoring: A dynamic scheduling method tuned to the rate of weight changes. In: Proceedings of the High Performance Computing Symposium (HPC). (2000) 122–129
8. Cariño, R.L., Banicescu, I.: A load balancing tool for distributed parallel loops. In: International Workshop on Challenges of Large Applications in Distributed Environments. (2003) 39–46
9. Evans, D.J.: Parallel SOR iterative methods. Parallel Computing **1** (1984) 3–18
10. Hageman, L.A., Young, D.M.: Applied Iterative Methods. Academic Press (1981)
11. Wolski, R., Spring, N.T., Hayes, J.: Predicting the CPU availability of time-shared unix systems on the computational grid. Cluster Computing **3** (2000) 293–301
12. Berman, F.D., Wolski, R., Figueira, S., Schopf, J., Shao, G.: Application-level scheduling on distributed heterogeneous networks. In: Proceedings of the 1996 ACM/IEEE conference on Supercomputing, ACM Press (1996) 39
13. Wolski, R.: Forecasting network performance to support dynamic scheduling using the Network Weather Service. In: HPDC. (1997) 316–325
14. Shum, K.H.: Adaptive distributed computing through competition. In: Proceedings of the International Conference on Configurable Distributed Systems, IEEE Computer Society (May 1996) 200–227

Load Balancing for Hierarchical Grid Computing: A Case Study

Chunxi Chen and Bertil Schmidt

School of Computer Engineering, Nanyang Technological University, Singapore
{pg03452644, asbschmidt}@ntu.edu.sg

Abstract. Hierarchical grid computing is a way to gain high compute power at low cost by combining existing computational resources instead of building a new one. It typically has heterogeneous characteristics, such as: (1) Resources have different computational power; and (2) Resources are shared among users; and (3) Resources are usually connected by networks with widely varying performance characteristics. This makes the development or adaptation of parallel applications on hierarchical grids challenging. In this paper, we study three load balancing techniques for hierarchical grids: static load balancing, master-slave and a new technique called "scheduler-worker". We evaluate the performance of these techniques for computing the alignment of long DNA sequences on a grid.

1   Introduction

Hierarchical grid computing describes the combination of several PC clusters within one architecture. Using PC clusters as in the Beowulf approach is currently one of the most efficient and simple ways to gain high compute power at a reasonable price. The development or adaptation of parallel applications for the hierarchical grid architecture is made challenging by the often heterogeneous nature of the resources involved. This establishes the need for new load balancing techniques for efficient hierarchical grid computing.

The grid application that we use in this paper is long DNA sequences alignment. Aligning long DNA sequences is a common and often repeated task in molecular biology. The need for speeding up this treatment comes from the rapid growth of the number of organisms whose genomes have been completely sequenced. Dynamic programming based algorithms can compute accurately the optimal alignment of a pair of sequences [12]. However, since their complexities are quadratic with respect to the length of the two sequences this approach leads to a high computing time. One effective approach to get high quality results in a short time is to use parallel processing. In this paper we present an efficient parallel implementation of the dynamic programming algorithm in linear space. In addition, our solution can compute near-optimal non-intersecting alignments since not only the optimal alignment but also sub-optimal alignments are biological significant. We show that this approach leads to significant runtime savings on a hierarchical grid system.

In order to map the application efficiently onto the hierarchical grid system, we are investigating static and dynamic load balancing approaches in this paper. On the basis

L. Bougé and V.K. Prasanna (Eds.): HiPC 2004, LNCS 3296, pp. 353–362, 2004.
© Springer-Verlag Berlin Heidelberg 2004

of this analysis, we propose a new dynamic load balancing approach named scheduler-worker technique, which can achieve better performance under disturbance and for low inter-cluster bandwidth.

The rest of this paper is organized as follows. In Section 2, we provide a description of hierarchical grid computing. Section 3 presents the sequence alignment algorithm. The mapping of the application onto the hierarchical grid architecture using three different techniques is explained in Section 4. The performance is evaluated in Section 5. Section 6 concludes the paper with an outlook to further research topics.

2   Hierarchical Grid Computing

The computational grid [3] is a new parallel and distributed computing paradigm that provides a resource for large scientific computing applications. It typically consists of heterogenous resources, e.g., clusters, that may reside in different administrative domains, run different softwares, be subject to different access policies, and be connected by networks with widely varying performance characteristics. Grid infrastructure, e.g., Globus Toolkits [16], is installed in the head nodes of these resources. Inside each resource, the resource is managed by the local resource management systems, such as SGE [18] or PBS [19]. Since Message Passing Interface (MPI) is the dominant programming paradigm for clusters, some Grid-enabled MPI implementations, such as PACX-MPI [4], MPICH-G2 [8], Stampi [7] and MagPIe [9], are widely used for hierarchical grid computing.

We have built an experimental testbed consisting of three Linux PC clusters. The three clusters are located at two different research centers (PDCC: parallel and distributed computing center and BIRC: bioinformatics research center) at Nanyang Technological University. Two clusters are in PDCC (8 nodes for every cluster: Intel PIII 733) and one cluster is in BIRC (8 notes: Intel Itanium-1). Each cluster is internally connected by a Myrinet switch (the intra-cluster connection) and an Ethernet switch is used as an inter-cluster connection. The normal application-level bandwidth inside each cluster is about 190 MByte/s. The normal application-level inter-cluster bandwidth is about 8 MByte/s. In order to evaluate different inter-cluster bandwidths, we run an application, which only sends or receives data packages between the clusters. We can control the sizes and frequencies of the data packages to get different levels of inter-cluster bandwidth. The experimental testbed is similar to a real wide-area grid system.

The software architecture is shown in Figure 1. It can be divided into two layers. The upper layer is the MPICH-G2 [8] layer that runs on the control node of each cluster. This allows (slow) inter-cluster communication. The lower one is the MPICH [17] layer that runs on all nodes within a cluster. This allows (fast) intra-cluster communication. Each cluster has SGE [18] installed. SGE is a Distributed Resource Management (DRM) software. It can allocate parallel tasks from the control node to execution nodes inside a cluster. Each parallel task is firstly distributed to the control nodes of each cluster. Secondly, SGE allocates the task from the control node to another execution node within the cluster, complying with its scheduling strategy. Parallel processes can communicate via MPICH-G2 (between clusters) and MPICH (within the same cluster).

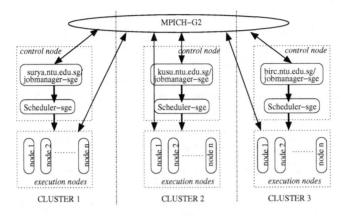

Fig. 1. The hierarchical parallel programming environment consisting of MPICH-G2, MPICH and Sun Grid Engine

3   Aligning Long DNA Sequences

3.1   Smith-Waterman Algorithm

The Smith-Waterman algorithm [12] finds the most similar subsequences of two sequences (the local alignment) by dynamic programming. The algorithm compares two sequences by computing a distance that represents the minimal cost of transforming one segment into another. Two elementary operations are used: substitution and insertion/deletion (also called a gap operation). Consider two strings $S1$ and $S2$ of length $l1$ and $l2$. To identify common subsequences, the Smith-Waterman algorithm computes the *similarity matrix H(i,j)* of the two sequences. The computation of $H(i,j)$ is given by the following recurrences:

$$H(i,j) = \max \begin{cases} 0 \\ E(i,j) \\ F(i,j) \\ H(i-1,j-1) + sbt(S1_i, S2_j) \end{cases}$$

$$E(i,j) = \max \begin{cases} H(i,j-1) - \alpha \\ E(i,j-1) - \beta \end{cases}$$

$$F(i,j) = \max \begin{cases} H(i-1,j) - \alpha \\ F(i-1,j) - \beta \end{cases}$$

where $1 \leq i \leq l1, 1 \leq j \leq l2$ and *Sbt* is a character substitution cost table. Initialization of these values are given by: $H(i,0) = E(i,0) = 0, 0 \leq i \leq l1$ and $H(0,j) = F(0,j) = 0, 0 \leq j \leq l2$.

Multiple gap costs are taken into account as follows: α the cost of the first gap; β is the cost of the following gaps. Each position of the matrix H is a similarity value. The two segments of $S1$ and $S2$ producing this value can be determined by a traceback procedure.

Although able to report all possible alignments between two sequences, the Smith-Waterman algorithm imposes challenging requirements both on computer memory and run-time. Considering comparing two sequences with length of l_1 and l_2, the memory and time complexity for Smith-Waterman algorithm is $O(l_1 \times l_2)$. For aligning two sequences of a few million base pairs in length this would lead to a memory requirement of several Terabytes. However, it is possible to reduce the memory space complexity from quadratic to linear by the algorithm described in the next subsection.

3.2   Optimal Alignments in Linear Space

The values in the *similarity matrix* can be computed in linear space as follows. The value of cell (i,j) in the *similarity matrix* only depends on values of the cells $(i - 1, j - 1)$, $(i - 1, j)$ and $(i, j - 1)$. Thus, the ith row in the *similarity matrix* can be computed by overwriting values for the $(i - 1)^{th}$ row in a left-to-right sweep. However, this approach will only find the maximal score, start and end points; it will not find the actual alignment. Hirschberg [5, 11] presented a recursive divide-and-conquer algorithm for computing this alignment in linear space.

3.3   Finding Near-Optimal Alignments in Linear Space

The detection of near-optimal (or high-scoring) non-intersecting local alignments is particular useful for the comparison of long DNA sequences. The Waterman-Eggert algorithm [14] finds a series of non-intersecting near-optimal local alignments that is widely used. Huang [6] has produced a linear-space version of this method using graph theory. Our approach is based on Huang's method with some modifications to make it more suitable for efficient parallelization.

4   Mapping the Application onto the Hierarchical Grid Architecture

Grid systems typically have a heterogeneous nature. Therefore, the following four aspects have to be taken into account when running a parallel application in a multi-clustered grid environment.

(1) Resources have different computational power.
(2) Resources are shared, i.e. there are several users' tasks running at the same time, therefore, the effective CPU time of an application depends on the number of jobs running on the node at that time.
(3) Resources in a grid system usually are connected by networks with widely varying performance characteristics. Furthermore, the inter-cluster connection is by one or two orders of magnitude slower than the intra-cluster connection.

In order to parallelize an application efficiently on a hierarchical grid architecture, the program should comply with the following rules:

(1) Reduction of inter-cluster data transfer, since the inter-cluster link is usually very slow.

(2) Amount of work allocated to a processor should depend on the computational power that processor allocates to the application at that time. This assures that no processor becomes the bottleneck.

4.1   Parallelization of Sequence Alignment

Parallelization of long DNA sequence alignment consists of two parts:

(1) Parallelization of the *similarity matrix* computation.
(2) Parallelization of the divide-and-conquer algorithm to calculate the actual alignments.

The parallelization of the *similarity matrix* calculation is based on the wavefront communication pattern. Each cell (i,j) of the *similarity matrix* is computed from the cells $(i - 1,j)$, $(i, j - 1)$, $(i - 1, j - 1)$. Therefore, On coarse-grained architectures like homogeneous PC clusters, it is more efficient to assign an equal number of adjacent columns to each processor as shown in Figure 2a. In order to reduce communication time further, matrix cells can be grouped into blocks. Processor i then computes all the cells within a block after receiving the required data from processor $i - 1$. Figure 2 shows an example of the computation for 4 processors, 8 columns and a block size of 2×2.

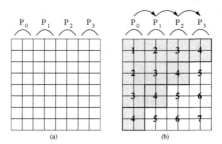

Fig. 2. (a) Column-based division of an 8×8 matrix using 4 processors; (b) Wavefront computation for 4 processors, 8 columns and a 2×2 block size. The complete 8×8 matrix can then be computed in 7 iteration steps

Calculation of the actual alignment is only parallelized if the distance between these points is reasonably large. The parallel solution for determining the optimal path on the hierarchical grid is as follows. Let us define special columns as the last columns of the parts of the *similarity matrix* allocated to each processor, except the last processor. If we can identify the intersection of an optimal path with the special columns, we can split the problem into sub-problems and solve them sequentially on each processor using linear space algorithm. The solutions of the sub-problems are then concatenated to get the optimal alignment.

4.2   Mapping onto the Hierarchical Grid Using Static Load Balancing

In this approach, the mapping has two levels of partitioning. Firstly, the matrix is divided into parts of adjacent columns equal to the numbers of clusters. Secondly, the part within each cluster is further partitioned. The computation is then performed in the same way as shown in Figure 2b. This reduces the inter-cluster data transfer to a single column per iteration step. In order to avoid bottlenecks on the heterogeneous hierarchical grid architecture, the number of columns assigned to each cluster depends on its computational capabilities. The order of the partitioning depends on the inter-cluster bandwidths.

The static load balancing approach can achieve good performance under the condition that there is no disturbance from other applications. Figure 5 shows the speedups of different number of processors and different inter-cluster bandwidths. If the execution of a job is disturbed by another application in a node, it might become the bottleneck for the whole system. An experiment is therefore designed to measure the performance degradation.

In order to scale the extent to which a disturbance affects the application performance, we define PDRD (*performance degradation ratio under disturbance*) as [15]:

$$PDRD = (\frac{T'-T}{T}) \times 100\%$$

where T' denotes the execution time under disturbance and T denotes the execution time without any disturbance. Smaller PDRD values indicate a better robustness of the application to disturbance.

Figure 5 shows how is the effect of disturbance by an application running on the CPU that uses around 50% CPU time.

4.3   Mapping onto the Hierarchical Grid Using Dynamic Load Balancing

A dynamic load balancing mechanism can be implemented using two approaches: *system-level dynamic load balancing* and *application-level dynamic balancing*. The former uses system state information to make decision at runtime on how to dispatch workloads. The latter implements dynamic load balancing as part of the application.

The dominant programming paradigm for clusters is message passing using Message Passing Interface (MPI) [17]. With the goal of coupling several clusters in a computational grid environment, grid-enabled MPI solutions are developed by some institutions. By now there are four most widely used MPI implementations for grid environment, which are PACX-MPI [4], MPICH-G2 [8], Stampi [7] and MagPIe [9]. Unfortunately all these grid-enabled MPI implementations do not support *system-level dynamic load balancing* [10]. Therefore, the dynamic load balancing in our paper is an *application-level dynamic balancing*, as means balancing mechanism and balancer are implemented by users.

Traditional Master-Slave Paradigm

The master-slave paradigm is a widely used technique to implement dynamic load balancing. It works like a server-client model as illustrated in Figure 3. Some parallel bioinformatics applications have achieved good performance using this technique, such as fastDNAml [13] and HMMER [15].

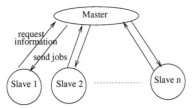

Fig. 3. Master-slave paradigm. Once a slave node finishes a job, it sends a request to the *master* process. The *master* responds by sending back a new job

The implementation of long DNA sequence alignment on the hierarchical grid using this approach works as follows: the *similarity matrix* is divided into rectangular blocks. The computation of a rectangular block is assigned by the *master* to an available *slave*. This requires the sending of the left column and upper row of the block to be computed to the *slave*. The *slave* then returns the right column and bottom row of the computed block to the *master*. Compared to static load balancing implementation, the advantage of this approach is its robustness under disturbance (see Figure 6). Unfortunately, it requires much more communication. For example, for a *similarity matrix* size of $100k \times 100k$ and a block size of 1250×1250, the overall data to be transferred sequentially through inter-cluster link is around 240MBytes. This makes the approach very sensitive to the inter-cluster bandwidth (see Figure 6).

New Scheduler-Worker Technique
We present a new technique named scheduler-worker in order to achieve good robustness under disturbance as well as high performance for low inter-cluster bandwidth. The *workers* report their computing performances to the *scheduler* every time they finish a piece of work. The *scheduler* then produces a new job allocation form depending on every node's performance and broadcasts it to each *worker*. The new job allocations are implemented by exchanging data among *workers*. In the case of DNA sequence alignment, the data transfer for rearranging jobs only happens between two neighboring processes (see Figure 4).

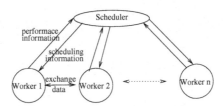

Fig. 4. Scheduler-worker technique

Initially, the *similarity matrix* is partitioned in the same way as in the static load balancing approach. During the computation each *worker* reports its performance to the *scheduler*. We define *NP* (*node performance*) as:

$$NP_i = \frac{SB_i}{T}$$

where SB_i denotes the size of the block assigned to $worker_i$ and T denotes the time for finishing the computation of this block. NP-values describe the currently available compute power in a node.

The *scheduler* judges whether a new job allocation is needed depending on all the NPs. If there exits a disturbance in one node, the sizes of blocks will be rearranged. The equation is as follows:

$$SB_i = \frac{Size_x}{\sum_{j=0}^{N} \frac{NP_i}{NP_j}}$$

where SB_i is the size of the block allocated to $worker_i$, NP_i is the node performance of $worker_i$ and $Size_x$ is the size of Sequence X.

The *workers* will receive the job allocation form produced by the *master*. If no new allocation is needed, the *workers* continue to compute without any interruption. If a new allocation is need, the *workers* implement the rearrangement by exchanging data between neighboring *workers*.

The scheduler-worker technique has several advantages. Its speedup without disturbance is only slightly slower than the static load balancing implementation (see Figure 7). However, it achieves much better PDRD values (see Figure 7). It has only slightly worse PDRD values compared to the implementation using the master-slave approach. However, because of the significantly reduced data transfer, it achieves much higher speedups for low inter-cluster bandwidths.

5   Performance Evaluation

In our experiments, intra-cluster application-level bandwidth is almost 190MByte/s and the two sequences are length 100,1000. We investigate the speedups and PDRDs of each implementation on the grid system with 4 different application-level inter-cluster bandwidths: 8MByte/s, 1MByte/s, 0.5MByte/s and 0.3MByte/s (shown in Figure 5, Figure 6 and Figure 7). How to construct this kind of test environment is described in Section 2. The method for computing speedup is:

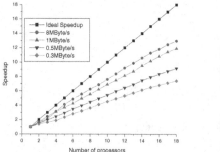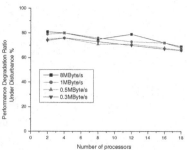

Fig. 5. The left part shows speedups of the implementation using static balancing and without disturbance; The right part shows the PDRCs when one application is running in 1 CPU and uses around 50% CPU time to disturb the execution of our application

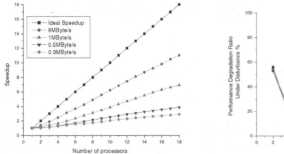
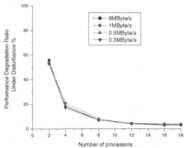

Fig. 6. The left part shows speedups of the implementation using master-slave dynamic balancing and without disturbance; The right part shows the PDRCs when one application is running in 1 CPU and uses around 50% CPU time to disturb the execution of our application

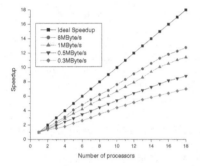
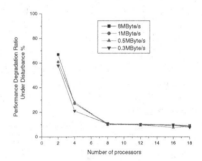

Fig. 7. The left part shows speedups of the implementation using scheduler-worker dynamic balancing and without disturbance; The right part shows the PDRCs when one application is running in 1 CPU and uses around 50% CPU time to disturb the execution of our application

$$speedup = \frac{RT_{cluster1(1)} + RT_{cluster2(1)} + RT_{cluster3(1)}}{3}$$

where $RT_{cluster1(1)}$, $RT_{cluster2(1)}$ and $RT_{cluster3(1)}$ are the runtimes for one processor in cluster1, cluster2 and cluster3.

6   Conclusions and Future Work

In this paper we have demonstrated that the computational grid concept can be efficiently applied to aligning long DNA sequences. We have presented three techniques to map the application onto a hierarchical grid system. We have studied the performance of the these techniques under disturbance and for different levels of application-level inter-cluster bandwidths. The results show that the scheduler-worker technique performs best with respect to these two parameters. Our future work in grid computing will include identifying more biology applications that profit from hierarchical grid systems and presenting more efficient parallel models to map these applications onto hierarchical grid systems.

References

1. Chen C.X., Schmidt B., "Computing Large-scale Alignments on a Multi-cluster", *Cluster 2003*, Hongkong, 2003.
2. Chen C.X.,Schmidt B., "Performance Analysis of Computational Biology Applications on Hierarchical Grid Systems", *CCGrid'04*, Chicago, 2004.
3. Foster, I., Kesselman, C., "The Grid 2: Blueprint for a New Computing Infrastructure," *Morgan Kaufmann*, 2004.
4. Gabriel E., Resch M., Beisel T., Keller R., "Distributed computing in a heterogenous computing environment", *Recent Advances in Parallel Virtual Machine and Message Passing Interface, Lecture Notes in Computing Scicence*. Springer, 1998.
5. Hirschberg, D.S., "A linear space algorithm for computing longest common subsequences," *Comm.ACM*, 18:341-343. 1975.
6. Huang, X., Miller, W., "A time efficient, linear-space local similarity algorithm," *Advances in Applied Mathematics*, 12:337-357, 1991.
7. Imamura T., Tsujita Y., Koide H., Takemiya H., "An architecture of Stampi: MPI library on a cluster of parallel computers", *Recent Advances in Parallel Virtual Machine and Message Passing Interface, Dongarra J., Kacsuk P., Podhorszki N., editors, volume 1908 of Lecture Notes in Computer Science*, Pages 200-207, Springer, Sep. 2000.
8. Karonis N.T., Toonen B., MPICH-G2 project: http://www3.niu.edu/mpi
9. Kielmann T., Hofman R.F.H., Bal H.E., Plaat A., Bhoedjang R.A.F., "MagPIe: MPI's collective communication operations for clustered wide area systems", *Ppopp'99*, pages 131-140, ACM, May, 1999.
10. Müller M., Hess M. Gaberel E., "Grid enabled MPI solutions for Clusters", *CCGrid'03*, Tokyo, Japan.
11. Myers, E., Miller, W., "Optimal alignments in linear space," *Computer Applications in the Biosciences*, 4:11-17, 1988.
12. Smith T.F., Waterman, M. S., "Identification of common molecular subsequences," *Journal of Molecular Biology*, 147:195–197, 1981.
13. Stewart C. A. , Hart D., Berry D.K. , Olsen G.J., Wernert E.A., Fischer W., "Parallel implementation and performance of fastDNAml: a program for maximum likelihood phylogenetic inference", *SC2001*. Denver, CO, USA. 2001.
14. Waterman, M. S., Eggert, M., "A new algorithm for best subsequence alignments with application to tRNA-rRNA comparisons," *Journal of Molecular Biology*, 197:723-728, 1987.
15. Zhu W.R., Niu Y.W., Lu J.Z., Shen C., Gao G.R., "A Cluster-Based Solution for High Performance Hmmpfam Using EARTH Execution Model", *Cluster 2003*, Hongkong, 2003.
16. GLOBUS project: http://www.globus.org
17. MPICH project: http://www-unix.mcs.anl.gov/mpi/mpich/
18. Sun grid engine project: http://gridengine.sunsource.net/
19. openPBS project: http://www.openpbs.org/

A-FAST: Autonomous Flow Approach to Scheduling Tasks

Sagnik Nandy, Larry Carter, and Jeanne Ferrante

Department of Computer Science and Engineering,
University of California at San Diego
{snandy, carter, ferrante}@cs.ucsd.edu*

Abstract. This paper investigates the problem of *autonomously* allocating a large number of independent, equal sized tasks on a distributed
heterogeneous grid-like platform, using only local information. We propose A-FAST (Autonomous Flow Approach to Scheduling Tasks), an
efficient, scalable, dynamic and generic (imposing no restrictions on the
topology) protocol for this purpose. Motivated by the idea of pressure
guiding the flow in fluid networks, A-FAST only uses parameters available *locally* to a node to guide scheduling decisions. Simulations show
that the protocol performs well over a variety of networks, averaging
more than 99.5% of the optimal performance and outperforms related
techniques like *RID* (Receiver Initiated Diffusion). We also show how
a modified use of local information can improve the performance of an
unreliable system. Preliminary results from implementing A-FAST on a
small but real-life distributed system show the performance of our protocol to be near the maximum throughput of the system. Such a protocol
has the potential to aid the efficient deployment of large, data intensive
applications on very large or dynamically changing heterogeneous peer-
to-peer computing platforms.

Keywords: Heterogeneous computing, peer-to-peer computing, network
flows, scheduling.

1 Introduction

The advent of collaborative computing efforts like SETI@home project [25],
GIMP [21] and Entropia [9] has given rise to a range of applications where
a large set of tasks can be distributed across a grid-like platform and solved
concurrently. These applications form the driving motivation of our work, which
aims to schedule a large number of *independent, equal-sized* tasks *online* across a
dynamic and *heterogeneous* computing platform. We seek a scheduling strategy
with the following properties:

- **Autonomous** - Uses minimal (or no) global information. In particular it
  should not require network-wide information.

---

* This work was supported in part by NSF grant ACI-0234233.

L. Bougé and V.K. Prasanna (Eds.): HiPC 2004, LNCS 3296, pp. 363–374, 2004.
© Springer-Verlag Berlin Heidelberg 2004

- **Generic** - Applies to all kinds of networks, regardless of topology.
- **Efficient** - Results in high overall throughput.
- **Scalable** - Applies to networks of very large size.
- **Dynamic** - Adjusts to systems where, due to contention or other reasons, the bandwidths and computation speeds change over time.
- **Practical** - Is easy to implement in real-life scenarios.

The autonomic behavior of fluid networks, using pressure as a guiding force, forms the key inspiration for our work. One can imagine the nodes in a grid as fluid reservoirs, and the links as pipes connecting these reservoirs. Tasks are analogous to the circulating fluid in this scenario. In case of the fluids, *pressure* helps in bringing the system to a steady state without the use of any centralized control. We propose a similar approach where nodes autonomously measures their own pressure. This pressure is then used to decide when to move a task to a neighboring node, eliminating the need for centralized control over scheduling. A-FAST shares similarities with well-known techniques like Cycle Stealing [5] and RID [22], but differs from these techniques by taking both computation and communication into account, which makes it better suited for a wider range of networks. We show how several important scheduling-related issues, including fairness, throughput and reliability, can be easily incorporated in our approach. Initial simulations show that the protocol achieves more than **99.5%** of the maximum throughput over a range of networks, while preserving the above-mentioned properties.

The rest of the paper is organized as follows - Section 2 discusses the related work in this area and Section 3 describes the protocol in detail. In Section 4 we present experimental results showing performance of the protocol under various conditions. We conclude in Section 5 with a summary of our findings and suggest future research directions.

2   Related Work

Scheduling independent tasks across heterogeneous sets of resources is a well known problem. We differ from many of these approaches [14, 1, 6, 24, 10, 17, 19, 12, 29] in that we are developing an *autonomous* scheduling strategy that does not require centralized control or knowledge for scheduling.

Several research efforts have formulated the problem of scheduling tasks across heterogeneous systems as a max-flow problem [7, 30]. However, the most popular max-flow algorithms, including Ford-Fulkerson [15] and Edmonds-Karp [8] use global information to make network-wide decisions. Golberg's algorithm [11] is closer to being autonomous but still requires a notion of *height* that depends on the total number of nodes in the network. In [26], the authors provide a parallel solution to the max-flow problem. However, their approach uses a notion of *timesteps* across the network. This involves network-wide synchronization and is difficult to achieve in large networks. Moreover, all these techniques were designed specifically for static systems. In practice, system properties, such as node speed and bandwidth, network topology, change over time, making these techniques unsuitable.

A-FAST shares similarities with the *RID* (Receiver Initiated Diffusion) [22, 13] and other similar *gradient-based approaches* [18, 27]. In these approaches nodes use some notion of gradient to balance their workload among their neighbors. However, they make their scheduling decisions completely based on the load at a node without taking its communication into account. A-FAST adopts a diffusion-like approach similar to these techniques, but requests tasks based on the supply rate of a node. This ensures that more tasks are received from nodes connected by faster link-speeds. It also makes the protocol applicable to both *computation and communication* dominated systems. Moreover, in A-FAST all communication decisions for a pair of nodes are done independently of their remaining neighbors, reducing the synchronization requirements among nodes. We also show later in the paper how A-FAST manages to capture other system properties like reliability in its notion of pressure.

In [5, 2, 28] variants of the *Cycle Stealing* technique addresses a similar problem as ours. In Cycle Stealing, a node that has exhausted all its work randomly asks its neighbors for additional work. While this approach is autonomous and works well for computation intensive applications, it requires the nodes to be arranged in a hierarchical fashion to avoid unnecessary transfer of tasks. Moreover, Cycle Stealing does not take communication time into account and does not differentiate between nodes connected by different connection speeds.

We only consider applications where there are far more tasks to be executed than nodes in the system, so throughput is more important than makespan, latency or response time. In our previous work [4], [16], we presented an autonomous algorithm that, when the network is a *tree*, achieves the optimum throughput for a static network. Our experiments showed that the protocol reacts quickly to changes in the network as well. However, it may not be desirable to impose a tree-structure on large networks. In [3], it is proven that the problem of finding the best tree from a given network is NP-complete, and even if one could find the best tree, there are networks for which the performance of the optimal tree is unboundedly worse than the whole network's performance. Thus, finding an autonomous solution for a generic network is still open.

3   The A-FAST Task Scheduling Protocol

We begin with a formal description of the problem. We are given a labeled, directed graph $G = (N, E, P, C)$ representing the network. $N = \{0, 1, ..., n-1\}$ is the set of computing resources. Each node $i \epsilon N$ has a computing speed $P(i)$ $(P : N \to R^+)$, denoting the number of tasks the node can complete in a unit time. $E = \{(i, j) : i, j \epsilon N\}$ is the set of links connecting the various nodes in this graph, and $C(i, j)$ $(C : N \times N \to R^+)$ denotes the number of tasks that can be sent from node i to node j in a unit time. All tasks are of equal size (both computationally and communication-wise)[1] and initially reside in the source node 0. We assume that node 0 has a large number of tasks, so we can ignore the start-up and wind-down times of the protocol. The graph G is dynamic in nature, i.e. *(N, E, P, C)* can evolve during execution. Nodes and edges can be added and deleted from N and E (except for node 0, which is always present) and $P(i)$ and $C(i,j)$ can also change. Our objective is to maximize the number of tasks completed per unit time.

The A-FAST protocol assumes that some number of incoming tasks can be buffered in a node. Nodes begin by advertising a quantity we will call their *pressure (p)* to their neighbors, requesting tasks. On receiving a request, a node compares the requester's p to its own to decide whether the request should be serviced. Such an approach allows us to do away with the need for a centralized scheduler, and instead make all scheduling decisions locally based on differences in pressure. If a node does not service a request, it informs the requestor of its decision. On being serviced by a neighbor, a node requests another task from the same neighbor. However, if its request is denied, it waits for a set length of time before making another request. Nodes thus periodically query their neighbors, requesting further tasks. To process a task, a node takes a task from its buffer. If the buffer is empty the node waits till it receives a task.

We now give two case studies that show how A-FAST can autonomously achieve two different system requirements — high throughput and improved reliability — by making suitable definition of pressure.

3.1   Task Scheduling in Dynamic Heterogeneous Systems

For each edge $(j,i)\epsilon E$, we assume there is an *incoming buffer* $IB_{j,i}$ on node i that holds the most recent response (either a denial or a task) sent by j to i. Additionally, each node i has a *task buffer* TB_i that has a capacity of m_i "slots", where each slot can hold one task. These slots can be in one of the following states:

- **S1**: the slot is "empty".
- **S2**: a task is being transferred into the slot from one of the IB_{ji}'s.
- **S3**: the task in the slot is getting executed by N_i.
- **S4**: the task in the slot is being sent to another node N_k i.e. it is being transferred from TB_i into IB_{ik}.
- **S5**: the slot holds a task and is currently not in any of the above states.

Task buffers can have multiple slots in states **S1**, **S2**, **S4** and **S5**, but for simplicity we will allow only one task at a time to be in state **S3**. We say "TB_i is full" when the number of slots e_i in state **S1** is zero. We define the *buffer occupancy*, b_i of a node to be the number of slots in state **S5** at the current time.

---

[1] We conjecture that if tasks are of different sizes, but have a constant computation-to-communication ratio, that the behavior of algorithms will be similar to the equal-size task problem. An interesting open question is how to make scheduling decisions when the ratio is non-constant but known.

For scheduling tasks in a heterogeneous system we set the pressure, p_i, of each node to its buffer occupancy.

The sub-protocols for responding to a request, processing a response and performing a task are shown in Figures 1 and 2. The highlighted sections of the protocols use the shared variables b_i and/or e_i, and must be synchronized to run correctly. This can be done by acquiring locks (if each protocol is a separate process), or by executing the shaded sections atomically (if a single buffer manager procedure handles all three protocols). The Wait primitive in Figure 2 should be implemented using a periodic polling mechanism that prevents livelock.

```
OnRecvReqest(j, bj) { // request from node j

    i = CurrentNode;
    pj = bj ; // pressure of node is equal to its buffer occupancy
    Pj = bj ;

    if (pi-1 > pj) { // node has more tasks than requesting node
        bi = bi - 1;
        send(task, Nj); // send single task to Nj
        ei = ei + 1;
    } else {
        send(refuseMsg, Nj ); // refuse Nj
    }

}
```

```
OnRecvData(j, rj) { // response from node j

    i = CurrentNode;

    if (rj is a task) {
        flag = true;
        while(flag) {
            if(ei > 0) { // there is an empty slot
                ei = ei - 1 ;
                transfer task from iBji to TBi ;
                bi = bi + 1 ;
                requestData(j, bi) ; // request more tasks from node j
                flag = false;
            } else {
                wait for a while;
            }
        }
    } else {
        wait for a while
        requestData(j, bi); // request tasks again
    }

}
```

Fig. 1. Protocol nodes follow on (a) receiving a task request (b) on receiving a response from a neighbor

```
ProcessTask() {
    i = CurrentNode;

    if (bi > 0) { // There exists some task
        dispatch task for processing;
        bi = bi - 1;
        perform task;
        ei = ei + 1;
    } else {
        Wait(till pi > 0);
    }

}
```

Fig. 2. Protocol nodes follow to perform a task

Intuitively, A-FAST should adapt to both a computation-dominated system as well as a communication-dominated one: faster nodes empty their buffers faster and their *pressure* decreases, making them likely to receive more tasks. Similarly if a link is fast, tasks will be delivered more quickly across it, making the receiver request more tasks along that link as compared to a slower link. We will verify these claims experimentally in Section 4.

3.2   Adding Reliability to the System

We now show how the idea of pressure can be modified to incorporate a measure of node reliability into the scheduling strategy.

We define an unreliability parameter, τ_i, for each node in the system, which reflects the average time a node remains online. A fair estimate of the value of τ_i can be computed completely independently by each node. This can be done by maintaining a three tuple of $<num\_of\_readings,\tau_i,last\_val>$ in the persistent storage of each node. On coming online, node i increments the value of $num\_of\_readings$, sets τ_i to $\frac{(\tau_i+last\_val)}{num\_of\_readings}$, saves these values, assigns $last\_val$ to 0 and then continues. $last\_val$ is periodically updated to the elapsed time and saved. The last recorded value of this variable can then be used as an estimate of how long the node remained online (the accuracy depends on the frequency of updates). τ_i thus gives an estimate of the expected duration node i is likely to remain online.

To incorporate reliability into A-FAST we modify our existing definition of pressure to $p_i = \frac{b_i}{\tau_i^K}$, where K is some real positive constant (we shall term it *Assurance Constant*) denoting the importance of reliability to the system. A node now sends the buffer occupancy and unreliability constant to its neighbors when requesting a task and the neighbor can calculate its value of p.

By doing this we make the pressure of a node inversely proportional to its chances of breaking down. Thus for two nodes with similar buffer occupancies, the node with a smaller value of τ (less reliable) will have a higher pressure, making tasks flow out of it towards a more "reliable" node. It must however be mentioned that giving too much importance to reliability might have adverse effects since slower but more reliable nodes will start getting more jobs assigned to them. This can be controlled by choosing an appropriate value of K and will be studied further in the experimental evaluations.

4   Experimental Results

We now present experimental results from simulations as well as real life systems to show how A-FAST works under different situations.

4.1   Experimental Setup

We tested A-FAST on two different networks topologies - internet-like graphs (G1), generated using the Network-Emulator package (NEM) [23] and cluster-like graphs (G2)[2]. For both topologies, we generated graphs of four different sizes (n = 200, 400, 600 and 800). Each node i in these graphs were assigned a random processing speed, $P(i)$ (ranging between 1 and MAX_P), representing the number of tasks node i can process in a minute of simulated time. The values

---

[2] For G2, we built k clusters of equal size. Nodes in these clusters were heavily connected (average connectivity of $k/2$). The clusters were then connected to each other in a random tree topology.

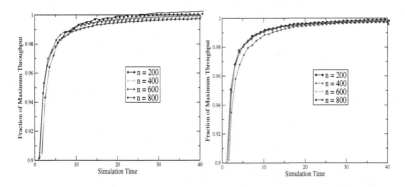

Fig. 3. (a)Performance of A-FAST on *Internet-like* graphs (G1) (b) Performance of A-FAST on *Cluster-like* graphs (G2)

of $C(i, j)$ were similarly assigned (ranging between 1 and MAX_C), denoting the number of tasks that can be sent along the link in one minute. We set both MAX_P and MAX_C to the same value (40) to allow the throughput of the system to be equally dependent on computation and communication. We assumed zero latency networks for our simulations. This might be unreasonable in certain scenarios where the frequent exchange of request messages and the single task transfer approach of the protocol might affect the performance of the system. We discuss in the concluding section how our ongoing work is addressing this issue for real systems (Note that by assuming zero latency the request/denial message transfer times were reduced to zero but the task transfer times were non-zero, depending on the bandwidth of the connecting links). Experiments were repeated multiple times and the average value over all the runs were reported.

4.2   Throughput in a Heterogeneous System

We compared the performance of the first variant of the protocol as a percentage of the maximum throughput of the system (calculated using the *maxflownet* package [20]). The results for the two types of topologies, are shown in Figures 3(a) and (b).

A-FAST performed very well, averaging over *99.5%* of the maximum throughput for both the topologies and the different sizes. Though our graphs were small, this showed A-FAST to be both generic and scalable. It can also be observed that the startup time of A-FAST is also small with almost all the simulations reaching 98% efficiency within 5 minutes of simulated time (5 minutes corresponded to approximately 750 completed tasks in our simulation setup).

We also implemented a version of the RID algorithm to compare its performance against A-FAST for communication-dominated systems. We generated these systems by generating the G1 type graphs where link speeds were less than the processing speed of the nodes joining them. The version of RID balanced the load every time the number of tasks in the Task Buffer fell below 5. The experiments were run for 60 minutes to allow the RID algorithm to reach steady state throughput. The results are shown in Figure 4. While A-FAST achieves

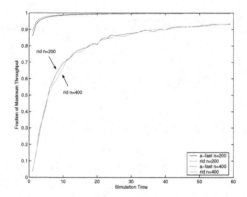

Fig. 4. Relative Performance of A-FAST vs RID on *Communication-dominated Graphs*

nearly 99.5% of the optimal throughput, RID only achieves around 95% of the optimal value. RID also takes a larger amount of time to reach the steady state throughput.

4.3   Adding Reliability to A-FAST

To test the reliability-aware A-FAST variant described in Section 3.2, each node was randomly assigned a value τ between the range $(5, 75)$. We conducted our experiments for 40 minutes of simulation time, giving nodes an equal chance of failing or surviving in the lifetime of the experiments. We then tested A-FAST with four different values of the unreliability constant K, denoting the importance of reliability for the experiments (note that for $K = 0$ the protocol reduces to the standard buffer-based pressure approach described in Section 3.1 and is provided as a base case)[3]. We measured the change in throughput and number of lost tasks (tasks that were assigned to nodes when they broke down). The results are shown in Figure 5 (a) and (b).

In all our experiments, the throughput of the reliability-aware version of A-FAST achieves better throughput when compared to the standard version. However, one cannot conclude anything definitive about the impact of K on throughput. This is because a smaller value of K reduces the importance of reliability and increases the chance of a potentially faulty node getting more tasks while a larger value of K might make slower and more reliable nodes get more tasks, thereby affecting performance. However, it is evident that the introduction of reliability as a parameter to pressure does pay off.

We also see a marked improvement in the reduction of task losses with A-FAST. A task loss might eventually require re-transmitting the task and by reducing the task loss one might eventually improve the system-throughput even further.

---

[3] We also tried the experiments with larger and smaller values of but for K>1.25 the results were similar to that of K=1.25 and therefore have not been shown in the graphs.

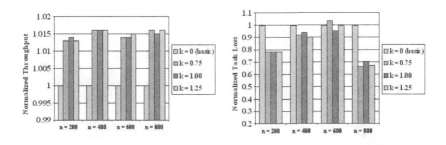

Fig. 5. Effect of adding reliability on (a) system throughput and (b) task loss

4.4   Practical Implementation

As a proof of concept we implemented a prototype version of A-FAST. This section presents some initial results from these experiments. The version of A-FAST described in Section 3.1 was implemented using Java RMI. The system was tested on a 9-node cluster of Pentium III (800 MHz) processors. One of these nodes was designated as a source with a large number of matrix multiplication tasks. We arranged the remaining 8 nodes in three *virtual* topologies - star, ring and a complete binary tree. As a yardstick for A-FAST's performance we also provide an approximate upper bound of the throughput of the system. The maximum throughput of the system is bounded by the sum of the maximum throughput of each participating node. To get an estimate of a node's maximum throughput, we ran the tasks on each target node separately (multiple times). The total of these values is provided in Figure 6 (a). The individual throughput of the nodes are shown in Figure 6 (b)[4].

Though preliminary in nature, these results show A-FAST performs efficiently on all three topologies. The three topologies produced comparable results with an additive difference (contributed mostly by their start-up times). Figure 6 (b) shows that for the star and ring topologies A-FAST does a very good job of autonomous load balancing. However, for the tree topology certain nodes outperform the others. This happens mainly because nodes in the tree do not have equal connectivity and the throughput of the communication-intensive nodes was affected negatively.

5   Conclusion and Future Work

This paper presents a new autonomous scheduling protocol based on the idea of pressure in fluid networks. Preliminary experiments show that the protocol is efficient and can scale and autonomously adjust in dynamic heterogeneous

---

[4] We ignored the first 100 tasks transferred by n0 (to avoid the effect of startup time) and tracked the number of results contributed by each node for the *subsequent* 250 tasks.

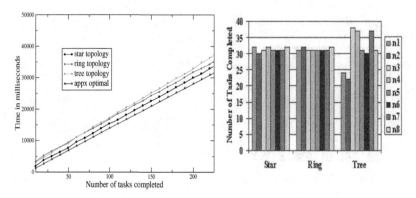

Fig. 6. (a) Execution time of A-FAST on a small cluster. (b) Individual throughput of nodes for the different topologies

networks. We showed how different parameters like throughput and reliability can also be captured. Simulations showed the protocol to be efficient, achieving more than 99.5% of the maximum throughput on average. The need for such protocols is likely to grow as we start using the world wide web not only as an information medium but also as a computing resource.

We are currently pursuing a number of different aspects of this problem. Some of these include: a theoretical bound on the performance of the protocol; the effect of unequal-sized tasks; the effect of dependency between tasks; and capturing other aspects of scheduling with pressure. We are also working on a version of A-FAST supporting *lazy* updates of pressure to reduce the effect of latency and periodic message transfers.

References

1. D. Andresen and T. McCune. Towards a Hierarchical Scheduling System for Distributed WWW Server Clusters. In *Proceedings of the Seventh International Symposium on High Performance Distributed Computing (HPDC-7)*, July 1998.
2. J. Baldeschwieler, R. Blumofe, and E. Brewer. ATLAS: An Infrastructure for Global Computing. *In Proceedings of the Seventh ACM SIGOPS European Workshop on System Support for Worldwide Applications*, 1996.
3. C. Baninio, O. Beaumont, L. Carter, J. Ferrante, A. Legrand, and Y. Robert. Scheduling Strategies for Master-Slave Tasking on Heterogeneous Processor Platforms. In *IEEE Transactions on Parallel and Distributed Systems*, April 2004.
4. O. Beaumont, L. Carter, J. Ferrante, A. Legrand, and Y. Robert. Bandwidth-centric Allocation of Independent Task on Heterogeneous Platforms. In *Proceedings of the International Parallel and Distributed Processing Symposium (IPDPS'02)*, Fort Lauderdale, Florida, April 2002.
5. R. Blumofe, C. Joerg, B. Kuszmaul, C. Leiserson, K. Randall, and Y. Zhou. Cilk: An Efficient Multithreaded Runtime System. In *Proceedings of the 5th Symposium on Principles and Practice of Parallel Programming*, 1995.

6. H. Casanova, A. Legrand, D. Zagorodnov, and F. Berman. Heuristics for Scheduling Parameter Sweep Applications in Grid Environments. In *Proceedings of the 9th Heterogeneous Computing Workshop (HCW'00)*, May 2000.
7. T. Cormen, C. Leiserson, and R. Rivest. Introduction to Algorithms, MIT Press. 1990.
8. Jack Edmonds and Richard M. Karp. Theoretical improvements in the algorithmic efficiency for network flow problems. *Journal of the ACM (JACM)*, 19, 1972.
9. Entropia Inc. http://www.entropia.com, 2001.
10. S. Flynn Hummel, J. Schmidt, R. Uma, and J. Wein. Load-Sharing in Heterogeneous Systems via Weighted Factoring. In *Proceedings of the 8th Annual ACM Symposium on Parallel Algorithms and Architectures (SPAA'96)*, Jun 1996.
11. Andrew V. Goldberg. *Efficient Graph Algorithms for Sequential and Parallel Computers*. PhD thesis, Department of Electrical Engineering and Computer Science, MIT, 1987.
12. T. Hagerup. Allocating Independent Tasks to Parallel Processors: An Experimental Study. *Journal of Parallel and Distributed Computing*, 47, 1997.
13. H.-U. Heil and M. Schmitz. Decentralized Dynamic Load Balancing: The Particles Approach. *Proc. 8th Int. Symp. on Computer and Information Sciences*, 1993.
14. O. H. Ibarra and C. E. Kim. Heuristic algorithms for scheduling independent tasks on non-identical processors. *Journal of the ACM (JACM)*, 24(2), 1997.
15. L. R. Ford Jr. and D. R. Fulkerson. Flow in Networks, Princeton University Press. 1962.
16. B. Kreaseck, H. Casanova L. Carter, and J. Ferrante. Autonomous Protocols for Bandwidth-Centric Scheduling of Independent-task Applications. In *Proceedings of the International Parallel and Distributed Processing Symposium (IPDPS'03), Nice, France*, April 2003.
17. C.P. Kruskal and A. Weiss. Allocating Independent Subtasks on Parallel Processors. *IEEE Transactions on Software Engineering*, 11, 1984.
18. Frank C. H. Lin and Robert M. Keller. The gradient model load balancing method. *IEEE Trans. Softw. Eng.*, 13(1):32–38, 1987.
19. M. Maheswaran, S. Ali, H. J. Siegel, D. Hensgen, and R. Freund. Dynamic Matching and Scheduling of a Class of Independent Tasks onto Heterogeneous Computing Systems. In *8th Heterogeneous Computing Workshop (HCW'99)*, pages 30–44, Apr. 1999.
20. Max-Flow-Solution. http://elib.zib.de/pub/Packages/mathprog/maxflow/index.html.
21. Mercenne Prime Search. http://www.mercenne.com
22. Willebeek-LeMair M.H. and A.P Reeves. Strategies for Dynamic Load Balancing on Highly Parallel Computers. In *Parallel and Distributed Systems, IEEE Transactions*, 1993.
23. Network Emulator. http://clarinet.u-strasbg.fr/nem/.
24. A. Rosenberg. Sharing Partitionable Workloads in Heterogeneous NOWs: Greedier Is Not Better. In *Proceedings of the IEEE International Conference on Cluster Computing (Cluster'01), Newport Beach, California*, October 2001.
25. SETI@home. http://setiathome.ssl.berkeley.edu, 2001.
26. Y. Shiloach and U. Vishkin. An $O(n^2 log\, n)$ parallel max-flow algorithm. *Journal of Algorithms*, (3), 1982.
27. W. Shu and L. V. Kale. A dynamic scheduling strategy for the chare-kernel system. In *Proceedings of the 1989 ACM/IEEE conference on Supercomputing*, pages 389–398. ACM Press, 1989.

28. Rob van Nieuwpoort, Thilo Kielmann, and Henri E. Bal. Satin: Efficient parallel divide-and-conquer in java. In *Proceedings from the 6th International Euro-Par Conference on Parallel Processing*, pages 690–699. Springer-Verlag, 2000.
29. B. Veeravalli, D. Ghose, and T. G. Robertazzi. Divisible load theory: A new paradigm for load scheduling in distributed systems. *Cluster Computing*, 6(1), January 2003.
30. Kevin Daniel Wayne. *Generalized Maximum Flow Algorithms*. PhD thesis, Cornell University, 1999.

Integration of Scheduling and Replication in Data Grids

Anirban Chakrabarti, R.A. Dheepak, and Shubhashis Sengupta

Software Engineering and Technology Laboratory,
Infosys Technologies Ltd, Bangalore (India)
Tel: 91 80 852 0261
{anirban_chakrabarti, dheepak_ra,
shubhashis_sengupta}@infosys.com

Abstract. Data Grids seek to harness geographically distributed resources for large-scale data-intensive problems. Such problems involve loosely coupled jobs and large data sets distributed remotely. Data Grids have found applications in scientific research fields of high-energy physics, life sciences etc. as well as in the enterprises. The issues that need to be considered in the Data Grid research area include resource management for computation and data. Computation management comprises scheduling of jobs, scalability, and response time; while data management includes replication and movement of data at selected sites. As jobs are data intensive, data management issues often become integral to the problems of scheduling and effective resource management in the Data Grids. The paper deals with the problem of integrating the scheduling and replication strategies. As part of the solution, we have proposed an Integrated Replication and Scheduling Strategy (IRS) which aims at an iterative improvement of the performance based on the coupling between the scheduling and replication strategies. Results suggest that, in the context of our experiments, IRS performs better than several well-known replication strategies.

1 Introduction

In an increasing number of scientific and enterprise applications, large data collections are emerging as important resources that need to be shared and accessed by research teams dispersed geographically. In domains as diverse as global climate change, high energy physics, and computational genomics, the volume of interesting data will soon total petabytes[1]. The combination of large data size, geographic distribution of users and resources, diverse data sources, and computationally intensive analysis results in complex and stringent performance demands that are not satisfied by any existing data management infrastructure. The literature offers numerous point solutions that address the issues of data management, data distribution and job scheduling (e.g., see [2,3]). However, no integrating architecture exists that allows one to identify requirements and components common to different systems and hence apply different technologies in a coordinated fashion to a range of data-intensive application domains. Motivated by these considerations, researchers have launched a collaborative effort called *Data Grids* to design and produce such an integrating architecture.

L. Bougé and V.K. Prasanna (Eds.): HiPC 2004, LNCS 3296, pp. 375–385, 2004.
© Springer-Verlag Berlin Heidelberg 2004

1.1 Motivation and Objectives

Most previous scheduling work has considered data locality/storage issues as secondary to job placement. Casanova et al. [4] describe an adaptive scheduling algorithm for parameter sweep applications that uses a centralized scheduler to compute an optimal placement of data prior to job execution. Banino et. al [5] talks about scheduling in a heterogeneous scenario. Work on data replication strategies for Grids includes [6], where the authors examined dynamic replica placement strategies in a hierarchical Grid environment. Recently, some work has been carried out which combines the scheduling and replication strategies to provide better overall performance in Data Grids [7]. Paper [8] talks about combining the replication and scheduling strategies in a more organized manner. The authors assumed three components: an External Scheduler (ES), which determines where (i.e. to which site) to send jobs that originate at that site; a Local Scheduler (LS), which determines the order in which jobs that are allocated to that site are executed; and a Data Scheduler (DS), responsible for determining if and when to replicate data and/or delete local files. The Grid architecture considered in this paper is similar to one proposed in [8].

In Data Grid, both scheduling and replication aim at reducing the latency for job execution. While scheduling does that by directing the jobs to certain sites so that the latency involved in data movement and job processing is reduced, replication moves the data around so that the data access time during scheduling is reduced. The key contribution of the paper lies in the idea of the possible integration between scheduling and replication called *Integrated Replication and Scheduling (IRS)* Approach. Most of the works in this field have concentrated either on replication or scheduling aspects of the problem. Though, some hybrid strategies have been proposed in [7], the first real effort to study the combination of these two strategies was first done in [8]. In [8], the authors have assumed that at a time each job will access only a single data resource like a file. However, in practical situations one job may require multiple files. In this paper, we propose a replication-scheduling algorithm which iteratively improves the performance of the Data Grids. The main objectives of the paper are to develop and evaluate an iterative replication and scheduling strategy.

The assumptions made are: (i) Data Grid is considered to be an undirected graph. Hence, the transfer cost is same both ways, (ii) a two-stage scheduling as mentioned in [8] is assumed, (iii) the Grid is more or less stable i.e., the chances of link and node failures and rare, (iv) the data is mostly handled in a read-only mode, (vi) the jobs are non-preemptable. The rest of the paper is organized as followed. In Section 2 we outline our IRS algorithms in detail with suitable examples. In Section 3, we present and discuss the performance test results vis-à-vis some other approaches. We conclude in Section 4 by pointing out the salient contributions and future work.

1.2 Data Replication (DR) and Job Scheduling (JS) Problem

We model a job request as a 3-tuple $J = <S, \tilde{F}, \tilde{c}>$, where Sj is the site at which the job is fired, $\tilde{F}$ is the list of files needed by the job and $\tilde{C}$ is the computation time required by the job J at site s_j. A site is modeled as a 3-tuple $S = <\hat{F}, V, P_s>$, where

$\hat{F}$ is the set of files stored in the site S, V is the storage capacity at that site and P_s is the computation capacity at that site. It is to be noted that P_s is expressed in sec/GB. In [8], the authors have stated that P_s varies between 10 sec/GB to 50 sec/GB. The *Job Scheduling (JS)* problem states that: Let J_i be a job, and $\hat{s} = \{ S_1, S_2 \ldots S_n \}$ be the set of sites, then the problem is to schedule the job J_i to a site S_j, where $s_j \in \hat{s}$, such that the latency between submitting the job and job execution is minimized. A Demand Matrix $D_{F_iS_j}{}^T \; \forall i = 1 \ldots n, j = 1 \ldots m$, is created based on a set of jobs J within a time interval T. The replication involves creation of identical copies of data files and their distribution over the nodes in a Grid. The Data Replication (DR) problem states that: Let $D_{F_iS_j}{}^T$ be a demand matrix and $\hat{s}$ be a set of sites; the aim is to distribute a set of files to the sites, so that the latency is minimized based on the demand matrix and the volume constraint at each site is maintained. In this paper, an *Integrated Replication and Scheduling (IRS)* approach is proposed which combines the replication and scheduling schemes. Data Replication (DR) algorithm is a centralized algorithm running at certain interval of time. After the arrival of jobs, each External Schedulers take the help of replication information and schedule so that the job scheduled has the least latency in terms of execution.

2 Integrated Replication and Scheduling (IRS) Approach

We start by defining some operational terms.

Normalized Demand (η_{F_i}): Ratio of the demand for file F_i to the demand of all files.

$$\eta_{F_i} = \frac{\sum_{j=1}^{n} D_{F_iS_j}}{\sum_{j=1}^{n}\sum_{i=1}^{m} D_{F_iS_j}} \tag{1}$$

File Latency (Δ^k_{ij}): Latency for a file F_k to be moved from site S_i to S_j.

Computational Latency (ω_{ij}): Latency for a job i to be executed at site S_j

$$\omega_{ij} = \frac{\sum_{i=1}^{m} \tau_i}{P_j} \tag{2}$$

Queuing Latency (Q_{ij}): Latency for a job i due to the queuing at the site S_j (queue size (q_i). In case of assumption that all the jobs take the same time for execution, then

$$Q_{ij} = \frac{q_j \cdot \sum_{i=1}^{m} \tau_i}{P_j} \tag{3}$$

Slots Available (γ_j): Average number of files that can be stored in site S_j. Thus,

$$\gamma_j = \frac{V_j}{\bar{\tau}}, \; \bar{\tau} = \text{average file size} \tag{4}$$

2.1 Job Scheduling (JS) Algorithm

The JS algorithm has two parts – (a) Job Scheduling and (b) Matrix Updating.

Job Scheduling Strategies: Two different Job Scheduling Strategies have been proposed: Matching based Job Scheduling (MJS) and Cost Based Job Scheduling (CJS).
Matching based Job Scheduling (MJS): In MJS, the jobs are scheduled to those sites which have the highest match in terms of data (maximum number of files for the job available at the site). Any tie is broken by reducing the latency involved in moving the data which is not present in the scheduled site from the site(s) containing the data. It is possible that MJS may distribute the jobs to the same site resulting in the queue size increase in that site. To distribute the jobs to different sites the scheduling is done based on $v = m.\dfrac{\overline{q}}{q_i}$ factor, where m is the maximum match and $\overline{q}$ is the average queue size and q_i is the queue size at the site. MJS schedules based on the maximum v value. Figure 1 shows the topology of a Data Grid. S1, S2, S3 and S4 are the different sites in the Data Grid. The numbers and the arrows show the latency to move a file from one data site to the other. The elements in each site indicate the files that are present in each of those sites. Let a job come which requires files D1, D3 and D6. According to the MJS algorithm, both S2 and S4 are candidate sites where the job can be scheduled. If the job is scheduled in S2, then it takes 7 seconds to move the file D6 from S3 (File Originating Site) to S2. On the other hand, if the job is scheduled onto S4, then it takes 4 seconds. Therefore, the job is scheduled onto site S4.

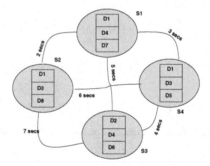

Fig. 1. Topology of an example Data Grid

Cost Based Job Scheduling (CJS): Another alternative to matching based job scheduling, a cost based job scheduling strategy is proposed. Cost (C_{ij}^{s}) of scheduling a job J_i onto a site S_j is defined as the combined cost of moving the data into the site S_j, latency to compute the job J_i in the site S_j and the wait time in the queue in the site S_j. The job is scheduled onto the site which has the minimum C_{ij}^{s}. Referring back to the example shown in Figure 1, we assume that in this case the computational time is 0 and queues at each site is also 0. Therefore C_{ij}^{s} is composed of only the data latency. The values of

C_{ij}^{s} for j=1,2,3,4 are: $C_{i1}^{s} = 7$ secs, $C_{i2}^{s} = 7$ secs, $C_{i3}^{s} = 8$ secs, $C_{i4}^{s} = 4$ secs. Therefore, the job is scheduled onto site S4, same as MJS. Though both the algorithms provide similar performance in this example, generally CJS will be better if instantaneous queue information is available. However, in case of partial information the comparison between these algorithms can be an interesting future study.

Updating the Demand Matrix: In this step, the Demand Matrix is updated as illustrated below. The example is based on the topology shown in Fige 1. Let the files required by job J_i be $\tilde{F}_i$. The data files required are: $\tilde{F}_1 = (D1, D3, D4)$, $\tilde{F}_2 = (D1, D2, D5)$, $\tilde{F}_3 = (D2, D3, D8)$, $\tilde{F}_4 = (D1, D3, D7)$, $\tilde{F}_5 = (D4, D5, D8)$, $\tilde{F}_6 = (D1, D5, D7)$, $\tilde{F}_7 = (D3, D4, D5)$, $\tilde{F}_8 = (D1, D7, D8)$. Based on the job requests, the given topology and the MJS; the following Demand Matrix can be constructed:

Table 1. Demand Matrix for the topology in Figure 1 based on a job pattern

Files\Sites	S1	S2	S3	S4	Total
D1	0	0	0	1	1
D2	0	0	4	0	4
D3	0	2	1	1	4
D4	0	1	1	0	2
D5	0	0	0	1	1
D6	0	1	3	2	6
D7	0	0	1	1	2
D8	0	2	2	0	4
Total	0	6	12	6	24

Scheduling of Jobs with Data Ordering: Till now we have assumed that each job requires data all at the same time i.e., at the time of starting the job. However, the cost of scheduling can be modified in case of order of data files. By order of data files we mean that say the job requires files (f_1, f_2, f_3) at the start of the job and requires (f_4, f_5) later. Then files f_4, f_5 can be obtained later than files f_1, f_2, f_3. Therefore,

$$Latency = \max(\sum_{j=1}^{3}{}^{s}\Delta_{ij} + \sum_{i=1}^{3}\omega_{ij}, \sum_{j=4}^{5}\Delta_{ij}^{s}) + \sum_{i=4}^{5}\omega_{ij}, \tag{5}$$

Let $f_{11}, f_{12} \cdots f_{1k_1}$ be the set of files required initially (Step 1), $f_{21}, f_{22} \cdots f_{2k_2}$ be the number of files required in Step 2, $f_{j1}, f_{j2} \cdots f_{jk_j}$ be the files required in Step j, and $f_{p1}, f_{p2} \cdots f_{pk_p}$ be the files required in Step p (last step), then

$$L(i) = \max(L(i-1), \sum_{j=1}^{K_1}\Delta_{ij}) + \sum_{j=1}^{K_1}\omega_{ij}$$

$$L(1) = \sum_{j=1}^{K_1}\Delta_{ij}^{s} + \sum_{j=1}^{K_1}\omega_{ij} \tag{6}$$

Where L(i) is the latency at the ith step, K files are required at step i and L(p) is the total latency.

2.2 Data Replication Strategy

Data Replication Strategy has two steps: (i) Allocation of Replication Limits to each file and (ii) Replication.

Allocation of Replication Limits: We define Replication Limit (χ_i) of file F_i as the number of sites where the file F_i should be replicated. It is to be noted that each file should be replicated at least once, therefore the χ_i is defined as:

$$\chi_i = \min(\#sites, 1 + (ceiling(\eta_{F_i}.(\sum_{j=1}^{n}\gamma_j - \Phi)))) \tag{7}$$

In the Equation 7, the minimum of the ceiling value and the #of sites is taken as a file could not be replicated more than the number of available sites. During the allocation of χ_i, priority is given to the files having the highest Normalized Demand (η_{F_i}). Based on the Demand Matrix shown in table 1, the following η_{F_i} values are calculated for each of the different files: $\eta_{D1} = 1/24, \eta_{D2} = 1/6, \eta_{D3} = 1/6, \eta_{D4} = 1/12, \eta_{D5} = 1/24, \eta_{D6} = 1/4, \eta_{D7} = 1/12, \eta_{D8} = 1/6.$ Based on the η_{F_i} values calculated above, the Replication Limit allocations are: $\chi_{D1} = 1, \chi_{D2} = 2, \chi_{D3} = 2, \chi_{D4} = 1, \chi_{D5} = 1, \chi_{D6} = 2, \chi_{D7} = 1, \chi_{D8} = 2.$

Data Replication: Data Replication Strategy is based on the principle of choosing sites based on the expected latency that the site is going to provide. We assume that the data is placed at various storage elements as a result of replication strategy and no further caching takes place. Let the probability of a job scheduled in site S_k requiring the file F_i be p_{ik}. Then the expected file latency (δ_{ij}) of the file F_i scheduled in site S_j is defined as:

$$\delta_{ij} = \sum_{k=1}^{n} p_{ik} l_{kj} \tag{8}$$

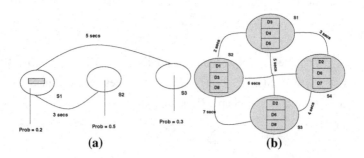

(a)   (b)

Fig. 2. Illustration of the (a) Expected Latency, (b) Final Config

Figure 2(a) illustrates the concept of expected latency. Let a file be replicated in Site 1. Probabilities of using the file by a job scheduled in site S1, S2 and S3 are 0.2, 0.5 and 0.3 respectively. Therefore the expected latency of the file in site S1 becomes (0*0.2 + 3*0.5 + 5*0.3) = 3 seconds. It is to be noted that the latency of the job requiring the file in site S1 is 0, as the site contains the file. Similarly, Expected Computation (θ_{ij}) and Queuing Latencies (α_{ij}) are given by:

$$\theta_{ij} = \sum_{k=1}^{n} p_{ij}.(\tau_j / P_j), \quad \alpha_{ij} = q_j.\theta_{ij} \tag{9}$$

Cost function for data replication strategy is $C^R{}_{ij} = \delta_{ij} + \theta_{ij} + \alpha_{ij}$, the algorithm aims at minimizing $C^R{}_{ij}$. It is to be noted that this is an NP-Complete problem as it can be reduced to a K-Median problem [9]. The Data Replication Strategy is based on a simple greedy approach. Based on the Demand Matrix, a Normalized Demand is calculated (η_{F_i}) (See Equation 1). Replication algorithm starts with the node having the maximum η_{F_i} value. Then Replication Limit (χ_i) is calculated, where χ_i indicates how many times the file F_i is replicated in the Data Grid. More the value of η_{F_i}, higher is χ_i. The best χ_i sites are selected among n sites which has the lowest $C^R{}_{ij}$. Then replication is carried out by the file having the second highest η_{F_i} value. This process is repeated until all the files are exhausted.

An Example: The data replication strategy is shown in Table1 and Figure 1. The cost of the strategy is provided in Table 2.

Table 2. Cost table based on the data replication strategy

Files\ Sites	S1	S2	S3	S4
D6	3.83	5.50	**2.50**	3.00
D8	3.50	**3.50**	3.50	5.00
D3	**3.00**	3.25	4.50	4.00
D2	5.00	7.00	**0.00**	4.00
D4	**3.50**	3.50	3.50	5.00
D7	4.00	6.50	2.00	**2.00**
D5	**3.00**	6.00	4.00	0.00
D1	3.00	**6.00**	4.00	0.00

The final replication is shown in Figure 2(b). Based on the replication the latency is improved from 5.38 seconds to 2.25 seconds, an improvement of 58%.

3 Performance Studies

To evaluate the performance of the proposed strategies, the OptorSim [10] simulator was used, which is a commonly used simulator for the simulation of Data Grid
environment. The simulation runs are taken with jobs arriving average exponential
inter-arrival time of 0.25 seconds, the processing speed at the nodes are considered
constant at 10 second/Gb of data. Number of jobs requesting a particular file
is distributed exponentially. This gives an elliptical file distribution per job
with an average of 7 and total files in the system (ϕ) as 20. The initial file

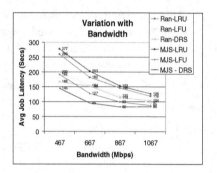
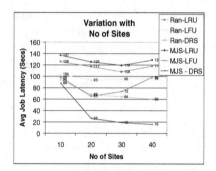

Fig. 3. Performance of schemes with (a) Bandwidth variation, (b) No. of Sites

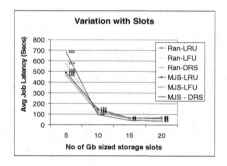
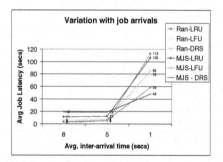

Fig. 4. Performance of schemes with (a) no. of slots, (b) Job Arrival

distribution in the Grid is random. The data replication was carried out at
pre-determined intervals during the runs. The parameters that were varied in simulation runs are network bandwidth (467 to 1067 Mbps); number of computing and
storage nodes (10 to 40); and storage capacity V_j (5 to 20 Gb). A separate set of
runs were taken by varying the average job inter-arrival times (1 seconds to 8 seconds). We have compared performance of our data replication strategy against no
replication, and commonly used Least Recently Used (LRU), Least Frequently
Used (LFU) replication strategies. For job scheduling, the MJS version of our

scheduling algorithm is used (with α value in the average queue size factor taken as 1). Essentially, during evaluation, the following combinations are considered : Ran-NoRep – random scheduling with no replication, Ran-LRU – random scheduling with LRU replication strategy, RAN- LFU – random with LFU, Ran-DRS – random with our replication strategy, MJS-NoRep – our scheduling with no replication, MJS-LRU,MJS-LFU and MJS-DRS respectively.

3.1 Discussions

Figures 3 and 4 highlight the performance of the schemes in terms of variations in average job latency with respect to the variations in bandwidth, number of sites, SE size, and job arrival distributions. It is clear that the MJS scheduling scheme with DRS performs best in most of the cases. The cases with no data replication strategies perform far worse than other cases, and therefore, the results are omitted. For example, average job latency with Ran-NoRep strategy with 10 sites is 1512 seconds – 10 to 12 times the average job latencies observed in other schemes.

The idle times for the processors were evenly distributed at high loads following MJS-DRS scheme with a standard deviation of 3. The average job queue sizes at high loads were also evenly distributed. The results show that with variation of bandwidth (figure 3(a)), the Ran-DRS scheme achieves a performance improvement of up-to 19% over Ran-LFU and up-to 54% over Ran-LRU schemes. The MJS-DRS scheme results in a job latency improvement of an average of 56% over Random scheduling schemes and of 23% over other data replication schemes with our scheduling. With increasing numbers of computing elements n, the MJS-DRS scheme performs significantly better than other schemes (figure 3(b)). At 40 nodes the MJS-DRS scheme is almost two times better than Ran-DRS scheme and even better than other replication schemes. The results of the job latency with variations of storage capacity, and hence, the number of gigabyte slots are interesting (figure 4(a)). While at lower storage capacity (5 GB), other schemes perform marginally better than MJS-DRS; with increase in capacity and γ_j , MJS-DRS fares better than other schemes by an average of 17%. As evident from figure 4(b), the MJS-DRS scheme scales up better than other schemes with increase in load. With a decrease of average job inter-arrival time from 5 seconds to 1 seconds, the average job latency increases by 2.3 times while in case of the other schemes the performance deteriorates by 10-12 times.

We have also compared our scheduling – replication scheme against economy based replication strategy proposed in [11]. The preliminary results, as given in Table 3, suggest that the MJS-DRS scheme improves job latency over EcoModel optimizer (EO) with ZipF file distribution.

Table 3. Performance comparison of MJS-DRS with EcoModel optimizer

Scheduler + Replication	RS+EO	RS+DRS	MJS+EO	MJS +DRS
Job Latency (secs)	106	104	69	48

4 Conclusions

In this paper an interaction between replication and scheduling strategy called the Integrated Replication and Scheduling (IRS) strategy, has been proposed. The data replication is carried out in an asynchronous timer-controlled process that takes into account history of jobs and data access patterns and is primarily based on the notion of expected data file latency and a greedy optimization approach. The scheduling is carried out in a matching-based or a cost-based manner with view of transient system-state data like queue length. The approach MJS-DRS has shown promising results with respect to the popular and commonly used data replication strategy, while the cost-based scheduling approach is yet to be tested. Contrary to [8], our experience shows that it is better to consider the interactions of replication and scheduling while scheduling in a Data Grid and replication strategy works well even if the jobs are not scheduled locally. In this paper, we have considered a centralized external scheduler which may prove costly with increase in Data Grid size. In subsequent works, we propose to extend this scheduling and replication scheme to a decentralized and hierarchical environment. Further, we propose to analyze the sensitivity of the schemes with respect to variations *viz* in file sizes, processor speeds, effect of data arrival pattern in job execution etc.

References

1. Chervenak, I. Foster, C. Kesselman, C. Salisbury, S. Tuecke, "The Data Grid: Towards an Architecture for the Distributed Management and Analysis of Large Scientific Datasets," *Journal of Network and Computer Applications*, vol. 23, pp. 187-200, 2001.
2. M. Beck and T. Moore, "The Internet2 distributed storage infrastructure project: An architecture for internet content channels," *Computer Networking and ISDN Systems*,1998.
3. Foster and C. Kasselman, "The Grid 2: Blueprint for a new Computing Infrastructure," *Morgan Kaufman*, 2004.
4. H. Casanova, G. Obertelli, F. Berman and R. Wolski, "The AppLeS Parameter Sweep Template: User-Level Middleware for the Grid," in *Proc. SuperComputing'00*, 2000.
5. C. Banino, O. Beaumont, L. Carter, J. Ferrante, A. Legrand, and Y. Robert, "Scheduling Strategies for Master-Slave tasking for Heterogeneous Processor Platforms," in *IEEE Trans. On Parallel and Distributed Systems*, vol. 15, no. 4, Apr. 2004.
6. K. Ranganathan and I. Foster, "Identifying Dynamic Replication Strategies for a High Performance Data Grid," in *Proc. Second IWGC*, 2001.
7. D. Thain, J. Bent, A. Arpaci-Dusseau, R. Arpaci-Dusseau and M. Livny, "Gathering at the Well: Creating Communities for Grid I/O," in *Proc. SuperComputing 2001*, 2001.
8. K. Ranganathan and I. Foster, "Simulation Studies of Computation and Data Scheduling Algorithms for Data Grids," in *Journal of Grid Computing*, vol. 1, no. 2, Apr. 2003.
9. R.R. Mettu and K.G. Plaxton, "The Online Median Problem", in *SIAM Journal on Computing, Vol. 32, No. 3*, pp 816- 832, 2003

10. W.H. Bell, D.G. Cameron et al., "Simulation of Dynamic Grid Replication Strategies in OptorSim," in *Proc. Third Int'l Workshop on Grid Computing*, 2002.
11. W.H. Bell, D.G. Cameron, R. Carvajal-Schiaffino, A. P. Millar, K. Stockinger and F. Zini, "Evaluation of an Economy-Based File Replication Strategy for a Data Grid", in *Proc. CCGrid*, May 2003.

Efficient Layout Transformation for Disk-Based Multidimensional Arrays

Sriram Krishnamoorthy[1], Gerald Baumgartner[2], Chi-Chung Lam[1],
Jarek Nieplocha[3], and P. Sadayappan[1]

[1] Department of Computer Science and Engineering,
The Ohio State University, Columbus, OH 43210, USA
{krishnsr, clam, saday}@cse.ohio-state.edu
[2] Department of Computer Science,
Louisiana State University, Baton Rouge, LA 70803, USA
gb@csc.lsu.edu
[3] Computational Sciences and Mathematics,
Pacific Northwest National Laboratory, Richland, WA 99352, USA
jarek.nieplocha@pnl.gov

Abstract. I/O libraries such as PANDA and DRA use blocked layouts for efficient access to disk-resident multi-dimensional arrays, with the shape of the blocks being chosen to match the expected access pattern of the array. Sometimes, different applications, or different phases of the same application, have very different access patterns for an array. In such situations, an array's blocked layout representation must be transformed for efficient access. In this paper, we describe a new approach to solve the layout transformation problem and demonstrate its effectiveness in the context of the Disk Resident Arrays (DRA) library. The approach handles re-blocking and permutation of dimensions. Results are provided that demonstrate the performance benefit as compared to currently available mechanisms.

1 Introduction

Many scientific and engineering applications need to operate on data sets that are too large to fit in the physical memory of the machine. Due to the extremely large seek time relative to the per-word transfer time for disk access, it is imperative that I/O be done using contiguous blocks of disk resident data. To optimize performance in collective I/O operations between arrays located on disk and in distributed main memory of parallel computers [1], I/O libraries like PANDA [2, 3] and DRA [4] use a blocked layout representation for the disk-based multidimensional arrays instead of the dimension-ordered representation used typically for the representation of multidimensional arrays in main memory. Thus, the disk-based multidimensional array is partitioned into a number of multidimensional blocks or "bricks", and the elements within a brick are linearized using some dimension order. Such a bricked representation of disk-based multidimensional arrays permits efficient access as long as the accessed regions mostly contain full bricks.

However, the access patterns to some disk-based multidimensional arrays in two successive phases (or the access pattern of the producer and the consumer) are so different that no choice of brick shape will allow for efficient access. An example is the out-of-core

L. Bougé and V.K. Prasanna (Eds.): HiPC 2004, LNCS 3296, pp. 386–398, 2004.
© Springer-Verlag Berlin Heidelberg 2004

2D Fast Fourier Transform (FFT), where the array is accessed by columns in one phase and by rows in the other. The multi-dimensional FFT [5, 6] can be implemented as a series of one-dimensional FFTs, one along each dimension. Another example illustrating very different access patterns is with image data in three and four (including time) dimensions. The production of data from scanning occurs plane by plane. However, examination of the time evolution of a 3D block of data requires a very different access pattern than that by which the data was generated. In isosurface construction in three and four dimensions, the data is typically produced in a row-major format by scanning or simulation. The amount of memory available determines the amount of data generated between writes to disk, and hence limits the blocking possible. To efficiently perform computations on the stored data in a parallel system, the data might have to transformed into a different blocked form [7]. Thus there are situations where performance can be greatly improved by transforming the layout of a multidimensional array on disk to match the application's access pattern.

Our primary motivation for addressing the layout transformation problem arises from the domain of electronic structure calculations using ab initio quantum chemistry models such as Coupled Cluster models. We are developing an automatic synthesis system called the Tensor Contraction Engine (TCE)[8], to generate efficient parallel programs from high level expressions, for a class of computations expressible as tensor contractions [9, 10, 11, 12, 13, 14]. Often the tensors (essentially multi-dimensional arrays) are too large to fit in memory and must be disk-based. The input tensors are often generated by other quantum chemistry packages such as NWChem [15], with a layout quite different from that needed for efficient processing by the TCE-generated code.

This paper describes an approach to efficient transformation of data between disk-based multidimensional arrays. Experimental results indicate that this approach delivers comparable or better performance than other techniques currently used in practice, that are based on reading data from one disk-based array to distributed main memory, in-memory data transformation, and then writing data to the destination disk array. For example, improvements exceeding 80 percent were observed on a Linux cluster.

The paper is organized as follows. Section 2 describes the DRA framework, within which we implement our solution to the layout transformation problem. The array re-blocking problem is explained is detail in Section 3. Section 4 presents the proposed approach for efficient layout transformation. In Section 5, experimental results are presented. Section 6 concludes the paper.

2   Disk Resident Arrays

The Global Arrays (GA) library [16] [17] provides a shared-memory programming model in which data locality is explicitly managed by the programmer. Explicit function calls are used to transfer data between global address space and local storage. It is similar to distributed shared-memory models in providing an explicit acquire-release protocol, but differs with respect to the level of explicit control in moving blocks of data in multidimensional arrays between remote global storage and local storage. The functionality provided by GA has proved useful in the development of large scale parallel quantum chemistry suites such as NWChem [15] (which contains over a million lines of code).

The Disk Resident Arrays (DRA) model [18] extends the GA NUMA programming model to secondary storage. It provides a disk-based representation for multidimensional arrays and functions to transfer blocks of data between global arrays and disk resident arrays. DRA, along with GA, provides a unified programming model for handling different levels of the memory hierarchy in which the user controls the location of data in the memory hierarchy. This has been shown to provide high performance while providing a programming model that is simpler than message passing.

Henceforth, we shall use GA and DRA to refer both to the library and the arrays handled by them. The reference will be clear from the context.

3   The Layout Transformation Problem

Internally, the data in a DRA is stored in a blocked fashion. When a DRA is created, a typical request shape/size can be specified. This is used to determine the shape of the basic layout block or "brick". The shape of the brick is chosen to match the specified access shape. The size of the brick is chosen as a compromise between two competing objectives: 1) optimize disk I/O bandwidth - this requires that the brick size be large enough to amortize the disk seek time and 2) minimize wastage of disk I/O - since I/O is done in units of the basic block (brick), small bricks imply less wastage at the boundaries of the DRA regions being read/written.

An application might have an access pattern that is very different from the organization of the DRA on disk. This can happen when an application uses the output of another program, or because different phases of the same program use different access patterns. This can be handled by creating another copy of the disk resident array to match the new request size and transformed dimensions.

We have implemented the copy routine, referred to as $NDRA\_Copy$, together with dimension permutation. The routine takes as input the source and target DRA handles and the dimension permutation to be performed. Henceforth, the data in the DRA corresponding to the dimensions of blocking in the source and target arrays are referred to as the source and target blocks respectively.

The disk array layout transformation problem we consider here is a generalization of the out-of-core matrix transposition problem. Out-of-core matrix transposition has been widely studied in the literature. The algorithms perform out-of-core transposition by making passes through the entire array a number of times. During each pass through the array, each element of the source array is read once and each element of the target array is written once. Each pass consists of a series of steps in which a portion of data from the source array is brought into memory, permuted and written to the target out-of-core array. Different steps in a pass operate on disjoint sets of data. The block transposition algorithm is a single-pass algorithm in which a 2-D tile of data is brought into memory, transposed and written to disk. Since the different row segments of a 2D tile are not contiguous on disk, this could be extremely inefficient unless the tile size is very large. Eklundh [19] proposed a multi-pass algorithm, in which the minimum unit of I/O is a row. The number of passes in the algorithm is proportional to the array dimensions. Kaushik et al. [20] reduced the number of read operations and increased the read block size compared to Eklundh's algorithm. Sun and Prasanna [21] proposed an algorithm that minimized the total number of I/O operations, while potentially increasing the total volume of I/O. Krishnamoorthy et al. [22] formulated these algorithms in a tensor product notation and derived a generic algorithm that attempts to minimizes the total execution time by taking into consideration the I/O characteristics of the system, and subsequently extended it to a multi-processor system, in which each processor has a local disk [23] .

Most of the above approaches assume the array dimensions and the memory size to be powers-of-2. This assumption, coupled with the fact that the required transformation is a transposition, allows different steps in the re-blocking process to operate on disjoint sets of data. In each step, the set of data read into memory form an integral number of write blocks, which are written out. So no data is retained across steps during the transposition. When arbitrary blocking, array dimensions and memory sizes are to be handled, it may not be possible to process and write out all the data read into memory in a given step. Some data either needs to be discarded and re-read, increasing the I/O cost, or needs to be retained, increasing the memory requirement. The memory cost for retaining the data unused from a step depends on the order of traversal of dimensions, and hence is not straight forward. The out-of-core transposition algorithms involve I/O of blocks of data at specific strides, which is fixed for a pass. This regularity allows accurate prediction of the I/O cost. The in-memory permutation of data can be modeled as a bit-permutation on the linear address space of the data stored in disk. This provides a regular structure to the in-memory computation. In the general case, in-memory permutation corresponds to a series of collect operations for combining portions of different read blocks to create a write block. The simplicity in the cost models for the power-of-2 transposition problem makes it amenable to mathematical treatment as done in [22].

In the next section, we detail our approach to solving the generalized re-blocking problem.

4   Algorithm Design

The disk array layout transformation problem is modeled as an I/O optimization problem. The total I/O cost is to be minimized, subject to the amount of physical memory available. The cost model and the algorithm to obtain the multi-pass solution are explained in this section. In the ensuing discussion, we shall consider an n-dimensional matrix of dimensions $< d_1, \ldots, d_n >$. The matrix is blocked in brick shape $< s_1, \ldots, s_n >$. The target matrix has the same ordering of dimensions as the source but is blocked using bricks of shape $< t_1, \ldots, t_n >$. The source and target bricks are assumed to be of size that is large enough for efficient access from/to disk. DRA typically uses a brick size of around 1 Mbyte. Reads from the source disk array are assumed to be in units of the source brick, and writes to the target disk array are done in units of the target brick.

4.1   Solution Approach

If feasible, a single-pass solution (in which each element is read and written exactly once) would provide the minimum I/O cost. But the memory requirement for a single-pass solution might exceed the physical memory available. In this case, we either need to choose a multi-pass solution or perform redundant I/O in one pass. In this sub-section, we present the intuition behind the design of our algorithm. We begin with a basic single-pass algorithm and determine its I/O and memory cost. We then incrementally improve

```
Input:   (1) Source and target block sizes [s] and [t],
         (2) Template size [templ]
Output:  (1) Total memory cost (2) Dimension traversal order [T]
1)   foreach dimension i
2)      L[i] = LCM(s[i], t[i])
3)      U[i] = min(s[i], t[i]) - gcd(s[i], t[i])
4)      M[i] = ceil(max(s[i], t[i])/s[i])*s[i]
5)   Sort dimensions into array T such that
        forall i<j => U[T[i]]*M[T[j]] + L[T[i]]*U[T[j]] <
        U[T[j]]*M[T[i]] + L[T[j]]*U[T[i]]
6)   memCost=0
7)   foreach dimension i
8)      pdt=U[T[i]]
9)      foreach j<i
10)        pdt *= L[T[j]]
11)     foreach j>i
12)        pdt *= M[T[j]]
13)     memCost += pdt
```

Fig. 1. Pseudo-code to determine the memory cost for a given template size

the single-pass algorithm to lower the memory requirement and/or the I/O cost. The multi-pass solution is discussed in a subsequent sub-section.

Consider the region $< 0 - LCM(s_1, t_1), \ldots, 0 - LCM(s_n, t_n) >$. This region contains an integral number of source and target blocks along all the dimensions. Thus the data in the source matrix from this region maps onto complete blocks in the target matrix. This region can be processed independent of other such blocks, without any redundant I/O. We shall refer to such regions as *LCM blocks*. If the amount of physical memory were large enough to hold an LCM block, then a single-pass solution is clearly possible - read in source blocks contained in an LCM block into memory, construct the target blocks corresponding to the data in memory, and write them into the target array. The I/O cost is defined as the I/O required per element of the source array. This algorithm has the minimum I/O cost of one read and one write per element of the source array. Assuming the read and write operations are equivalent the I/O cost is two units per element.

The memory cost is the size of the LCM block. Since arbitrary re-blocking needs to be supported, the source and target block sizes could have arbitrary dimensions (provided their total size corresponds to a reasonable block size for I/O on the target file system). Hence the LCM block can be arbitrarily large and might not fit in physical memory. We can improve the single-pass algorithm to handle this scenario without increasing the I/O cost. Instead of reading entire LCM blocks into memory, the algorithm reads in a set of blocks of data from the source matrix and writes out those target blocks that can be completely constructed from the data available in memory. Any data in memory that cannot be used to construct a complete target block is retained in memory. Any source block in an LCM block contributes to target blocks within the same LCM block. Hence no data needs to be retained across LCM blocks. The algorithm processes all the data in one LCM block before processing any other LCM block. The algorithm requires enough memory to retain unused data and read in additional data for processing. The additional data read into memory for processing must be enough to write at least one target block to disk. This is referred to as the Max block and corresponds to $< M_1, \ldots, M_n >$ where $Max_i = \lceil (\max(s_i, t_i)/s_i) \rceil * s_i$. The algorithm traverses each LCM block along each of the dimensions and processes data in units of the Max block. The buffer to store the unused data is partitioned into one buffer per dimension. Unused data from a Max block along a dimension needs to be retained until the adjacent Max block along that dimension is processed. Thus the amount of unused data to be retained depends on the order of traversal of dimensions. Along the dimension traversed first, only data unused from the last processed Max block needs to be stored. Other dimensions require more data to be retained. A static memory cost model is used, in which the sizes of buffers used to store data is determined before the transformation begins. The maximum memory required to perform the transformation is the sum of the size of the Max block and the sizes of the buffers.

$$\text{MemCost} = \sum_{i=1}^{n} \text{bsize}_i + \prod_{i=1}^{n} Max_i$$

where bsize_i represents the size of buffer to store unused data along the i-th dimension.

Let $< T_1, \ldots, T_n >$ be the order of traversal of dimensions. The unused data along a dimension (say T_i) is an n-dimensional region. For a given dimension i, the size of this region along dimension j can be as much as $\text{LCM}(s_{T_j}, t_{T_j})$ for $j < i$, but is bounded above by Max_{T_j} for $j > i$. Hence, the size of the buffer to store the unused data along a dimension T_i is bounded by

$$\text{bsize}_{T_i} = \prod_{j=1}^{n} S_j$$
$$S_j = \begin{cases} \text{LCM}(s_{T_j}, t_{T_j}) & \text{if } j < i \\ U_{T_j} & \text{if } j = i \\ Max_{T_j} & \text{if } j > i \end{cases}$$

where U_i be the maximum unused data that needs to be stored along dimension i. Since U_i must be smaller than both s_i and t_i, and for every s_i elements along dimension i brought into memory, at least $\gcd(s_i, t_i)$ elements must be written out, we have

$$U_i = \min(s_i, t_i) - \gcd(s_i, t_i)$$

As can be seen from the above formulae, the sizes of the unused buffers is proportional to the LCM block dimensions. This could lead to situations in which the memory requirement still exceeds the available memory. In this case, there are two options to be considered. A multi-pass solution could be determined, which is discussed later, or a single-pass solution that performs redundant read of data can be designed.

We propose a single-pass algorithm that differs from the discussion above in one respect. Instead of traversing an entire LCM block, a smaller template is chosen. No unused data is stored across templates. A template is an integral number of write blocks along all dimensions. There is no redundant read within a template. But unlike LCM blocks, templates might have source blocks on their boundaries that straddle across two templates. This results in redundant reads across templates, increasing the I/O cost. The memory cost is reduced and is given by:

$$\text{MemCost} = \sum_{i=1}^{n} \text{bsize}_i + \prod_{i=1}^{n} Max_i$$
$$\text{bsize}_{T_i} = \prod_{j=1}^{n} S_j$$
$$S_j = \begin{cases} \text{templ}_{T_j} & \text{if } j < i \\ U_{T_j} & \text{if } j = i \\ Max_{T_j} & \text{if } j > i \end{cases}$$

where templ_i represents the size of the template along the i-th dimension.

For a two-dimensional array, the memory cost due to the unused buffers is $U_1 * Max_2 + \text{LCM}(s_1, t_1) * U_2$ if dimension 1 is traversed first; otherwise, it is $U_2 * Max_1 + \text{LCM}(s_2, t_2) * U_1$. In an n-dimensional array, the traversal order is determined by sorting the dimensions by comparing these expressions.

The minimum template size corresponds to a target block. In this case, the memory requirement is reduced to a Max block. Thus the necessary condition for the existence of a single-pass solution is that the Max block fit in memory.

The I/O cost is multiplicative along the dimensions. Within an LCM block, the number of source blocks that need to be reread is the number of templates minus one, which is $(\text{LCM}(s_i, t_i) - \text{templ}_i) - 1$. Therefore, the I/O cost of re-blocking is given by templ_i is

$$\text{IOCost} = \prod_{i=1}^{n} \text{IOCost}_i$$
$$\text{IOCost}_i = \frac{(s_i * (\frac{\text{LCM}(s_i, t_i) - \text{templ}_i}{\text{templ}_i}) + \text{LCM}(s_i, t_i))}{\text{LCM}(s_i, t_i)}$$

In reality, the LCM along a dimension might be larger than the length of the array along the dimension, in which case we replace the LCM by the array dimension. Note that the array dimensions are not considered while determining U_i. Hence, U_i does not provide an exact estimate, but only an upper bound on the memory requirement. Note that though the I/O cost for the single-pass solution is increased, the total I/O cost could be reduced due to a decrease in the number of passes.

4.2   Template Determination for Single-Pass Solution

Both the I/O cost and the memory cost are affected by the choice of the template. In this section, we discuss the algorithm used to determine the template sizes. The template is a set of write blocks along all the dimensions. It can range in size from one write block, to an LCM block. For re-blocking an n-dimensional array, the template needs to be determined from an n-dimensional solution space. A template is a feasible solution if its processing does not require more memory than available. The algorithm exploits the characteristics of the solution space and the optimization function.

Consider a template A. An enclosing template is defined as a template that is at least as large as the given template in all the dimensions. Let B be an enclosing template of A. From the memory cost equations, it can be seen that the memory required to process A cannot exceed that required to process B. Conversely, processing B requires at least as much memory as processing A. This implies that once a template has been determined to require more memory than available (an infeasible solution), no enclosing templates needs to be considered. This relation separates the solution space into a feasible and an infeasible solution space (where the surface of separation approximates to a hyperbola when $n = 2$).

```
Input:   Source and target block sizes [s] and [t],
Output:  Template size for single-pass solution, if it exists.
Support Routines: MemCost(templ) - Memory cost for processing
                                   the given template
                  DiskCost(templ) - I/O cost for processing
                                    the given template
                  MemoryExceeded(templ) - returns true if the
                                          template is infeasible
1) Initialize template to LCM block
2) Reduce template size along along all dimensions equally
      (in units of write block size) until the template is a
      feasible solution.
3) If no feasible solution is found return "No solution exists"
4) Adjust the template size so that increasing the template size
      along any dimension makes it infeasible.
5) Repeat the following steps
6)    Among adjacent template sizes choose the one that has the
         maximum rate of decrease in I/O cost to increase
         in memory cost.
7)    Determine a feasible solution that leads to the least
         increase in disk I/O cost from the chosen template.
8)    If the feasible solution found has lesser I/O cost than the
         current template, choose that as the current template.
      Otherwise return the current template as the solution.
```

Fig. 2. Algorithm to determine template size for a single-pass solution

The I/O cost has a similar characterization. The I/O cost equation shows that decreasing the template size along any dimension increases the I/O cost. Thus the I/O cost of template A is at least as much as that of template B. This implies that when searching through the solution space, no template that is enclosed by a feasible template needs to be considered. Thus the optimal solution resides on the surface separating the feasible and infeasible solution spaces.

Our algorithm to determine the template for a single-pass solution involves three phases. The algorithm begins with the LCM block as the template and tests for feasibility. If an LCM block is the feasible solution, it is chosen as the template. Otherwise, a solution is chosen that is just feasible, i.e. , increasing the template size along any dimension violates the memory constraint. This is a solution on the boundary between the feasible and infeasible solution spaces and hence is a candidate solution. From this solution, we perform a steepest descent to arrive at a local minimum in the search space. Note that other optimization algorithms that can optimize on a surface can be used. The algorithm used is shown in Fig.2.

4.3   Multi-pass Solution Determination

When a single-pass solution does not exist or is too expensive, a multi-pass solution is chosen by determining intermediate block sizes. An intermediate disk-based array is used to store the intermediate results. Hence, additional disk space equal to the size of the arrays is required. The multi-pass solution proceeds as repeated execution of the single-pass algorithm, for the source and target block sizes determined for that pass. The source block size of the first pass is the block size of the source array. The target block size of the last pass if the block size of the target array. The skew between the source and target block sizes decreases as the multi-pass solution proceeds from one pass to the next. The intermediate block size are chosen to effect the maximum re-blocking possible with the available memory.

A simple heuristic is used to determine the intermediate tile sizes for the multi-pass solution. Two candidate intermediate block sizes are considered. The first candidate intermediate block size is the geometric mean of the source and target block sizes. This block size is "equidistant" from the source and target block sizes. This can be an effective intermediate block size of for solutions with an even number of passes. The second intermediate block size is, in fact, a pair of block sizes. Let s_i and t_i be the source and target block sizes along dimension i. The intermediate block sizes chosen are $s_i^{2/3} * t_i^{1/3}$ and $s_i^{1/3} * t_i^{2/3}$. This pair of intermediate block sizes can be effective for solutions with an odd number of passes. These two options allow a more refined search for intermediate block sizes. Without the second choice, any solution that requires an odd number of passes, each transforming to an intermediate block "equidistant" from the previous one, might be harder to achieve. Higher order intermediates were not considered as solutions with a larger number of passes seldom occur in practice and can be handled by a combination of these choices.

```
Input:  Source and target block sizes [s] and [t],
Output: Sequence of intermediate block sizes [seqB], order of
            traversal of dimensions for each pass I/O cost
1)  Determine the cost (a) of single-pass solution
2)  foreach dimension i
3)      B1[i] = floor(sqrt(s[i]*t[i]))
4)      B2[i] = (s[i]^(2/3)*t[i]^(1/3))
5)      B3[i] = (s[i]^(1/3)*t[i]^(2/3))
6)  Determine the cumulative cost (b) of multi-pass solutions
        for re-blocking from s to B1 and B1 to t by recursively
        calling this routine.
7)  Determine the cumulative cost (c) multi-pass solutions
        for re-blocking from s to B1, B1 to B2 and B2 to t
        by recursively calling this routine.
8)  If no multi-pass solution exists
        return "no solution exists"
9)  Choose solution with least I/O cost from (a), (b) and (c)
10) If the single-pass solution has the minimum cost
11)     Return the solution with the order of traversal
            determined by invoking memCost
12) else
13)     Concatenate the sequence of solutions returned by the
            two parts of the solution with the minimum I/O cost.
```

Fig. 3. Pseudo-code to determine a multi-pass solution

Once the intermediate block(s) are determined, the multi-pass solution is determined recursively for transforming from source to intermediate, and intermediate to target block sizes. In the case of two intermediate blocks, the transformation between the intermediate blocks is determined as well. The algorithm for determining the multi-pass solution is shown in Fig. 3.

Consider an instance of the matrix re-blocking problem in which the source and target arrays are blocked as $< 32, 9 >$ and $< 5, 16 >$, respectively. The array dimensions are much larger than the blocking and hence are not considered. The Max block is $< 32, 16 >$ and the unused data along each dimension is bounded by $< 4, 8 >$. The solution to the re-blocking problem depends on the memory available. An LCM block contains $LCM(s_1, t_1) * LCM(s_2, t_2)$=23040 elements. When enough memory is available to hold an LCM block, the re-blocking can be performed by reading in an entire LCM block and writing out the target blocks. But if the memory can hold $U_2 * Max_1 +$ LCM$(s_2, t_2) * U_1 + Max_1 * Max_2$=1344 elements, it is sufficient to hold all unused data when an LCM block is processed. The second dimension is traversed first in the re-blocking procedure. If the memory available is lesser, say enough to hold just 900 elements, a single-pass solution with a template size of $< 120, 6 >$ elements is used for the re-blocking. When the memory size is 800, a two-pass solution with an intermediate tile size of $< 12, 12 >$ is determined. The template for the first pass is $< 96, 12 >$, and that for the second pass is $< 60, 48 >$.

5   Experimental Results

In order to evaluate the effectiveness of the proposed approach, we compared the time for layout transformation using our implementation with the time for transformation using currently available mechanisms. The present interface to DRA is through Global Arrays. When a DRA is to be copied to another DRA with different blocking, the source array is read into a GA one section at a time, and written into the same section of the target array. This is a single-pass solution. The basic unit of access, i.e. the shape and size of the

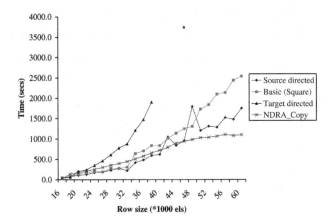

Fig. 4. Execution time to transform set-of-rows blocking to set-of-columns blocking

GA needs to be determined. The size is determined independent of the blocking of the source and target arrays to equal the amount of available physical memory. We evaluated three options for the shape of the GA used. One option was to use the largest square tile that fits within the available memory. If the blocking of the DRAs is known, the GA can be chosen to be a multiple of the source block size or the target block size. These three options are labeled Basic(square), Source Directed and Target Directed, respectively. The implementation of the new approach is labeled NDRA_Copy.

We evaluated the mechanisms on the OSCBW machine at the Ohio Supercomputer Center [24]. Each node in the cluster has Dual AMD Athlon MP processors (1.533 GHz) and 2GB of memory. The PGI pgcc 4.0-2 compiler was used to generate the executables. Two sets of experiments were conducted. In one, a set of rows form the blocks in the source array. The target array is blocked as a set of columns. The corresponding results are shown in Fig. 4. The second experiment involved the reverse - transforming from a set-of-columns blocking into a set-of-rows blocking, and its results are shown in Fig. 5. The number of rows (or columns) in a block was chosen such that the block size was greater than 1MByte, the typical brick size chosen by DRA for this system. For example, for a $< 4096, 4096 >$ array, where each element is of size four bytes, set-of-rows blocking corresponds to a block size of 1 Mb, with each brick holding a $< 64, 4096 >$ block of data; and a set-of-columns layout corresponds to a 1 Mb brick holding a $< 4096, 64 >$ block of DRA data.

In both the experiments, the array size was increased from 16000 to 60000 in steps of 2000 and all four mechanisms were evaluated. For our approach, the template size is determined automatically using the algorithms described in Section 4. The x-axis in the graphs shows the array dimension in number of elements. The y-axis shows the transformation time in seconds. We were unable to run larger experiments due to the limited amount of disk space available on the local disks (around 60GB).

In transforming the set-of-rows bricks into a set-of-columns bricks, the target directed method performs significantly worse than other approaches. This is because the data to be read in is not contiguous on disk. The DRA reads in entire blocks of data to 'collect'

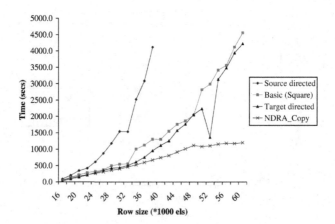

Fig. 5. Execution time to transform set-of-columns blocking to set-of-rows blocking

the data into the global array. This leads to exponential increase in cost. Due to this obvious trend, this approach was evaluated with only certain sample array dimensions. The source directed approach performs better, as DRA implementation allows writes of partial blocks, if it is contiguous on disk. Though the unit of write is small, it still performs better than the target directed approach. With larger array dimensions, both the source directed and basic (square) approach increase in cost.

Our implementation performs better than the alternatives. The relative performance benefit of our new approach increases with the size of the array. It starts with a single-pass solution and then uses a two-pass solution for arrays with dimensions larger than 32,000. But the execution time increases gradually and is not drastically affected by the exact problem instance at hand. Unlike the other three approaches, our implementation performs comparably for both the transformations evaluated.

6   Conclusions

In this paper we proposed a new approach to efficient transformation of the blocked layout of multidimensional disk-based arrays. The proposed approach was implemented as a new copy primitive within the DRA I/O library. Experimental results demonstrated the benefit of the new approach over existing mechanisms. The extension of this approach to the parallel context is being pursued.

Acknowledgments

We acknowledge the support from the National Science Foundation through the Information Technology Research program (CHE-0121676) and Pacific Northwest National Laboratory. In addition, We would like to thank the Ohio Supercomputer Center (OSC) for the use of their computing facilities.

References

1. Chen, Y., Foster, I., Nieplocha, J., Winslett, W.: Optimizing collective I/O performance on parallel computers: A multisystem study. In: 11th ACM Intl. Conf. on Supercomputing. (1997)
2. Seamons, K.E., Winslett, M.: Multidimensional array I/O in Panda 1.0. The Journal of Supercomputing **10** (1996) 191–211
3. The Panda Project – Data Management for High-Performance Scientific Computation. (http://drl.cs.uiuc.edu/panda/)
4. Foster, I., Nieplocha, J.: Disk Resident Arrays: An array-oriented I/O library for out-of-core computations. In Buyya, R., Jin, H., Cortes, T., eds.: Disk Arrays and Parallel I/O: Theory and Practice. IEEE Computer Society Press (2001)
5. Anderson, G.L.: A stepwise approach to computing the multidimensional fast Fourier transform of large arrays. IEEE Transactions on Acoustics and Speech Signal Processing **28** (1980) 280–284
6. Bailey, D.H.: FFTs in external or hierarchical memory. Journal of Supercomputing **4** (1990) 23–35

7. Kazhiyur-Mannar, R., Wenger, R., Crawfis, R., Dey, T.K.: Adaptive resolution isosurface construction in three and four dimensions. Technical Report OSU-CISRC-7/03–TR38, School of Computer and Information Science, The Ohio State University (2003)
8. Tensor Contraction Engine – Synthesis of High-Performance Algorithms for Electronic Structure Calculations. (http://www.cse.ohio-state.edu/~saday/TCE/)
9. Baumgartner, G., Bernholdt, D., Cociorva, D., Harrison, R., Hirata, S., Lam, C., Nooijen, M., Pitzer, R., Ramanujam, J., Sadayappan, P.: A high-level approach to synthesis of high-performance codes for quantum chemistry. In: Proceedings of Supercomputing 2002. (2003)
10. Cociorva, D., Gao, X., Krishnan, S., Baumgartner, G., Lam, C., Sadayappan, P., Ramanujam, J.: Global communication optimization for tensor contraction expressions under memory constraints. In: 17th International Parallel & Distributed Processing Symposium (IPDPS). (2003)
11. Cociorva, D., Baumgartner, G., Lam, C., Sadayappan, P., Ramanujam, J., Nooijen, M., Bernholdt, D., , Harrison, R.: Space-time trade-off optimization for a class of electronic structure calculations. In: Proc. of ACM SIGPLAN PLDI 2002. (2002)
12. Cociorva, D., Wilkins, J., Baumgartner, G., Sadayappan, P., Ramanujam, J., Nooijen, M., Bernholdt, D., Harrison, R.: Towards automatic synthesis of high-performance codes for electronic structure calculations: Data locality optimization. In: Proc. of the Intl. Conf. on High Performance Computing. (2001)
13. Krishnan, S., Krishnamoorthy, S., Baumgartner, G., Cociorva, D., Lam, C., Sadayappan, P., Ramanujam, J., Bernholdt, D., Choppella, V.: Data locality optimization for synthesis of efficient out-of-core algoritms. In: Proc. of the Intl. Conf. on High Performance Computing. (2003)
14. Krishnan, S., Krishnamoorthy, S., Baumgartner, G., Lam, C., Ramanujam, J., Choppella, V., Sadayappan, P.: Efficient synthesis of out-of-core algorithms using a nonlinear optimization solver. In: Proc. of 18th Intl. Parallel & Distributed Processing Symposium (IPDPS). (2004)
15. High Performance Computational Chemistry Group: NWChem, A Computational Chemistry Package for Parallel Computers, Version 4.6. Pacific Northwest National Laboratory, Richland, Washington 99352–0999, USA. (2004)
16. Nieplocha, J., Harrison, R.J., Littlefield, R.J.: Global arrays: a portable programming model for distributed memory computers. In: Supercomputing. (1994) 340–349
17. Nieplocha, J., Harrison, R.J., Littlefield, R.J.: Global arrays: A nonuniform memory access programming model for high-performance computers. The Journal of Supercomputing **10** (1996) 169–189
18. Nieplocha, J., Foster, I.: Disk resident arrays: An array-oriented I/O library for out-of-core computations. In: Proceedings of the Sixth Symposium on the Frontiers of Massively Parallel Computation, IEEE Computer Society Press (1996) 196–204
19. Eklundh, J.O.: A fast computer method for matrix transposing. IEEE Trans. on Computers **20** (1972) 801–803
20. Kaushik, S.D., Huang, C.H., Johnson, R.W., Sadayappan, P., Johnson, J.R.: Efficient transposition algorithms for large matrices. In: Proceedings of the 1993 ACM/IEEE conference on Supercomputing, ACM Press (1993) 656–665
21. Suh, J., Prasanna, V.K.: An efficient algorithm for out-of-core matrix transposition. IEEE Trans. on Computers **51** (2002) 420–438
22. Krishnamoorthy, S., Baumgartner, G., Cociorva, D., Lam, C., Sadayappan, P.: On efficient out-of-core matrix transposition. Technical Report OSU-CIRSC-9/03-T52, School of Computer and Information Science, The Ohio State University (2003)
23. Krishnamoorthy, S., Baumgartner, G., Cociorva, D., Lam, C.C., Sadayappan, P.: Efficient parallel out-of-core matrix transposition. In: Proceedings of the International Conference on Cluster Computing, IEEE Computer Society Press (2003) to appear.
24. The Ohio Supercomputer Center. (http://www.osc.edu)

Autonomic Storage System Based on Automatic Learning[*]

Francisco Hidrobo[1] and Toni Cortes[2]

[1] Universidad de Los Andes, Mérida 5101, Venezuela,
[2] Universitat Politécnica de Catalunya, Barcelona 08034, Spain
`hidrobo@ula.ve, toni@ac.upc.es`

Abstract. In this paper, we present a system capable of improving the I/O performance in an automatic way. This system is able to learn the behavior of the applications running on top and find the best data placement in the disk in order to improve the I/O performance. This system is built by three independent modules. The first one is able to learn the behavior of a workload in order to be able to reproduce its behavior later on, without a new execution. The second module is a drive modeler that is able to learn how a storage drive works taking it as a "black box". Finally, the third module generates a set of placement alternatives and uses the afore mentioned models to predict the performance each alternative will achieve. We tested the system with five benchmarks and the system was able to find better alternatives in most cases and improve the performance significantly (up to 225%). Most important, the performance predicted where always very accurate (less that 10% error).

1 Introduction

One of the main trends in the computing world is the increasing needs for I/O capacity and performance shown by applications.

Many approaches to solve this problem have been proposed in the last decades. One of the most promising consists of configuring the storage system and the placement of data to maximize the storage-system performance for a specific workload. In general, this approach consists of finding the optimal configuration and data placement for the I/O system given a specific workload. Currently, these optimizations are usually done by experts who use their experience and intuition to make this configuration and placement. A tool that could perform this tuning in an automatic way would be a great step in making this technique available to a wider range of sites. Furthermore, this tool becomes even more useful if the optimal configuration and placement varies throughout the time making it more difficult to keep the right placement up to date.

---

[*] This work was supported in part by a grant from FONACIT (Venezuela) which is gratefully acknowledged, by the Ministry of Science and Technology (Spain), and by FEDER funds of the European Union under grants TIC2001-0995-C02-01.

L. Bougé and V.K. Prasanna (Eds.): HiPC 2004, LNCS 3296, pp. 399–409, 2004.
© Springer-Verlag Berlin Heidelberg 2004

Our objective is to design a storage system capable of extracting all potential performance and capacity available in a heterogeneous environment with as little human interaction as possible. We envision the system as an advanced data-placement mechanism that analyzes the workload to decide the best distribution of data among all available devices, and the best placement within each device.

The aim of this paper is to present the global design of our proposal. This design includes a disk model based on neural networks, an approach to model the workload based on reduced traces and some sample strategies to generate different placements.

2   Autonomic Storage System: A Global Picture

In order to place the work in the right context, we will first give a global description of how the whole system works, and then we will describe in detail each of the parts that build it.

In the starting point, when the system is new, we have a set of disks attached to the system. The first thing we have to do is to model them. This model should have two main properties. First it should be able to predict the performance of a given workload, without having to run it. Second, it should treat the disk as a "black box".

Once we have all disks modeled, and thus we can predict the performance of any possible workload, we start learning the workload behavior. This step mainly consists of tracing the requests done to the disk and keeping them in file-system internal data structures.

Periodically, which could be once a day, once a week, or any other period depending on the needs, the system uses the workload behavior learned to generate different placement alternatives that may (or may not) improve the performance of the applications running on the system. As these new placements cannot be implemented and tested, we will use the disk models to predict the performance each new placement would achieve.

After the performance of all proposed placements is predicted, we pick the best one and compare it with the performance of the current workload (which has been learned at the same time as the workload). If the new placement is better that 10% the performance of the current one, then we take the effort of moving the blocks to implement the new placement and thus improve the performance of the applications using these disks.

It is also important to see, that whenever a new disk is added, it has to be modeled and then it will be used by the generator of placement alternatives to place blocks in it.

Once described the module as a whole, it is important to notice that the objective of this paper is to test that all modules are possible and that the final system works well. Our current version is not integrated in the file system, but it is implemented as separated modules that work off-line, but placing them in the kernel and running them on-line it is just an implementation problem that lies as future work.

3   Disk Model

The main objective of this module, and also the main difference compared to other approaches, is to design a model that has no previous knowledge about the drive to model.

In our system, we propose a general model based on a mathematical function. We assume that we can find a function (M) that approximates the service time St for each request. Thus, our general model can be expressed by:

$$St \approx M(R)$$

where: R is the input vector with components:
R_{Addr}: address of the first-requested block,
R_{Jump}: difference (in blocks) between this request and the previous one.
R_{Size}: request size.
St is the request service time.

Our experience has shown us that it is better to have one model for each operation (Read and Write). Thus, we will use one model for each request type.

In [1], we studied different approaches using several applications and drives, and we found that neural network are the best mechanism to model drives.

Neural Networks have a high capacity for function approximation, and this is exactly our objective. We have tested a simple architecture based on feed-forward network to resolve the function approximation problem because it has been proved to be a simple and effective approach.

We use a feed-forward neural net with the following configuration: 3 neurons in input layer, 25 neurons in the hidden layer, and one neuron in output layer (service time). To resolve the problem, we use a Levenberg-Marquardt back-propagation algorithm.

As we can see, this approach behaves as a "black box" and does not need any previous knowledge about the device, it can learn the behavior by running a synthetic trace. This learning process was described in a previous paper [1].

4   Workload Model

We focused the workload model on files because it allows us to focus the effort toward the most important (i.e. used) files of the application to reduce the size of the model and the complexity of the data-placement module.

Our file model is based on the segments accessed and their relationship. We define a segment as a set of blocs requested in a single disk operation. Please keep in mind that our model works at disk-drive level and thus segments are not necessarily what users request, but what the operating system (or file system) sends to the disk drive after being filtered by the buffer cache and some possible reordenation done during the process. Thus, our file model stores, for each segment, the following information:

- The first logical block requested. Using logical blocks makes our model independent of the physical location of the block.
- The total number of requests made to this direction. Used in order to keep some information about the importance of different segments.
- Request type (read/write). This value is expressed as read probability.
- The average size of the requests made to this segment.
- The list of possible predecessor requests. This list contains information about the requests that preceded it in the trace. Each entry in the list has:
  - the file id. This field is used for two reasons. The first one is to identify the physical location of the last accessed block. Second, it is used to find relationships among different files.
  - the last logical block of the previous request. This field is used to find the number of blocks the disk has "jumped" from one request to the other.
  - the number of times that this element preceded the request. This field is used to know the percentage of times this sequence occurred.

As we can see, this model fulfills all our requirements. It can be generated on-line because new requests can be added very easily. No physical information is kept, and thus it is block-location independent. It does not keep any time information and thus it is device independent. It does not grow unnecessarily but it has enough information (even inter-file relationships) for the data placement module to take decisions. Finally, building a trace with the information required by the drive model previously described (initial block, R/W, jump and size of the request) is trivial.

5   Prediction and Data Placement

This module implements its functionality through three independent steps: generation of placement alternatives, evaluation of the generated alternatives, and application of the best placement found.

5.1   Generation of Placement Alternatives

In order to generate a placement, we first decide a logical order in which blocks should be placed and then we decide in which disk blocks we map them. Finally, we may refine the solution to take into account specific drive characteristics.

Logical Ordering. The main idea is to encourage spatial locality. This approach can reduce the seek distance and can increase the probability that the disk read-ahead improves the general performance. We can follow two different approaches:

a) **Based on the Access Pattern.** We use the access pattern represented in the file model, to find the probabilistic access sequence. We can generate a directed graph $G(V, E)$, where the vertex set (V) represents the accessed blocks. The set of edges (E) is defined as follows: E_{ij} belongs to E if block i is acceded after block j. Furthermore, each edge E_{ij} has a weight (edge weight), which represents the number of times that block j is preceded by i.

In order to generate the logical ordering, we eliminate edges, taking only the edge that has the largest weight, until each node only has one input edge and one output edge, except the first and the last one.

b) **Based on Sequential Structure.** We directly use the logical order of the file regardless of the real accesses. This means that we use the pointers to data blocks to make the list.

Physical Placement. Once we have the logical order, we need to find a spot (or spots) in the disk to place them sequentially according to the logical order. To do it, we have several alternatives regarding the sets of disk blocks we consider to relocate the file. The three studied alternatives are presented here:

a. **All disk blocks.** It uses all disk blocks except those that contain metadata such as superblock copies. This option may move blocks from other files.
b. **File-used blocks.** It uses all blocks used by the defined as important files.
c. **File-used blocks plus free blocks.**

Refinement: Special Drive Conditions. Depending on the drive type, it may have some special conditions that could be taken into account to refine the physical placement. In this work we have only tested one that deals with the different density of disk zones.

Hot Blocks. If the disk has a faster area, we can try to place the most used blocks in this area. We will divide the previous list in two sections. One, as big as the fast zone, with the most used blocks. A second one, with the rest of the file blocks. Both lists will be ordered as explained before. This refinement looses locality, but gains bandwidth, a trade off that has to be evaluated. To simplify things, we will only apply this refinement when all disk blocks are used.

5.2   Evaluation of Alternatives

For each proposed placement, the system generates a new allocation map and takes the file model for each "important" file to create an estimate trace for these files. Then, the disk model is used to predict the performance.

5.3   Application of the Proposed Placement

Here, the module applies the placement proposed by the "winning alternative". In order to take some action we demanded that the placement improves the performance, at least, 10%. We decided to use 10% as threshold, because smaller gains may not be enough for the trouble of data movement and because our predictions have a potential error of up to 10%.

6   Experiments and Results

In this section, we are going to show how our global system works. In order to show it, we are going to use different workloads and to run the workload in a file system artificially aged according to [2] (We also executed the tests in a new file system; but for space reasons will not be presented).

6.1   Workloads

Synthetic Benchmarks

1. **SSA:** Simple Sequential Access. The application reads one file sequentially. In each read it requests 64KB. It will show us the behavior in the simplest case: accessing a single file (but no necessarily using regular intervals).
2. **FSA:** Fixed Stride Access. In this case, we used more complex pattern access than sequential. It uses a stride of 512KB and reads 8KB in each requests.
3. **MSA:** Multiple Sequential Access. Here, the application reads simultaneous three files. It requests a segment of the first file, then a segment of second one and a segment of third file. In each read it requests 64KB. These operations are repeated until reading the files completely. It will allow us to observe what happens when several files are involved, but with a reproducible behavior.

The inter-arrival time was uniformly distributed (0 and 100 ms) in all case. The files used were 300Mbytes each and we called them: $file_1$, $file_2$ and $file_3$.

Sequence of Access to WEB Pages. In this case, we took a personal WEB page generated by a standard tool. This information is organized in many directories and files. In the test, we accessed a sequence of pages, where each requested page implied the access to a set of related files, which are normally small in size. Therefore, in this application the interrelation between files is more important that the access pattern within each file.

TPC-H Benchmark. Finally, we have tested the TPC-H benchmark, which is a decision support benchmark for Databases. We created a database of 1 Gbytes and used *Postgres SQL 7.2.1* for LINUX. We ran all queries, except queries number 7, 9, 20 and 21 because these queries contain functions not supported by our database manager system.

6.2   Results

Following the procedure described in the section 5.1, we created 10 possible placements for each list of important files. Table 1 presents all possible combinations and the names we have given them.

Table 1. Possible placements implemented

		Based on Access Pattern	Based on Sequential Structure
		APAD	SSAD
All disk blocks	Hot blocks (all file blocks fit)	APAD_HF	SSAD_HF
	Hot blocks (not all file blocks fit)	APAD_HNF	SSAD_HNF
Used blocks		APUB	SSUB
Used+free blocks		APUF	SSUF

Synthetic Workload on an Aged File System. If we apply the SSAD placement alternative for the SSA benchmark created on an aged file system, we could improve the performance in 44.5% as Figure 1 shows. In this experiment it is important to observe that, although the access pattern should be sequential, APAD does not achieve as good performance as SSAD. The reason behind this unexpected behavior, is that some requests were reordered when the access pattern was learned, but the real access is sequential. If more repetitions had been done to learn the pattern, as will probably be in a real system, this reordenation will only appear in the pattern if they occur very frequently and not just once.

For the FSA benchmark, Figure 2, the improvement could reach up to 225% with APAD placement.

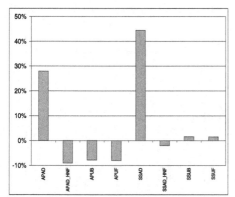
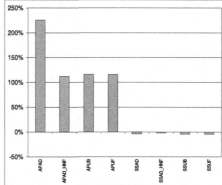

Fig. 1. SSA on on file system aged   **Fig. 2.** FSA on file system aged

With regard to MSA benchmark, Figure 3 shows that the best placement, APAD, could improvement the performance over 100% for three files.

When we applied the winning placement and measured the real performance (Figure 4), we can observe that the relative errors were again very small, but for file1 of MSA where the relative error is over 100%, which makes the global error for the MSAA benchmark grow unexpectedly.

The problem in MSA benchmark is that the buffer cache policies generate a different behavior when the blocks are remapped. Therefore, the access pattern learned and used to make the prediction is very different from the one obtained once the new placement has been applied. To prove this fact, we executed MSA benchmark turning off the read-ahead mechanism in the buffer cache. In this new experiment, the relative error was, once again, below 10% for all files. Obviously, the proposal is not to eliminate the read-ahead in the buffer cache to obtain reliable results; rather the idea is to develop a integrated approach between our system and the file system to guarantee that such changes of behavior will not be produced or will be taken into account by the prediction system.

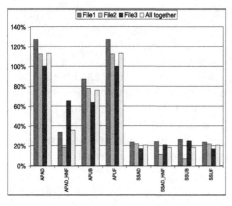

Fig. 3. MSA on file system aged

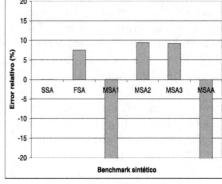

Fig. 4. Best placement (Relative error)

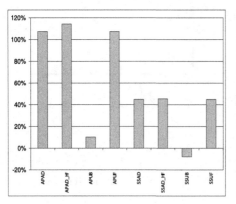

Fig. 5. Sequence of access to WEB pages

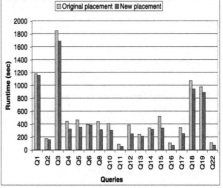

Fig. 6. Runtime reduction for TPC-H

Sequence of Access to Pages WEB on a File System Aged. As we can see in figure 5, the values show us that the alternatives based on the access pattern could improve the performance near to 100%. The reason for this behavior is that file were spread throughout all the disk while they need to be close.

TPC-H on a File System Aged. The best placement alternative for the TPC-H on an aged file system is SSAD, and could improve the performance in 142% for access to partsupp table, 73 % for order table, 32% for lineitem table and 39% if we take all tables together.

To give a final idea of the potential improvements, we computed the real runtime improvement for each query and then all queries together. Figure 6 shows the runtime (in seconds) obtained for each query with the original placement and when the placement SSAD was applied. The global reduction is 14.64%.

In a real system, we would not first learn the behavior, then move data, and finally reexecute the application. The normal behavior is that we learn during some time, move the data, and then continue the execution (not repeating exactly what we learned). To test how our system would behave in such an environment we tested WEB and TPC-H benchmarks. In the first case we learned from one sequence of pages (one possible session) and tested with a different one (another session by another user). In the TPC-H we learned from half of the queries and tested the other half.

The results showed that the performance improvement obtained with the new placements and the "not learned" part of application were similar to the ones presented in the previous sections. This means that the system is able to do accurate predictions and it can be used for autonomic computing.

7   Related Work

The first kind of storage-drive models we find in the bibliography are based on simulation techniques. In this group we found specially interesting the proposals done by Ruemmler and Wilkes [3] and the one done by Ganger et al. [4, 5]. Another possibility is the usage of analytical models where the input is not a trace (as in the previous group), but a characterization of the load [6] Both, simulations and analytical models need prior knowledge of the disks, which is not always easy to find, while our neural network approach does not. Finally, there is also another group of proposals that treat the drive as a "black box" and learn the behavior after a training period [7, 8] but they either need huge data strutres or work at a very high level (not request level as we need).

With regard to application model, some works have been proposed in characterizations and modeling of I/O access pattern [9–14]. Most of these studies were made at the file system level. Gómez and Santonja [15] developed an approach to analyze and model disk access pattern but cannot be learned and used on-line. In some aspect our workload model is similar to the presented by Madhyastha and Reed in [16], and by Oly in [17]. They used Hidden Markov Model (HMM) for classification and prediction of I/O access patterns. Basically, They use the model to predict the next action for the file system or to propose an allocation strategy according to the access pattern classification.

Finally, some work has been done on block replacement to improve I/O performance [18, 19], but are not integrated in an autonomic system as ours.

Regarding our global system, there are some tools similar to our system such as MINERVA [20] and Hippodrome [21] development in the HP Laboratories. However, they target their work to the configuration of RAID systems.

8   Conclusions

In this paper, we have presented an autonomic storage system based on automatic learning of previous behavior. This system uses a modular design to explore different placement possibilities and decides which one is better and whether the new placement is worth the effort of making the changes.

In addition, to achieve the autonomic storage system, we have proposed novel mechanisms to model disks and applications that may be of use to the research community in different environments.

The tests done showed that the performance prediction is reliable. Furthermore, we have also shown that this prediction can be used to improve the performance significantly when the data is not correctly placed.

References

1. Hidrobo, F., Cortes, T.: Towards a Zero-Knowledge Model for Disk Drives. In: Proceedings of the AMS, Seattle, WA, USA, IEEE Computer Society Press (2003)
2. Smith, K.A., Seltzer, M.I.: File System Aging Increasing the Relevance of File System Benchmark . In: Proceedings of SIGMETRICS. (1997)
3. Ruemmler, C., Wilkes, J.: An introduction to disk drive modeling. IEEE Computer **27** (1994) 17–28
4. Ganger, G., Worthington, B., Patt, Y.: The DiskSim Simulation Environment (Version 2.0). http://www.ece.cmu.edu/~ganger/disksim/ (2004)
5. Schindler, J., Ganger, G.R.: Automated Disk Drive Characterization. In: Proceedings of SIGMETRICS, Santa Clara, CA, USA, ACM Press (2000) 112–113
6. Shriver, E., Merchant, A., Wilkes, J.: An analytic behavior model for disk drives with readahead caches and requests reordering. In: SIGMETRICS, Madison, Wisconsin, USA, ACM Press (1998) 182–191
7. Thornock, N.C., Tu, X.H., kelly Flanagan, J.: A STOCHASTIC DISK I/O SIMULATION TECHNIQUE. In: Proceedings of Winter Simulation Conference, Atlanta, GA, USA, ACM Press (1997) 1079–1086
8. Anderson, E.: Simple table-based modeling of storage devices. Technical Report HPL-SSP-2001-04, HP Laboratories (2001) http://www.hpl.hp.com/SSP/papers/.
9. Kotz, D., Nieuwejaar, N.: Dynamic File-Access Characteristics of a Production Parallel Scientific Workload. In: Proceedings of Supercomputing, Washington, DC, USA, IEEE Computer Society Press (1994) 640–649
10. Kroeger, T.M., Long, D.D.: The Case for Efficient File Access Pattern Modeling. In: Proceedings of HotOS, AZ, USA, IEEE Computer Society (1999) 14–19
11. Ware, P.P., Jr., T.W.P., Nelson, B.L.: Modeling File-system Input Traces via a Two-level Arrival Process. In: Proceedings of the 28th Conference on Winter Simulation, Coronado, California, United States, ACM Press (1996) 1230–1237
12. Ware, P.P., Thomas W. Page, J.: Automatic Modeling of File System Workloads Using Two-Level Arrival Processes. ACM TOMACS **8** (1998) 305–330
13. Ruemmler, C., Wilkes, J.: *UNIX Disk Access Patterns.* In: Proceedings of Winter USENIX Conference, San Diego, CA, USA (1993) 405–420
14. Ganger, G.R.: Generating Representative Synthetic Workloads. An Unsolved Problem. In: Proceedings of 21st International Computer Measurement Group Conference, Nashville, TN, USA, Computer Measurement Group (1995) 1263–1269
15. Gómez, M.E., Santoja, V.: A new approach in the analysis and modeling of disk access pattern. In: Proceedings of ISPASS, Austin, Texas, IEEE Press (2000)
16. Madhyastha, T., Reed, D.: Input/Output Access Pattern Classification Using Hidden Markov Models. In: IOPADS, San Jose, CA, USA, ACM Press (1997)
17. Oly, J.: Markov Model Prediction of I/O requests for Scientific Applications. Master's project, University of Illinois at Urbana-Champaign (2000)

18. Vongsathorn, P., Carson, S.: A system for adaptive disk rearrangement. Software - Practice and Experience **20** (1990) 225–242

19. Akyürek, S., Salem, K.: Adaptive Block Rearrangement. ACM TOCS **13** (1995) 89–121

20. Alvarez, G., Borowsky, E., Go, S., Romer, T., Becker-Szendy, R., Golding, R., Merchant, A., Spasojevic, M., Veitch, A., Wilkes, J.: *MINERVA: an automated resource provisioning tool for large-scale storage systems*. TOCS **19** (2001) 483

21. Anderson, E., Hobbs, M., Keeton, K., Spence, S., Uysal, M., Veitch, A.: Hippodrome: running circles around storage administration. In: Proceedings of FAST, Monterey, CA, USA, USENIX, Berkeley, CA. (2002) 175–188

Broadcast Based Cache Invalidation and Prefetching in Mobile Environment

Narottam Chand, Ramesh Joshi, and Manoj Misra

Electronics and Computer Engineering Department,
Indian Institute of Technology,
Roorkee – 247 667, India
{narotdec, joshifcc, manojfec}@iitr.ernet.in

Abstract. Caching at mobile client is an important technique for improving the performance in wireless data dissemination. It can reduce the number of uplink requests, server load, query latency and can increase the data availability. Battery energy and limited bandwidth are two major constraints imposing challenges to the realization of caching at mobile clients. A cache invalidation strategy ensures that the data items cached in a mobile client are consistent with those stored on the server. In our invalidation scheme, to minimize uplink requests, all recently updated or requested items are broadcast immediately following the invalidation report. To further improve the caching performance, prefetching is used. Three update report strategies are presented to reduce average access time and minimize the client energy for cache invalidation. Simulation experiments show that the LCF strategy dramatically improves the average access time for mobile clients, reduces the number of uplink requests and conserves the client energy.

1 Introduction

Users of mobile devices wish to access dynamic data, such as stock quotes, news items, current traffic conditions, weather reports, email and video clips via wireless networks. Caching at mobile client can relieve bandwidth constraints imposed on wireless and mobile computing. Copies of remote data can be kept in the local memory of the mobile client to substantially reduce user requests for retrieval of data from origin server. This not only reduces the uplink and downlink bandwidth consumption but also the average query latency. Caching frequently accessed data in mobile client can also save the energy used to retrieve the repeatedly requested data at client side. A *cache invalidation* strategy is used to ensure that the data items cached in a mobile client are consistent with those stored on the server.

Barbara and Imielinski [14] provide a caching solution where the server periodically broadcasts an *invalidation report* (IR) in which the changed data items are indicated. An IR based strategy has long query latency and it makes poor utilization of available wireless bandwidth. To minimize uplink requests and downlink broadcasts,

L. Bougé and V.K. Prasanna (Eds.): HiPC 2004, LNCS 3296, pp. 410–419, 2004.
© Springer-Verlag Berlin Heidelberg 2004

we propose a broadcast strategy, called *update report (UR)*, where all the recently updated or requested items are broadcast immediately after the invalidation report (IR).

One important issue in broadcast strategies is to determine an optimal broadcast schedule i.e. the order in which items are transmitted. This paper presents broadcast based invalidation techniques that are favorable in terms of access time and tuning time. The objective is to provide an efficient and effective update report, so that mobile clients can validate their cache and download updated or requested data from the broadcast channel with minimum of energy expenditure such that average access time associated with each data item is minimized. To further improve the caching performance, prefetching at mobile clients is used.

The rest of the paper is organized as follows. Section 2 gives a description of the related work. Section 3 presents the caching model and preliminaries. In section 4, we present the update report based caching scheme. In section 5, we describe simulation experiments for establishing the performance of our methods. Conclusion is given in section 6.

2 Related Work

The caching in mobile environment is complicated by the fact that the caches need to be kept consistent. A number of broadcast based cache invalidation strategies have been proposed for mobile environments. Barbara and Imielinski [14] provided three cache invalidation schemes, namely Broadcasting Timestamps (TS), Amnesic Terminals (AT) and Signatures (SIG), which use different invalidation reports for a stateless server. Jing et al. [10] proposed a Bit-Sequence (BS) scheme that uses a hierarchical structure of binary bit sequences with an associated set of timestamps to represent clients with different disconnection times. Tan [7] reexamined the BS method and studied different organizations of the invalidation report. These new organizations facilitate clients to selectively tune to the portion of the report that are of interest to them. Hue et al. [12] proposed a scheme to reduce the false invalidation rates based on BS reports. Wu et al. [6] proposed a scheme which modifies the TS or AT algorithms to include cache validity checks after reconnection. Hu and Lee [3] have proposed a family of invalidation algorithms. The essence of these algorithms is that the type of invalidation report to be sent is determined dynamically based on system status such as disconnection frequency and duration as well as update and query pattern. G. Cao in [4], [5] addresses the problem of long query latency with a UIR based approach. In this approach, a small fraction of the essential information (called *updated invalidation report (UIR)*) related to cache invalidation is replicated several times within an IR interval, and hence the client can answer a query without waiting until next IR. However, if there is a cache miss, the client still needs to wait for the data to be delivered. In [1], [15], author addresses various problems associated with the IR based cache invalidation strategies. To improve the query latency and cache hit ratio, clients intelligently prefetch the data that are most likely used in the future. Kahol et al. [2], [9] present an asynchronous stateful (AS) scheme to maintain cache consistency. Each mobile client maintains its own Home Location Cache (HLC) to deal with the problem of disconnections. Yuen et al. [8]

proposed a cache invalidation scheme based on absolute validity interval (AVI) for each data item. Although a number of broadcasting algorithms that aim to minimize the access time for data items [20], [21], [22] have been proposed, but most of them are only based on access frequency of data items and do not take into consideration the clients who have cached the data items. In our broadcast based caching strategies, data items cached by different clients are also considered while selecting the contents and schedule of broadcast.

3   Background and Preliminaries

This section describes our caching model and some of the preliminaries.

3.1   Caching Model in Mobile Environment

A mobile computing environment consists of two distinct sets of entities: *Mobile Hosts (MHs)* and *Fixed Hosts (FHs)*. Some of the fixed hosts called *Mobile Support Stations (MSSs)*, are augmented with a wireless interface in order to communicate with the mobile hosts, which are located within a radio coverage called a *cell*. Each MSS stores a complete copy of the database and hence acts like a database server. Henceforth, we use the terms MSS and server interchangeably. An MH communicates with a server over a wireless communication link. The wireless channel is logically separated into two sub-channels: downlink and uplink channel [17]. The communication is asymmetric (i.e. the uplink bandwidth is much less than that of downlink).

It is assumed that the database is updated only by the server. The servers themselves form a wired distributed system in which a fully replicated database resides. The database comprises a set D of N data items, d_1, d_2, ..., d_N. An important property of data items is that their values can be highly dynamic as they are used to maintain some real-world information e.g., a news update. Each item d_j is of same size S_{data} (in bits). As the wireless communication within a cell is independent of other cells, henceforth we concentrate on a single cell. Inside a cell, there are M mobile hosts, MH_1, MH_2, ..., MH_M.

An application program runs on the mobile host as a client process and communicates with the server through message, i.e., the client sends an uplink request (query) for the data it needs to the server and the server responds by sending the requested data on the downlink [2]. In order to minimize the number of uplink requests, the client caches a portion of the database in its local nonvolatile memory. To ensure cache consistency, each server periodically broadcasts update reports. All active clients listen to the reports and invalidate their cache contents accordingly.

3.2   Definitions

Following are some definitions and notations that are used during further discussion:

Access Time (AT). It indicates the time elapsed between the query submission and receipt of the response. We denote AT_i as the access time associated with data item d_i.

Tuning Time (TT). It indicates the total time that a mobile client spends actively listening to the channel in a complete access period. TT determines the energy consumption of MHs because they could slip into doze mode when they are not actively listening on the channel. As mobile clients have limited battery energy, the reduction of TT is an important issue in data broadcast technology [18], [23].

D_i. Actual data of an item with id d_i.

t_i^r. The *latest request time* of item d_i. It represents the most recent time when d_i has been requested by a client.

n_i. The *number of mobile clients* who have cached the data item d_i.

TAT_i. The *total access time* of data item d_i indicates the sum of access times for d_i by different clients. If n_i is the number of clients who have cached the item d_i, then $TAT_i = n_i.AT_i$. We need to concentrate on the order of data items in one broadcast interval, so as to minimize $TAT = \sum TAT_i$. TAT gives a measure of total access time experienced by all the MHs in one broadcast cycle.

AAT. The *average access time* to download the actual data for an item from the UR.

$$AAT = \frac{TAT}{\sum n_i}.$$

4   Update Report (UR) Based Caching

To reduce the number of uplink requests and downlink broadcasts, we introduce the concept of update report (UR). URs are broadcast synchronously with period L. The structure of a UR is shown in Fig. 1.

IR	INDEX	DATA

Fig. 1. Structure of a UR

At interval T_i:

$IR_i = \{(d_x, t_x) | (d_x \in D) \wedge (n_x > 0) \wedge (T_i - w.L < t_x \leq T_i)\}$

$INDEX_i = \{d_x | ((T_{i-1} < t_x \leq T_i) \wedge (n_x > 0)) \vee (T_{i-1} < t_i^r \leq T_i)\}$

$DATA_i = \{D_x | d_x \in INDEX_i\}$

$INDEX_i$ defines the order in which data appears in $DATA_i$. IR_i contains the update history of past w broadcast intervals, whereas $DATA_i$ contains actual data value for the items which have been updated or requested during the previous IR interval. UR can handle the broadcast of variable size items by including size of each item as part of INDEX, i.e., $INDEX = \{(d_x, l_x)\}$, where l_x is the size (bytes) of item d_x. Supporting variable size data, increases the size of INDEX and hence the client tuning time.

4.1 Broadcast UR to Utilize Bandwidth

In most IR based algorithms [2], [3], [6], [7], [8], [10], [11], [13], [14], [16], [19], updating a cached item, may generate many uplink requests and downlink broadcasts, and thus make poor utilization of available wireless bandwidth. Similar to [1], [4], [5], [15], [23], we address the problem by asking the server to broadcast all the recently updated/requested data items which are cached by one or more clients. If a client observes that the server is broadcasting an item which is an invalid entry in its local cache, it will download the item, thus, saving on uplink and downlink bandwidth.

4.2 Saving Energy

Clients may wake up during the UR broadcast time and selectively tune-in to the channel to save energy. After broadcasting IR, the server broadcasts INDEX followed by actual data. Every client listens to the IR if not disconnected. At the end of IR, the client downloads INDEX and locates the interesting item that will come, and listens to the channel at that time to download the data. This strategy saves energy since the client selectively tunes to the channel and can stay in doze mode most of time.

4.3 Scheduling Algorithms

Given various data items with their cache count, problem is to decide in what sequence to schedule the updated/requested items in INDEX, which are cached by one or more clients, so as to minimize the average access time and tuning time to download the actual data broadcast in DATA. Formally the problem may be stated as:

The Input: The data items d_i, i=1, 2, ..., N, are defined by their cache counts n_i.

The Schedule: A schedule INDEX of $d_{i(1)}$, $d_{i(2)}$, ..., $d_{i(n)}$ is a finite sequence $d_{s(1)}$, $d_{s(2)}$, ..., where $1 \leq s(i) \leq n$, $n_{i(r)} > 0$, r=1, 2, ..., n. Here, n is the total number of items, cached by one or more clients and have been updated/requested in the recent interval. Schedule defines the order in which actual data for recently updated/requested items is to be broadcast.

Objective Function: We are interested in minimizing the metrics AAT and TT.
Now we present three scheduling strategies to broadcast update reports.

Algorithm 1 (IOI)

for each recently updated/requested data item d_i with $n_i > 0$
Item[d_i].priority=N-d_i;
select data items in decreasing order of priority from Item[];
add these items into INDEX to prepare update report;

Increasing Order of Items id (IOI) Strategy. In this strategy, the server chooses n data items to be broadcast in the increasing order of their item id i.e. if d_i and d_{i+1} are to be broadcast then d_i is scheduled before d_{i+1}. Algorithm 1 shows the detailed steps.

Broadcast by Random Order (BRO) Strategy. In this strategy, each data item to be broadcast is assigned unique random priority and items are scheduled in decreasing order of priority. Algorithm 2 describes this scheduling strategy.

Largest Count First (LCF) Strategy. This strategy schedules the recently updated/requested items according to their cache count. The data for an item with higher count is broadcast earlier than for an item with lower count. Ties are resolved by scheduling an item with lower id earlier. Details are shown in Algorithm 3.

Algorithm 2 (BRO)

for each recently updated/requested data item d_i with $n_i>0$
        Item[d_i].priority=unique random value not assigned earlier to any item;
select data items in decreasing order of priority from Item[];
        add these items into INDEX to prepare update report;

Algorithm 3 (LCF)

for each recently updated/requested data item d_i with $n_i>0$
        Item[d_i].priority=n_i;
select data items in decreasing order of priority from Item[];
        add these items into INDEX to prepare update report;

An example. We explain an example to demonstrate the above scheduling strategies. Consider 10 items to be broadcast due to invalidation during a particular UR interval as shown in Table 1. For the understanding purpose, we consider AT_i as the time elapsed from the start of DATA broadcast in UR to the time when D_i has been downloaded. It is assumed that it takes a unit time to download actual data for an item.

Table 1. An example

Item d_i	Count n_i	AT_i	IOI		BRO		LCF	
			INDEX	TAT_i	INDEX	TAT_i	INDEX	TAT_i
2	5	1	2	5	9	12	23	25
5	2	2	5	4	7	2	30	32
6	1	3	6	3	21	18	9	36
7	1	4	7	4	23	100	41	36
9	12	5	9	60	5	10	21	30
21	6	6	21	36	2	30	2	30
23	25	7	23	175	41	63	5	14
24	2	8	24	16	6	8	24	16
30	16	9	30	144	30	144	6	9
41	9	10	41	90	24	20	7	10
Total	79			537		407		238

AAT (IOI) = 6.79, AAT (BRO) = 5.15, AAT (LCF) = 3.01

4.4 Prefetching

In mobile environment, data item requested by one client is available for all other clients in the cell, thus prefetching such data item into the cache may improve cache hit ratio and reduce further access latency [24], [25]. However, prefetching the data items may result in the replacement of some recently accessed useful data items thus degrading cache performance. We use a prefetching scheme where the items are preferred for prefetching based on high access probability. A client takes the decision to prefetch a hot (most frequently) item if either it has sufficient empty space or there is an item in the cache that has not been used for a time period longer than some threshold (α). Thus, LRU with time threshold is used as replacement policy to evict an item from the cache and replace it with the item being prefetched. As the proposed UR strategy broadcasts items in nonincreasing order of hotness, a client willing to prefetch k hottest items may download first k items from the broadcast.

In the broadcast based prefetching, the mobile client does not send uplink request, thus reducing the burden on the server while improving cache performance at the mobile client and utilization of scarce wireless bandwidth. With a proper selection of time threshold (α), a performance trade off between energy consumption and access latency may be maintained. Simulation experiments show an improvement in hit ratio due to prefetching.

5 Performance Evaluation

We have performed simulation to evaluate the performance of the proposed schemes. Table 2 shows most of the system parameters. The simulation model consists of a single server per cell serving multiple clients. The database can only be updated by the server, while the queries are generated by the clients following an exponential distribution. The mean inter-arrival time of queries generated by all clients is T_q. The inter-arrival time of updates at the server is distributed exponentially with a mean of T_u. The server periodically broadcasts update reports (UR) every L seconds. The IR field of UR report covers a broadcast window of w broadcast intervals.

In the experiment, we study the average access time (AAT) of the three proposed scheduling strategies for UR broadcasting under different update arrival times. The results are shown in Fig. 2. The AAT falls with the increase of update time as the number of invalidations decreases, hence requiring less downloads. LCF algorithm

Table 2. Simulation parameters

Parameter	Value	Parameter	Value
Server database size (N)	1000 items	UIR replicate times (m-1)	4
Item size (S_{data})	4096 bits	Broadcast window (w)	10 L
Client cache size (C)	30 items	Mean query generate time (T_q)	2 sec
Maximum clients per cell (M)	30	Mean update arrival time (T_u)	1 - 10000 sec
Item id size (S_{id})	32 bits	Uplink bandwidth (B_{up})	19.2 Kbps
Timestamp size (T_{data})	32 bits	Downlink bandwidth (B_{down})	100 Kbps
Broadcast interval (L)	20 sec	Time threshold (α)	30 sec

outperforms the IOI and BRO algorithms because an item with higher cache count is given priority so that most of the client requests are satisfied earlier, thus reducing the AAT. The LCF scheduling is orthogonal to other IR based methods, it can be used in conjunction with any caching schemes where the server aggregates data requests from all its clients and broadcasts the requested data after each IR. We demonstrate the effectiveness of LCF in combination with the UIR scheme [1]. Fig. 3 shows that UIR+LCF scheme reduces the query delay.

To evaluate the effectiveness of UR strategy to bandwidth utilization and conserve client energy, we study the effect of update arrival time over number of uplink requests and tuning time, and compare the results with TS based invalidation strategy. In Fig. 4, the mean number of uplink requests per 100 queries is plotted against the mean update arrival time. At low update arrival time (i.e. high update rate), the number of uplinks is higher because more items are invalidated. The UR scheme performs better than TS scheme as in UR, most of the uplink requests are generated because of cache miss whereas in TS both the cache miss and invalidation cause the uplink requests.

As shown in Fig. 5, the UR scheme which uses selective tuning, performs better under all update rates than TS scheme, and thus, conserves the client energy. Fig. 6 shows how prefetching improves the cache hit ratio.

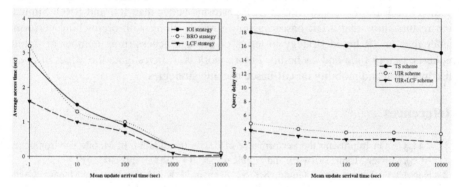

Fig. 2. Average access time as a function of mean update arrival time

Fig. 3. Query delay as a function of mean update arrival time

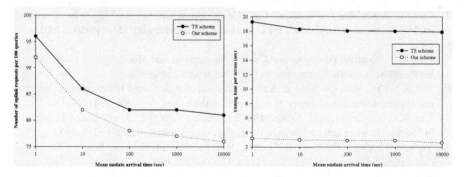

Fig. 4. Uplink requests as a function of mean update arrival time

Fig. 5. Tuning time as a function of mean update arrival time

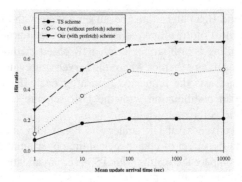

Fig. 6. Cache hit ratio as a function of mean update arrival time

6   Conclusions

This paper considers the cache invalidation and broadcasting in integration and studies the problem of scheduling update reports. Three scheduling schemes have been
studied and it has been shown that LCF performs better than IOI and BRO. Simulation results show that a UR based caching strategy along with prefetching performs
better than an IR based strategy in terms of average access time, number of uplink
requests, tuning time and cache hit. Future work is to investigate the effect of client
disconnection and mobility on UR based caching strategies.

References

1. Cao, G.: On Improving the Performance of Cache Invalidation in Mobile Environments.
   ACM/Kluwer Mobile Network and Applications, 7(4). (2002) 291-303.
2. Kahol, A., Khurana, S., Gupta, S.K.S., Srimani, P.K.: A Strategy to Manage Cache
   Consistency in a Disconnected Distributed Environment. IEEE Transaction on Parallel
   and Distributed Systems, 12(7). (2001) 686-700.
3. Hu, Q., Lee, D.K.: Cache Algorithms Based on Adaptive Invalidation Reports for Mobile
   Environments. Cluster Computing, 1(1). (1998) 39-50.
4. Cao, G.: A Scalable Low-Latency Cache Invalidation Strategy for Mobile Environments.
   ACM International Conference on Computing and Networking (Mobicom), (2001) 200-
   209.
5. Cao, G.: Proactive Power-Aware Cache Management for Mobile Computing Systems.
   IEEE Transactions on Computers, Vol. 51, No. 6 (2002) 608-621.
6. Wu, K.L., Yu, P.S., Chen, M.S.: Energy-Efficient Mobile Cache Invalidation. Distributed
   and Parallel Databases, Kluwer Academic Publishers, Vol. 6. (1998) 351-372.
7. Tan, K.L.: Organisation of Invalidation Reports for Energy-Efficient Cache Invalidation
   in Mobile Environments. Mobile Networks and Applications, 6 (2001) 279-290.
8. Yuen, J.C., Chan, E., Lam, K., Lueng, H.W.: Cache Invalidation Scheme for Mobile
   Computing Systems with Real-Time Data. SIGMOD, (2000) 34-39.
9. Kahol, A., Khurana, S., Gupta, S., Srimani, P.: An Efficient Cache Maintenance Scheme
   for Mobile Environment. 20th Int. Conf. on Distributed Computing Systems, (2000) 530-
   537.

10. Jing, J., Elmagarmid, A., Helal, A., Alonso, R.: Bit-Sequences: An Adaptive Cache Invalidation Method in Mobile Client/Server Environments. Mobile Networks and Applications (1997) 115-127.
11. Yao, J.F., Dunham, M.H.: Caching Management of Mobile DBMS. Journal of Integrated Computer-Aided Engineering, Vol. 8, No. 2 (2001).
12. Hou, W.C., Su, M., Zhang, H., Wang, H.: An Optimal Construction of Invalidation Reports for Mobile Databases. In Proceedings of CIKM, (2001) 458-465.
13. Nam, S.H., Chung, Y., Cho, S.H., Hwang, C.S.: Asynchronous Cache Invalidation Strategy to Support Read-Only Transactions in Mobile Environments. IEICE Trans. Inf. and Syst. Vol. E85-D, No. 2 (2002).
14. Barbara, D., Imielinski, T.: Sleepers and Workaholics: Caching Strategies in Mobile Environments. Proceedings of the ACM SIGMOD Conference on Management of Data, (1994) 1-12.
15. Cao, G.: A Scalable Low-Latency Cache Invalidation Strategy for Mobile Environments. IEEE Transactions on Knowledge and Data Engineering, Vol. 15, No. 5 (2003) 1251-1265.
16. Lee, S.K.: Caching and Concurrency Control in a Wireless Mobile Computing Environment. IEICE Trans. Inf. and Syst. Vol. E85-D, No. 8 (2002).
17. Fong, C.C.F., Lui, J.C.S., Wong, M.H.: Distributed Caching and Broadcast in a Wireless Mobile Computing Environment. The Computer Journal, Vol. 42, No. 6 (1999) 455-472.
18. Cai, J., Tan, K.L.: Energy-Efficient Selective Cache Invalidation. Wireless Networks, (1999) 489-502.
19. Tan, K.L., Cai, J., Ooi, B.C.: An Evaluation of Cache Invalidation Strategies in Wireless Environments. IEEE Transactions on Parallel and Distributed Systems, Vol. 12, No. 8 (2001).
20. Sun, W., Shi, W., Shi, B.: A Cost-Efficient Scheduling Algorithm of On-Demand Broadcasts. Kluwer Wireless Networks, No. 9 (2003) 239-247.
21. Chen, C., Lee, C., Ke, C.: Compression-based Broadcast Strategies in Wireless Information Systems. In Proceedings of the AINA, (2003) 13-18.
22. Kenyon, C., Schabanel, N.: The Data Broadcast Problem with Non-Uniform Transmission Times. Algorithmica, (2003) 146-175.
23. Lai, K.Y., Tari, Z., Bertok, P.: Cost Efficient Broadcast Based Cache Invalidation for Mobile Environments. SAC, (2003) 871-877.
24. Shen, H., Kumar, M., Das, S., Wang, Z.: Energy-Efficient Data Caching and Prefetching of Mobile Devices Based on Utility. ACM/Kluwer Journal of Mobile Networks and Applications (MONET), Special Issue on Mobile Services.
25. Yin, L., Cao, G.: Adaptive Power-Aware Prefetch in Wireless Networks. IEEE Transactions on Wireless Communication (to appear).

Efficient Algorithm for Energy Efficient Broadcasting in Linear Radio Networks

Gautam K. Das, Sandip Das, and Subhas C. Nandy

Indian Statistical Institute, Kolkata - 700 108, India

Abstract. Given a set S of n radio-stations located on a d-dimensional space, a source node $s(\in S)$ and an integer h ($1 \leq h \leq |S| - 1$), the *h-hop broadcast range assignment* problem deals with finding the range assignments for the members in S so that s can communicate with all other members in S in at most h-hops, and the total power consumption is minimum. The problem is known to be NP-hard for $d \geq 2$. We propose an $O(n^2)$ time algorithm for the one dimensional version ($d = 1$) of the problem. This is an improvement over the existing results on this problem by a factor of h [5].

1 Introduction

While designing radio network, several interesting and difficult problems arise due to the shared nature of wireless medium, limited transmission power (range) of wireless devices, node mobility, and battery limitations. Here we consider the problem of assigning transmission ranges to the nodes of the radio-network to minimize power consumption while ensuring broadcasting from a dedicated node (called source) to all other nodes in the network.

A radio-network is a finite set S of *radio-stations* located on a geographical region which can communicate each other by transmitting and receiving radio signals. Each radio-station $s \in S$ is assigned a range $\rho(s)$ (a positive real number) in order to communicate with other stations. A radio-station s can communicate (i.e., send a message) directly (i.e., in *1-hop*) to any other station t, if the Euclidean distance between s and t is less than or equal to $\rho(s)$. If s can not communicate directly with t due to its assigned range, then communication between them can be achieved using *multi-hop* transmissions. If the maximum number of hops allowed (h) is small, then communication between a pair of radio-stations happen very quickly, but the power consumption of the entire radio-network becomes high. On the other hand, if h is large then the power consumption decreases, but communication delay takes place. The tradeoff between the power consumption of the radio-network and the maximum number of hops needed between a communicating pair of radio-stations are studied extensively in [7, 8, 9]. The power $power(s)$ required by a radio station s to transmit a message to another radio-station s' satisfies $\frac{power(s)}{\delta(s,s')^\beta} > \gamma$, where $\delta(s, s')$ is the Euclidean distance between s and s', β is referred as the distance-power gradient and $\gamma(\geq 1)$ is the transmission quality of the message [10]. We assume that

L. Bougé and V.K. Prasanna (Eds.): HiPC 2004, LNCS 3296, pp. 420–429, 2004.
© Springer-Verlag Berlin Heidelberg 2004

$\beta = 2$ and $\gamma = 1$, and so $power(s) = (\rho(s))^2$. The total cost of a range assignment $\mathcal{R} = \{\rho(s) \mid s \in S\}$ is written as $cost(\mathcal{R}) = \sum_{s \in S} power(s) = \sum_{s \in S} (\rho(s))^2$.

The objective of *h-hop broadcast range assignment* problem is to assign transmission ranges $\rho(t)$ to the radio-stations $t \in S$ so that a dedicated radio-station (say $s \in S$) can transmit message to all other radio-stations in the network using at most *h-hops*, and the total power consumption of the entire network is minimum. If $h = 1$, the problem becomes trivial. Here, $\rho(s) = Max_{t \in S} \delta(s, t)$, and $\rho(t) = 0$ for all $t \in S \setminus \{s\}$. For arbitrary h, the problem is known to be NP-hard even in 2D [2, 4]. In [5], the one dimensional version of the problem is considered, and a dynamic programming based algorithm is proposed. It runs in $O(hn^2)$ time. We improve the time complexity result of the 1D version of the problem. Our algorithm is simple and it runs in $O(n^2)$ time and $O(hn)$ space.

In spite of the fact that the model considered in this paper is simple, it is very much useful in studying road traffic information system where the vehicles follow roads and messages are to be broadcasted along lanes. Typically, the curvature of the road is small in comparison to the transmission range so that we can consider that the vehicles are moving on a line [3]. Linear radio networks have been observed to be important in several recent studies [3, 5–8]

2   Structure of Optimal Broadcast Range Assignment

Let us assume that the radio-stations $S = \{s_1, s_2, \ldots, s_n\}$ are ordered on the x-axis from left to right. The radio-station s_1 is positioned at the origin. The position of s_i will be denoted by $x(s_i)$. We will use $\mathcal{R}(S, s, h)$ to denote the optimum *h-hop* broadcast range assignment for broadcasting message from the source station $s(\in S)$ to all other radio-stations in S in at most *h-hops*.

Definition 1. *In a h-hop broadcast range assignment $\mathcal{R}(S, s, h)$ a right-bridge $\overleftarrow{s_\ell s_r}$ corresponds to a pair of radio-stations (s_ℓ, s_r) such that s_ℓ is to the left of s, s_r is to the right of s, and $\delta(s_\ell, s_r) \leq \rho(s_r) < \delta(s_{\ell-1}, s_r)$. In other words, s_r can communicate with s_ℓ in 1-hop due to its assigned range, but it can not communicate with $s_{\ell-1}$.*

Definition 2. *In a h-hop broadcast range assignment $\mathcal{R}(S, s, h)$, a right-bridge $\overleftarrow{s_\ell s_r}$ (if it exists) is said to be functional in $\mathcal{R}(S, s, h)$, if there exists a radio-station $s_i \in S$ such that the minimum number of hops among all paths from s to s_i that avoids the direct (1-hop) communication $\overleftarrow{s_\ell s_r}$, is greater than h.*

Similarly, define a *left-bridge* $\overrightarrow{s_\ell s_r}$ and a *functional left-bridge* in a range assignment $\mathcal{R}(S, s, h)$. Here s_ℓ and s_r are respectively to the left and right of s.

Theorem 1. *[5] Given a set of radio-stations $S = \{s_1, s_2, \ldots, s_n\}$, a source node $s \in S$, and an integer h ($1 \leq h \leq n-1$), the optimal h-hop broadcast range assignment $\mathcal{R}(S, s, h)$ contains at most one functional bridge.*

The dynamic programming based algorithm proposed in [5] solves the problem in three phases. It computes optimal solutions having (i) no functional (left/right) bridge, (ii) one functional left-bridge only, and (iii) one functional right-bridge only. Finally, the one having minimum total power requirement is reported. Our algorithm is based on the same principle as in [5], but it considers the geometry of the range assignment for obtaining the optimal solution, and this leads to an algorithm with improved time complexity.

3   Geometric Properties

Lemma 1. *For a set of radio-stations* $S = \{s_1, s_2, \ldots, s_n\}$, $cost(\mathcal{R}(S, s_1, \mu)) = cost(\mathcal{R}(S, s_n, \mu))$.

Lemma 2. *In the optimum μ-hop broadcast range assignment $\mathcal{R}(S, s_1, \mu)$, if the range assigned to s_1 is $\rho(s_1) = \delta(s_1, s_j)$ for some $j > 1$, then in $\mathcal{R}(S \backslash \{s_1\}, s_2, \mu)$, $\rho(s_2) \geq \delta(s_2, s_j)$.*

Proof. In $\mathcal{R}(S, s_1, \mu)$, $\rho(s_1) = \delta(s_1, s_j)$ implies that $\rho(s_2) = \ldots = \rho(s_{j-1}) = 0$. Thus, if $cost(\mathcal{R}(S, s_1, \mu)) = C$ then $cost(\mathcal{R}(S \setminus \{s_1, \ldots, s_{j-1}\}, s_j, \mu - 1)) = C - (\delta(s_1, s_j))^2$. Now, let us assume that the range assigned to s_2 in $\mathcal{R}(S \backslash \{s_1\}, s_2, \mu)$ is $\rho(s_2) = \delta(s_2, s_k)$. We need to prove $k \geq j$.

Let $cost(\mathcal{R}(S \setminus \{s_1\}, s_2, \mu)) = C'$.

This implies, $cost(\mathcal{R}(S \setminus \{s_1, s_2, \ldots, s_{k-1}\}, s_k, \mu - 1)) = C' - (\delta(s_2, s_k))^2$.

Thus, $\{\delta(s_1, s_k), \mathcal{R}(S \backslash \{s_1, s_2, \ldots, s_{k-1}\}, s_k, \mu - 1)\}$ is a feasible range assignment (may not be optimum) for the μ-hop broadcast from s_1 to all the nodes in $S \backslash \{s_1\}$, and its cost is equal to $C' - (\delta(s_2, s_k))^2 + (\delta(s_1, s_k))^2 \geq C$. This implies $C - C' \leq (\delta(s_1, s_2))^2 + 2\delta(s_1, s_2)\delta(s_2, s_k)$.

By a similar argument, $\{\delta(s_2, s_j), \mathcal{R}(S \backslash \{s_1, s_2, \ldots, s_{j-1}\}, s_j, \mu - 1)\}$ is a feasible range assignment (may not be optimum) for the μ-hop broadcast from s_2 to all the nodes in $S \backslash \{s_1, s_2\}$, and its cost is equal to $C - (\delta(s_1, s_j))^2 + (\delta(s_2, s_j))^2 \geq C'$. This implies, $C - C' \geq (\delta(s_1, s_2))^2 + 2\delta(s_1, s_2)\delta(s_2, s_j)$. Combining these two inequalies, we have $k \geq j$.   □

In the following lemma, we prove that if we increase the number of hops for broadcasting from a fixed vertex, say s_1, to all the vertices to its right, then the gain in the cost obtained in two consecutive steps are monotonically decreasing.

Lemma 3. $cost(\mathcal{R}(S, s_1, \mu)) - cost(\mathcal{R}(S, s_1, \mu + 1)) \geq cost(\mathcal{R}(S, s_1, \mu + 1)) - cost(\mathcal{R}(S, s_1, \mu + 2))$.

Proof. Let $A = \{a_0 = s_1, a_1, a_2, \ldots a_{\mu-1}\}$ denote the (ordered) subset of radio-stations of S having non-zero ranges in $\mathcal{R}(S, s_1, \mu)$. We use a_μ to denote the radio-station s_n, and $cost(A)$ to denote $cost(\mathcal{R}(S, s_1, \mu))$. Here, the range assigned to $a_i \in A$ is $(x(a_{i+1}) - x(a_i))$ for $i = 0, 1, 2, \ldots, \mu - 1$. Again, let $B = \{b_0 = s_1, b_1, b_2, \ldots b_{\mu+1}\}$ denotes the set of radio stations having non-zero ranges in $\mathcal{R}(S, s_1, \mu + 2)$, i.e., $cost(B) = cost(\mathcal{R}(S, s_1, \mu + 2))$. As earlier, s_n is denoted by $b_{\mu+2}$, and the ranges assigned to $b_i (\in B)$ are $(x(b_{i+1}) - x(b_i))$ for $i = 0, 1, 2, \ldots, \mu + 1$. The two range assignments (A and B) are shown in Figure 1(a) using dashed and solid lines.

Observe that, $x(a_0) - x(b_1) < 0$, and $x(a_\mu) - x(b_{\mu+1}) > 0$. This implies, there exists at least one $i \in \{1, 2, \ldots, \mu - 1\}$ such that $x(a_i) - x(b_{i+1}) \leq 0$ and $x(a_{i+1}) - x(b_{i+2}) \geq 0$. We consider the smallest $i \geq 0$ such that $x(a_{i+1}) - x(b_{i+2}) \geq 0$, and construct two (ordered) subsets of radio stations of length $\mu + 1$, namely C and D, where $C = \{a_0 = b_0, a_1, \ldots, a_i, b_{i+2}, b_{i+3}, \ldots, b_{\mu+1}\}$ and $D = \{a_0 = b_0, b_1, b_2, \ldots, b_{i+1}, a_{i+1}, a_{i+2}, \ldots, a_{\mu-1}\}$. The ranges assigned to the members in C and D are respectively

- $\{x(a_1) - x(a_0), \ldots, x(a_i) - x(a_{i-1}), x(b_{i+2}) - x(a_i), x(b_{i+3}) - x(b_{i+2}), \ldots, x(b_{\mu+2}) - x(b_{\mu+1})\}$ (see Figure 1(b)), and
- $\{x(b_1) - x(b_0), \ldots, x(b_{i+1}) - x(b_i), x(a_{i+1}) - x(b_{i+1}), x(a_{i+2}) - x(a_{i+1}), \ldots, x(a_\mu) - x(a_{\mu-1})\}$ (see Figure 1(c)).

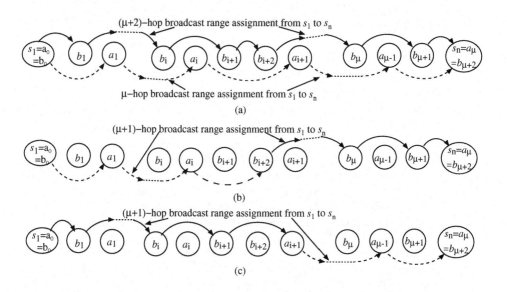

Fig. 1. Proof of Lemma 3

The corresponding costs of the range assignments are

$$cost(C) = \sum_{j=0}^{i-1}(x(a_{j+1}) - x(a_j))^2 + (x(b_{i+2}) - x(a_i))^2 + \sum_{j=i+2}^{\mu+1}(x(b_{j+1}) - x(b_j))^2,$$
$$cost(D) = \sum_{j=0}^{i}(x(b_{j+1}) - x(b_j))^2 + (x(a_{i+1}) - x(b_{i+1}))^2 + \sum_{j=i+1}^{\mu-1}(x(a_{j+1}) - x(a_j))^2.$$

Thus, $cost(C) + cost(D) = cost(A) + cost(B) + 2(x(a_i) - x(b_{i+1}))(x(a_{i+1}) - x(b_{i+2})) \leq cost(A) + cost(B)$ (due to the choice of i as mentioned above).

Let O indicate the optimal choice of the subset of $\mu + 1$ radio-stations having non-zero ranges such that s_1 can send message to s_n in $\mu + 1$ hops and the cost of range assignment is minimum, (i.e., $cost(O) = cost(\mathcal{R}(S, s_1, \mu + 1))$). Thus we have, $2 \times cost(O) \leq cost(C) + cost(D) \leq cost(A) + cost(B)$. □

4   Algorithms

Let $s_\alpha \in S$ be the given source station (not necessarily the left-most/right-most in the ordering of S). Our algorithm for broadcasting from s_α to all other radio-stations $s_j \in S$ consists of three phases. In Phase 1, we prepare four initial matrices. These are used in Phases 2 and 3 for computing optimal solution with no functional bridge, and exactly one functional bridge.

Lemma 4. *If in a linear ordered set of radio-stations* $\{s_a, s_{a+1}, \ldots, s_b\}$*, the source station is at* s_a *(one end of the above set), then for any* $1 \leq m < b - a - 1$*, the optimum* m*-hop broadcast range assignment* $\mathcal{R}(\{s_a, s_{a+1}, \ldots, s_b\}, s_a, m)$ *should satisfy* $\sum_{k=a}^{b-1} \rho(s_k) = x(s_b) - x(s_a)$.

For notational convenience, if the source radio-station (say s_a) is at one end of the linearly ordered destination stations, then we use $R(s_b, s_a, \mu)$ to denote the optimal range assignment $\mathcal{R}(\{s_a, s_{a+1}, \ldots, s_b\}, s_a, \mu)$.

4.1   Phase 1

In this phase, we prepare the following four initial matrices. These will be extensively used in Phases 2 and 3. Recall that s_α is the source station.

M_1: It is a $h \times \alpha - 1$ matrix. Its (m, j)-th element $(1 \leq j < \alpha)$ indicates the optimum cost of sending message from s_j to s_α using m hops. In other words, $M_1[m, j] = cost(R(s_\alpha, s_j, m))$, where $1 \leq m \leq h$ and $1 \leq j < \alpha$.

M_2: It is a $h \times \alpha - 1$ matrix. Its (m, j)-th element $(1 < j \leq \alpha)$ indicates the optimum cost of sending message from s_j to s_1 using m hops. In other words, $M_2[m, j] = cost(R(s_1, s_j, m))$, where $1 \leq m \leq h$ and $1 < j \leq \alpha$.

M_3: It is a $h \times n - \alpha$ matrix. Its (m, j)-th element $(\alpha < j \leq n)$ indicates the optimum cost of sending message from s_j to s_α using m hops. In other words, $M_3[m, j] = cost(R(s_\alpha, s_j, m))$, where $1 \leq m \leq h$ and $\alpha < j \leq n$.

M_4: It is a $h \times n - \alpha$ matrix. Its (m, j)-th element $(\alpha \leq j < n)$ indicates the optimum cost of sending message from s_j to s_n using m hops. In other words, $M_4[m, j] = cost(R(s_n, s_j, m))$, where $1 \leq m \leq h$ and $\alpha \leq j < n$.

Note that, the columns of M_1 are indexed as $[1, 2, \ldots, \alpha - 1]$, whereas those in M_2 are indexed as $[2, 3, \ldots, \alpha]$. Similarly, the columns of M_3 are indexed as $[\alpha + 1, \alpha + 2, \ldots, n]$, whereas those in M_4 are indexed as $[\alpha, \alpha + 1, \ldots, n - 1]$. We use incremental approach (in terms of hops) for constructing the matrices. Let us consider the construction of the matrix M_1. Similar procedures work for constructing the other three matrices.

Each entry of the matrix M_1 contains a tuple (χ, ptr); the χ field contains $cost(R(s_\alpha, s_j, m))$, and its ptr field is an integer which contains the index of the first radio-station (after s_j) on the m-hop path from s_j to s_α. We interchangably use $M_1[m, j]$ and $M_1[m, j].\chi$ to denote $cost(R(s_\alpha, s_j, m))$. After computing up to row m of the matrix M_1, the elements in the $m + 1$-th row can easily be obtained as follows:

Consider an intermediate matrix $A = ((A[j, k]))$ where $A[j, k] =$ the cost of $(m + 1)$-hop communication from s_j to s_α with first hop at $s_k = (\delta(s_j, s_k))^2 +$

$M_1[m, k].\chi$. Thus, $M_1[m + 1, j].\chi = Min_{k=j+1}^{\alpha-1} A[j, k]$, and $M_1[m + 1, j].ptr$ will contain the value of k for which $A[j, k]$ is contributed to $M_1[m + 1, j].\chi$.

Straight forward application of the above method needs $O(n^2)$ time. But, Lemma 1 says that, if in the optimum $(m + 1)$-hop path from s_j to s_α first hops at node s_k, then for any node $s_{j'}$ with $j' > j$, the optimum $(m + 1)$-hop path from s'_j to s_α first hops at node $s_{k'}$ with $k' \geq k$. A simple method for computing the minimum of every row in the matrix A works in $O(n\log n)$ time as follows:

Compute all the entries in the $\frac{\alpha}{2}$-th row of the matrix A, and find the minimum. Let it corresponds to $A[\frac{\alpha}{2}, \beta]$. Next, compute the minimum entry in the $\frac{\alpha}{4}$-th row of A by considering $\{A[\frac{\alpha}{4}, j], j = 1, 2, \ldots, \beta\}$, and compute the minimum entry in the $\frac{3\alpha}{4}$-th row of A by considering $\{A[\frac{3\alpha}{4}, j], j = \beta, \beta+1, \ldots, \alpha-1\}$. The process continues until all the rows of A are considered.

We now describe an efficient method of computing the minimum entry in each row of matrix A using *monotone matrix searching* [1].

Definition 3. *[1] A matrix M is said to be* monotone *if for every j, k, j', k' with $j < j'$, $k < k'$, if $M[j, k] \geq M[j, k']$ then $M[j', k] \geq M[j', k']$.*

Lemma 5. *The matrix A is a monotone matrix.*

Proof. Given $A[j, k] \geq A[j, k']$, where $A[j, k] = (\delta(s_j, s_k))^2 + M_1[m, k]$ and $A[j, k'] = (\delta(s_j, s_{k'}))^2 + M_1[m, k']$. Thus, $M_1[m, k] - M_1[m, k'] \geq (\delta(s_j, s_k))^2 - (\delta(s_j, s_{k'}))^2$. Now, $A[j', k] - A[j', k'] = (\delta(s_{j'}, s_k))^2 - (\delta(s_{j'}, s_{k'}))^2 + M_1[m, k] - M_1[m, k'] \geq (\delta(s_{j'}, s_k))^2 - (\delta(s_{j'}, s_{k'}))^2 + (\delta(s_j, s_{k'}))^2 - (\delta(s_j, s_k))^2 \geq 0$ (on simplification). ∎

A recursive algorithm is described in [1]. It can compute the minimum entry in each row of a $\alpha \times \alpha$ monotone matrix in $O(\alpha)$ time provided each entry of the matrix can be computed in $O(1)$ time. Using that algorithm, the matrix M_1 can be computed in $O(\alpha \times h)$ time.

Lemma 6. *Phase 1 needs $O(nh)$ time.*

Proof. Follows from the fact that M_1, M_2 can be constructed in $O(\alpha \times h)$ time, and M_3, M_4 needs $O((n - \alpha) \times h)$ time. ∎

4.2   Phase 2

In this phase, we compute the optimal functional bridge-free solution for broadcasting message from s_α to the other nodes in S. Note that, the range to be assigned to s_α is at least $Max(\delta(s_\alpha, s_{\alpha-1}), \delta(s_\alpha, s_{\alpha+1}))$.

Without loss of generality, assume that $\delta(s_\alpha, s_{\alpha-1}) \leq \delta(s_\alpha, s_{\alpha+1})$. Thus, $\rho(s_\alpha)$ is initially assigned to $\delta(s_\alpha, s_{\alpha+1})$, and let s_k $(k < \alpha)$ be the farthest radio-station such that s_α can communicate with s_k in 1-hop (i.e., $\delta(s_k, s_\alpha) \leq \delta(s_\alpha, s_{\alpha+1}) < \delta(s_{k-1}, s_\alpha))$. If we use $\mathcal{R}(S, s_\alpha, h | \rho(s_\alpha) = d)$ to denote the optimum range assignment for the h-hop broadcasting from s_α to all the nodes in S subject to the condition that the range assigned to s_α is d, then

$\mathcal{R}(S, s_\alpha, h|\rho(s_\alpha) = \delta(s_\alpha, s_{\alpha+1})) =$
$\{\mathcal{R}(\{s_1, \ldots, s_k\}, s_k, h - 1), \delta(s_\alpha, s_{\alpha+1}), \mathcal{R}(S \setminus \{s_1, \ldots, s_\alpha\}, s_{\alpha+1}, h - 1)\},$
$= \{R(s_1, s_k, h - 1), \delta(s_\alpha, s_{\alpha+1}), R(s_n, s_{\alpha+1}, h - 1)\}$
and its cost is $C = cost(\mathcal{R}(S, s_\alpha, h|\rho(s_\alpha) = \delta(s_\alpha, s_{\alpha+1}))$
$= (\delta(s_\alpha, s_{\alpha+1}))^2 + M_2[h - 1, k] + M_4[h - 1, \alpha + 1].$

This can be computed in $O(1)$ time using the matrices M_2 and M_4.

We use two temporary variables TEMP_Cost and TEMP_id to store C and $s_{\alpha+1}$.

Next, we increment $\rho(s_\alpha)$ to $Min(\delta(s_\alpha, s_{k-1}), \delta(s_\alpha, s_{\alpha+2}))$, and apply same procedure to calculate the optimum cost of the h-hop broadcast from s_α under this circumstance. This may cause updating of TEMP_Cost and TEMP_id. The same procedure is repeated by incrementing $\rho(s_\alpha)$ to its next choice in the set $\{\delta(s_\alpha, s_k), k = 1, 2, \ldots, \alpha - 1\} \bigcup \{\delta(s_\alpha, s_j), j = k + 1, \ldots, n\}$ so that it can communicate direcetly with one more node than its previous choice. At each step, the TEMP_Cost and TEMP_id are adequately updated. Thus, the procedure is repeated for $O(n)$ times, and the time complexity of this phase is $O(n)$.

4.3   Phase 3

In this phase, we compute an optimal range assignment for the h-hop broadcasting from s_α to all other nodes in S where the solution contains a functional *right-bridge only*. Similar method is adopted to compute the optimal solution with one functional *left-bridge only*. The one having minimum cost is chosen as the optimal solution obtained in this phase.

Let us consider a range assignment which includes a right-bridge $\overleftarrow{s_i s_j}$, $i < \alpha < j$. It is realized in the following ways:

Scheme 1: Assign $\rho(s_j) = \delta(s_j, s_i)$.
Scheme 2: If $\delta(s_j, s_k) < \delta(s_j, s_i) < \delta(s_j, s_{k+1})$ and $\delta(s_j, s_{k+1}) < \delta(s_j, s_{i-1})$ then $\rho(s_j) = \delta(s_j, s_{k+1})$.

We assume that s_j is reached from s_α using m hops. Thus, using Scheme 1, h-hops connection from s_α to all the nodes in S is achieved by (i) reaching s_1 from s_i in $(h - m - 1)$ hops, and (ii) reaching s_n from s_k in $(h - m - 1)$ hops. Here the cost of range assignment is $B_1 = cost(R(s_j, s_\alpha, m)) + (\delta(s_i, s_j))^2 + cost(R(s_1, s_i, h - m - 1)) + cost(R(s_n, s_k, h - m - 1))$.

In Scheme 2, as s_j can directly communicate to s_{k+1} to the right and s_i to the left, the h-hop connection from s_α to all the nodes in S is established by (i) reaching s_1 from s_i in $(h - m - 1)$ hops, and (ii) reaching s_n from s_{k+1} in $(h - m - 1)$ hops. Here the cost of range assignment is $B_2 = cost(R(s_j, s_\alpha, m)) + (\delta(s_j, s_{k+1}))^2 + cost(R(s_1, s_i, h - m - 1)) + cost(R(s_n, s_{k+1}, h - m - 1))$.

Denoting by $B(\overleftarrow{s_i s_j}, m)$ the cost of range assignment with a right bridge $\overleftarrow{s_i s_j}$ where s_j is reached from s_α in m hops, we have $B(\overleftarrow{s_i s_j}, m) = Min(B_1, B_2)$.

Apart from identifying s_k, $B(\overleftarrow{s_i s_j}, m)$ can be calculated in $O(1)$ time, because $cost(R(s_j, s_\alpha, m)) = cost(R(s_\alpha, s_j, m)) = M_3[m, j]$ (by Lemma 4),

$cost(R(s_1, s_i, h - m - 1)) = M_2[h - m - 1, i],$
$cost(R(s_n, s_k, h - m - 1)) = M_4[h - m - 1, k],$

and all these matrices are already available. In order to get an optimal solution with a right-bridge, we need to find $Min_{i=1}^{\alpha-1} Min_{j=\alpha+1}^{n} Min_{m=1}^{h} B(\overleftarrow{s_i s_j}, m)$.

In our algorithm, we fix each s_i, and compute $Min_{j=\alpha+1}^{n} Min_{m=1}^{h} B(\overleftarrow{s_i s_j}, m)$ with the help of the following lemma.

Lemma 7. *Let $s_j \in S_\alpha = \{s_{\alpha+1}, s_{\alpha+2}, \dots, s_n\}$ and $R(s_j, s_\alpha, \mu)$ indicates the optimal range assignment for sending message from s_α to s_j using μ hops, then $cost(R(s_j, s_\alpha, \mu-1)) - cost(R(s_j, s_\alpha, \mu)) \le cost(R(s_{j+1}, s_\alpha, \mu-1)) - cost(R(s_{j+1}, s_\alpha, \mu)).$*

Proof. Similar to the proof of Lemma 3.   □

Lemma 8. *While using the bridge $\overleftarrow{s_i s_j}$, $i < \alpha < j$, if $B(\overleftarrow{s_i s_j}, \mu) \le B(\overleftarrow{s_i s_j}, \mu+1)$ then $B(\overleftarrow{s_i s_j}, \mu + 1) \le B(\overleftarrow{s_i s_j}, \mu + 2).$*

Proof. The reduction in cost for increasing the number of hops from μ to $\mu+1$ to reach from s_α to s_j is $a_1 = cost(R(s_j, s_\alpha, \mu)) - cost(R(s_j, s_\alpha, \mu + 1)) \ge 0$. In order to maintain h-hop reachability from s_α to s_1 and s_n, we need to reach both from s_i to s_1 and from s_k to s_n using at most $h - \mu - 2$ hops instead of $h - \mu - 1$ hops. Thus, the amount of increase in the corresponding costs are $a_2 = cost(R(s_1, s_i, h - \mu - 2)) - cost(R(s_1, s_i, h - \mu - 1)) \ge 0$ and $a_3 = cost(R(s_n, s_k, h - \mu - 2)) - cost(R(s_n, s_k, h - \mu - 1)) \ge 0$.

As stated in the lemma, $B(\overleftarrow{s_i s_j}, \mu) - B(\overleftarrow{s_i s_j}, \mu+1) \le 0$ implies $a_1 - a_2 - a_3 \le 0$.

Now, the gain in cost for increasing the number of hops from $\mu + 1$ to $\mu + 2$ to reach from s_α to s_j is $a_1' = cost(R(s_j, s_\alpha, \mu + 1)) - cost(R(s_j, s_\alpha, \mu + 2)) \ge 0$ This causes the reduction in number of hops to reach from s_i to s_1 and s_k to s_n from $h - \mu - 2$ to $h - \mu - 3$. The loss in the corresponding costs are $a_2' = cost(R(s_1, s_i, h - \mu - 3)) - cost(R(s_1, s_i, h - \mu - 2)) \ge 0$ and $a_3' = cost(R(s_n, s_k, h - \mu - 3)) - cost(R(s_n, s_k, h - \mu - 2)) \ge 0$.

By Lemma 3, $a_1' \le a_1$, $a_2' \ge a_2$ and $a_3' \ge a_3$.

Thus, $B(\overleftarrow{s_i s_j}, \mu + 1) - B(\overleftarrow{s_i s_j}, \mu + 2) = a_1' - a_2' - a_3' \le a_1 - a_2 - a_3 \le 0.$ □

Lemma 8 implies that while using the right-bridge $\overleftarrow{s_i s_j}$, we vary the number of hops m to reach from s_α to s_j, and compute the corresponding cost $B(\overleftarrow{s_i s_j}, m)$. As soon as $B(\overleftarrow{s_i s_j}, \mu) < B(\overleftarrow{s_i s_j}, \mu + 1)$ is observed, there is no need to check the costs by increasing m beyond $\mu + 1$.

After computing the optimum range assignment with the right-bridge $\overleftarrow{s_i s_j}$, we proceed to compute the same with right-bridge $\overleftarrow{s_i s_{j+1}}$. The following lemma says that if the optimum $B(\overleftarrow{s_i s_j}, m)$ is achieved for $m = \mu$ then while considering the right-bridge $\overleftarrow{s_i s_{j+1}}$, the optimum $B(\overleftarrow{s_i s_{j+1}}, m)$ will be achieved for some $m \ge \mu$. It needs to mention that, we could not explore any relationship among the optimum costs of range assignments using $\overleftarrow{s_i s_j}$ and $\overleftarrow{s_i s_{j+1}}$.

Lemma 9. *For a given $s_i \in S$, $i < \alpha$, if $Min_{m=1}^{h} B(\overleftarrow{s_i s_j}, m)$ and $Min_{m=1}^{h} B(\overleftarrow{s_i s_{j+1}}, m)$ are achieved for $m = \mu$ and ν respectively, then $\nu \ge \mu$.*

Proof. As s_i is fixed, we compute optimal range assignment $R(s_1, s_i, h - m - 1)$ to reach from s_i to s_1.

While using $\overleftarrow{s_i s_j}$, $\rho(s_j) = \delta(s_j, s_i)$, and this enables s_j to reach s_k to its right (i.e. $\delta(s_j, s_i) \geq \delta(s_j, s_k)$). Similarly, while using $\overleftarrow{s_i s_{j+1}}$, $\rho(s_{j+1}) = \delta(s_{j+1}, s_i)$, and this enables s_{j+1} to reach s_ℓ to its right (i.e. $\delta(s_{j+1}, s_i) \geq \delta(s_{j+1}, s_\ell)$). Here $k \leq \ell$. In order to prove the lemma, we need only to show that $B(\overleftarrow{s_i s_{j+1}}, \mu - 1) \geq B(\overleftarrow{s_i s_{j+1}}, \mu)$.

By Lemma 8, this will automatically imply $B(\overleftarrow{s_i s_j}, m - 1) \geq B(\overleftarrow{s_i s_j}, m)$ for all $m \leq \mu$. Thus, if $Min(B(\overleftarrow{s_i s_{j+1}}, m))$ is achieved for $m = \nu$, then $\nu \geq \mu$. To prove the above inequality, let

$a_1 = cost(R(s_j, s_\alpha, \mu - 1)) - cost(R(s_j, s_\alpha, \mu))$,
$a_1' = cost(R(s_\alpha, s_{j+1}, \mu - 1)) - cost(R(s_\alpha, s_{j+1}, \mu))$,
$a_2 = cost(R(s_1, s_i, h - \mu - 1)) - cost(R(s_1, s_i, h - \mu))$,
$a_3 = cost(R(s_n, s_k, h - \mu - 1)) - cost(R(s_n, s_k, h - \mu))$ and
$a_3' = cost(R(s_n, s_\ell, h - \mu - 1)) - cost(R(s_n, s_\ell, h - \mu))$.

As $B(\overleftarrow{s_i s_j}, \mu - 1) > B(\overleftarrow{s_i s_j}, \mu)$, we have $a_1 - a_2 - a_3 > 0$. By Lemma 7, $a_1' \geq a_1$ and $a_3' \leq a_3$. Now, $B(\overleftarrow{s_i s_{j+1}}, \mu - 1) - B(\overleftarrow{s_i s_{j+1}}, \mu) =$ the amount of gain in cost for increasing the number of hops from $\mu - 1$ to μ to reach s_α to s_{j+1} while using the bridge $\overleftarrow{s_i s_{j+1}} = a_1' - a_2 - a_3' \geq a_1 - a_2 - a_3 > 0$.   □

Given a source-station s_α and another station s_i ($i < \alpha$), the optimal range assignment of the members in S consisting of a functional right-bridge incident at s_i, can be computed using the following algorithm:

Algorithm Range_Assign_Using_Right_Bridge(s_i)

Step 1: We initialize $\mu = 1$, $OPT\_j = \alpha$, $OPT\_cost = \infty$ and $k\_store = \alpha$; Next we start the execution with $m = 1$ and $j = \alpha + 1$. (* The role of k_store will be clear in the procedure **compute** invoked from this algorithm. *)

Step 2: At each j, we execute **compute**($B(\overleftarrow{s_i s_j}, m), k\_store$) by incrementing m from its current value upwards until
  (i) $B(\overleftarrow{s_i s_j}, m) > B(\overleftarrow{s_i s_j}, m - 1)$ is achieved (see Lemma 8) or
  (ii) m attains its maximum allowable value $Min(h - 2, j - \alpha)$.

Step 3: Update OPT_cost and OPT_j observing the value of $B(\overleftarrow{s_i s_j}, m - 1)$ or $B(\overleftarrow{s_i s_j}, m)$ depending on whether Step 2 has terminated depending on Case (i) or Case (ii).

Step 4: For the next choice of j, update μ by $m - 1$ or m depending on whether Case (i) or (ii) occured in Step 2 (see Lemma 9).

Procedure **compute**($B(\overleftarrow{s_i s_j}, m), k\_store$)

- Initialize $k = k\_store$.
- Increment k to identify the rightmost radio-station such that $\delta(s_j, s_k) \leq \rho(s_j)(= \delta(s_j, s_i))$.
- Set $k\_store = k$ for further use.
- Compute $B(\overleftarrow{s_i s_j}, m) = (\rho(s_j))^2 + R(s_j, s_\alpha, m) + R(s_1, s_i, h - m - 1) + R(s_n, s_k, h - m - 1)$; the last three terms are available in $M_3[m, j]$, $M_2[h - m - 1, i]$ and $M_4[h - m - 1, k]$ respectively.

Theorem 2. *For a given s_i ($i < \alpha$), algorithm Range_Assign_using_Right_Bridge needs $O(n - \alpha + h)$ time in the worst case.*

Proof. Follows from Lemmata 8 and 9, and the role of k_store in the procedure **compute** for locating rightmost s_k such that $\delta(s_j, s_k) \geq \rho(s_j)$.   □

4.4   Complexity Analysis

Theorem 3. *Given a set of radio stations S and a source station $s_\alpha \in S$, the optimum range assignment for broadcasting message from s_α to all the members in S using at most h-hops can be computed in $O(n^2)$ time and using $O(nh)$ space.*

Proof. Phase 1 and Phase 2 can be executed in $O(nh)$ and $O(n)$ respectively. Finally in Phase 3, we fix s_i to left of s_α and identify the optimum solution with a functional right-bridge incident at s_i in $O(n - \alpha + h)$ time (see Theorem 2). For $(\alpha - 1)$ such s_i's, the total time required in this phase is $O(\alpha \times (n - \alpha + h))$. Similarly, the worst case time required for finding the optimum range assignment with one functional left-bridge is $O((n - \alpha) \times (\alpha + h))$. Thus, the result follows.

□

References

1. A. Aggarwal and M. Klawe, *Applications of generalized matrix searching to geometric algorithms*, Discrete Applied Mathematics, vol. 27, pp. 3-23, 1990.
2. A.E.F. Clementi, P. Crescenzi, P. Penna, P. Rossi, P. Vocca, *On the Complexity of Computing Minimum Energy Broadcast Subgraph*, 18th Annual Symp. on Theoretical Aspects of Computer Science(STACK'01), Lecture Notes in Computer Science, vol. 1770, pp. 651-660, 2000.
3. A. E. F. Clementi, A. Ferreira, P. Penna, S. Perennes, R. Silvestri, *The minimum range assignment problem on linear radio networks*, Algorithmica, vol. 35, pp. 95-110, 2003.
4. M. Zagalj, J. P. Hubaux, C. Enz, *Minimum-Energy broadcast in All-Wireless Networks : NP-Completeness and Distribution Issues*, MOBICOM, pp. 172-182, 2002.
5. A.E.F. Clementi, M Di Ianni, R. Silvestry, *The Minimum Broadcast Range Assignment Problem on Linear Multi-Hop wireless Networks*, Theoretical Computer Science(TCS), vol. 299, pp. 751-761, 2003.
6. C. Gaibisso, G. Proietti, R. Tan, *Efficient Management of Transient Station Failures in Linear Radio Communication Networks with Bases*, Proc. 2nd Internatonal Workshop on Approximation and Randomized Algorithms in Communication Networks (ARACNE), Carleton Scientific, pp. 37-54, 2001.
7. L. Kirousis, E. Kranakis, D. Krizanc and A. Pelc, *Power consumption in packet radio networks*, Theoretical Computer Science, vol. 243, pp. 289-305, 2000.
8. R. Mathar and J. Mattfeldt, *Optimal transmission ranges for mobile communication in linear multihop packet radio networks*, Wireless Networks, vol. 2, pp. 329-342, 1996.
9. P. Piret, *On the connectivity of radio networks*, IEEE Trans. on Information Theory, vol. 37, pp. 1490-1492, 1991.
10. K. Pahlavan, A. Levesque, *Wireless Information Networks*, John Wiley, New York, 1995.

Characterization of OpenMP Applications on the InfiniBand-Based Distributed Virtual Shared Memory System*

Inho Park[1], Seon Wook Kim[1], and Kyung Park[2]

[1] Department of ECE, Korea University, Seoul, Korea
[2] ETRI, Taejeon, Korea

Abstract. For the past years, architectures and programming models about distributed virtual shared-memory (DVSM) systems have been extensively studied. The DVSM needs communication between distributed processing nodes in order to maintain memory consistency, therefore the communication-related overhead determines the overall performance. Recently many advanced hardware-based interconnection technologies have been introduced, and one of them is the InfiniBand Architecture (IBA) which supports shared-memory programming semantics by means of remote direct-memory access (RDMA) and atomic operations. In this paper, we describe the implementation of our InfiniBand-based DVSM system, and evaluate its performance using SPEC OMP benchmarks. We show that our DVSM system to use full features of the IBA can improve the performance significantly over the IPoIB-based traditional system on the IBA, and furthermore the performance of one application on the IBA-based DVSM system is better than on the hardware-based shared-memory system.

1 Introduction

For the past years, in order to build high-performance large-scale computer systems we have connected distributed processing nodes through the high speed interconnection network instead of building shared-memory multiprocessor (SMP) systems. One of the popularly used distributed-memory multiprocessor systems is a cluster due to easy maintainability, expandability, and scalability at low cost [1]. But it is difficult for a programmer to parallelize sequential programs on the distributed-memory system, since he must know when to communicate, where data exist, and who receives or sends data.

In order to provide easy programming environment on the distributed-memory system, a distributed virtual shared-memory (DVSM) system has been intro-

---

* This work was supported in part by the Ministry of Information & Communications, Korea, under the Information Technology Research Center (ITRC) Support Program, and by the University Research Program by the Electronics and Telecommunications Research Institute, Taejeon, Korea.

L. Bougé and V.K. Prasanna (Eds.): HiPC 2004, LNCS 3296, pp. 430–439, 2004.
© Springer-Verlag Berlin Heidelberg 2004

duced [2–4]. The DVSM system allows processes to access physically distributed memory spaces through one virtual shared memory space, therefor a programmer is able to use the shared-memory parallel programming APIs on the system, such as OpenMP [5] and PThread. However, the performance of applications on the DVSM system, especially when executing parallel applications, heavily depends on the network performance and network programming semantics. For example, a coherence mechanism to support memory consistency through networks occurs many kinds of overheads, such as interrupting communicating processes, sending the most recent copies, communicating and maintaining the execution orders, and so on. In order to overcome the performance limitation, various versions of DVSM systems and code optimization techniques have been proposed. [2, 3, 4, 6, 7, 8, 9].

Recently many advanced interconnection technologies to connect multiple processing nodes have been proposed to support Gbit bandwidth, such as AM, VMMC, FM, U-Net, LAPI, and so on [10]. These techniques are usefully used to build high-performance distributed systems due to low communication latency. One of the proposed architectures is the InfiniBand Architecture [11] that implements in hardware legacy software protocol tasks to support Remote Direct Memory Access (RDMA) and atomic operations. These functionalities can remove many DVSM-related overheads because a process is able to receive and send data without interrupting other processes. The InfiniBand standard is proposed to overcome the PCI bus speed. The InfiniBand Architecture (IBA) consists of processing nodes, I/O nodes, and System Area Network (SAN) to connect nodes. InfiniBand supports switched fabric to guarantee high stability and wide bandwidth. IBA can be used to construct from a small size cluster to a parallel supercomputer. The detailed description is shown in [11].

There are a few research activities about programming environment on top of the InfiniBand. The research in [12] describes the implementation of the OS-layered DVSM on the IBA. This approach needs to modify kernels in order to integrate the InfiniBand primitives with OS kernels. Our approach is the application-layered InfiniBand-based DVSM, and therefore we do not need any modification inside kernels. And while our framework uses a lazy release consistency, the DVSM in [12] uses a sequential consistency. Also the InfiniBand-based MPI implementation has been done extensively [13].

In this paper, we describe briefly the three variants of implementation (a base, an ownership assignment, and a page prefetching) for the IBA-based DVSM system based on the lazy release consistency (LRC) model. And we evaluate their performance by using realistic SPEC OMP benchmarks. We show that our DVSM to use full features of the IBA (to use Verbs API) can improve the performance significantly over using the IPoIB protocol stack-based DVSM system on the IBA. Furthermore, the performance of one application on our DVSM system is higher than the hardware SMP machine.

2   InfiniBand-Based Distributed Virtual Shared-Memory Systems

In this section, we briefly describe three implementations of our page-based DVSM systems on the InfiniBand [14]. One is a basic implementation which does not include any optimization scheme, i.e. a minimal implementation to preserve memory consistency. Another implementation assigns an ownership to each page in order to reduce page segmentation violation overhead, and the third implementation uses a page prefetching technique at serial and parallel section boundaries to hide page copy latency. All implementations are based on the lazy release consistency model [2].

2.1   Basic Implementation

In the traditional implementation of the DVSM system to use socket programming a process should talk with other processes in order to send and receive the data through process interruption. But on the InfiniBand we use RDMA Infini-Band primitives without interrupting processes. Figure 1 explains the difference between two schemes.

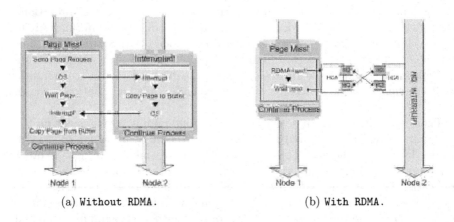

(a) Without RDMA.   (b) With RDMA.

Fig. 1. The difference to access data on remote nodes

In order to record read and write operations per page, we maintain an interval table whose entry has two bits to mark read and write operations and whose number of entries is equal to the number of pages in globally allocated virtual shared-memory. When a page segmentation violation occurs, the related bit is set. The interval table is cleared after all necessary diff operations are performed to other processes.

Each synchronization includes two internal barriers, b_0 and b_1. At the first barrier (b_0) each process broadcasts the interval table to all the participant processes. Since the size of the interval table is small (only 6.4KB for 100MB virtually allocated shared-memory) and we use RDMA write operations (i.e. does not need to wait completion message), the related communication overhead is negligible. After the first barrier, the diff operations are applied to other processes if necessary by considering all the other processes' interval tables. After all the diff operations are performed, a page to have the most recent copy is write-memory protected. If a page has an old copy, the page is both read and write memory-protected for receiving the most recent copy from other processes at the first read and write operations.

2.2   Ownership Assignment

It is well known that the overhead associated with the page segmentation violation is very large, and it degrades the performance seriously. In order to reduce the occurrence of segmentation violations, we assign an ownership to each page. The basic idea is that if a page is completely localized to one process, then the page does not need to be communicated with other processes.

The page owner information is maintained inside the interval table by extending one bit to mark a page ownership. By using the owner information, page owners are not memory-protected. The page owner is write memory-protected after only other processes perform read and write operations. If this situation happens one more time, then the page ownership is assigned to the other process at the next synchronization. The next new owner is determined by each process locally.

2.3   Page Prefetching

We consider the history about page movement across processes between serial and parallel sections in OpenMP execution. There are two region boundaries: from a serial to a parallel, and from a parallel to a serial sections. We maintain fork and join prefetch tables per parallel region by using the start address of the parallel region as a reference. The fork prefetch table records the page segmentation violation due to data movement from a serial to a parallel regions, and the joint prefetch table does from a parallel to a serial regions. Our DVSM system examines these two tables at each execution boundary, and performs diff operations ahead of page segmentation faults, i.e. page prefetch.

3   Performance Evaluation

3.1   Experiment Methods

For performance evaluation of our IBA-based DVSM system, we used four benchmarks (swim, mgrid, wupwise, and applu) from SPEC OMP3.51 suite [15]. We *did not* optimize any code for performance improvement from the officially delivered benchmarks from SPEC, but we scaled down the number of iterations in SPEC OMP for short execution (it does not change any program characteristics).

We implemented the DVSM system on the InfiniBand to use Verbs API in order to use full features of the InfiniBand semantics. We used four processing nodes connected with the 4X Mellanox InfiniBand, and each node includes a Intel 2.0GHz Xeon processor, 512MB memory, and 133MHz 4X PCI-X 128MB memory HCA board. Also for comparing with the performance on the hardware SMP machine, we measured the SPEC applications on the bus-based SMP to use four 2.8GHz Xeon processors, 512MB memory.

There are the following five measurements: Execution on the hardware-based SMP machines (SMP), TreadMarks execution to use IPoIB protocol stack on the InfiniBand (TMK), execution on the basic implementation to use the Verbs on the InfiniBand (IBA-B), execution to use ownership assignment with IBA-B (IBA-O), and execution to use page prefetching with IBA-O (IBA-P). We used TreadMarks DVSM for performance comparison, since it also uses the lazy release consistency model [2].

3.2   Performance

Overall. Figure 2 shows the speedup of the OpenMP benchmarks on 2 and 4 processors with respect to one processor execution. In applu and swim there is no speedup in the DVSM executions. In mgrid there is speedup at the execution of our DVSM system, but less than the SMP execution. In wupwise, the speedup of our InfiniBand-based DVSM system is higher than the SMP execution. The figure shows that there is no speedup in the TreadMarks execution, and the result is different from that in [6]. The reason of the performance difference is that our processor's performance is much higher than those in [6], so the communication overhead dominates the overall execution time. The ownership assignment implementation performs well in most applications, but the page prefetching technique does not work well in mgrid and wupwise.

Figure 3 shows the memory size in the RDMA transactions on the IBA-based DVSM system on P0 in 4 process execution. In each application about 100MB virtual shared-memory is used in each HCA. In applu the ownership assignment increases the page read transactions, but decreases the diff transactions. It implies that the page ownership is assigned correctly, but many pages are shared across processes. The page prefetching technique increases the diff transactions significantly without increasing diffs and performance gain in mgrid. It implies that the page ownership is assigned correctly, but the program behavior is changing and the history is ineffective.

Applu. Figure 4 shows the overhead analysis in applu execution. The barrier overhead in the TreadMarks execution is considered only as BAR1 in the legend. The execution in the application itself is marked as COMP. In the SMP execution, it is impossible to distinguish operations in the legend, so we consider as COMP only.

The barrier overhead on the TreadMarks is significantly reduced on the InfiniBand-based approach. But the second barrier is still large, and there is no performance difference in three variants of the IBA-based DVSM system.

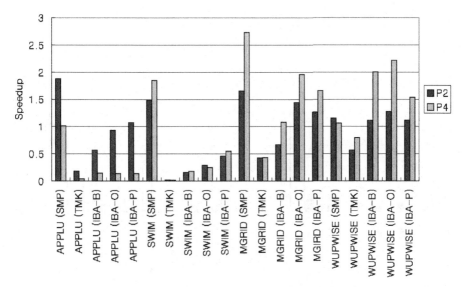

Fig. 2. Speedup of the OpenMP applications

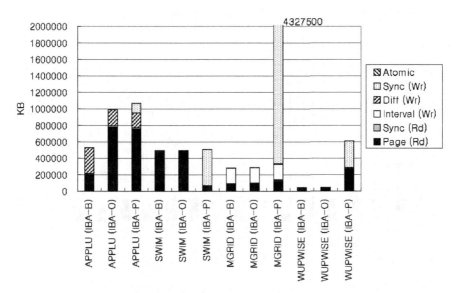

Fig. 3. Sizes (KB) of the RDMA transactions

It is found that the performance degradation occurs in one of the major parallel regions, ssor_do_2_2. The parallel region includes two barriers per loop iteration, and it results in a large number of synchronizations which incurs significantly large memory consistency transactions on the DVSM system. There is also performance degradation from 2 to 4 processors in the SMP execution.

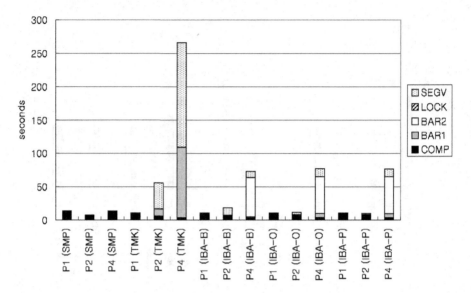

Fig. 4. The overhead analysis in `applu`

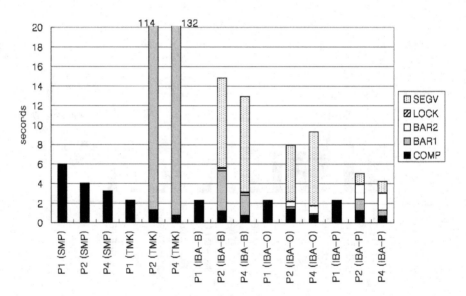

Fig. 5. The overhead analysis in `swim`

When we serialized this parallel section, we could achieve linear speedup in total execution.

Swim. Figure 5 shows execution time in `swim`, and it shows that our ownership assignment and page prefetching technique are very effective. The ownership assignment approach reduces segmentation violation occurrences, and it reduces

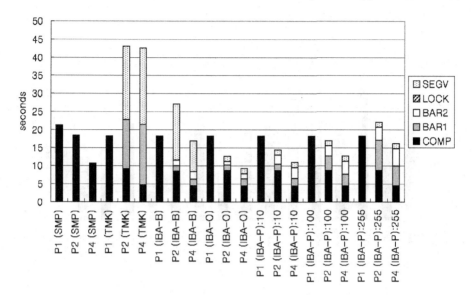

Fig. 6. The overhead analysis in `mgrid`

the load imbalance at the first barrier. The page prefetching technique reduces the number of segmentation violations by about 90%. But the IBA-based execution is still slower than the SMP. There is small amount data movement between serial and parallel regions in `swim`. On the SMP machine, small data movement is very fast since the granularity of communication between processors is a cache-line. But on the DVSM, the granularity is a page, and the related transaction overhead is much larger than on the SMP.

Mgrid and Wupwise. Figures 6 and 7 show the execution time in `mgrid` and `wupwise` respectively. There is speedup in both applications, and furthermore the performance in an ownership assignment approach is higher than the SMP execution.

We update the segmentation violation history at the synchronization regularly with a certain interval number which is shown in the x-axis legend of Figure 6. The ownership assignment is a special case of the page prefetch approach and the interval number is one. In two applications, the prefetch technique does not work positively. In `mgrid` the program behavior is changing, and therefore, when we check the history infrequently, the number of segmentation violation increases. In `wupwise`, each page is almost completely localized to one process, i.e. there is few data movement between processes.

4   Conclusion

In this paper, we presented three variants of implementations for the distributed virtual shared-memory system based on one of the next generation of intercon-

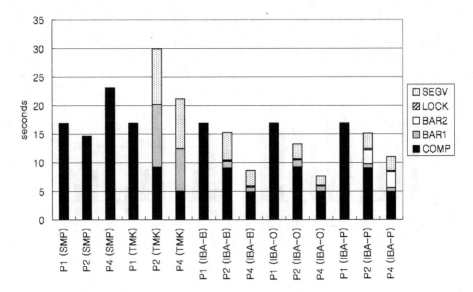

Fig. 7. The overhead analysis in `wupwise`

nection technologies, the InfiniBand Architecture: 1) base implementation to support memory consistency, 2) ownership assignment to reduce the overhead associated with page faults, and 3) page prefetching to hide page copy latency. Our system is the first approach to use the LRC protocol on the InfiniBand Architecture. In order to use full features of the InfiniBand architecture, we used the hardware-based Verbs API in implementation.

From experiment to use SPEC OMP applications, we showed that our implementation improved the speedup significantly higher than the IPoIB protocol stack-based TreadMarks on the InfiniBand. Also in `wupwise` the performance on the IBA-based DVSM system is better than the hardware SMP machine. The performance gain results from using the remote DMA features instead of send/receive communications between processes.

Our final research goal is to find out the software and hardware co-architecture to build low-cost shared-memory multiprocessor systems by using the InfiniBand.

References

1. ASCI pathforward. http://www.llnl.gov/asci/pathforward_trilab/overview.html.
2. C. Amza, A. L. Cox, S. Dwarkadas, P. Keleher, H. Lu, R. Rajamony, W. Yu, and W. Zwaenepoel. Treadmarks: Shared memory computing on networks of workstations. *IEEE Computer*, 29(2):18–28, 1996.
3. L. Kontothanassis, G. Hunt, R. Stets, N. Hardavellas, M. Cierniak, S. Parthasarthy, W. Meira, S. Dwarkadas, and M. Scott. VM-based shared memory on low-latency, remote-memory access networks. In *The 24th International Symposium on Computer Architecture (ISCA-24)*, June 1997.

4. R. Samanta, A. Bilas, L. Iftode, and J. P. Singh. Home-based SVM protocols for SMP clusters: Design, simulations, implementation and performance. In *The Fourth IEEE Symposium on High-Performance Computer Architecture (HPCA-4)*, January 1998.

5. OpenMP Forum, http://www.openmp.org/. *OpenMP: A Proposed Industry Standard API for Shared Memory Programming*, October 1997.

6. S. J. Min, A. Basumallik, and R. Eigenmann. Supporting realistic OpenMP applications on a commodity cluster of workstations. *Lecture Notes in Computer Science (WOMPAT2003)*, 2716:170–179, 2003.

7. J. Bircsak, P. Craig, R. Crowell, Z. Cvetanovic, J. Harris, C. Nelson, and C. Offner. Extending OpenMP for NUMA machines. In *Proceedings of the IEEE/ACM Supercomuting*, pages 49–55, Dallas, TX, 2000.

8. V. Schuster and D. Miles. Distributed OpenMP, Extensions to OpenMP for SMP Clusters. 2000.

9. A. Basumallik, S. J. Min, and R. Eigenmann. Towards OpenMP execution on software distributed shared memory systems. *Lecture Notes in Computer Science (WOMPEI'2002)*, 2327:457–468, 2002.

10. D. Dunning, G. Regnier, G. McAlpine, D. Cameron, B. Shubert, F. Berry, A. Merritt, E. Gronke, and C. Dodd. The virtual interface architecture. *IEEE Micro*, pages 66–76, 1998.

11. InfiniBand Trade Association. *InfiniBand Architecture Specification, Release 1.0*. October 2000.

12. T. Birk, L. Liss, and A. Schuster. Efficient exploitation of kernel access to InfiniBand: a software DSM example. In *Hot Interconnects*, Stanford, CA, August 2003.

13. J. Liu, W. Jiang, P. Wyckoff, D. K. Panda, D. Ashton, D. Buntinas, W. Gropp, and B. Toonen. Design and implementation of MPICH2 over infiniband with RDMA support. In *Proceedings of the IEEE International Parallel and Distributed Processing Symposium (IPDPS)*, 2004.

14. I. Park and S. W. Kim. Implementation of infiniband-based software distributed shared-memory systems. Technical Report ECE-CL-20040101, 2004.

15. V. Aslot, M. Domeika, R. Eigenmann, G. Gaertner, W. B. Jones, and B. Parady. SPEComp: A new benchmark suite for measuring parallel computer performance. *Lecture Notes in Computer Science (WOMPEI2001)*, 2104:1–10, 2001.

Fast and Scalable Startup of MPI Programs in InfiniBand Clusters*

Weikuan Yu, Jiesheng Wu, and Dhabaleswar K. Panda

Network-Based Computing Lab
Dept. of Computer Science and Engineering
The Ohio State University
{yuw, wuj, panda}@cse.ohio-state.edu

Abstract. One of the major challenges in parallel computing over large scale clusters is fast and scalable process startup, which typically can be divided into two phases: process initiation and connection setup. In this paper, we characterize the startup of MPI programs in InfiniBand clusters and identify two startup scalability issues: serialized process initiation in the initiation phase and high communication overhead in the connection setup phase. To reduce the connection setup time, we have developed one approach with data reassembly to reduce data volume, and another with a bootstrap channel to parallelize the communication. Furthermore, a process management framework, Multi-Purpose Daemons (MPD) system is exploited to speed up process initiation. Our experimental results show that job startup time has been improved by more than 4 times for 128-process jobs, and the improvement can be more than two orders of magnitude for 2048-process jobs as suggested by our analytical models.

1 Introduction

The MPI (Message Passing Interface) Standard [12] has evolved as a *de facto* parallel programming model for distributed memory systems. Traditional research over MPI has been largely focusing on the high performance communication between processes. As cluster computing becomes a prominent platform of high performance computing, scalable process management of MPI applications becomes an active research topic [3, 1]. One of the major challenges in process management is the fast and scalable startup of large-scale applications [2, 6, 10, 4, 9]. This issue becomes even more pronounced in the large scale systems with thousands of nodes. A parallel job is usually launched by a process manager, which is often referred to as the *process initiation phase*. These initiated processes usually require assistance from the process manager to set up peer-to-peer connections before starting communication and computation. This is referred to as the *connection setup phase*.

InfiniBand Architecture (IBA) [8] has been recently standardized in industry to design next generation high-end clusters for both data-center and high performance computing. Large cluster systems with InfiniBand are being deployed. For example, in the Top500 list released in November 2003 [15], the 3rd, 111th, and 116th most powerful supercomputers use InfiniBand as their parallel application communication interconnect. These

---

* This research is supported in part by a DOE grant #DE-FC02-01ER25506, NSF Grants #CCR-0204429 and #CCR-0311542, and a grant from Los Alamos National Laboratory.

L. Bougé and V.K. Prasanna (Eds.): HiPC 2004, LNCS 3296, pp. 440–449, 2004.
© Springer-Verlag Berlin Heidelberg 2004

three systems have 2200, 256, and 512 processors, respectively. The startup of MPI applications in InfiniBand clusters at such a large scale is a challenging issue. It may take more than ten minutes to go through the above mentioned process initiation and connection setup phases for an application with 1000 processes without scalable and high performance startup support.

In this paper, we have taken on the challenge to support a scalable and high performance startup of MPI programs over InfiniBand clusters. With MVAPICH [13] as the platform of study, we have analyzed the startup bottlenecks. Accordingly, different approaches have been developed to speed up the connection setup phase, one with data reassembly at the process manager and another using pipelined all-to-all broadcast over a ring of InfiniBand queue pairs (referred to as a bootstrap channel). In addition, we have exploited a process management framework, Multi-Purpose Daemons (MPD) system to further speed up the startup. The bootstrap channel is also utilized to reduce the impact of communication bottlenecks in MPD, including multiple process context switches and quadratically increasing data volume over the MPD management ring. Over 128 processes, our work improves the startup time by more than 4 times. Scalability Models derived from these results suggest that the improvement can be more than two orders of magnitude for the startup of 2048-process jobs.

The rest of the paper is structured as follows. Section 2 gives an overview of Infini-Band. Section 3 describes the challenge of scalable startup faced by parallel programs over InfiniBand and related work on process management. Section 4 describes the design of startup with different approaches to improve the connection setup time and the process initiation phase. Experiments results are provided in 5. Finally, we conclude the paper in Section 6.

2   Overview of InfiniBand Architecture

The InfiniBand Architecture (IBA) [8] defines a System Area Network (SAN) for interconnecting computing nodes and I/O nodes. In an InfiniBand network, a switched communication fabric is defined to allow many devices to communicate concurrently at high bandwidth and low latency. Processing nodes are connected as end-nodes to the fabric with Host Channel Adapters (HCAs).

InfiniBand provides four types of transport services: Reliable Connection (RC), Reliable Datagram (RD), Unreliable Connection (UC), and Unreliable Datagram (UD). The often used service is RC in the current InfiniBand products and software. It is also our focus in this paper. To support RC, a connection must be set up between two QPs before any communication. In the current InfiniBand SDK, each QP has a unique identifier, called *QP-ID*. This is usually an integer. For network identification, each HCA also has a unique 16-bit local identifier (*LID*). To make a connection, a pair of QPs must exchange their QP IDs and LIDs.

3   Problem Statement and Related Work

This section first characterizes the scalability constraints of the startup of MPI programs in InfiniBand clusters. It then provides a brief discussion of related work and motivates the study for a scalable startup scheme.

3.1   Startup of MPI Applications using MVAPICH

MVAPICH [13] is a high performance implementation of MPI over InfiniBand. Its design is based on MPICH [5] and MVICH [11]. The current implementation of MVAPICH utilizes the Reliable Connection (RC) service for the communication between processes. The connection-oriented nature of IBA RC-based QPs requires each process to create at least one QP for every peer process. To form a fully connected network of N processes, a parallel application needs to create and connect at least $N \times (N - 1)$ QPs during the initialization time. Note that it is possible to have these QPs be allocated and connected in an on-demand manner [16], however this requires that the connection management subsystem of IBA can handle either peer-to-peer or client-server model connection establishment, which is not mature yet in the current IBA software. Another reason for the fully-connected connection model is simplicity and robustness. Therefore, this connection model has been used in many MPI implementations, including MVAPICH.

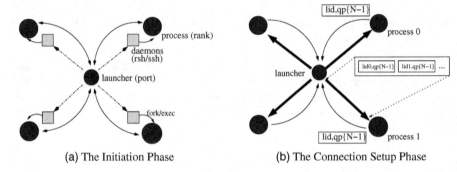

(a) The Initiation Phase   (b) The Connection Setup Phase

Fig. 1. The Startup of MPI Applications in Current MVAPICH

The startup of an MPI application using MVAPICH can also be divided into two phases. As shown in Fig. 1(a), an MPI application using MVAPICH is launched with a simple process launcher iterating over UNIX remote shell (rsh) or secure shell (ssh) to start individual processes. Each process connects back to the launcher via a port exposed by the launcher. Except the rank of the process, each process has no global knowledge about the parallel program. In the second phase of connection setup, as shown in Fig. 1(b), each process creates $N - 1$ QPs, one for each peer process, for an N-process application. Then, these processes exchange their local identifiers (LIDs) and corresponding QP identifiers (QP-IDs), as mentioned in Section 2 for connection setup. Since each process is not connected to its peer processes, the data exchange has to rely on the connections that are created to the launcher in the first phase. The launcher collects data about LIDs and QP-IDs from each process, and then sends the combined data back to each process. Each process in turn sets up connections over InfiniBand with the received data. A parallel application with fully connected processes is then created.

3.2   The Scalability Problem

The startup paradigm described above is able to handle the startup of small scale parallel applications. However, as the size of an InfiniBand cluster goes to 100s–1000s, the limitation of this paradigm becomes pronounced. For example, launching a parallel application with 2000 processes may take tens of minutes. There are two main scalability bottlenecks, one in each phase. The first bottleneck is *rsh/ssh-based startup* in the process initiation phase. This process startup mechanism is simple and straightforward, but its performance is very poor on large systems. The second bottleneck is the communication overhead for exchanging LIDs and QP-IDs in the connection setup phase. To launch an N-process MPI application, the launcher has to receive data containing $(N - 1)$ QP-IDs from each process. Then it returns the combined data with $N \times (N - 1)$ QP-IDs to each process. In total, the launcher has to communicate data in the amount of $O(N^3)$ for an N-process application. Each QP-ID is usually a four-byte integer, for a 1024-process application the launcher will receive almost 4 MegaBytes data and sends almost 4 Gigabytes of data. This communication typically goes through the management network which is normally Fast Ethernet or Gigabit Ethernet. This incurs significant communication overhead and slowdown to the application startup.

3.3   Related Work

Numerous work have been done to provide resource management framework for collections of parallel processes, ranging from basic iterative rsh/ssh-based process launch in MVICH [11] to more sophisticated packages like MPD [3], Cplant [2], PBS [14], LoadLeveler/POE [7], to name a few. Compared to the rsh/ssh-based iterative launch of processes, all these packages can provide more scalable startup and retain better monitoring and control of parallel programs. However, they typically lack efficient support for complete exchange of LIDs and QP-IDs as required by parallel programs over Infini-Band clusters. In this paper, we focus on providing an efficient support for the complete exchange of LIDs and QP-IDs, and applying such a scheme to one of these package, MPD, in order to obtain efficient process initiation support. We choose to study MPD [3] because it is one of the systems widely distributed along with MPICH [5] releases and has a large user base.

4   Designing Scalable Startup Schemes

This section describes the design of scalable startup schemes in InfiniBand clusters. We first describe different approaches used to enhance the connection setup phase while the processes are still launched via rsh/ssh daemons. Then we exploit the advantages of *MPD* [3], to replace the rsh/ssh based scheme and achieve efficient process initiation. We also characterize some MPD features and their limitations to the scalable startup of MPI applications in InfiniBand clusters. We also introduce the concept of a bootstrap channel which can be used to overcome these limitations.

4.1   Efficient Connection Setup

As mentioned in the previous section, because the launcher has to collect, combine and broadcast QP IDs, the volume of these data scales up in the order of $O(N^3)$, which leads to prolonged connection setup time. One needs to consider two directions in order to reduce the connection setup time. The first direction is to reduce the volume of data that needs to be communicated. The other direction is to parallelize communication for the exchange of QP IDs.

Approach 1: Reducing the Data Volume with Data Reassembly (DR). To have processes fully connected over InfiniBand, each process needs to connect with another peer process via one QP. This means that each process needs to obtain $N - 1$ QP IDs, one for each peer. That is to say, out of the combined data of $N \times (N - 1)$ QP IDs in the launcher, each process only needs to receive $N - 1$ QP IDs that is specifically targeted for itself. This requires a centralized component, i.e., the launcher, to collect and reassembly QP IDs. The biggest advantage of this data reassembly (DR) scheme is that the data volume exchanged can be reduced down to an order of $O(N^2)$. But there are several disadvantages associated with this scheme. First, the entire set of QP IDs need to be reassembled before sending them to each client processes. This constitutes another performance/scalability bottleneck at the launcher. Second, the whole procedure of receive-reassembly-send is also serialized at the launcher.

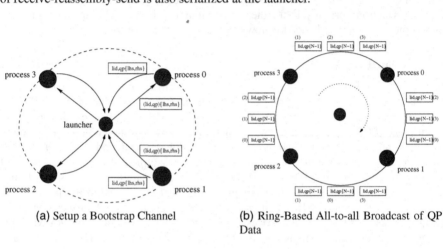

(a) Setup a Bootstrap Channel

(b) Ring-Based All-to-all Broadcast of QP Data

Fig. 2. Parallelizing the Total Exchange of InfiniBand Queue Pair Data

Approach 2: Parallelizing Communication with a Bootstrap Channel (BC). More insights can be gained on the possible parallelism with further examination of the startup. Essentially, what needs to be achieved at the startup time is an all-to-all personalized exchange of QP IDs, i.e., each process receives the specific QP IDs from other processes. In the original startup scheme as shown in Fig. 1, the launcher performs a gather/broadcast to help the all-to-all broadcast of their QP data. On top of that, the DR scheme in Section 4.1 reassembles and "personalizes" QP data to reduce the data volume. Both do not exploit the parallelism of all-to-all personalized exchange. Algorithms that parallelize an all-to-all personalized exchange can be used here. These algorithms are usually based on a ring-, hypercube- or torus-based topology, which requires more connections to be provided among processes. With the initial star topology in the original startup scheme, providing these connections has to be done through the launcher. However, since a parallel algorithm can potentially overlap both sending and receiving QP data, it promises better scalability over clusters with larger sizes.

Among the three possible parallel topologies, the ring-based topology requires the least number of additional connections, i.e., 2 per process. This would minimize the impact of the ring setup time. Another design option to be considered is that which type of connections should be provided. Either TCP/IP- or InfiniBand-based connections can be used. Since the communication over InfiniBand is much faster than that over TCP/IP (see [17] for detail latency comparison between them), we choose to use a ring of InfiniBand QPs as a further boost to the parallelized data exchange.

The second approach works as follows. First, each process creates two QPs for its left hand side (lhs) and right hand side (rhs) processes, respectively. We call these QPs *bootstrap QPs*. Second, the DR scheme mentioned in Section 4.1 is used to set up connections between these bootstrap QPs as shown in Figure 2(a). Thus, a ring of connections over InfiniBand is created, as shown by the dotted line in Figure 2(a). We refer to this ring as a *bootstrap channel* (BC). After this channel is set up, each process initiates a broadcast of its own QP IDs through the channel in the clockwise direction as shown in Fig. 2(b) with four processes. Each process also forwards what it receives to its next process. In this scheme, we take advantage of both communication parallelism and high performance of InfiniBand QPs to reduce the communication overhead.

4.2   Fast Process Initiation with MPD

MPD [3] is designed to be a general process manager interface that provides the needed support for MPICH, from which MVAPICH is developed. It mainly provides fast startup of parallel applications and process control to the parallel jobs. MPD achieves its scalable startup by instantly spreading a job launch request across its ring of daemons, then launches one ring of manager and another ring of application processes in a parallel fashion (see [17] for detailed description of MPD systems). For processes to exchange individual information MPD system also exposes a BNR interface with a put/fence/get model. A process stores (puts) a (key,value) pair at its manager process, a part of the MPD database, then another process retrieves (gets) that value by providing the same key after a synchronization phase (fence).

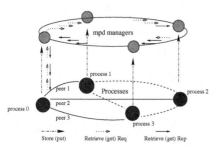
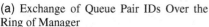
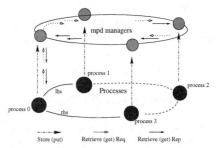

(a) Exchange of Queue Pair IDs Over the Ring of Manager

(b) Setting up Bootstrap Channel within Processes

Fig. 3. Improving the Scalability of MPD-Based Startup

Although this fast and parallelized process startup from MPD solves the process initiation problem, the significant volume of QP data still poses a great challenge to the MPD model. As shown in Fig. 3(a), the database is distributed over the ring of manager processes when each process stores (puts) their process-specific data to its manager. To collect the data from every peer process, one process has to send a request and get the reply back for the target process. At the completion of these data exchanges, each process then sets up connections with all the peers, as shown with process 0 in Fig. 3(a). Together, messages for the request and the reply make a complete round over the manager ring. For a parallel job with N processes, there are $N \times (N - 1)$ message exchanges in total. Each of these messages is in the order of $O(N)$ bytes and has to go through the ring of manager processes. In addition, since application processes store and retrieve data through their corresponding manager processes at each node, process context switches are very frequent and they further degrade the performance of ring-based communication. Furthermore, the message passing is over TCP/IP sockets, which delivers lower performance than InfiniBand-based connections (see [17] for latency comparisons).

There are different alternatives to overcome these limitations. One way of doing that is to replace the connections for the MPD manager ring with VAPI connections to provide fast communications. In addition, copies of QP data can be saved at each manager process as the first copy of QP data passes through the ring. Then further retrieve (get) requests can get the data from the local manager directly instead of the MPD manager ring. This approach will improve the communication time, however, the process context switches still exist between the application processes and manager processes. In addition, retrieve requests made before QP data reaches the local manager process still has to go through the manager ring. Last but not least, this approach necessitates a significant amount of instrumentation of MPD code and has only limited portability to InfiniBand-ready clusters.

Instead of exchanging all the QP data over the ring of MPD manager processes, we propose to exchange QP IDs over the bootstrap channel described in Section 4.1. Though setting up the bootstrap channel still needs help from the ring of manager processes. As shown in Fig. 3(b), each process first creates and stores QP IDs for its left side (lhs) and right hand side (rhs) processes to the local manager. Then, from the database, they retrieve QP IDs for its left hand side and right hand side processes, and set up Infini-Band connections accordingly. Eventually a ring of such connections are constructed and together form a bootstrap channel. This bootstrap channel is utilized to perform a complete exchange of QP IDs as described in Section 4.1. Since this bootstrap channel is provided within the application processes and over InfiniBand, this approach will not only provide fast communication and eliminate the process context switches, but also reduce the number of communications through each manager process.

5   Performance Evaluation

Our experiments were conducted on a 256-node cluster of 4GB DRAM dual-SMP 2.4GHz Xeon at the Ohio Supercomputing Center. For fast network discovery with data reassembly (DR) or the bootstrap channel (BC), we used ssh to launch the parallel processes. Performance comparisons were provided against MVAPICH 0.9.1 (Original). Since Networked File System (NFS) performance could be a big bottleneck in a large cluster and mask out the performance improvement of startup, all binary executable files were duplicated at local disks to eliminate its impact.

5.1   Experimental Results

Table 1 shows the startup time for parallel jobs of different number processes using different approaches. SSH-DR represents ssh-based startup with QP data assembly (DR)

Table 1. Comparisons of Parallel Job Startup Time over MVAPICH with Different Approaches

Number of Processes	4	8	16	32	64	128
Original (sec)	0.59	0.92	1.74	3.41	7.3	13.7
SSH-DR (sec)	0.58	0.94	1.69	3.37	6.77	13.45
SSH-BC (sec)	0.61	0.95	1.70	3.38	6.76	13.3
MPD-BC (sec)	0.61	0.63	0.64	0.84	1.58	3.10

at the process launcher. SSH-BC represents ssh-based startup using the bootstrap channel (BC) to exchange QP IDs. MPD-BC represents MPD-based startup with a bootstrap channel for the exchange of QP IDs.

As the number of processes increases, both SSH-DR and SSH-BC reduce the startup time, compared to the original approach. This is because data reassembly can reduce the data volume by an order of $O(N)$ and the bootstrap channel can parallelize the communication time. Note that the BC-based approach performs slightly worse than the the original and DR-based approach for small number of processes. This is due to the overhead from setting up the additional ring over InfiniBand. As the number of processes increases, the benefits become greater. Both SSH-BC and SSH-DR will be able to provide more scalable startup for a job with thousands of processes since they remove the major communication bottleneck imposed by potentially large volume of QP data. In contrast, the MPD-based approach with a bootstrap channel provides the most scalable startup. On one hand, MPD-BC provides efficient parallelized process initialization, compared to the ssh-based schemes. On the other hand, it also pipelines the QP data exchange over a ring of VAPI connections, hence this approach speeds up the connection setup phase. Compared to the original approach, the MPD-BC approach reduces the startup time for a 128-process job by more than 4 times.

5.2   Analytical Models and Evaluations for Large Clusters

As indicated by the results from Section 5.1, the benefits of the designed schemes will be more pronounced for parallel jobs with larger number of processes. In this section, we further analyze the performance of different startup schemes and provide parameterized models to gain insights about their scalability over large clusters. The total startup time $T_{startup}$ can be divided into the process initiation time and the connection setup time, denoted as T_{init} and T_{conn} respectively. Based on the scalability analysis, we use the following model to describe the startup time of the original scheme (Original), ssh-based scheme with data reassembly (SSH-DR) and the MPD-based scheme with the bootstrap channel (MPD-BC). Each of the models shows the time for the startup of N processes, and the last component describes the time for other overheads that are not quantified in the models, for example, process switching overhead.

Original: $T_{startup} = (O_0 * N) + (O_1 * N * (W_N + W_{N^2})) + O_2$

The process initiation phase time T_{init} scales linearly as the number of processes increases with ssh/rsh-based approaches, while during the connection setup there are $2N$ messages communicated over TCP/IP. Half of them are gathered by the launcher, each being in the order of $O(N)$ bytes; the other half are scattered by the launcher, each of $O(N^2)$ bytes .

SSH-DR: $T_{startup} = (D_0 * N) + (D_{comp} * N^3 + D_1 * 2N * W_N) + D_2$
The process initiation time T_{init} scales linearly with ssh/rsh. During the connection setup phase, the amount of computation scales in the order of $O(N^3)$ (the constant D_{comp} can be very small, being the time for extracting one QP Id), and there are 2*N message communicated over TCP/IP. Half of them are gathered by the launcher, each being in the order of $O(N)$ bytes; The other half are scattered by the launcher, each of them is only $O(N)$ bytes due to reassembly.

MPD-BC: $T_{startup} = (M_0 + N * W_{req}) + (M_{ch\_setup} * N + M_1 * N * W_N) + M_2$
The process initiation time T_{init} scales constantly using MPD, however there is a small fractional increase of communication time for the request message W_{req}. During the connection setup phase, the time to setup a bootstrap channel increases in the order of $O(N)$. Each process also handles N message in the pipeline, each in the order of $O(N)$ bytes.

Original: $T_{startup}$ (sec) $= (0.100 * N) + (10.5 * N * (W_N + W_{N^2})) + 0.12$
SSH-DR: $T_{startup}$ (sec) $= (0.100 * N) + (8.5e^{-9} * N^3 + 10.5 * N * W_N) + 0.12$
MPD-BC: $T_{startup}$ (sec) $= (0.20 + 0.0010 * N) + (0.0180 * N + 2.5 * N * W_N) + 0.30$

The above scalability models are parameterized based on our analytical modeling. As shown in Fig. 4, the experiment results confirm the validity of these models for jobs with 4 to 128 processes. Fig. 5 shows the scalability of different startup schemes when applying the same models to larger jobs from 4 to 2048 processes. Both SSH-DR and MPD-BC improves the scalability of job startup significantly. Note that MPD-BC scheme improves the startup time by about two orders of magnitudes for 2048-process jobs.

6   Conclusions and Future Work

In this paper, we have presented schemes to support scalable startup of MPI programs in InfiniBand clusters. With MVAPICH as the platform of study, we have characterized the startup of MPI jobs into two phases: process initiation and connection setup. To speed up connection setup phase, we have developed two approaches, one with queue pair data reassembly at the launcher and the other with a bootstrap channel. In addition, we have

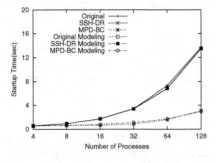

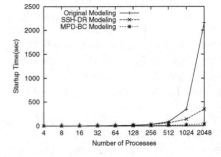

Fig. 4. Performance Modeling of Different Startup Schemes

Fig. 5. Scalability Comparisons of Different Startup Schemes

exploited a process management framework, Multi-Purpose Daemons (MPD) system, to improve the process initiation phase. The performance limitations in the MPD's ring-based data exchange model, such as exponentially increased communication time and numerous process context switches, are eliminated by using the proposed bootstrap channel. We have implemented these schemes in MVAPICH [13]. Our experimental results show that, for 128-process jobs, the startup time has been reduced by more than 4 times. We have also developed an analytical model to project the scalability of the startup schemes. The derived models suggest that the improvement can be more than two orders of magnitudes for the startup of 2048-process jobs with the MPD-BC startup scheme.

In future, we want to provide a file broadcast mechanism to MPD system to achieve efficient loading of jobs [10]. Furthermore, we intend to provide a hypercube-based scalable startup over really large systems, e.g., future Peta-scale clusters with tens of thousands of processors.

References

1. M. Baker, G. Fox, and H. Yau. Cluster Computing Review, November 1995.
2. R. Brightwell and L. A. Fisk. Scalable parallel application launch on Cplant. In *Proceedings of Supercomputing, 2001*, Denver, Colorado, November 2001.
3. R. Butler, W. Gropp, and E. Lusk. Components and interfaces of a process management system for parallel programs. *Parallel Computing*, 27(11):1417–1429, 2001.
4. E. Frachtenberg, F. Petrini, J. Fernandez, S. Pakin, and S. Coll. STORM: Lightning-Fast Resource Management. In *Proceedings of the Supercomputing '02*, Baltimore, MD, November 2002.
5. W. Gropp, E. Lusk, N. Doss, and A. Skjellum. A High-Performance, Portable Implementation of the MPI Message Passing Interface Standard. *Parallel Computing*, 22(6):789–828, 1996.
6. E. Hendriks. Bproc: The beowulf distributed process space. In *Proceedings of the International Conference on Supercomputing*, New York, New York, June 2002.
7. IBM. Using the Parallel Operating Environment, Version 4, Release 1, 2004.
8. Infiniband Trade Association. http://www.infinibandta.org, 2000.
9. M. Jette and M. Grondona. SLURM: Simple Linux Utility for Resource Management. In *Proceedings of the International Conference on Linux Clusters*, San Jose, CA, June 2003.
10. A. Kavas, D. Er-El, and D. G. Feitelson. Using Multicast to Pre-Load Jobs on the ParPar Cluster. *Parallel Computing*, 27(3):315–327, 2001.
11. Lawrence Berkeley National Laboratory. MVICH: MPI for Virtual Interface Architecture. http://www.nersc.gov/research/FTG/mvich/index.html, August 2001.
12. Message Passing Interface Forum. MPI: A message-passing interface standard. *The International Journal of Supercomputer Applications*, 8(3–4):159–416, 1994.
13. Network-Based Computing Laboratory. MVAPICH: MPI for InfiniBand on VAPI Layer. http://nowlab.cis.ohio-state.edu/projects/mpi-iba/index.html.
14. OpenPBS Documentation. http://www.openpbs.org/docs.html, 2004.
15. TOP 500 Supercomputers. http://www.top500.org/, 2003.
16. J. Wu, J. Liu, P. Wyckoff, and D. K. Panda. Impact of On-Demand Connection Management in MPI over VIA. In *Proceedings of the International Conference on Cluster Com puting*, 2002.
17. W. Yu, J. Wu, and D. K. Panda. Fast and Scalable Startup of MPI Programs in InfiniBand Clusters. Number OSU-CISRC-5/04-TR33, Columbus, OH 43210, May 2004.

Parallel Performance of Hierarchical Multipole Algorithms for Inductance Extraction[*]

Hemant Mahawar[1,**], Vivek Sarin[1], and Ananth Grama[2]

[1] Department of Computer Science, Texas A&M University,
College Station, TX, U.S.A.
{mahawarh, sarin}@cs.tamu.edu
[2] Department of Computer Science, Purdue University, West Lafayette, IN, U.S.A.
ayg@cs.purdue.edu

Abstract. Parasitic extraction techniques are used to estimate signal delay in VLSI chips. Inductance extraction is a critical component of the parasitic extraction process in which on-chip inductive effects are estimated with high accuracy. In earlier work [1], we described a parallel software package for inductance extraction called *ParIS*, which uses a novel preconditioned iterative method to solve the dense, complex linear system of equations arising in these problems. The most computationally challenging task in *ParIS* involves computing dense matrix-vector products efficiently via hierarchical multipole-based approximation techniques. This paper presents a comparative study of two such techniques: a hierarchical algorithm called Hierarchical Multipole Method (HMM) and the well-known Fast Multipole Method (FMM). We investigate the performance of parallel MPI-based implementations of these algorithms on a Linux cluster. We analyze the impact of various algorithmic parameters and identify regimes where HMM is expected to outperform FMM on uniprocessor as well as multiprocessor platforms.

1 Introduction

The design and testing phases in the development of VLSI chips rely on accurate estimation of the signal delay. Signal delay in a VLSI chip is due to the parasitic resistance (R), capacitance (C), and inductance (L) of the interconnect segments. At high frequencies, the physical proximity of interconnect segments leads to strong inductive coupling between neighboring conductors. This coupling arises because a magnetic field is created when current flows through a conductor. This magnetic field opposes any change in the current flow within the conductor as well as in the neighboring conductors. Self-inductance is the

---

[*] Support for Mahawar and Sarin was provided by NSF-CCR 9984400, NSF-CCR 0113668, and Texas ATP 000512-0266-2001 grants. Grama's research was supported by NSF-EEC 0228390 and NSF-CCF 0325227 grants. Computational resources were acquired through NSF-DMS 0216275 grant.
[**] Corresponding author.

L. Bougé and V.K. Prasanna (Eds.): HiPC 2004, LNCS 3296, pp. 450–461, 2004.
© Springer-Verlag Berlin Heidelberg 2004

resistance offered to change in current within the conductor. Mutual inductance refers to the resistance offered to change in current in a neighboring conductor. Inductance extraction refers to the process of estimating self and mutual inductance between interconnect segments of a chip.

To estimate inductance between a set of conductors in a particular configuration, one needs to determine current in each conductor under appropriate equilibrium conditions. The surface of each conductor is discretized using a uniform two-dimensional grid whose edges represent current-carrying filaments. The potential drop across a filament is due to its own resistance and due to the inductive effect of other filaments. Kirchoff's current law is enforced at the grid nodes. This results in a large dense system of equations that is solved using iterative methods such as the generalized minimum residual method (GMRES) [2]. Each iteration requires a matrix-vector product with the coefficient matrix, which can be computed without explicitly forming the matrix itself. Matrix-vector products with the dense matrix are computed approximately via multipole algorithms such as the Fast Multipole Method (FMM) [3, 4].

In an earlier paper [1], we described an object-oriented parallel inductance extraction software called *ParIS*. The software uses a formulation in which current is restricted to the subspace satisfying Kirchoff's law through the use of solenoidal basis functions. The reduced system of equations is solved by a preconditioned iterative solver in which products with the dense coefficient matrix and the preconditioner are computed via FMM. Improved formulation and the associated preconditioning is responsible for significant reduction in computational and storage requirements [5]. *ParIS* achieves high parallel efficiency on a variety of multiprocessors with shared-memory, distributed-memory, and hybrid architectures.

In this paper, we present a comparative study of multipole-based methods for computing dense matrix-vector products. We consider the well-known FMM algorithm and a hierarchical algorithm, called Hierarchical Multipole Method (HMM), which can be considered as a variant of the FMM based on particle-cluster multipole evaluations only (related to a Barnes-Hut type approach [6]). We present parallel formulations of these methods and discuss their performance on a Beowulf cluster. We analyze the impact of parameters such as the multipole degree (d), the multipole acceptance criterion threshold (α), and the maximum number of particles allowed in a leaf box (s) on these methods. Since these parameters influence both accuracy and cost, it is important to develop a framework to select the optimal method for a given set of parameters. The experimental results presented in this paper can be used to identify the optimal method for difference parameter subspaces.

The paper is organized as follows – Section 2 outlines the inductance extraction problem, the solenoidal basis method, and the software design of *ParIS*; Section 3 describes HMM and FMM algorithms and outlines their parallel formulations; Section 4 presents a set of experiments on an AMD cluster to illustrate the performance of these methods for a range of parameters; and Section 5 presents concluding remarks.

2   Background

2.1   Inductance Extraction Problem

For a set of t conductors, we need to determine an $t \times t$ impedance matrix that represents pairwise mutual inductance among the conductors at a given frequency. The element (l, k) of the matrix equals the potential drop across conductor l when there is zero current in all the conductors except conductor k that carries unit current. The kth column is computed by solving an instance of the inductance extraction problem with the right hand side denoting unit current flow through conductor k. The impedance matrix can be computed by solving t instances of this problem with different right hand sides.

The current density $\mathbf{J}$ at a point r is related to potential ϕ by the following equation [7]

$$\rho \mathbf{J}(\mathbf{r}) + j\omega \int_V \frac{\mu}{4\pi} \frac{\mathbf{J}(\mathbf{r}')}{\|\mathbf{r} - \mathbf{r}'\|} dV' = -\nabla\phi(\mathbf{r}), \tag{1}$$

where μ is magnetic permeability of the material, ρ is the resistivity, r is position vector, ω is frequency, $\|\mathbf{r} - \mathbf{r}'\|$ is the Euclidean distance between $\mathbf{r}$ and $\mathbf{r}'$, and $j = \sqrt{-1}$. The volume of the conductor is denoted by V and incremental volume with respect to r' is denoted by dV'.

To obtain a numerical solution for (1), each conductor is discretized into a mesh of n filaments $f_1, f_2, \ldots, f_n$. Current is assumed to flow along the filament length. The current density within a filament is assumed to be constant. Filament currents are related to the potential drop across the filaments according to the linear system

$$[\mathbf{R} + j\omega\mathbf{L}]\mathbf{I}_f = \mathbf{V}_f, \tag{2}$$

where $\mathbf{R}$ is an $n \times n$ diagonal matrix of filament resistances, $\mathbf{L}$ is a dense inductance matrix denoting the inductive coupling between current carrying filaments, $\mathbf{I}_f$ is the vector of filament currents, and $\mathbf{V}_f$ is the vector of potential difference between the ends of each filament. The kth diagonal element of $\mathbf{R}$ is given by $\mathbf{R}_{kk} = \rho l_k / a_k$, where l_k and a_k are the length and cross-sectional area of the filament f_k, respectively. Let $\mathbf{u_k}$ denote the unit vector along the kth filament. The elements of the inductance matrix $\mathbf{L}$ are given by

$$\mathbf{L}_{kl} = \frac{\mu}{4\pi} \frac{1}{a_k a_l} \int_{r_k \in f_k} \int_{r_l \in f_l} \frac{\mathbf{u}_k \cdot \mathbf{u}_l}{\|\mathbf{r}_k - \mathbf{r}_l\|} dV_k dV_l.$$

Kirchoff's current law states that the net current flow into a mesh node must be zero. These constraints on current lead to additional equations

$$\mathbf{B}^T \mathbf{I}_f = \mathbf{I}_s, \tag{3}$$

where $\mathbf{B}^T$ is a sparse $m \times n$ branch index matrix and $\mathbf{I}_s$ is the known branch current vector of length m with non-zero values corresponding to the source currents. The branch index matrix defines the connectivity among filaments and nodes. The (k, l) entry of the matrix is -1 if filament l originates at node k, 1 if filament l terminates at node k, and 0 otherwise. Since the unknown filament potential drop $\mathbf{V}_f$ can be represented in terms of node potential $\mathbf{V}_n$ by the relation $\mathbf{B}\mathbf{V}_n = \mathbf{V}_f$, one needs to solve the following system of equations to determine the unknown filament current $\mathbf{I}_f$ and node potential $\mathbf{V}_n$

$$\begin{bmatrix} \mathbf{R} + j\omega\mathbf{L} & -\mathbf{B} \\ \mathbf{B}^T & 0 \end{bmatrix} \begin{bmatrix} \mathbf{I}_f \\ \mathbf{V}_n \end{bmatrix} = \begin{bmatrix} 0 \\ \mathbf{I}_s \end{bmatrix}. \tag{4}$$

For systems involving a large number of filaments, it is not feasible to compute and store the dense matrix $\mathbf{L}$. These linear systems are typically solved using iterative techniques such as GMRES. The matrix-vector products with $\mathbf{L}$ are computed using fast hierarchical methods such as the FMM. The main hurdle in this matrix-free approach is the construction of effective preconditioners for the coefficient matrix.

2.2   The Solenoidal Basis Method

We present a brief overview of the solenoidal basis method for solving (4) (see, e.g., [5] for details). Consider the discretization of a ground plane shown in Fig. 1. Current flowing through the filaments must satisfy Kirchoff's law at each node in the mesh. The bold line indicates a path for current that satisfies boundary conditions. Current is made up of two components: constant current along the bold line shown on the left and a linear combination of mesh currents as shown in the partial mesh on the right. This converts the system in (4) into the following system with a different right hand side

$$\begin{bmatrix} \mathbf{R} + j\omega\mathbf{L} & -\mathbf{B} \\ \mathbf{B}^T & 0 \end{bmatrix} \begin{bmatrix} \mathbf{I} \\ \mathbf{V}_n \end{bmatrix} = \begin{bmatrix} \mathbf{F} \\ 0 \end{bmatrix}. \tag{5}$$

The main difference between matrix representations in (4) and (5) is that the former uses current boundary conditions and the later uses potential boundary conditions.

Solenoidal functions are a set of basis functions that satisfy conservation laws automatically. Figure 1 shows how to construct unit circular flows on mesh cells that automatically satisfy Kirchoff's law at the grid nodes. The unknown filament currents can be expressed in the solenoidal basis: $\mathbf{I} = \mathbf{P}x$, where x is vector of unknown mesh currents and $\mathbf{P}$ is a sparse matrix whose columns denote filament current in each mesh. A column of $\mathbf{P}$ consists of four non-zero entries that have the value 1 or -1 depending on the direction of current flow in the filaments of the cell.

The system (5) is converted to a *reduced* system

$$\mathbf{P}^T \left[\mathbf{R} + j\omega\mathbf{L} \right] \mathbf{P}x = \mathbf{P}^T \mathbf{F}, \tag{6}$$

which is solved by a preconditioned iterative method. The preconditioning step involves product with a dense matrix that represents the inductive coupling among filaments placed at the cell centers. This preconditioning scheme can be implemented using FMM as well, and leads to rapid convergence of the iterative

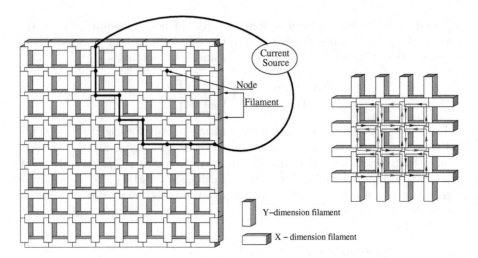

Fig. 1. Discretization of a ground plane with a mesh of filaments (left) and solenoidal current flows in each mesh cell (right) (Reproduced from [5])

method. On a set of benchmark problems, a serial implementation of this software is up to 5 times faster than FastHenry [7], a commonly available induction extraction software, with only one-fifth of memory requirements [5].

2.3   *ParIS*: Parallel Inductance Extraction Software

We have developed an object-oriented parallel implementation of the solenoidal basis algorithm for inductance extraction [1]. This software combines the advantages of the solenoidal basis method, fast hierarchical methods for dense matrix-vector products, and a highly effective preconditioning scheme to provide a powerful package for inductance extraction. In addition, the software includes an efficient parallel implementation to reduce overall computation time [8] on multiprocessors.

The building blocks of *ParIS* are conductor elements. Each conductor is uniformly discretized with a mesh of filaments. Kirchoff's law constraints on the filament currents of a conductor contribute a block in the system matrix. The most time-consuming step in the solution of the reduced system involves matrix-vector products with the impedance matrix, which is the sum of a diagonal matrix **R** and a dense inductance matrix **L**. Since the preconditioning step involves matrix-vector product with a dense matrix, which is similar to **L**, it is worthwhile to reduce the cost of the matrix-vector product with **L**.

3   Hierarchical Multipole-Based Algorithms

The computational complexity of a matrix-vector product with a dense $n \times n$ matrix is $O(n^2)$. This can be reduced significantly through the use of hierarchical approximation techniques. These algorithms exploit the decaying nature of the $\frac{1}{r}$ kernel for the matrix entries to compute approximations with acceptable error. Higher accuracy can be achieved at the expense of more computation. Well-known techniques such as the Barnes-Hut [6] method compute particle-cluster interactions to achieve $O(n \log n)$ complexity, whereas the Fast Multipole Method (FMM) [4] computes cluster-cluster interactions in addition to particle-cluster interactions to achieve $O(n)$ complexity.

3.1   Hierarchical Multipole Method

The hierarchical multipole method (HMM) can be viewed either as an augmented version of the Barnes-Hut method or as a variant of FMM that uses only particle-cluster multipole evaluations. The method works in two phases: the tree construction phase and the potential computation phase. In the tree construction phase, a spatial tree representation of the domain is derived. At each step in this phase, if the domain contains more than s particles, where s is a preset constant, it is recursively divided into eight equal parts. This process continues until each part has at most s elements. The resulting tree is an unstructured oct-tree. Each internal node in the tree computes and stores an approximate multipole series representation of the particles contained in its subtree. The multipole series of a node is computed from the series of its children through an up-traversal of the nodes from the leaves to the root. Once the tree has been constructed, the potential at each particle can be computed as follows: a *multipole acceptance criterion* is applied to the root of the tree to determine if an interaction can be computed; if not, the node is expanded and the process is repeated for each of the eight children. The multipole acceptance criterion computes the ratio of the distance of the point from the center of the box to the dimension of the box. If this ratio is greater than α, where α is a constant greater than $\sqrt{3}/2$, an interaction can be computed.

3.2   Fast Multipole Method

ParIS uses a variant of FMM to compute approximate matrix-vector products with dense matrices. FMM is used to compute the potential at each filament due to the current flow in all filaments. The algorithm divides the domain into eight equal non-overlapping subdomains, and continues the process recursively until each subdomain has at most s filaments, where s is a parameter that is chosen to maximize computational efficiency. A subdomain is represented by a subtree whose leaf nodes contain the filaments in the subdomain. These subdomains are distributed across processors. The potential evaluation phase consists of two traversals of the tree. During the up-traversal, multipole coefficients are computed at each node. These coefficients can be used to compute potential due to all the filaments within the node's subdomain at a *far away* point. The multipole computation does not require any communication between processors. During the down-traversal, local coefficients are computed at each node from the multipole coefficients. The local coefficients can be used to compute potential due to *far away* filaments at a point within the node's subdomain. Potential due to *near by* filaments is computed directly.

3.3   Parallel Formulation

To exploit parallelism at the conductor level, each conductor is assigned to a different processor. The data structures native to a conductor are local to its processor. This includes the filaments in a conductor and the associated oct-tree. With the exception of matrix-vector products with the inductance matrix, all other computations are local to each conductor.

The matrix-vector product with the inductance matrix involves two types of filament interactions. Interactions among the filaments of the same conductor are computed locally by the associated processor. To get the effect of filaments in other conductors, a processor needs to exchange multipole coefficients with other processors. Since matrix-vector products with the dense inductance matrix and the preconditioner are computed at each iteration, *ParIS* identifies those nodes in a conductor's tree that are required by other conductors during a pre-processing step. The cost of this step is amortized over the number of iterations of the solver. While computing the dense matrix-vector product, communication is needed for the translation of the multipole coefficients of these nodes to nodes on other processors. Communication is also needed between adjacent nodes that belong to different subtrees when computing direct interactions. This type of communication is proportional to the number of filaments on the subdomain boundary.

Additional parallelism is available within each conductor. By assigning different processes or threads to all the nodes at a specific level in the oct-tree, we are able to partition the computation for subdomains among processes. Fewer processes can be assigned to the top part of the oct-tree to further improve parallel efficiency. With different sized conductors, one can have more processes associated with larger conductors. This scheme allows load balancing to a certain extent. A variety of parallel implementations are discussed in [9–12].

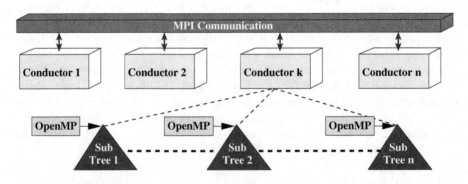

Fig. 2. Two-tier parallelization scheme implemented in *ParIS*

A two-tier parallelization approach shown in Fig. 2 simplifies the implementation in hybrid or mixed mode using both MPI and OpenMP. The software can be executed on a variety of platforms ranging from shared-memory multiprocessors to workstation clusters seamlessly [1].

4   Experiments

To investigate the performance of hierarchical multipole algorithms used in *ParIS* we considered the *cross-over* benchmark problem. Figure 3 shows two layered cross-over of interconnect segments called buses. The problem consists of determining the impedance matrix of these buses. Each bus is assumed to be 2cm long and 2mm wide. Buses within a layer are separated by 300μm while the layers are separated by 3mm. This problem leads to a non-uniform point distribution for the dense matrix-vector multiplication algorithm.

The main goal of this study is to analyze the performance of HMM and FMM codes within *ParIS*. Instead of solving the full inductance extraction problem, we observed the performance of the codes for a fixed number of GMRES iterations. Each iteration involved dense matrix-vector products with the coefficient matrix as well as the preconditioner. The results are identical to the case when the full inductance extraction problem is solved because the dense matrix-vector products account for over 98% of the execution time (see, e.g., [5]).

A generalized notion of efficiency is used to provide a uniform basis to compare different experiments. We compute *scaled efficiency* as shown below:

$$E_s = \frac{BOPS}{p}, \qquad (7)$$

where p is the number of processors and $BOPS$ is the average number of *base operations* executed *per second*. A base operation equals the cost of computing a direct interaction between a pair of filaments. In principle, $BOPS$ should remain unchanged when the number of conductors and filaments per conductor are varied. With this definition of efficiency, it possible to compare the performance of the code on a variety of benchmarks that require different number of interactions. The experiments were conducted on the *Tensor* cluster at Texas A&M University. The cluster consists of 1.4GHz 64-bit AMD Opteron processors running LAM/MPI on SuSE-Linux, connected via Giga-bit ethernet. GNU compilers were used on *Tensor* for compiling the code.

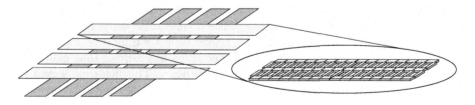

Fig. 3. The cross-over benchmark

4.1   Impact of Parameters

The performance of the hierarchical multipole algorithms depends on the choice of multipole degree (d), the multipole acceptance criterion determined by α, and the maximum number of particles allowed per leaf box (s). Since d and α parameters influence accuracy of the approximate dense matrix-vector product, a fair comparison is possible only when the impedance error is bounded. In these experiments, the impedance error was always within 1% of a reference value that was calculated by FMM with $d = 8$.

The dominant computation in FMM consists of multipole-to-local translations (M2Ls) with computational cost proportional to $(d + 1)^4$. The dominant computation in HMM consists of multipole evaluations at particles (M2Ps) with computational cost proportional to $(d + 1)^2$. Table 1 shows that with increase in d, the FMM time increases proportional to $(d + 1)^4$, while the HMM time increases proportional to $(d+1)^2$. For HMM experiments, α was chosen to be 1.

Table 1. Effect of the multipole degree (d) on the execution time, in secs, for different choices of maximum particles per leaf box (s)

	FMM code				HMM code			
d	$s{=}2$	$s{=}8$	$s{=}32$	$s{=}128$	$s{=}2$	$s{=}8$	$s{=}32$	$s{=}128$
1	49.5	18.3	12.7	29.9	25.7	21.5	21.3	34.8
2	225.8	62.5	25.3	32.8	46.8	36.5	31.3	41.9
4	1513.3	398.2	110.8	50.7	110.8	84.5	63.0	61.9

The execution time for both methods decreases when s is increased due to a decrease in the number of M2Ls and M2Ps. The cost of direct interactions is proportional to s^2 and is negligible for small values of s. Direct interactions begin to dominate the overall cost for large values of s, resulting in higher execution time. Table 1 shows that when s is increased, the FMM execution time reduces rapidly due to reduction in M2Ls, until the direct interactions begin to dominate the computational cost. Similarly, the HMM execution time decreases due to reduction in M2Ps, until the direct interactions begin to dominate. The decrease in the HMM case is not as rapid due to the lower complexity of M2Ps compared to M2Ls. For a given problem, one can identify (d, s) pair that minimizes the execution time for each method.

HMM has an additional parameter for the multipole acceptance criteria. A large value of α improves the accuracy of the approximate dense matrix-vector product at additional computational cost. Larger values of α increase the number of direct interactions as well as the number of M2Ps by ensuring that multipole evaluations at particles are computed for smaller boxes. This behavior is clear in Tables 2 and 3. The increase in time with α can be estimated from the increase in the number of direct interactions. A choice of $s = 8$ is used in Table 2 and $d = 2$ is used in Table 3.

Table 2. Effect of the multipole acceptance criterion threshold (α) on the execution time, in secs, of the HMM code for different choices of multipole degree (d)

α	$d=1$	$d=2$	$d=4$
1	21.5	36.5	84.5
1.5	40.1	70.6	158.2

Table 3. Effect of the multipole acceptance criterion threshold (α) on the execution time, in secs, of the HMM code for different choices of maximum particles per leaf box (s)

α	$s=2$	$s=8$	$s=32$
1	46.8	36.5	31.3
1.5	89.3	70.5	59.5

4.2   Parallel Performance

The parallel performance of FMM and HMM codes is primarily determined by the ratio of computation to communication. To compute M2L between a pair of oct-tree nodes residing on different processors, multipole coefficients must be exchanged. This requires communication of $(d + 1)^2$ data units followed by M2L computation, which is proportional to $(d + 1)^4$. Thus, the computation-to-communication ratio grows rapidly with increase in d. On the other hand, computing M2P between a node and a particle requires communication of $(d+1)^2$ data units followed by M2P computation, which is proportional to $(d + 1)^2$. In this case, there will be limited effect of d on the parallel performance as long as the multipole coefficients received by a processor q are stored and reused by the particles on q.

The use of scaled efficiency E_s defined in (7) allows us to scale the problem linearly with processors. A cross-over problem with p conductors was chosen for experiments that used p processors. This benchmark is characterized by proximity between pairs of conductors on different layers. Thus, the number of M2Ls and M2Ps requiring communication grows linearly with the number of conductors p. Similarly, the number of direct interactions that require communication between processors also grows linearly with p. This is observed in Table 4 for the HMM code with $\alpha = 1$.

The computation in FMM is varied, with M2Ls forming the dominant component. Table 5 shows the parallel execution time for the FMM code for $s = 8$ and $s = 32$. The execution time grows much faster with p for the case when $s = 32$ because of reduced M2Ls and increased direct interactions. This behavior is consistent with the observation that the FMM code achieves higher parallel efficiency for larger d.

The scaled efficiency allows us to compare the performance if the two methods. Table 6 shows the efficiency of the HMM and FMM codes on the cross-over problem with $s = 8$ and $\alpha = 1$. The codes maintain high efficiency as p increases. The efficiency also increases when d is increased, and the effect is more pronounced in the FMM code.

A comparison of the parallel execution time of the two methods for different values of d is also instructive. Table 7 shows the ratio of parallel execution times

Table 4. Impact of multipole degree (d) on the execution time, in secs, of the HMM code on p processors for two different choices of maximum particles per leaf box (s)

	$s = 8$				$s = 32$			
d	$p{=}1$	$p{=}2$	$p{=}4$	$p{=}8$	$p{=}1$	$p{=}2$	$p{=}4$	$p{=}8$
1	21.5	26.5	50.9	105.8	21.3	24.4	48.8	94.1
2	36.5	46.5	96.5	184.3	31.3	38.3	77.9	157.5
4	84.5	101.9	220.9	436.8	63.0	78.2	169.6	347.9

Table 5. Impact of multipole degree (d) on the execution time, in secs, of the FMM code on p processors for two different choices of maximum particles per leaf box (s)

	$s = 8$				$s = 32$			
d	$p{=}1$	$p{=}2$	$p{=}4$	$p{=}8$	$p{=}1$	$p{=}2$	$p{=}4$	$p{=}8$
1	18.3	25.7	34.5	59.2	12.7	13.9	40.4	94.4
2	62.5	72.5	87.5	131.3	25.3	26.6	58.0	126.3
4	398.2	431.4	470.9	683.3	110.8	113.4	165.7	277.8

Table 6. Efficiency of the extraction codes on p processors for different choices of multipole degree (d)

	HMM code				FMM Code			
d	$p{=}1$	$p{=}2$	$p{=}4$	$p{=}8$	$p{=}1$	$p{=}2$	$p{=}4$	$p{=}8$
1	0.99	0.93	0.94	0.86	0.98	0.74	0.87	0.87
2	1.00	0.92	0.90	0.92	0.99	0.86	0.97	0.98
4	1.00	0.98	0.93	0.94	1.00	0.93	1.04	0.98

Table 7. Ratio of the execution time of FMM and HMM codes on p processors for different choices of multipole degree (d)

	$s = 8$				$s = 32$			
d	$p{=}1$	$p{=}2$	$p{=}4$	$p{=}8$	$p{=}1$	$p{=}2$	$p{=}4$	$p{=}8$
1	0.9	1.0	0.7	0.6	0.6	0.6	0.8	1.0
2	1.7	1.6	0.9	0.7	0.8	0.7	0.7	0.8
4	4.7	4.2	2.1	1.6	1.8	1.4	1.0	0.8

of FMM and HMM codes for $d = 1, 2, 4$ and $s = 8, 32$. It is clear that HMM is superior to FMM when a larger value of d is used. The comparative advantage of HMM is diminished for $s = 32$ due to improved performance of FMM.

5   Conclusions

This paper presents a comparison of multipole-based methods for computing dense matrix-vector products in inductance extraction problems. The Fast Multipole Method is compared with a hierarchical multipole method on a set of benchmark problems. Numerical experiments are conducted on an AMD cluster for range of parameters such as the multipole degree (d), the multipole acceptance criterion threshold (α), and the maximum number of particles allowed in a leaf box (s). The results provide insight into the relative merits of these methods and suggest ways to determine the optimal method for a given set of parameters.

References

1. Mahawar, H., Sarin, V.: Parallel software for inductance extraction. In: Proceedings of the International Conference on Parallel Processing, Montreal, Canada (2004)
2. Saad, Y.: Iterative Methods for Sparse Linear Systems. PWS Publishing Company, Boston (1996)
3. Greengard, L.: The Rapid Evaluation of Potential Fields in Particle Systems. The MIT Press, Cambridge, Massachusetts (1988)
4. Greengard, L., Rokhlin, V.: A fast algorithm for particle simulations. Journal of Computational Physics **73** (1987) 325–348
5. Mahawar, H., Sarin, V., Shi, W.: A solenoidal basis method for efficient inductance extraction. In: Proceedings of the IEEE Design Automation Conference, New Orleans, Louisiana (2002) 751–756
6. Barnes, J., Hut, P.: A hierarchical O($n \, log \, n$) force calculation algorithm. Nature **324** (1986) 446–449
7. Kamon, M., Tsuk, M.J., White, J.: FASTHENRY: A multipole-accelerated 3D inductance extraction program. IEEE Transaction on Microwave Theory and Techniques **42** (1994) 1750–1758
8. Mahawar, H., Sarin, V.: Parallel iterative methods for dense linear systems in inductance extraction. Parallel Computing **29** (2003) 1219–1235
9. Grama, A., Kumar, V., Sameh, A.: Parallel hierarchical solvers and preconditioners for boundary element methods. SIAM Journal on Scientific Computing **20** (1998) 337–358
10. Sevilgen, F., Aluru, S., Futamura, N.: A provably optimal, distribution-independent, parallel fast multipole method. In: Proceedings of the International Parallel and Distributed Processing Symposium, Cancun, Mexico (2000) 77–84
11. Singh, J.P., Holt, C., Totsuka, T., Gupta, A., Hennessy, J.L.: Load balancing and data locality in hierarchical n-body methods. Journal of Parallel and Distributed Computing **27** (1995) 118–141
12. Teng, S.H.: Provably good partitioning and load balancing algorithms for parallel adaptive N-body simulation. SIAM Journal of Scientific Computing **19** (1998) 635–656

A New Adaptive Fault-Tolerant Routing Methodology for Direct Networks[*]

M.E. Gómez[1], J. Duato[1], J. Flich[1], P. López[1], A. Robles[1],
N.A. Nordbotten[2], T. Skeie[2], and O. Lysne[2]

[1] Dept. of Computer Engineering, Universidad Politécnica de Valencia,
Camino de Vera, 14, 46071–Valencia, Spain
megomez@disca.upv.es
[2] Simula Research Laboratory,
P.O. Box 134, N-1325 Lysaker, Norway

Abstract. Interconnection networks play a key role in the fault tolerance of massively parallel computers, since faults may isolate a large
fraction of the machine containing many healthy nodes. In this paper,
we present a methodology to design fully adaptive fault-tolerant routing
algorithms for direct interconnection networks that can be applied to different regular topologies. The methodology is mainly based on the selection of an intermediate node (if needed) for each source-destination pair.
Packets are adaptively routed to the intermediate node and, from this
node, they are adaptively forwarded to their destination. This methodology requires only one additional virtual channel, even for tori. Evaluation
results show that the methodology is 7-fault tolerant, and for up to 14
faults, more than 99% of the combinations are tolerated, also without
significantly degrading performance in the presence of faults.

1 Introduction

There exist many compute-intensive applications that require continued research
and technology development to deliver computers with steadily increasing computing power. The required levels of computing power can only be achieved with
massively parallel computers, such as the Earth Simulator [8] and the Blue-
Gene/L [1]. The long execution times of these applications requires keeping such
systems running even in the presence of failures. However, the huge number of
processors and associated devices (memories, switches, links, etc.) significantly
increases the probability of failure. In particular, failures in the interconnection network may isolate a large fraction of the machine, wasting many healthy
processors that otherwise could have been used. Although network components
are robust, they are usually working close to their technological limits and are
therefore prone to failures. Increasing clock frequencies leads to a higher power
dissipation, which again could lead to premature failures. Hence, fault-tolerant
mechanisms for interconnection networks are becoming a critical design issue for
large massively parallel computers.

[*] This work was supported by the Spanish Ministry of Science and Technology under
Grant TIC2003-08154-C06-01.

L. Bougé and V.K. Prasanna (Eds.): HiPC 2004, LNCS 3296, pp. 462–473, 2004.
© Springer-Verlag Berlin Heidelberg 2004

Faults can be classified as transient or permanent. Transient faults are usually handled by communication protocols, using CRCs to detect faults and retransmitting packets. In order to deal with permanent faults in a system, two fault models can be used: static or dynamic. In a static fault model, all the faults are known in advance when the machine is (re)booted. This fault model relies on checkpoints in order to be effective. In a dynamic fault model, once a new fault is found, actions are taken in order to appropriately handle the faulty component.

There exist several approaches to tolerate faults in the interconnection network. Most of them are based on fault-tolerant routing algorithms. However, these strategies require a significant amount of extra hardware resources (e.g., virtual channels), to route packets around faulty components [4], [12]. Alternatively, there exist some fault-tolerant routing strategies that use none or very few extra resources at the expense of providing a lower fault-tolerance degree [4], [9], disabling a certain number of healthy nodes (either in blocks (fault regions) [2], [3] or individually [5], [6]), preventing packets from being routed adaptively [10], or drastically increasing the latencies for some packets [14].

What is really needed is a fault-tolerant strategy for the interconnection network that does not degrade performance at all in the absence of faults, does not significantly decrease network performance in the presence of faults, and tolerates a reasonably large number of faults. This should be achieved without disabling any healthy node and without requiring too many extra resources.

In this paper, we take on this challenge and propose a fault-tolerant routing methodology that satisfies the properties mentioned above. The methodology relies on a static fault model with checkpointing. It allows the use of fully adaptive routing in most cases and it does not sacrifice any healthy node. In order to avoid faults, packets are sent adaptively to an intermediate node[1]. From that node, the packet will be sent adaptively to the destination. The methodology requires the use of at least three virtual channels. Note that two virtual channels are already required to provide fully adaptive routing [13].

It is important to highlight the main differences between the proposed methodology and similar approaches in the literature. Unlike [14], in no case the proposed methodology requires ejecting/reinjecting the packet at the intermediate node, thus drastically reducing latency. Moreover, unlike [10], it does not need to deactivate any *lamb* node to achieve high fault-tolerance. Furthermore, the proposed methodology allows packets to be routed using adaptive routing, instead of deterministic routing, thus increasing the overall network throughput.

The rest of the paper is organized as follows. In Section 2, the proposed methodology is presented. In Section 3, the fault-tolerant routing algorithm resulting from applying the methodology is evaluated. Finally, in Section 4, some conclusions and future work are drawn.

2   Description of the Methodology

The methodology provides fault-tolerance both in n-dimensional mesh and torus networks. For the sake of clarity, the description will mainly be based on a 3D

---

[1] Intermediate nodes were also used by Valiant [15] for traffic balance.

torus network, with some particular cases explained on a 2D torus network. The methodology tolerates both link and node failures. A node failure can be modeled by the failure of all the links connected to it. Therefore, we will focus only on link failures. When a link fails, we assume that it fails in both directions. The methodology will assume a static fault model, thus, it will know in advance where the failures are located. The proposed methodology will be focused only in the computation of the new routing info for every source-destination pair[2].

The methodology assumes that the initial (i.e., without faults) routing algorithm routes packets by using fully adaptive routing with at least two virtual channels (at least one adaptive and one escape) per physical channel. The adaptive channel(s) enables routing through any minimal path whereas the escape channel guarantees deadlock freedom based on the bubble flow control [13]. A fault-free path is computed by the methodology for each source-destination pair. In the presence of faults, those paths that may use some faulty components are not valid. The methodology avoids those faults by using intermediate nodes for routing. Packets are first forwarded to an intermediate node, and later, from this node to the destination node. Minimal adaptive routing is used in both subpaths. Notice that packets are not ejected from the network at the intermediate node.

Next, we will describe how the intermediate nodes are selected.

2.1   Intermediate Nodes

We denote the source node of a path as S and the destination node as D. A link connects two nodes whose coordinates only differ in one of the dimensions. Faulty links (F_f) are represented by using the identifier of the node (the faulty node F) with the lowest coordinate in the faulty dimension, or the node with the highest coordinate if the link is the wraparound, and the faulty dimension (f). The three coordinates of a given node N are denoted as (X_N, Y_N, Z_N). Intermediate nodes are denoted as I.

If a faulty link F_f can be reached by packets sent from S to D, then an intermediate node I is selected (if possible) in order to avoid F_f. Thus, an intermediate node is used only if there is at least one failure along any possible minimal path between S and D, that is, in a 3-D torus, if the failure is within the cube defined by S and D. On the contrary, if the failure F_f can not be reached for a particular $S - D$ pair, normal routing is used.

The intermediate node I is selected inside the minimal adaptive cube defined by S and D. Thus, both subcubes defined by S and I, and by I and D, are inside this cube, but they are smaller and avoid the failure. Figure 1 shows these subcubes for a given S, D, and I. If the packet is first sent to I and then to D, the possible paths are reduced to the shaded areas, thus, avoiding the failure. Packets are adaptively routed inside each stretch (S-I and I-D).

At the intermediate node, some action must be performed in order to avoid deadlocks. We simply propose the use of two different escape channels. One of them will be used as escape channel for the S-I stretch and the other one for the I-D stretch. Therefore, we define two virtual networks. Each one relies on

---

[2] Detection of faults, checkpointing, and distribution of routing info is out of the scope of the methodology.

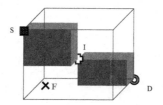

Fig. 1. A minimal adaptive path for a *S-D* pair using an intermediate node

a different escape channel, but both use the same adaptive channel(s). That is, if a packet is on its way to I, then it uses (if required) the first escape channel. From I, the packet uses (if required) the second escape channel.

2.2   Selecting Intermediate Nodes

There may be several possible I nodes that can be used for each $S - D$ pair. The methodology computes the set of possible I nodes and then selects one of them. Given a faulty link F_f, in order to compute the set of I nodes, some properties can be deduced from the relative positions of S, D, and F in a 3D torus network[3]:

- In one of the three dimensions, the I node must be placed between F and D. That is, I must overcome or leave behind the failure in one of the dimensions. This allows overriding the failure in the path between I and D.
- In another of the dimensions, the coordinate of the I node must lie between S and F. That is, in one of the dimensions I must not overcome the failure. This allows overriding the failure in the cube defined between S and I.
- Finally, in the remaining dimension, the coordinate of the I node can vary between the coordinates of S and D.

Following the previous rules, I will avoid the failure, also providing a minimal path. Figure 2 shows all the possible I nodes in a 3D torus when X is overcome and Y is not overcome and vice versa. Notice that the two first aforementioned rules can also be applied to the Z-dimension, leading to a possibly larger set of possible intermediate nodes.

There are some cases where the previous rules must be done more precise:

- If the S coordinate is equal to the F coordinate in one dimension, then this dimension must be overcome by the I nodes. This case is shown in Figure 3.(a) for a 2D torus, where Y is overcome and X is not, since $Y_S = Y_F$.
- If the D coordinate is equal to the F coordinate in one dimension, then the coordinate of the I node in this dimension must lie between the S and the F coordinates. That is, another dimension should be overcome. In Figure 3.(b), $Y_D = Y_F$, so the I nodes do not overcome Y.

---

[3] For the sake of simplicity, the dimension f where the faulty link is located has not been taken into account.

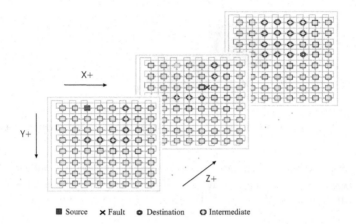

Fig. 2. All the possible intermediate nodes for a particular $S - D$ pair

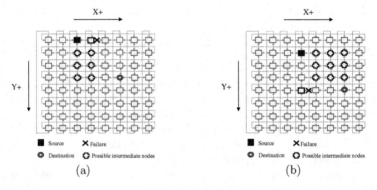

Fig. 3. Possible I nodes for two particular cases: (a) $Y_S = Y_F$. (b) $Y_D = Y_F$

- Another special case arises when S, F, and D are in the same plane. That is, the S, D, and F coordinates are the same for one dimension (for instance, $Z_F = Z_S = Z_D$). In this situation, if we want to follow a minimal path, then the coordinate of I in this dimension must be also the same (i.e., I must also be located in the same plane). Regarding the remaining coordinates, one of them must overcome the fault and the the other one must not overcome the fault. Such situations can be seen in Figures 3.(a) and 3.(b).
- If S, D, and F are in the same ring and the failure is between S and D through the minimal path, it is impossible to find an I node in the minimal path from S to D.

   A possible solution is to travel in the opposite ring direction in order to avoid the failure, selecting an intermediate node along this path. This case can be seen in Figure 4.(a). A possible I node is the one that is halfway between S and D through the non-minimal path. Notice that both paths $S - I$ and $I - D$ are minimal, but the resulting $S - D$ path is non-minimal.

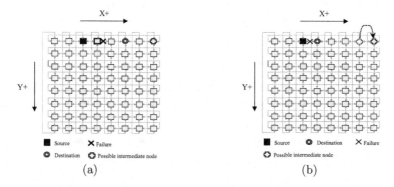

Fig. 4. Situations when S, F, and D are aligned in the same ring

- If S and D share the faulty link (see Figure 4.(b)) and the number of nodes in the dimension is even, it is impossible to find a valid I node in the non-minimal path from S to D as in the previous case.

  The problem is that distance from S to the computed I is the same both using the positive (faulty) and the negative directions of the dimension. Hence, the routing algorithm may select the wrong path. However, the solution is simple. All we have to do is to move I nearer to S. Notice that selecting I in this way also solves the previous case.

As the methodology uses at most one I node for every path, a selection of the final I node is required. Although this selection may affect system performance, at the current stage of this research, the selection is performed randomly.

2.3 Extension to More than One Failure

With the previous rules, all the 1-fault combinations are tolerated. In order to support more than one failure, we define a forbidden zone where all the failures are confined and use I nodes to avoid this zone. Moreover, for a given $S-D$ pair, only those failures located along any minimal path between them will be considered. The I node must be computed following the rules presented in Section 2.2. Figure 5.(a) shows the forbidden zone and the possible intermediate nodes for a 2D Torus network.

However, in a scenario with more than one failure it may be necessary to use additional mechanisms. E.g., with two faults located in the same ring, it is impossible for some $S - D$ pairs to find an adequate I. Figure 5.(b) shows a case, where a combination of seven faults can not be tolerated with one I node and adaptive routing in both stretches.

The figure shows a possible path that could be used, using the node located at $X_S - 1, Y_S + 1$ as I node. If minimal adaptive routing were used from S to I, then some faults could be encountered. In order to properly reach I from S, misrouting and switching off adaptive routing will be used. Misrouting will force routing packets several hops along different directions. Once misrouting is completed, then normal routing (or deterministic routing if adaptive routing

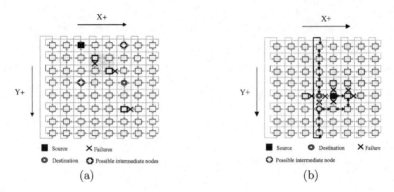

Fig. 5. 2D tori with several failures. (a). A faulty region is defined and I nodes are computed. (b). Misrouting and disabling adaptive routing is required

is switched off) will be applied to the packet. In order to be deadlock-free, the directions to misroute a packet must be used according to the order established by the deterministic routing. In particular, the methodology will use the $X+Y+Z+X-Y-Z-$ direction-order routing, which is deadlock-free and adds routing flexibility (it allows routing packets in both directions of the same dimension).

In the example shown in Figure 5.(b), by using direction-order routing, misrouting, and switching off adaptive routing, the packets will be misrouted one hop in the $X+$ direction, and then forwarded deterministically (by using direction-order routing) to the I node along the $Y+$ and $X-$ direction. In order to reach D from I, misrouting must also be used in order to avoid faults. The packet will be misrouted seven hops in the $Y+$ direction to reach the D node. In each subpath, we assume that packets are misrouted in at most three directions.

To sum up, in a scenario with more than one failure, the methodology will try to override the faults by using all the strategies we have shown so far. Notice that by applying the three mechanisms, it will be possible to compute different fault-free paths for a given $S-D$ pair, thus, being necessary to select among them. In particular, for every $S-D$ pair the methodology will first try to get a minimal path. If no minimal path is found then it will try non-minimal paths by switching off adaptivity and using misrouting in both stretches. The methodology will always pick the path that provides the shortest path.

2.4   Required Resources and Complexity

A packet routed through intermediate nodes requires two subheaders. The first one is used for routing the packet towards the intermediate node, and the second one for routing the packet towards the final destination. At the intermediate node, the first header is removed. However, nodes must select the proper escape channel (if required). If the packet has two headers, then it must select the first escape channel. If it has only one header, the second escape channel must be selected. Packet subheaders also include control fields about misrouting (direction and hops) and switching off adaptive routing (one bit).

The methodology also requires routing info to be stored at each source node. For every destination, this info includes the possible intermediate node (if required) and info about misrouting and switching off adaptive routing. Notice that the amount of required memory is low. In a large system with 65,536 nodes and 5 bytes for routing purposes, the memory size required will be 320KB.

The computational cost of the proposed methodology is low, especially if we take into account that routing info is computed off-line. Most of the affected paths only use an intermediate node to avoid faults, and the cost of computing each intermediate node is $O(1)$. Thus, for all the paths, the computational cost is $O(n^2)$, where n represents the number of nodes. However, in a few cases, misrouting has to be used, thus increasing the computational complexity. The algorithm has to explore all the possible hops along each dimension (up to the network radix, k) until a fault-free path is found. For a 3D Torus (with $n = k^3$ nodes), in the worst case, the computational cost is $O(k^3) = O(n)$. Hence, the computational cost for all the paths by using the proposed methodology is $O(n^3)$.

3   Evaluation of the Methodology

In this section, we evaluate the proposed methodology. First, we are interested in analyzing its fault-tolerance. The methodology is $n-$fault tolerant, if it is able to tolerate any combination of n failures. A given combination of failures is tolerated if every $S - D$ pair in the network can communicate avoiding the failures. Some fault combinations may physically disconnect some nodes from the network. This is not considered as a not tolerated combination.

We are also interested in analyzing how the methodology influences network performance. For this purpose, we compare the performance degradation experienced by our methodology against the one experienced by a mechanism similar to the one used in the BlueGene/L supercomputer. This system is chosen because it represents a state-of-the-art system, and uses adaptive routing and a direct network (3-D torus). Thus, the methodology could be applied to this system.

3.1   BlueGene/L Supercomputer

BlueGene/L [1] is configured as a $64 \times 32 \times 32$ torus of computing nodes constructed with point-to-point serial links between the routers. It uses virtual cut-through [11] and provides both adaptive and deterministic minimal-path routing. Physical channels are multiplexed into up to four virtual channels. Virtual channels are divided into two groups. Two of the virtual channels are used for adaptive minimal routing [7], and two for deterministic minimal routing. One of the deterministic channels is used as escape channel. The bubble flow control [13] is used in order to guarantee deadlock-freedom. The last deterministic virtual channel is reserved for high-priority packets.

The BlueGene/L supercomputer uses a static fault model with checkpointing. Fault-tolerance is achieved by marking healthy nodes as faulty in order to preserve topology and routing, which is extremely convenient when it is hardwired in each router. All the nodes included in four planes (4,096 or 8,192 nodes) that contain the faulty node/link are marked as faulty. A special hardware bypasses the four planes.

3.2   Simulation Model

A detailed event-driven simulator has been used which models a direct interconnection network with point-to-point bidirectional serial links. Each router has a non-multiplexed crossbar with queues only at the input ports. Each physical input port uses four virtual channels, each providing buffering resources in order to store up to two packets. A round-robin policy has been chosen to select among packets contending for the same output port.

Packets are adaptively routed with minimal paths by using the two adaptive virtual channels. In the two escape channels, packets are deterministically routed following the $X + Y + Z + X - Y - Z-$ order. The two escape channels are used according to the bubble flow control mechanism. When a packet arrives at an input port, the escape queue will be used only if the adaptive queues are full. The output port selected for each routed packet will take into account the information located in the packet header, the status of available output ports, and the status of the neighbor nodes queues.

In all the presented results, the network topology is a 3D torus. Each node has four internal ports connected to the processing node. We present results for $3 \times 3 \times 3$ (27 nodes) and $8 \times 8 \times 8$ (512 nodes) tori. Although actual systems are built with larger topologies (e.g., a $32 \times 32 \times 64$ torus for BlueGene/L), smaller networks can be evaluated exhaustively from a fault-tolerant point of view and the results can easily be extended to larger networks.

For each simulation run, the traffic has the following features. Packet generation rate is constant and the same for all the nodes. The destination of a message is chosen randomly with the same probability for all the nodes. The packet length is set to 128 bytes.

3.3   Evaluation Results

First, we have analyzed all the fault combinations for up to 5 faults in a $3 \times 3 \times 3$ torus. With 5 faults, 25,621,596 fault combinations have been analyzed, and all of them are tolerated by the methodology.

As the number of faults increases, the number of possible fault combinations exponentially increases. Therefore, from a particular number of faults (six faults for the $3 \times 3 \times 3$ torus), it is impossible to explore all the fault combinations in a reasonable amount of time. We will tackle this problem with two approaches. First, we will focus on faults confined in a limited region of the network. Notice that, the worst combinations of faults to be solved by the methodology are those where they are closely located. As the number of fault combinations within such a region is much lower than for the entire network, all the fault combinations can be evaluated. This gives us an approximation of the effectiveness of the methodology in the worst case. Secondly, a statistical analysis is performed, analyzing a subset of the fault combinations, where the faults are randomly located over the entire network. From the obtained results, statistical conclusions are extracted about the fault-tolerance degree of the proposed methodology.

For the first study, the faults are located over a region which will be formed by all the links in the positive direction (in each dimension) of the nodes that are one hop away from a node (the center node) randomly selected[4]. Such a region will be referred to as a *distance 1* region, and will be formed by 21 links (3 links · 6 neighbours + 3 links of the center node) in a $3 \times 3 \times 3$ Torus. Notice that for a high number of faults, the center node will be hardly accessible, as very few links will be not faulty. We have defined the *distance 1* region in such a way that only the center node can become disconnected.

All the fault combinations for up to 14 faults have been analyzed in the *distance 1* region. The methodology is 7-fault tolerant for the *distance 1* region, since the 116,280 fault combinations are tolerated. From 8 faults upwards, some cases are not handled by the methodology. However, the methodology is able to tolerate up to 11 faults in more than 99% of the analyzed combinations. For a greater number of faults, the percentage of supported combinations progressively decreases. Nevertheless, it is hard to imagine a system working for a long time with such a high number of faults, without being repaired.

Next, we present a more realistic scenario, where the faults are randomly located over the entire network. We generate random combinations of n faults and analyze them in order to know if they can be tolerated by our mechanism. We have analyzed 28,400,000 fault combinations in a $3 \times 3 \times 3$ torus with different numbers of link failures (up to 14). For all the cases analyzed, all the fault combinations were solved by the methodology with an error always lower than 0.00074. This error represents the maximum probability that a fault combination is not tolerated by the methodology. Therefore, the mechanism handles faults very efficiently, even for more than 7 link faults.

It must be noticed that the fault tolerance analysis has been performed in a $3 \times 3 \times 3$ Torus. However, because the faults will be at the same or greater distance in a larger network, it is reasonable to expect an equal or even better fault-tolerance degree in larger networks.

Following, we focus on the performance analysis. In order to make the results independent of the relative positions of the failures, we have run 50 simulations (for each number of failures), each of them corresponding to a different randomly-selected failure combination. Figure 6.(a) shows the mean overall network throughput achieved for different numbers of failures in a $8 \times 8 \times 8$ Torus. The confidence intervals are always lower than ± 1.6. As can be seen, the network performance is not seriously affected by the presence of failures (throughput decreases by 5.5% for 6 faults and by 10% for 14 faults).

Finally, we compare the performance degradation when using our methodology against the performance degradation that would be obtained by a fault-tolerant mechanism similar to the one used in the BlueGene/L supercomputer. The BlueGene/L system disables four planes of nodes in order to deal with a fault. As we are using a smaller torus network, we model the mechanism of the BlueGene/L system by only disabling one plane. Figure 6.(b) shows the network

---

[4] The selection of the center node will not affect results due to the symmetry property of the torus network.

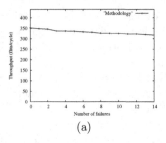 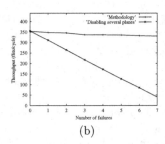

(a)   (b)

Fig. 6. (a). Mean overall network throughput (flits/cycle) degradation in a $8 \times 8 \times 8$ torus network. (b). Mean overall network throughput degradation for the proposed methodology and for the BlueGene/L like mechanism, in an $8 \times 8 \times 8$ Torus network

throughput obtained with the two methodologies when there are up to 7 faults in a $8 \times 8 \times 8$ Torus. Error bars are not shown as they are too small.

Notice that this is a worst case for the fault-tolerant mechanism similar to the BlueGene/L mechanism, as we assume that the seven faults are located in seven different planes. If the seven faults were located in the same plane, only one plane would be disconnected. As shown in Figure 6.(b), when seven faults are present, our proposed mechanism achieves better performance results than the BlueGene/L-like mechanism does with only one fault present. Thus, even if all the seven faults were in the same plane, our proposed mechanism would still outperform the BlueGene/L-like mechanism. Network throughput degrades only by up to a 6.4% with 7 random faults when using our mechanism, whereas when using the BlueGene/L mechanism, the network performance drops by 88% when disabling seven planes. These results must be put in context. That is, they are obtained in a $8 \times 8 \times 8$ Torus. For larger networks, in particular for the $32 \times 32 \times 64$ Torus used in the BlueGene/L supercomputer, a fault would disconnect four planes of at least 32×32 nodes. That is, 4,096 out of 65,536 nodes. So, in the presence of seven faults, performance would decrease at least 6.25% (if the seven faults are in the same four planes). For our mechanism, in a larger network, performance degradation in the presence of seven faults should be significantly lower than the 6.4% obtained in the $8 \times 8 \times 8$ Torus. This is because the traffic unbalance introduced by the faulty links will be lower in a larger network.

4   Conclusions

In this paper, we have proposed a fully adaptive fault-tolerant methodology valid for n-dimensional mesh and torus networks with a static fault model. The methodology relies mainly on the use of intermediate nodes in order to avoid faults. However, to deal with particular fault configurations, some source-destination pairs communicate through non minimal paths. Also, for some pairs of nodes, adaptive routing is disabled in the subpaths in order to tolerate more faults. Unlike other fault-tolerant approaches, the proposed methodology does not need to disable any healthy node, only requires one additional virtual channel, and does not degrade performance in the absence of failures.

Evaluation results on a 27-node tori show that the proposed methodology is 7-fault tolerant. Additionally, the percentage of tolerated fault combinations is greater than 99.9% when up to 14 failures are considered. Also, network throughput degrades less than 10% when injecting 14 random failures in a 512-node Torus. In contrast, a fault-tolerant mechanism similar to the one used in the BlueGene/L may degrade network throughput by 88%.

References

1. IBM BG/L Team, An Overview of BlueGene/L Supercomputer, *ACM Supercomputing Conference*, 2002.
2. A.A. Chien and J.H. Kim, Planar-adaptive routing: Low-cost adaptive networks for multiprocessors, *Proc. of the 19th Int. Symp. on Computer Architecture*, pp. 268-277, May 1992.
3. S.Chalasani and R.V. Boppana, Communication in multicomputers with nonconvex faults, *IEEE Trans. on Computers*, vol. 46, no. 5, pp. 616-622, May 1997.
4. W.J. Dally and H. Aoki, Deadlock-free adaptive routing in multicomputer networks using virtual channels. *IEEE Trans. on Parallel and Distributed Systems*, vol. 4, no 4. pp. 466-475, April 1993.
5. W. J. Dally et al., The Reliable Router: A Reliable and High-Performance Communication Substrate for Parallel Computers, *Proc. Parallel Computer Routing and Communication Workshop*, 1994.
6. J. Duato, A theory of fault-tolerant routing in wormhole networks, *Proc. of the Int. Conf. on Parallel and Distributed Systems*, pp. 600-607, Dec. 1994.
7. J. Duato, A Necessary and Sufficient Condition for Deadlock-Free Outgoing in Cut-Through and Store-and-Forward Networks, *Proc. of IEEE Trans. on Parallel and Distributed Systems*, vol. 7, no. 8, pp. 841-854, August 1996.
8. Earth Simulator Center, http://www.es.jamstec.go.jp/esc/eng/index.html.
9. G.J. Glass, and L.M. Ni, Fault-Tolerant Wormhole Routing in Meshes without Virtual Channels, *IEEE Trans. on Parallel and Distributed Systems*, vol. 7, no. 6, pp. 620-636, 1996.
10. C.T. Ho and L. Stockmeyer, A New Approach to Fault-Tolerant Wormhole Routing for Mesh-Connected Parallel Computers, *Proc. of 16th Int. Parallel and Distributed Processing Symp.*, April 2002.
11. P. Kermani and L. Kleinrock, Virtual cut-through: A new computer communication switching technique, *Computer Networks*, vol. 3, pp. 267-286, 1979.
12. D.H. Linder and J.C. Harden, An Adaptive and fault tolerant wormhole routing strategy for k-ary n-cubes, *IEEE Trans. Computers*, vol. C-40 no. 1, pp. 2-12, 1991.
13. V. Puente et al., Adaptive Bubble Router: A Design to Balance Latency and Throughput in Networks for Parallel Computers, *Proc. of the 22nd Int. Conf. on Parallel Processing*, September 1999.
14. Y.J. Suh, B.V. Dao, J. Duato, and S.Yalamanchili, Software-based rerouting for fault-tolerant pipelined communication, *IEEE Trans. on Parallel and Distributed Systems*, vol. 11, no. 3, pp. 193-211, 2000.
15. L.G. Valiant, A Scheme for Fast Parallel Communication, *SIAM Journal on Computing*. vol. 11, pp. 350-361, 1982.

Fast and Efficient Submesh Determination in Faulty Tori

R. Pranav and Lawrence Jenkins

Department of Electrical Engineering, Indian Institute of Science, Bangalore
{pranav, lawrn}@ee.iisc.ernet.in

Abstract. In a faulty torus/mesh, finding the maximal fault-free submesh is the main problem of reconfiguration. Chen and Hu [1] proposed a distributed method to determine the maximal fault-free submesh in a faulty torus. In this paper, we show that it is sufficient to apply the distributed algorithm proposed by Chen and Hu [1] to only few nodes of a torus. The time for determination of the maximal fault free submesh/submeshes (MFSS) is considerably reduced, by reduction in the number of messages needed for determination of MFSS. In addition, it also reduces the congestion in the network. We present an algorithm to determine the smallest submesh containing all faulty nodes in a torus. The proposed algorithm has a time complexity of $O(n(m+k))$ for a k-ary n-cube with m faults. Intensive simulation study reveals that number of messages is significantly reduced compared to Chen and Hu's [1] method.

1 Introduction

In recent years, researchers have become increasingly interested in mesh, torus, and hypercube based distributed systems. There are many commercial mesh, torus and hypercube based architectures, such as Cray T3D[2], Intel Paragon XP/S[3], nCUBE[4], Caltech Cosmic[5] Intel/DARPA Touchstone Delta[6] and the recent IBM BlueGene/L [7], which aims at delivering target peak processing power of 360 teraflops. IBM Bluegene is a 3-dimensional torus with 64x32x32 nodes i.e. total of 65,536 processing elements.

Fault tolerance in such a huge system becomes one of the critical aspects of interest, so as to maintain the high availability of the system. When the nodes/links in such a system become faulty, there is a need to detect, diagnose and reconfigure the system to a working state, and to recover from the errors. Reconfiguration is one of the main tasks. There are two traditional ways of reconfiguring the system. One adds redundant nodes/links, and the other uses the graceful degradation technique. Redundancy techniques [8], [9], [10], [11] use spare nodes and spare links. Whenever some node/link goes faulty, the system is reconfigured in such a way that it includes all fault-free nodes/links. This technique is inefficient, because many fault-free nodes/links are not used.

In the graceful degradation technique, faulty nodes/links are discarded, and the rest of the fault free subsystem is used, with small performance degradation. The

L. Bougé and V.K. Prasanna (Eds.): HiPC 2004, LNCS 3296, pp. 474–483, 2004.
© Springer-Verlag Berlin Heidelberg 2004

main problem in reconfiguring is to identify the MFFS in the faulty torus. Many researchers like Ozguner and Aykanat [12], Sridhar and Raghavendra [13], Latifi [14], and Chen and Tzeng [15],[16] have studied reconfiguration by the graceful degradation technique in hypercubes, and Seong-Moo and Yong-Youn [17] have studied reconfiguration by the graceful degradation technique in 2-dimensional meshes. All the above techniques are either not scalable and/or suffer from high time complexity. Chen and Hu [1] proposed a distributed algorithm ('Algorithm A') to determine the MFFS, which is applied to all nodes to determine the MFFS locally, and then in the next phase the information is passed, in a non-redundant way, to find the global MFFS.

In this paper, we show that, to determine MFFS it is sufficient to apply 'Algorithm A' to a few nodes, rather than applying it to all nodes of the torus. So, in the second stage of information gathering, there will be fewer messages passed. As message passing is the bottleneck in most of the distributed algorithms, the time required for determining the MFFS is reduced. The reduction in the number of messages in the network also reduces congestion.

The smallest submesh containing all faulty nodes is used to determine the nodes to which 'Algorithm A' is to be applied. We present a Cyclic Binary String Based (CBSB) algorithm to determine the smallest submesh containing all faulty nodes. The CBSB algorithm has a time complexity of $O(n(m+k))$ for a k-ary n-cube with m faulty nodes.

The reminder of the paper is divided into 6 sections. Section 2 and 3 explain the notation and definitions used in this paper. In section 4, we brief the Chen and Hu [1] method. The proposed strategy is delineated in section 5. Section 6 provides simulation results, comparing the proposed strategy with that of Chen and Hu's [1] method. Conclusions are discussed in Section 7.

2 Notation

An n-dimensional torus is denoted by $T_n(k_{n-1}, k_{n-2}, \ldots, k_0)$, where n is the number of dimensions and k_i is the number of nodes in the ith dimension, where $k_i > 1$. Each node is identified by an n-dimensional vector address, $<l_{n-1}, l_{n-2}, \ldots, l_0>$, where $0 \leq l_i < k_i$, for $0 \leq i < n$. Let $l_i(x)$ denote the sub-label corresponding to dimension i, of node x. Node y is the positive neighbor of node x along dimension j only if $l_i(x) = l_i(y)$ for all i, $0 \leq i < n$, except for $i = j$, where $l_i(y) = (l_i(x)+1) \bmod k_i$, and we say that the connection from node x to node y is in the positive direction. Similarly, node y is the negative neighbor of node x along dimension j only if $l_i(x) = l_i(y)$ for all i, $0 \leq i < n$, except for $i = j$, where $l_i(y) = (l_i(x)-1) \bmod k_i$, and we say that the connection from node x to node y is in the negative direction.

An n-dimensional mesh is an n-dimensional torus without wraparound connections. A corner node in an n-dimensional mesh is a node with exactly n neighbors. Thus, an n-dimensional mesh has 2^n corners. The corner, which has all n neighbors in positive (negative) direction, in an n-dimensional mesh is defined as the base (end) node of the mesh. A submesh in an n-dimensional torus specifying the base node

$<b_{n-1},b_{n-2},...,b_0>$ and the end node $<e_{n-1},e_{n-2},...,e_0>$ is denoted by $\{<b_{n-1},b_{n-2}, ...,b_0>: <e_{n-1},e_{n-2},...,e_0>\}$.

In the following, we apply set theory to represent submeshes in the torus. Let '·' denote intersection and '+' denote Union of two sets and Complement is denoted by '‾'. Let $d_i(b:e)$ be an n-dimensional (or (n−1)-dimensional, if b = e) submesh in an n-dimensional torus, consisting the set of nodes $<l_{n-1},l_{n-2},...,l_0>$ such that $0 \le l_j < k_j$ for all $j \ne i$ and l_i is one of b, (b+1)mod k_i, (b+2)mod k_i, ... , (e-1) mod k_i and e. The complement of $d_i(b:e)$, denoted by $\overline{d_i(b:e)}$, consists the set of nodes $<l_{n-1},l_{n-2}, ...,l_0>$ such that $0 \le l_j < k_j$ for all $j \ne i$ and l_i can be one of (e+1)mod k_i, (e+2)mod k_i, ..., (b-2)mod k_i and (b-1)mod k_i, i.e. d_i((e+1)mod k_i : (b-1)mod k_i). The complement of union/intersection of submeshes can be simplified using simple DeMorgan's laws.

Let '$\prod$' denote the intersection of set of submeshes and '$\bigcup$' denote the union of set of submeshes. A submesh with the base node $<b_{n-1},b_{n-2},...,b_0>$ and the end node $<e_{n-1},e_{n-2},...,e_0>$ is represented by $\prod_{i=0}^{n-1} d_i(b_i:e_i)$. Note that a submesh can be represented by a product term (minterm).

3 Definitions

Definitions used in this paper are given as follows:

Definition 1. *Given a node X, the candidate-submesh in a torus with respect to node X is the submesh whose base node is X and which contains all the torus nodes.*

Definition 2. *Given a node X, the antipodal-node of X in a torus is the farthest node from X in the candidate-submesh with respect to X.*

Definition 3. *Given a node X, a reject-region in a faulty torus with respect to node X is the smallest submesh which contains faulty node of the torus and the antipodal node of X in the candidate-submesh with respect to node X.*

Definition 4. *A prime-submesh with a given node, say X, as the base node is a fault-free submesh which involves X but is not contained entirely in any other fault-free submesh involving X as the base node.*

Note that the antipodal-node is the node which is the negative neighbor along all dimensions of the given node. A reject-region is addressed simply by using the labels of a faulty node and the antipodal node as the base and the end nodes respectively. The prime-submesh is defined with respect to the given node, and prime-submeshes with respect to different nodes could be of different sizes. The prime submesh is MFFS with respect to the given node.

Definition 5. *Let B be any submesh in the torus; let B be called a base-submesh. Then antipodal-submesh B* of B is the submesh formed by the antipodal-nodes of all nodes in B.*

Since an antipodal-node is the node which is the negative neighbor of the given node along all dimensions. For a base-submesh with the base node $<b_{n-1}, b_{n-2},...,b_0>$ and the end node $<e_{n-1}, e_{n-2},...,e_0>$ is given by expression $\prod_{i=0}^{n-1} d_i(b_i : e_i)$, and the antipodal-submesh is given by $\prod_{i=0}^{n-1} d_i((b_i - 1) \bmod k_i : (e_i - 1) \bmod k_i)$ in an n-dimensional torus $T_n(k_{n-1}, k_{n-2},...,k_0)$. Similarly, for a given antipodal–submesh we can determine a base-submesh.

4 Chen and Hu's Work

Chen and Hu[1] proposed a boolean expression based approach to determine the MFFS in a torus consisting of faulty nodes and/or faulty links. They proposed a two-phase approach. In the first phase, the distributed 'Algorithm A' which has a time complexity of $O(k^n.m.(m+k))$, is applied to all nodes of the torus (except the faulty nodes). The 'Algorithm A' simplifies the complement of the union of reject-regions expression from product-of-sum form to sum-of-product form. The MFFS with respect to a given node (prime-submesh) are extracted from the simplified expression. In the next phase the global MFFS is determined by passing messages between the nodes in a non-redundant manner. It is assumed that the set of faulty nodes is uncovered in a distributed manner by the fault-free nodes, following the diagnostic algorithm introduced by Armstrong and Gray[18]. For further explanations refer [1].

5 Proposed Approach

In the proposed approach, we first show that it is sufficient to apply 'Algorithm A' of Chen and Hu to a part of the torus, rather than to all the nodes of the submesh. This reduces the number of messages passed and the time required to determine the MFFS.

Theorem 1. *To determine MFFS, it is sufficient to apply 'Algorithm A' to the base-submesh of any submesh containing all faulty nodes.*

Proof. Let the base and the end nodes of the submesh containing the set of m (m>0) faulty nodes, say S, be $<b_{n-1}, b_{n-2},...,b_0>$ and $<e_{n-1}, e_{n-2},...,e_0>$, respectively, in an n-dimensional torus $T_n(k_{n-1}, k_{n-2},...,k_0)$ i.e. $S = \prod_{i=0}^{n-1} d_i(b_i : e_i)$.

Let the base-submesh formed by S be $B = \prod_{i=0}^{n-1} d_i((b_i + 1) \bmod k_i : (e_i + 1) \bmod k_i)$ Let $X = \{ X^1, X^2,..., X^m \}$ be the set of all m faulty nodes with labels $< x_{n-1}^1, x_{n-2}^1,..., x_0^1 >$, $< x_{n-1}^2, x_{n-2}^2,..., x_0^2 >$, ... , $< x_{n-1}^m, x_{n-2}^m,..., x_0^m >$ respectively. Depending on number of faults and pattern of fault distribution, there will be two cases.

Case 1:- If $\overline{B}$ is null, then B contains all the nodes of the torus, so 'Algorithm A' should be applied to all the nodes of the torus (like in Chen and Hu's method), to get global MFFS.

Case2:- If $\overline{B}$ is not null.

Now, apply 'Algorithm A' to any node

$$Y =< (y_{n-1}+1)\bmod k_{n-1},...,(y_0+1)\bmod k_0 > \text{ in } \overline{B} \text{ i.e.}$$

$$\overline{B} = \overline{\prod_{i=0}^{n-1} d_i((b_i+1)\bmod k_i : (e_i+1)\bmod k_i)}$$

$$= \bigcup_{i=0}^{n-1}\overline{d_i((b_i+1)\bmod k_i : (e_i+1)\bmod k_i)}$$

$$= \bigcup_{i=0}^{n-1} d_i((e_i+2)\bmod k_i : b_i)$$

Since Y lies in $\overline{B}$, the antipodal-node of node Y i.e. $< y_{n-1}, y_{n-2},..., y_0 >$ lies in

$$\bigcup_{i=0}^{n-1} d_i((e_i+1)\bmod k_i : (b_i-1)\bmod k_i)$$

$$= d_{n-1}((e_{n-1}+1)\bmod k_{n-1} : (b_{n-1}-1)\bmod k_{n-1}) + d_{n-2}((e_{n-2}+1)\bmod k_{n-2}$$

$$: (b_{n-2}-1)\bmod k_{n-2}) + ... + d_0((e_0+1)\bmod k_0 : (b_0-1)\bmod k_0)$$

$$= \overline{S}.$$

Therefore, in the node $< y_{n-1}, y_{n-2},..., y_0 >$, there exists at least a dimension i such that y_i lies between $(e_i+1)\bmod k_i$ and $(b_i-1)\bmod k_i$.

Let us consider a case such that in the node $< y_{n-1}, y_{n-2},..., y_0 >$ there exists only one dimension i $= p, 0 \le p \le n-1$ such that y_i lies between $(e_i+1)\bmod k_i$ and $(b_i-1)\bmod k_i$ and for $i \ne p$, y_i lies between b_i and e_i and the reject-region formed by faulty node X^j, say $R_Y^j = \prod_{i=0}^{n-1} d_i(x_i^j : y_i)$.

Now apply 'Algorithm A' to the node

$$E =< (y_{n-1}+1)\bmod k_{n-1},...,(y_{p+1}+1)\bmod k_{p+1},(e_p+1)\bmod k_p,$$

$$(y_{p-1}+1)\bmod k_{p-1},...,(y_0+1)\bmod k_0 >$$

where $< y_{n-1},..., y_{p+1}, e_p, y_{p-1},..., y_0 >$ is the antipodal-node of E, the reject-region formed by a faulty node X^j, say $R_E^j = \prod_{i=0}^{n-1} d_i(x_i^j : e_i)$. Since for any dimension i, and for any faulty node X^j, x_i^j lies between b_i and e_i, the node $< y_{n-1},..., y_{p+1}, e_p, y_{p-1},..., y_0 >$ lies in R_y^j. Therefore $R_E^j \subset R_y^j \ \forall j \ 1 \le j \le m$, and the prime submesh found by applying 'Algorithm A' to node Y will be the subset of the prime submeshes found by applying 'Algorithm A' to node E ($\because$ A prime - submesh does not contain any nodes inside reject-regions and a node outside all reject regions is contained in at least one prime submesh[1]). Hence there is no need to apply 'Algorithm A' to node Y.

Similarly, we can prove that there is no need to apply 'Algorithm A' to node Y, for all nodes $< y_{n-1}, y_{n-2}, ..., y_0 >$ such that there is more than one dimension along which y_i lies between $(e_i + 1) \bmod k_i$ and $(b_i - 1) \bmod k_i$ □

According to the above theorem, it is sufficient to apply 'Algorithm A' to the nodes of the base-submesh of a submesh containing all faulty nodes. Let us consider the smallest submesh containing all faulty nodes. Then the corresponding base-submesh will be the smallest one. This is because the number of nodes in the base submesh of a submesh, say S, will be equal to the number of nodes in S. Therefore the number of nodes to which the 'Algorithm A' is applied gets reduced, as compared with Chen and Hu's [1] method, and hence the number of messages passed will get reduced, and that in turn reduces the time required to determine the MFFS. Message passing is one of the bottlenecks of many distributed method of problem solving. The reduction in the number of messages passed will reduce the major overhead. It will also reduce the congestion in the network.

Now the problem is to find a smallest submesh containing all faulty nodes. In this paper we propose CBSB Algorithm to find the smallest submesh containing all faulty nodes. For an n-dimensional torus $T_n(k_{n-1}, k_{n-2}, ..., k_0)$ with m faulty nodes, the CBSB Algorithm finds n submeshes, where for each dimension j there is a submesh, which optimizes only along dimension j. In other words each submesh is optimized along one of the n different dimensions. To determine each of the n submeshes, we use the concept of cyclic binary string. The intersection of the n submeshes will have all m faulty nodes and will be optimized along all the n dimensions; it will be the smallest submesh that contains all faulty nodes. The CBSB algorithm is as follows:

CBSB Algorithm
Input: Set of all m faulty nodes in a torus T_n $(k_{n-1,...,} k_1, k_0)$ where m>0
Output: Smallest submesh containing all m faulty nodes.
For each dimension i=0 to n-1 begin
Step1: For j=0 to k_i-1
   If there is faulty node in submesh $d_i(j:j)$ Flag[j]=1;
   Else   Flag[j]=0;
// above step can performed like this: first initialize Flag[1...k_i-1]=0;
// for each faulty node $<X_{n-1}, X_{n-2}, ... , X_0>$ Flag[X_i]=1;
Step2: Now the array Flag is a string containing 0s and 1s. Consider the string to be a cyclic string of length k_i-1. Find two indices to Flag <s,e> such that the number of 0's between Flag[s] and Flag[e] is minimum and all 1s lie between Flag[s] and Flag[e] (inclusive).
  Submesh [i]= d_i(s:e).
end
Minsubmesh = submesh[0] ∩ submesh[1] ∩ ... ∩ submesh[n-1]
return Minsubmesh

The above algorithm has two major steps, both of which are repeated for n times. During each iteration i, we find a submesh which is optimized along the dimension i. The first step is initialization of Flag array (cyclic binary string), and as there are m faults, this takes a time of $O(m)$. The second step is to find the largest sequence of 0s in the flag array (considering Flag to be cyclic sequence of 0s and 1s of length k_i-1) and to find the complement of this sequence. This step takes a time of $O(k_i)$. As these two steps are repeated n times, the total time complexity of the algorithm is $O(n.m + \sum_{i=0}^{n-1} k_i)$. For a k-ary n-cube, there are k nodes across each of the n dimensions. Therefore the CBSB algorithm has a time complexity of $O(n(m+k))$ for k-ary n-cube.

It is assumed that the torus system has one host node, which is connected directly to every node in the system, and which is in charge of reconfiguration. This host node is similar to that assumed in Chen and Hu's [1] method. All faulty nodes are diagnosed and determined at the host node using the techniques described in [18] or [19]. Then the host node applies the CBSB Algorithm to find the smallest submesh, say S, containing all faulty nodes. From S, the base-submesh of S, say B, is calculated. The host node broadcasts a message describing all faulty nodes and the submesh B. All the healthy nodes which lie in B execute 'Algorithm A' and determine the local MFFS. The base node of B executes 'Algorithm A' and sends a message describing its local MFFS to all its neighbors in the positive direction which lie in B. Any healthy node in B other than the base node of B executes 'Algorithm A' and waits for the message from all the healthy neighbors in the negative direction that lie in B. After getting all the messages, the node chooses the maximal submeshes among the submeshes described in the messages and its local MFFS. Then the node sends a message describing the chosen maximal submeshes to all the healthy neighbors in positive direction that lie in B. In case the node is the end node of B or there is no healthy neighbor in the positive direction that lies in B, the message is directly sent to the host node. In a special case, if for any node there are no healthy neighbors in the negative direction that lie in B, the node need not wait for any message, and therefore it sends a message describing its local MFFS to all healthy neighbors that lie in B. The host node may get more than one message, so it chooses the maximal submeshes among the submeshes described in messages, which is the required global MFFS of the torus.

To extend the method to faulty links, we use the method described in [1], for finding the reject region formed by a faulty link. There are two nodes connecting each link, so a preferred node [1] can be one of the two nodes connecting the faulty link. As the preferred node is chosen based on a given node and is different from one given node to another [1], any of the nodes connecting faulty link can be the preferred node. Therefore, in the above method, S should consist all the faulty nodes, and all the nodes connecting faulty links. Now base-submesh of S, (i.e. B) is calculated and the rest of the process for determining MFFS is same as described above.

6 Performance Study

In this section, we have compared the proposed method of determining MFFS with that of Chen and Hu. In section 6.1 system models and performance parameters are discussed. The simulation results are presented in section 6.2.

6.1 System Models and Performance Measures

The model simulated is a 3 dimensional torus $T_3(10,10,10)$ with 1000 nodes. We conducted simulations for 100 different values of m (number of faulty nodes) starting from 1 to 100 and generated 1000 random fault patterns for each m.

The time taken to determine MFFS in a faulty torus by these two distributed methods depends on the number of nodes executing 'Algorithm A' and the number of messages spread in the network. Therefore, for each fault pattern, the number of nodes that execute 'Algorithm A' and number of messages that will be spread in the network were measured. Then the average of 1000 values for each m is calculated. So we use the average number of nodes participating in determination process and the average number of messages used as the performance measures.

6.2 Simulation Results

Fig. 1 shows the graph of average number of nodes that participate (executes 'Algorithm A') in MFFS determination Vs number of faults. In Chen and Hu's [1] method, 'Algorithm A' is applied to all healthy (fault-free) nodes of the torus. Initially when there is only one faulty node, the 'Algorithm A' is applied to 999 healthy nodes, as depicted in the graph. As the number of faults increases, the number of healthy nodes keeps decreasing. Therefore, there is a decreasing trend in the number of nodes to which the 'Algorithm A' is applied.

In the proposed approach, 'Algorithm A' is applied to base-submesh (B) of the smallest submesh containing all faulty nodes (S). Initially, when the number of faulty nodes is one, the submesh S contains only one node, and therefore submesh B contains only one node. Hence, only one node executes the 'Algorithm A'. Initially as the number of faults increases and occurs in a random fashion, the probability of scattering of faults also increases. This rapidly increases the number of nodes in S and B, especially the healthy nodes. Hence, there is a fast increasing trend in the number of nodes that participate in MFFS determination.

The size of S and B cannot exceed the size of the torus. As the number of faults increases (for m=20 and onwards) the sizes of S and B starts getting saturated due to the size restriction, leading to a slowly increasing trend in the number of nodes that participate in MFFS determination. For m=60 and onwards, the sizes of S and B will reach the limit (size of torus) in most of the fault patterns, and now the number of nodes that participate in MFFS determination in proposed approach will be equal to that of Chen and Hu's method, so both the curves converges. This also indicates that in the worst case the proposed method works as good as Chen and Hu's method. As the number of messages is proportional to the number of nodes that participate in the submesh determination, a similar trend is seen in the Fig 2.

These graphs clearly depict a significant difference in the number of messages and nodes that participate in submesh determination between Chen and Hu's [1] and the proposed approach for low number of faults. As the time required in determining the maximal submesh is proportional to the number of nodes participating in the determination and the number of messages, it also decreases compared to Chen and Hu's [1] method. In the worst case, when number of faults is high, our scheme performs as good as Chen and Hu's method. But with high number of faults, the average size of the fault-free submesh will be very small [1] and it is highly unrealistic to work in such a condition. Therefore in a practical system the number of faults is small. Hence in a practical system, where number of faults is less, our scheme outperforms Chen and Hu's method.

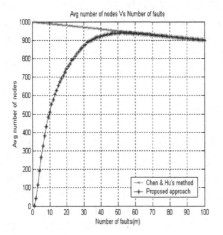

Fig. 1. Average number of nodes Versus Number of faults(m)

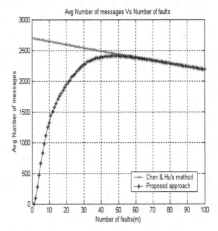

Fig. 2. Average number of messages Versus Number of faults(m)

7 Conclusions

In this paper we have showed that "it is sufficient to apply 'Algorithm A', proposed by Chen and Hu in [1] to only some part of the torus when there are few faulty elements". It is also shown that the part of the network to which 'Algorithm A' is to be applied is related to the smallest submesh containing all faulty nodes. In the proposed method, the number of messages passed during determination of MFFS is reduced (compared to Chen and Hu's method). The reduction in the number of messages will reduce the time taken for determination of MFFS and the congestion in the network.

This paper also presents an algorithm based on cyclic binary string to find a smallest submesh containing all faulty nodes in faulty torus. The CBSB Algorithm has a time complexity of $O(n(m+k))$ for a k-ary n-cube. The simulation results show that proposed strategy outperforms that of Chen and Hu's method.

This paper also extends the method to determine MFFS in the presence of faulty links. The proposed approach to determine MFFS in a torus containing faulty nodes and faulty links is fast and efficient method. This MFFS determination procedure could be useful for systems designed to operate in a gracefully degraded manner after faults occur.

References

1. H L Chen and S H Hu, "Submesh Determination in Faulty Tori and Meshes", IEEE Trans. Parallel and Distributed Systems, Vol. 12, no. 3, pp. 272-282, Mar. 2001.
2. R.E. Kessler and J.L. Schwarzmeier, "CRAY T3D: A New Dimension for Cray Research", Proc. 1993 Compcon Spring, pp. 176-182, 1993.
3. Intel Corporation, Paragon XP/S Product Overview, 1991.
4. NCUBE Corp., NCUBE/ten : An overview. Beaverton, Ore, Nov 1985.
5. C.L.Seitz, "The Cosmic Cube", Comm, ACM, Vol 28, no. 1, pp 22-23 Jan 1985.
6. "A Touchstone DELTA System Description ", Intel Corp. 1991.
7. The BlueGene/L Team, IBM and Lawrence Livermore National Laboratory, "An overview of the BlueGene/L Supercomputer", Proc. SuperComputing, Baltimore, Nov.16-22, 2002.
8. J. Bruck, R. Cypher, and C.-T. Ho, "Efficient Fault-Tolerant Mesh and Hypercube Architectures", Proc.22nd Int'l Symp. Fault-Tolerant Computing, pp. 162-169, July 1992.
9. T.A. Varvarigou, V.P. Roychowdhury, and T. Kailath, "Reconfiguring Processor Arrays Using Multiple-Track Models: The 3-Track-1-Spare-Approach", IEEE Trans. Computers, vol. 42, no. 11, pp. 1281-1293, Nov. 1993.
10. J. H. Kim and P. K. Rhee, "The Rule-Based Approach to Reconfiguration of 2-D Processor Arrays", IEEE Trans. Computers, vol. 42, no.11, pp. 1403-1408, Nov. 1993.
11. A. Chandra and R. Melhem, "Reconfiguration in 3D Meshes", Proc. 1994 Int'l Workshop Defect and Fault Tolerance in VLSI Systems, pp. 194-202, 1994.
12. F. Ozguner and C. Aykanat, "A Reconfiguration Algorithm for Fault Tolerance in a Hypercube Multiprocessor", Information Processing Letters, vol.29, pp.247-254, Nov 1988.
13. M.A. Sridar and C.S. Raghavendra, "On Finding Maximal Subcubes in Residual Hypercubes", Proc. Second IEEE Symp. Parallel and Distributed Processing, pp. 870-873, Dec. 1990.
14. S. Latifi, "Distributed Subcube Identification Algorithms for Reliable Hypercubes", Information Processing Letters, vol. 38, pp. 315-321, June 1991.
15. H.-L. Chen and N.-F. Tzeng, "Subcube Determination in Faulty Hypercubes", IEEE Trans. Computers, vol. 46, no. 8, pp. 871-879, Aug. 1997.
16. H.-L. Chen and N.-F. Tzeng, "A Boolean Expression-Based Approach for Maximum Incomplete Subcube Identification in Faulty Hypercubes," IEEE Trans. Parallel and Distributed Systems, vol. 8, no.11 pp. 1171-1183, Nov. 1997.
17. Seong-Moo Yoo, Hee Young Youn, "Finding Maximal Submeshes in Faulty 2D Mesh in the Presence of Failed Nodes", Proc. Second Aizu International Symp. Parallel Algorithms/ Architecture Synthesis, pps 97 –103, 17-21 March 1997.
18. J R Armstrong and F G Gray, "Fault Diagnosis in a Boolean n Cube Array of Microprocessors", IEEE Trans. Computers, Vol. 30, no. 8, pp. 587-590, Aug. 1981.
19. Stefano Chessa and Piero maestrini, "Correct and Almost Complete Diagnosis of Processor Grids", IEEE Trans. Computers, vol. 50, no. 10, pp. 1095-1102, Oct. 2001.

High Performance Cycle Detection Scheme
for Multiprocessing Systems

Ju Gyun Kim

Department of Computer Science, Sookmyung Women's University, Seoul, Korea
jgkim@sookmyung.ac.kr

Abstract. This paper presents a non-blocking deadlock detection scheme with immediate cycle detection in multiprocessing systems. It assumes an expedient state and a special case where each type of resource has one unit and each request is limited to one resource unit at a time. Unlike the previous deadlock detection schemes, this new method takes O(1) time for detecting a cycle and O(n+m) time for blocking or handling resource release where n and m are the number of processes and that of resources in the system. The deadlock detection latency is thus minimized and is constant regardless of n and m. However, in a multiprocessing system, the operating system can handle the blocking or release in parallel on a separate processor, thus not interfering with user process execution. To some applications where deadlock is concerned, a predictable and zero-latency deadlock detection scheme could be very useful.

1 Introduction

Researches on parallel, widely multiprocessing, systems have been carried out over many years. However, relatively little interest has been devoted to algorithms for deadlock problem even though its possibility of occurrence becomes larger. This paper presents a non-blocking deadlock detection algorithm which can detect a cycle immediately after its creation under multiprocessing systems.

If the system state is such that all satisfiable requests have been granted, there is a simple sufficient condition for deadlock. This situation occurs if the resource allocators do not defer satisfiable requests but grant them immediately when possible; most allocators follow this policy [2]. The resulting states are called *expedient*. If a system state is expedient, then a knot in the corresponding reusable resource graph is a sufficient condition for deadlock. Assume that a process may request only one unit at a time; that is, at most, one request edge can be connected to any process node. Then, for an expedient state, a knot becomes a necessary and a sufficient condition for deadlock [1,2,3,4]. An O(1) time knot detection scheme for this case has already been proposed in [9,11].

Moreover, in many situations, resources can be limited to have only one unit. With this additional restriction, existence of a cycle becomes a sufficient and necessary condition for deadlock so we can implement a cycle detection scheme with simpler and less resource waste way. This paper presents a revised version of cycle detection schemes, already presented by author [8,9], in order to get an algorithmic consistency and gives some generalities to the data structures used.

L. Bougé and V.K. Prasanna (Eds.): HiPC 2004, LNCS 3296, pp. 484–493, 2004.
© Springer-Verlag Berlin Heidelberg 2004

In order to detect a cycle, almost all existing algorithms start the detection procedure after an active process requests an unavailable resource. The process cannot progress further until the system completes its deadlock inspection, and this is particularly inefficient under multiprocessing environment.

Suppose that we can use an available processor for deadlock detection, then running the detection process and user process in parallel is impossible under existing algorithms. The user process must wait until it is informed of the result of the detection process, because one of two different procedures is required according to the detection result; that is, process switching is sufficient when the process is merely blocked, otherwise a proper deadlock recovery procedure must be run. The detection process is idle while there are no such requests for detection. This means that the busy periods of the two processes are not overlapped at all, so we can not achieve the enhanced efficiency of parallel computation, even if we can run the existing algorithms under multiprocessing systems.

For another example of poor CPU usage, suppose that several cooperating processes, communicating with each other, run in parallel and one of them is blocked. If we know immediately that this blocking leads to deadlock, the deadlocked and other affected processes are rolled back from this point soon, and process switching may be performed at the earliest possible time. But if we must spend $O(n+m)$ time to know that the blocking eventually leads to deadlock, other cooperating processes still running in this interval, but already affected before the blocking point, must be rolled back together. So, the work performed during the interval, otherwise useful for increasing the throughput under parallel processing environment, is wasted.

Thus, the nature of existing algorithms leads to loss of parallelism. They are invoked to find the existence of deadlock whenever a process is blocked after requesting an unavailable resource, and this requires $O(n+m)$ time, where n and m are the number of processes and resources in the system. If we could instead utilize the $O(n+m)$ time amounts to prepare for the next detection, we could use the prepared information to determine the existence of a cycle immediately after a process's request action. The scheme proposed in this paper takes $O(1)$ time to detect a cycle, and $O(n+m)$ time for handling the detection information - updating the data structures - for the next detection.

For $O(1)$ time cycle detection, proposed scheme makes several assumptions and defines system state with tree structure. Section 2 reviews related works, and section 3 describes the system model, and section 4 presents algorithms and analysis. Concluding remarks are given in section 5.

2   Related Works

The schemes of Shoshani and Holt[4,5] are valuable to refer to them because they present basic idea for the treatment of deadlock problems. Leibfried represented a system state with an adjacency matrix and proposed a scheme which can detect deadlock with repeated multiplications of the matrix [7], but it requires at least $O(m^3)$ time complexity to find a knot.

Any algorithms based on so-called WFG(Wait-For-Graph) method make an exhaustive search through reachable paths in the graph to find the existence of cycle

[1,2,5,6], and have $O(n+m)$ time complexity in its worst case [2,3,6]. Several drawbacks common to such algorithms are easily seen. First, deadlock detection may take longer as n and m increase. Second, the time required to detect deadlock may vary depending on the situation, and thus becomes unpredictable. Third, we cannot expect performance enhancement under multiprocessor systems because the algorithms have not been designed originally for such systems. Reduced parallelism and unpredictable deadlock detection time could complicate their implementation on some systems. Idea in [10] proposed $O(1)$ cycle detection but it is impossible to be extended to knot problem. As mentioned above, [8] also contains cycle detection but the data structures and algorithms are inconsistently different from that of [11], and thus can not be extended to knot detection.

Unlike the existing schemes, proposed scheme takes $O(1)$ time for detecting deadlock, and $O(n+m)$ time for updating data structures after detection or release. The deadlock detection latency is thus minimized, and is constant, regardless of n and m. Release operation takes longer with this scheme. However, in multiprocessor and distributed systems, the operating system can handle the release on the fly, running on a separate processor, and thus does not interfere with user process execution. If the process releasing the resource is a high priority job, then it may merely notify the release to the operating system but does not necessarily have to wait for the response from the operating system. This can maximizes the possibility of parallelism and resource utilization.

3 System Model

As mentioned earlier, we make several simplifying assumptions about the system. First, we assume that each resource has one unit. With this restriction, a cycle becomes a sufficient condition for deadlock. Single unit resources are not uncommon in present systems; for example, the ENQ/DEQ macros of IBM OS/360 deal exclusively with resources of this type [2]. Second, we assume that operating system does not defer satisfiable requests but grant them immediately when possible. The resulting states are called expedient. Finally, we assume that processes request only one unit of resource at a time. This assumption can be applicable because multiple requests can be serialized without severely restricting the process's request behavior.

By these assumptions, We can find that every process node in WFG can have at most one outgoing arc since we assumed single unit request. Also, every resource node can have at most one outgoing arc, since we have restricted to serially reusable resource and we assumed that each resource has one unit. Since WFG is bipartite, this implies that every pair of vertices in our WFG is connected by only one edge, or not connected at all. As for the incoming arc, both process and resource node can have more than one incoming arcs directed to them. Therefore, with the proposed system, the WFG becomes a collection of trees, when we ignore the direction of the edges.

In Fig. 1, we represent trees with root node at the top and all arcs directed upward, leaving the arrowheads of the arc implicit. Two resources R1 and R2 have been allocated to the process P1. The process P2 and P3 are blocked at resource R1. Two resources R5 and R6 have been allocated to process P3. All the processes and resources in Fig. 2 are connected, yet do not contain any cycle. Among the processes in Fig. 1,

P1 is unique in that it does not have any request edge. Thus, P1 is the only active process and all the other processes in Fig. 1 are in blocked states. We call P1 the root of the tree. Each active process in WFG owns a tree, in the sense that it is the root of respective tree. Resource nodes belonging to a tree have been allocated to some process. The sons of a root node represent the resources currently allocated to the root process. The sons of a resource node represent the processes blocked at its father resource node. In order to distinguish such trees in a WFG, we name each tree by the process number of its respective root process. Thus the tree illustrated in Fig. 1 will be identified by the process number of P1, number of its root process.

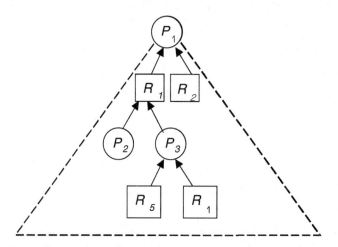

Fig. 1. An example of a tree from WFG. Its identity is equal to the root process number. Dotted line means that a subgraph in the WFG forms a tree

Clearly, state of the system can be changed only by the root nodes because only an active process can either request a new resource or release one of the resources it has been using. In the following, we show how deadlock detection could be accomplished in a fixed time regardless of the number of processes and the number of resources.

In the proposed system, the WFG consists of several disjoint trees and only the root nodes of the trees could request a new resource. A root node could form a new cycle only by requesting a resource node already belonging to the same tree it is in. In other words, a process Pi requesting resource Rj becomes deadlocked, if and only if Rj already belongs to the same tree rooted by Pi itself, thus creating a new cycle. Otherwise, Pi will either get Rj immediately or be blocked but not deadlocked.

Therefore, in order to detect a cycle, checking to which tree Rj belongs is enough. If Rj does not belong to any tree, Rj is free and can be immediately allocated. If Rj belongs to the same tree as the requesting process Pi, then it is a deadlock state. Otherwise, Pi will simply be blocked but not deadlocked (i.e., Rj and Pi currently belong to different trees). Such detection process could be performed in O(1) time if we associate each resource with the tree identifier it belongs to all the time.

4 Algorithms and Analysis

Request edges from a root to resources in Fig. 2 illustrate three possible scenarios for requests. A request labeled (a) in Fig. 2 is a trivial case. R_a is available and can be allocated to P_i immediately. In type (b) request, R_b is not available and P_i will have to be blocked. But this request does not cause a cycle because P_i and R_b belong to different trees i and j.

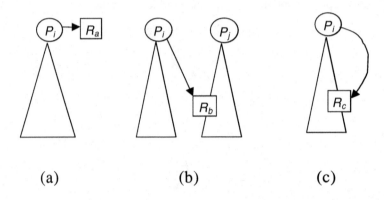

(a)   (b)   (c)

Fig. 2. Three possible scenarios of a process's request

When a process is blocked (but not deadlocked), the two trees are merged to become a larger one. The tree i rooted by blocked process P_i is now part of the tree j which contains R_b. That is, tree i originally rooted by P_i now becomes a subtree of tree j rooted by P_j. In (c), a cycle is formed. A proper recovery operation is now required.

When a process releases a resource, three types of scenario are possible. In Fig. 3, P_i is a root and is releasing a resource it has been using. In Fig. 3 (a), P_i releases a resource which is not wanted by any other processes (thus R_j is a leaf), The release is trivial in this case - simply delete allocation edge. In Fig. 3 (b), releasing R_j causes operating system to wake up P_k. Now P_k becomes a new root and all its descendents which previously had tree number i must now change their tree number to a new tree number k. In Fig. 3 (c), many processes are blocked at R_j. Suppose that P_k is chosen as a receiver of R_j. As a result of this, R_j is allocated to P_k and now P_k becomes an active process and roots a new tree. The tree rooted by P_k is now splitted from the tree rooted by P_i. All the descendants under R_j must be departed from the tree i and change their tree number to k while all the resource nodes of left subtree of P_i keep their tree number unchanged.

Data structures used in the proposed scheme are explained below.

① Matrix *History*(i,j) contains all status of block, allocation and root information. where rows represent processes and columns represent resources, and *History*(i,j)=1, i=1,...,n, j=1,...,m, means that resource, R_j, is allocated to process, P_i. One extra row is appended for providing root information and also one extra column is appended for providing blocking factor. For example, *History*(n+1,j) contains a root (process) number of resource j and *History*(i,m+1) contains a resource number by which process i has been blocked.

② Two set variables, *child* and *dsdnt* are used for readability and algorithm's purpose.

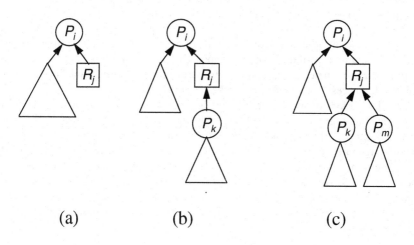

(a)   (b)   (c)

Fig. 3. Three possible scenarios of a process's release

Following algorithms perform allocation and release operations. Algorithm *split* called in *release* divides a tree into two smaller ones.

```
Algorithm Request(i, j)
      // Root i requests a resource j //
 1: begin
 2:    Key := History(n+1, j);
 3:    case
 4:    Key = null: // j is an available resource //
              History(i, j) := 1;
              History(n+1, j) := i;
 5:    Key = i : // Deadlocked! A proper recovery action
                    is needed //
 6:    Key • i : // Blocked //
 7:         History(i, m+1) := j;
 8:         for L = 1 to m do
              begin
                 if History(n+1, L) = i then
 9:                  History(n+1, L) := History(n+1, j);
                    // change root from i to j's root //
                 endif
              end
           endfor
       endcase
10: end
```

The state resulting from statement 4 is evidently not deadlocked as explained in Fig. 2 (a). Statement 6, as explained in Fig. 2 (b), means that process i requests an unavailable resource j and then blocked. This case causes tree merging and requires $O(m)$ time complexity in order to prepare for the next detection but these operations can be processed in the background, in parallel with other system activities, given an extra processor.

Case of statement 5 is the condition for deadlock and requires $O(1)$ time complexity. This is a major contribution of the proposed scheme. An $O(1)$ deadlock detection is more important when several coordinating processes run in parallel. Immediate decision of one process's state can allow other coordinating processes to continue, without any delay or loss of work which might otherwise be needed in order to know the result of deadlock inspection. Moreover, such immediate response make it possible to get better resource utilization and suitable for realtime systems.

When a resource is released, deadlock can not occur. Statement 2 of algorithm *Split* can be done by any of well known tree traversal algorithms and the complexity is known to be equivalent to the number of edges in the tree. By assumption, A tree in this paper can have $n + m$ edges at most. Therefore, *Release* operation takes $O(n+m)$ time in the worst case and this also can be processed in background like merging in *Request* operation.

```
Algorithm Release(i, j)

//Root i releases a resource j//
1: begin
2:    for L = 1 to n do
         begin
            if History(L, m+1) = j then
3:             child := History(L, m+1);
               // child is a set variable and contains proc-
                  esses who has been blocked by j //
            endif
         end
      endfor
4:    case
5:    child = null:   // Resource j is a leaf node //
               History(i, j) := 0;
               History(n+1,   j) := null;
6:    child = 1 : // Only one process, say process k, is
                     blocked by j and it becomes to a new
                     root by allocating resource j //
               History(i, j) := 0;
               History(k, j) := 1;
               History(n+1, j) := k;
               History(k, m+1) := null;
               // Data adjustment for future O(1) cycle
                  detection //
               call  Split(i, j, k);
```

```
7:    otherwise : // Several processes have been blocked
by resource j. Let process k is now selected to become
a new root among them //
                History(i, j) := 0;
                History(k, j) := 1;
                History(n+1, j) := k;
                History(k, m+1) := null;
                call  Split(i, j, k);
        endcase
8: end
```

```
Algorithm Split(i,j,k) // Divide a tree into two
                              smaller ones //

1: begin
2:    Find all descendants resource nodes from process k
      and put them into dsdnt.
3:    for all nodes in dsdnt do
                Set their (n+1)ᵗʰ row to k;
      endfor
4: end
```

The proposed scheme requires $O(n+m)$ time - same as existing schemes - to treat the deadlock problem, but makes it possible to detect deadlock within a constant time. In Fig. 4, B denotes a point of time at which a process is blocked. and D is a decision point whether it is deadlocked or not. All the operations related to deadlock resolution caused by event B are completed at time F. So, the time interval [B,F] is equivalent to $O(n+m)$. The time required for interval [B,D] of (a) and [D,F] of (b) are constant. During the interval [D,F] of (a), data structure adjustment for next detection is processed. In (b), interval [B,D] is consumed by a graph search to find a cycle.

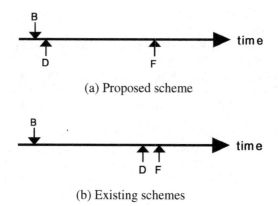

(a) Proposed scheme

(b) Existing schemes

Fig. 4. Time comparison between proposed scheme and existing schemes

Under multiprocessor systems, operations for release or blocking can be performed on the fly, running on a separate processor, and thus maximizing the inherent parallelism of the systems. More than one processor can be assigned for detection, if available. Non-blocking detection causes immediate response from the operating system, thus minimizing the waste of resources, and is beneficial for realtime systems. Although the proposed scheme is designed for multiprocessor systems, it is also advantageous on single CPU systems.

5 Concluding Remarks

This paper presents a new deadlock detection scheme for a special case system where each type of resource has one unit and each request is limited to one unit at a time. It needs $O(1)$ time for detecting deadlock and $O(n+m)$ time for blocking a process or releasing a resource. But the $O(n+m)$ time operations can be performed in the background or can run on separate processor. The proposed scheme is non-blocking and guarantees an immediate response on multiprocessor systems. Immediate detection also allows us to use the proposed scheme for deadlock avoidance if a process that causes deadlock is selected to be killed. Future work may include applying similar methods to less restrictive systems.

Acknowledgements. This research was supported by the Sookmyung Women's University Research Grants 2004.

References

1. A. Silberschatz, P. Galvin, Operating System Concepts, 4th Ed., Addison-Wesley, 1994.
2. L. Bic and A. C. Shaw, The Logical Design of Operating Systems, 2nd Ed., Prentice-Hall, 1988.
3. M. Maekawa, A. E. Oldehoeft and R. R. Oldehoeft, Operating Systems - Advanced concepts, Benjamin-Cummings Pub., 1987.
4. R. C. Holt, "Some Deadlock Properties of Computer Systems", ACM Computing surveys, Vol. 4, No. 3, Sep., 1972.
5. A. Shoshani and E. G. Coffman, "Prevention Detection and Recovery from System Deadlock", Proc. 4th annual Princeton Conf. on Information Sciences and System, Mar., 1970.
6. S. S. Isloor and T. A. Marsland, "The Deadlock Problem: An Overview", IEEE Computer, Sep., 1980.
7. T. F. Leibfried Jr., "A Deadlock Detection and Recovery Algorithm Using the Formalism of a Directed Graph Matrix", Operating System Review, Vol. 23, No. 2, Apr., 1989.
8. J. G. Kim and K. Koh, "An O(1) Time Deadlock Detection Scheme in Single Unit and Single Request Multiprocessor System", IEEE TENCON'91, Delhi, India, Aug., 1991, Vol. 2, pp. 219-223.
9. J. G. Kim, "A Non-blocking Deadlock Detection Scheme for Multiprocessor Systems", Ph.D Thesis, SNU, SEOUL, Feb., 1992.

10. Y. S. Ryu and K. Koh, "A Predictable Deadlock Detection Technique for a Single resource and single Request System", Proc. of the 14th IASTED Int'l Conf. on Applied Informatics, Innsbruck, Austria, Feb., 1996, pp. 35-38,

11. J. G. Kim, "An Algorithmic Approach on Deadlock Detection for Enhanced parallelism in Multiprocessing Systems", Proc. of the 2nd Aizu int'l Symp. on Parallel Algorithms/Architecture Synthesis, IEEE Computer Society Press PR07870, pp. 233-238, Mar., 1997.

Improved Quality of Solutions for Multiobjective Spanning Tree Problem Using Distributed Evolutionary Algorithm

Rajeev Kumar, P. K. Singh, and P. P. Chakrabarti

Department of Computer Science and Engineering,
Indian Institute of Technology Kharagpur,
Kharagpur, WB 721 302, India
{rkumar, pksingh, ppchak}@cse.iitkgp.ernet.in

Abstract. The problem of computing spanning trees along with specific constraints has been studied in many forms. Most of the problem instances are NP-hard, and many approximation and stochastic algorithms which yield a *single* solution, have been proposed. Essentially, such problems are multi-objective in nature, and a major challenge to solving the problems is to capture possibly all the (representative) equivalent and diverse solutions at convergence. In this paper, we attempt to solve the generic multi-objective spanning tree (MOST) problem, in a novel way, using an evolutionary algorithm (EA). We consider, without loss of generality, edge-cost and diameter as the two objectives, and use a multiobjective evolutionary algorithm (MOEA) that produces diverse solutions without needing *a priori* knowledge of the solution space. We employ a distributed version of the algorithm and generate solutions from multiple tribes. We use this approach for generating (near-) optimal spanning trees from benchmark data of different sizes. Since no experimental results are available for MOST, we consider two well known diameter-constrained spanning tree algorithms and modify them to generate a Pareto-front for comparison. Interestingly, we observe that none of the existing algorithms could provide good solutions in the entire range of the Pareto-front.

1 Introduction

Computing a minimum spanning tree (MST) from a connected graph is a well-studied problem and many fast algorithms and analytical analyses are available [1, 2, 3, 4, 5, 6, 7]. However, many real-life network optimization problems require the spanning tree to satisfy additional constraints along with minimum edge-cost. For example, communication network design problem for multicast routing of multimedia communication requires constructing a minimal cost spanning/Steiner tree with given constraints on diameter. VLSI circuit design problems aim at finding minimum cost spanning/Steiner trees given delay bound constraints on source-sink connections. Analogously, there exists the problem of degree/diameter-constrained minimum cost networks in many other engineering applications too (see [3] and the references therein).

Many such MST problem instances having a bound on the degree, a bound on the diameter, capacitated trees or bounds for two parameters to be satisfied simultaneously are

L. Bougé and V.K. Prasanna (Eds.): HiPC 2004, LNCS 3296, pp. 494–503, 2004.
© Springer-Verlag Berlin Heidelberg 2004

listed in [3]. Finding spanning trees of sufficient generality and of minimal cost subject to satisfaction of additional constraints is often NP-hard [3, 4]. Many such design problems have been attempted and approximate solutions obtained using heuristics. For example, the research groups of Deo et al. [5, 6, 7] and Ravi et al. [3, 4] have presented approximation algorithms by optimizing one criterion subject to a budget on the other. In recent years, evolutionary algorithms (EAs) have emerged as powerful tools to approximate solutions of such NP-hard problems. For example, Raidl & Julstorm [8, 9] and Knowles & Corne [10, 11] attempted to solve diameter and degree constrained minimum spanning tree problems, respectively using EAs. All such approximation and evolutionary algorithms yield a *single* optimized solution subject to satisfaction of the constraint(s).

We argue that such constrained MST problems are essentially multiobjective in nature. A multiobjective optimizer yields a set of all representative equivalent and diverse solutions rather a single solution; the set of all optimal solutions is called the Pareto-front. Secondly, extending this constraint-optimization approach to multi-criteria problems (involving two or more than two objectives/constraints) the techniques require improving upon more than one constraints. Thirdly and more importantly, such approaches may not yield all the representative optimal solutions. For example, most conventional approaches to solve network design problems start with a minimum spanning tree (MST), and thus effectively minimize the cost. With some variations induced by ϵ-constraint method, most other solutions obtained are located near the minimal-cost region of the Pareto-front, and thus do not form the complete (approximated) Pareto-front.

In this work, we try to overcome the disadvantages of conventional techniques and single objective EAs. We use multiobjective EA to obtain a (near-optimal) Pareto-front. For a wide-ranging review, a critical analysis of evolutionary approaches to multiobjective optimization and many implementations of multiobjective EAs, see [12, 13]. These implementations achieve diverse and equivalent solutions by some diversity preserving mechanism but they do not talk about convergence. Any explicit diversity preserving method needs prior knowledge of many parameters and the efficacy of such a mechanism depends on successful fine-tuning of these parameters. In a recent study, Purshouse & Fleming [14] extensively studied the effect of sharing, along with elitism and ranking, and concluded that while sharing can be beneficial, it can also prove surprisingly ineffective if the parameters are not carefully tuned.

Kumar & Rockett [15] proposed use of Rank-histograms for monitoring convergence of Pareto-front while maintaining diversity without any *explicit* diversity preserving operator. Their algorithm is demonstrated to work for problems of *unknown* nature. Secondly, assessing convergence does not need *a priori* knowledge for monitoring movement of Pareto-front using rank-histograms. Some other recent studies have been done on combining convergence with diversity. Laumanns et al. [16] proposed an ϵ-dominance for getting an ϵ-approximate Pareto-front for problems whose optimal Pareto-set is *known*.

In this work, we use the Pareto Converging Genetic Algorithm (PCGA) [15] which has been demonstrated to work effectively across complex problems and achieves diversity without needing *a priori* knowledge of the solution space. PCGA excludes any explicit mechanism to preserve diversity and allows a natural selection process to maintain diversity. Thus multiple, equally good solutions to the problem, are provided. Another major challenge to solving unknown problems is how to ensure convergence. We use a distributed version of PCGA and generate solutions using multiple tribes and merge them to ensure convergence. PCGA assesses convergence to the Pareto-front which, by definition, is unknown in most real search problems of multi-dimensionality, by use of rank-histograms.

We consider, without loss of generality, edge-cost and tree-diameter as the two objectives to be minimized, though the framework presented here is generic enough to include any number of objectives to be optimized. The rest of the paper is organized as follows. In Section 2, we include few definitions and a brief review of the multiobjective evolutionary algorithm (MOEA) along with the issues to be addressed for achieving quality solutions along a Pareto-front in the context of a MOEA. We describe, in Section 3, the representation scheme for the spanning tree and its implementation using PCGA. Then, we present results in Section 4 along with a comparison with other approaches. Finally, we draw conclusions in Section 5.

2   Multiobjective Evolutionary Algorithms: A Review

EAs have emerged as powerful black-box optimization tools to approximate solutions for NP-hard combinatorial optimization problems. In the multiobjective scenario, EAs often find effectively a set of mutually competitive solutions without applying much problem-specific information. However, achieving proper diversity in the solutions while approaching convergence is a challenge in multiobjective optimization, especially for unknown problems.

Mathematically, a general multiobjective optimization problem containing a number of objectives to be maximized/minimized along with (optional) constraints for satisfaction of achievable goal vectors can be written as:

$$Minimize/\ Maximize\ Objective\ f_m\left(\mathbf{X}\right),\quad m=1,2,...,M$$
$$subject\ to\ Constraint\ g_k\left(\mathbf{X}\right)\leq c_k,\quad k=1,2,...,K$$

$where\ \mathbf{X}=\{x_n:n=1,2,...,N\}\ is\ an\ N-tuple\ vector\ of\ variables$
$and\ \mathbf{F}=\{f_m:m=1,2,...,M\}\ is\ an\ M-tuple\ vector\ of\ objectives$

In a maximization problem of m objectives, an individual objective vector F_i is partially less than another individual objective vector F_j (symbolically represented by $F_i \prec F_j$) iff:

$$(F_i \prec F_j)=(\forall_m)(f_{mi}\leq f_{mj})\wedge(\exists_m)(f_{mi}<f_{mj})$$

Then F_j is said to dominate F_i. If an individual is not dominated by any other individual, it is said to be non-dominated. A relationship is termed as weakly-dominance if, either both the individuals are equal or one is better than the other. The notion of Pareto-optimality was introduced to assign equal probabilities of regeneration to all the individuals in the population. We use the notion of Pareto-optimality if $F=(f_1,....,f_m)$ is a vector-valued objective function.

Definition 1. Pareto Optimal Set : *A set $A\subseteq Y$(where Y denotes the entire decision space) is called a Pareto optimal set iff*

$\forall\, a \in A$: *there does not exist* $x \in Y$: $a \prec x$.

The primary goal of a multiobjective optimization algorithm is to attain the Pareto-optimal set. However, in most practical cases it is not possible to generate the entire Pareto-optimal set. This might be the case when the size of the set is exponentially large. Thus, we confine our goal to attain an approximate set. This approximate set is usually polynomial in size. Since in most cases the objective functions are not bijective, there are a number of individuals in the decision space which may be mapped to the same objective function.

Definition 2. Approximate Set : *A set* $A_p \subseteq A$ *(Pareto-optimal Set) is called an approximate set if there is no indiviual in* A_p *which is weakly dominated by any other member of* A_p.

Another strategy that might be used to attain an approximate set is to try and obtain an inferior Pareto front. Such a front may be inferior with respect to the distance from the actual front in the decision space or the objective space. If the front differs from the actual optimal front by a distance of ϵ in the objective space, then, the dominance relation is called a $(1 + \epsilon)$-dominance.

The major achievement of the Pareto rank-based research is that a multiobjective vector is reduced to a scalar fitness without combining the objectives in any way. Further, the use of fitness based on Pareto ranking permits non-dominated individuals to be sampled at the same rate thus according equal preference to all non-dominated solutions in evolving the next generation. The mapping from ranks to fitness values however is an influential factor in selecting mates for reproduction.

Almost all the multiobjective evolutionary algorithms/implementations have ignored the issue of convergence and are thus, unsuitable for solving *unknown* problems. Another drawback of most of these algorithms/implementations is the explicit use of parameterized sharing, mating restriction and/or some other diversity preserving operator. Apart from its heuristic nature, the selection of the domain in which to perform sharing (variable (genotype) or objective (phenotype)) is also debatable. Any explicit diversity preserving mechanism method needs prior knowledge of many parameters and the efficacy of such a mechanism depends on successful fine-tuning of these parameters. It is the experience of almost all researchers that proper tuning of sharing parameters is necessary for effective performance, otherwise, the results can be ineffective if parameters are not properly tuned [14]. In particular to MOST problem where we use a special encoding [9], incorporation of such knowledge is not an easy task.

Many metrics have been proposed for quantitative evaluation of the quality of solutions [12, 13]. Essentially, these metric are divided into two classes:

– Diversity metrics : Coverage and sampling of the obtained solutions across the front, and
– Convergence metrics : Distance of the obtained solution-front from the (known) optimal Pareto-front.

Some of these metrics (e.g., generational distance, volume of space covered, error ratio measures of closeness of the Pareto-front to the true Pareto front) are only applicable where the solution is known. Other metrics (e.g. ratio of non-dominated individuals, uniform distribution) quantify the Pareto-front and can only be used to assess diversity. The MOST problem is an NP-hard problem, the actual Pareto-front is not known. Therefore, we can not find the distance of the obtained solutions from the actual (which is unknown) Pareto-front. In Section 4, we will show that we could obtain superior solutions using our EA implementation than the solutions obtained from best-known heuristics for the problem; therefore, distance metrics becomes ineffective for the MOST problem.

3   Design and Implementation

Evolutionary algorithm operators, namely, mutation and crossover imitate the process of natural evolution, and are instrumental in exploring the search space. The efficiency of the evolutionary search depends how a problem (in this case, a spanning tree) is represented in a chromosome and the reproduction operators. There are many encoding schemes to represent spanning trees – see [9] for a detailed review and comparison. For example, one classic representation scheme is Prüfer encoding which is used by Zhou & Gen [17]. Raidl & Julstorm [9] and Knowles & Corne [11] have pointed out that Prüfer numbers have poor locality and heritability and are thus unsuitable for evolutionary search. Deo et al. suggested use of other variants of Prüfer mappings [7]. Recently, Raidl & Julstorm [9] proposed spanning trees to be represented directly as sets of the edges and have shown locality, heritability and computational efficiency of the edge sets for evolutionary search. In this work, we use edge-set scheme for representing spanning trees to exploring the search space.

Initial Population: We generate initial population based on random generation of spanning trees; we do not choose the cheapest edge from the currently eligible list of edges (as per Prim's algorithm) rather we select a random edge from the eligible list. The other variants of generating initial trees are based on One-Time-Tree Construction (OTTC) [6] and Randomized Greedy Heuristics (RGH) [8] algorithms.

Fitness Evaluation: We use Pareto-rank based EA implementation. The Pareto-rank of each individual is equal to one more than the number of individuals dominating it in the multiobjective vector space. All the non-dominated individuals are assigned rank one. The values of the two objectives to be minimized (cost and diameter) are used to calculate rank of the individual. Based on the two objectives rank of the individual is calculated. In this work, we calculate fitness of an individual by an inverse quadratic function of the Pareto-rank.

Other Genetic Operator: We select crossover operation to provide strong habitability such that the generated trees consist of the parental edges as far as possible. For generating valid trees, we include non-parental edges into the offspring tree. The mutation operator generates valid spanning trees. We use the Roulette wheel selection for selecting the parents.

Ensuring Convergence: We compute *Intra*-Island Rank-histogram for each epoch of the evolutionary evolution and monitor the movement of the Pareto-front. Since, this is a hard problem, it is likely that the improvements may get trapped in local minima. To ensure a global (near-) optimal Pareto-front, we use a multi-tribal/island approach and monitor the Pareto-front using *Inter*-Island Rank histogram. See [15] for details of computation of Intra-/Inter- Rank histogram.

Algorithm: The PCGA algorithm [15] used in this work is a steady-state algorithm and can be seen as an example of $(\mu + 2)$ – Evolutionary Strategy (ES) in terms of its selection mechanism [12, 13]. In this algorithm, individuals are compared against the total population set according to a tied Pareto-ranking scheme and the population is selectively moved towards convergence by discarding the lowest ranked individuals in each evolution. In doing so, we require no parameters such as size of the sub-population in tournament selection or sharing/niching parameters. Initially, the whole population of size N is ranked and fitness is assigned by interpolating from the best individual (rank = 1) to the lowest (rank $\leq N$) according to some simple monotonic function. A pair of mates is randomly chosen biased in the sizes of the roulette wheel segments and crossed-over and/or mutated to produce offspring. The offspring are inserted into the population set according to their ranks against the whole population and the lowest ranked two individuals are eliminated to restore the population size to N. The process is iterated until a convergence criterion based on rank-histogram is achieved [15].

If two individuals have the same objective vector, we lower the rank of one of the individual by one; this way, we are able to remove the duplicates from the set of nondominated solutions without loss of generality. For a meaningful comparison of two real numbers during ranking, we restrict the floating-point precision of the objective values to a few units of precision. This algorithm does not explicitly use any diversity preserving mechanism, however, lowering the rank of the individual having the identical objective vector (with restricted units of precision) is analogous in some way to a sort of sharing/niching mechanism (in objective space) which effectively controls the selection pressure and thus *partly* contributes to diversity (For other factors that contribute to diversity, see [15]).

4   Results

We tested generation of dual objective spanning tree using our MOEA framework and selected benchmark data taken from Beasley's OR library[1]. The OR-Library is a collection of test data sets for a variety of Operations Research (OR) problems. We considered the Euclidean Steiner problem data which was used by previous researchers, e.g., Raidl-SAC. We considered datasets of up to 250 nodes for this work, and few representative results are included in rest of this Section.

For comparison, we also include results obtained from two well-known diameter constrained algorithms, namely, One-Time Tree Construction (OTTC) [6] and Randomized Greedy Heuristics (RGH) [8] algorithms. Both the algorithms have been demonstrated for Beasley's OR data and few results included in their respective papers. Both algorithms are single objective algorithms and generate a single tree subject to the diameter

---

[1] http://mscmga.ms.ic.ac.uk/info.html

constraint. Our MOST algorithm simultaneously optimizes both the objectives and generates a (near-optimal) Pareto-front which comprises a set of solutions. Therefore, we iteratively run both the OTTC and RGH algorithms by varying the value of the diameter constraint and generate sets of solutions to form the respective Pareto-fronts, for comparison with the Pareto-front obtained from the proposed multiobjective evolutionary algorithm.

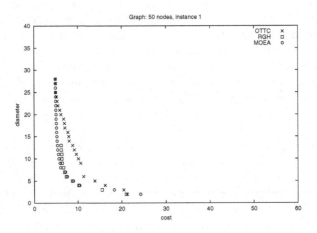

Fig. 1. Pareto front generated from evolutionary algorithm for a 50 node data. Other fronts generated from OTTC and RGH algorithms are also shown in the plot

For randomized algorithms, evolutionary and RGH, we have repeated experiments ten times and include here a single set of representative result obtained from the runs. We have included results obtained from 50 and 100 node data in Figures 1 and 2, respectively.

It can be observed from Figures 1 and 2, that this is indeed difficult to find the solutions in the higher range of diameter. In fact, RGH algorithm could not find any solutions in this range of diameter; we generated multiple sets of solutions with multiple runs of RGH algorithm with different initial values but none of the run could generate any solution in this range of diameter. It can also be observed from Figures 1 and 2 that the solutions obtained form OTTC algorithm are good in lower and higher range of diameter, however, the results obtained from RGH are good only in the lower range of the diameter. Contrary to this, EA is able to locate solutions in the higher range of the diameter with almost comparable quality of the solutions obtained by OTTC. The solutions obtained by OTTC in the middle range are much sub-optimal and are inferior to the solutions obtained by EA. In the upper-middle range of diameters, RGH could not locate solutions at all, and the solutions located in this range by the OTTC are much inferior to the solutions obtained by EA. Thus, the quality of solutions obtained by EA is much superior in this range, and comparable in higher range to those of OTTC.

The solutions obtained by EA are further improved by running the algorithm multiple times, and merging the obtained solutions to form a single Pareto-front. Results obtained from three randomly initialized runs of evolutionary algorithm for 100 node data to form

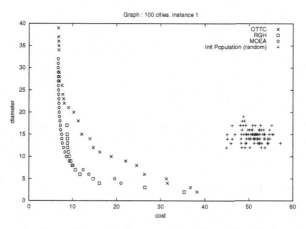

Fig. 2. Pareto front generated from evolutionary algorithm for a 100 node data. Initial population is also shown. Other fronts from OTTC and RGH algorithms are also shown in the plot

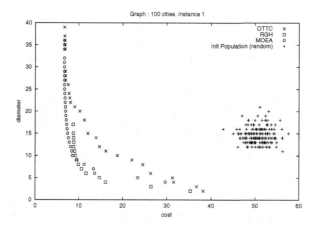

Fig. 3. Improved Pareto front generated from three tribes of evolutionary algorithm for the 100 node data; improvement in the lower and higher ranges of diameters are clearly visible. Initial population of one single tribe is shown. Other fronts from OTTC and RGH algorithms are also shown in the plot

an improved Pareto-front are included in Figure 3. It can be seen from Figure 3 that the solutions are improved in the lower and higher ranges of diameters. (Solutions obtained from EA are marginally sub-optimal compared to RGH algorithm in very low-range of diameter; this is obvious because EA is generic for any diameter values while RGH is tuned to the specific values. If EA is run with such known values, EA is guaranteed to give superior solutions.) This is possible because the solution points obtained from two tribes of EAs were distinct and diverse in lower and higher ranges of diameter. This is the clear advantage of using a distributed version of the evolutionary algorithm; the results could still be improved with a few more tribes. Additionally, such a multi-tribal approach is a test on convergence too.

These are interesting observations, and are partly contrary to those reported by Raidl & Julstorm [8]. Raidl & Julstorm have shown that their technique works the best over all the other such techniques including OTTC. We reiterate that their conclusions were based on the experiments which they did for a particular value of the diameter and they could not observe the results over the entire range of diameter. We are currently investigating the empirical behavior shown by these three algorithms, and how this knowledge can be used to further improve the solution-set.

5   Discussion and Conclusions

In this work, we demonstrated generating spanning trees subject to their satisfying the twin objectives of minimum cost and diameter. The obtained solution is a set of (near-optimal) spanning trees that are non-inferior with respect to each other. A network designer having a range of network cost and diameter in mind, can examine several optimal trees simultaneously and choose one based on these requirements and other engineering considerations.

To the best of our knowledge, this is the first work which attempts obtaining the complete Pareto front. Zhou & Gen [17] also obtained a set of solutions, they did not experiment on any benchmark data and, therefore, could not compare the quality of the solutions. It is shown by Knowles & Corne [11] that the front obtained by Zhou & Gen [17], was sub-optimal. We attribute the sub-optimality due to their use of an EA implementation which was unable to assess convergence. Knowles & Corne [11] used a weighted sum approach and could get the comparable solutions but their approach is sensitive to the selection of weight values.

The work presented in this paper presents a generic framework which can be used to optimize any number of objectives simultaneously for spanning tree problems. The simultaneous optimization of objectives approach has merits over the constrained-based approaches, e.g., OTTC and RGH algorithms. It is shown that the constrained-based approaches are unable to produce quality solutions over the entire range of the Pareto-front. For example, the best known algorithm of diameter-constrained spanning tree is RGH which is shown to be good for smaller values of diameters *only*, and is unable to produce solutions in the higher range. Similarly, the other well-known OTTC algorithm produces sub-optimal solutions in the middle range of the diameter. EA could obtain superior solutions in the entire range of the objective-values. The solutions obtained by EA may further be improved marginally by proper tuning of evolutionary operators for the specific values of the objectives by introducing problem specific knowledge while designing evolutionary operators; such type of improvement, is however, difficult with an approximation algorithm.

References

1. Garey, M.R., Johnson, D.S.: Computers and Interactability: A Guide to the Theory of NP-Completeness. San Francisco, LA: Freeman (1979)

2. Hochbaum, D.: Approximation Algorithms for NP-Hard Problems. Boston, MA: PWS (1997)

3. Marathe, M.V., Ravi, R., Sundaram, R., Ravi, S.S., Rosenkrantz, D.J., Hunt, H.B.: Bicriteria Network Design Problems. J. Algorithms **28** (1998) 142 – 171

4. Ravi, R., Marathe, M.V., Ravi, S.S., Rosenkrantz, D.J., Hunt, H.B.: Approximation Algorithms for Degree-Constrained Minimum-Cost Network Design Problems. Algorithmica **31** (2001) 58 – 78

5. Boldon, N., Deo, N., Kumar, N.: Minimum-Weight Degree-Constrained Spanning Tree Problem: Heuristics and Implementation on an SIMD Parallel Machine. Parallel Computing **22** (1996) 369 – 382

6. Deo, N., Abdalla, A.: Computing a Diameter-Constrained Minimum Spanning Tree in Parallel. In: Proc. 4th Italian Conference on Algorithms and Complexity (CIAC 2000), LNCS 1767. (2000) 17 – 31

7. Deo, N., Micikevicius, P.: Comparison of Prüfer-like Codes for Labeled Trees. In: Proc. 32nd South-Eastern Int. Conf. Combinatorics, Graph Theory and Computing. (2001)

8. Raidl, G.R., Julstrom, B.A.: Greedy Heuristics and an Evolutionary Algorithm for the Bounded-Diameter Minimum Spanning Tree Problem. In: Proc. 18th ACM Symposium on Applied Computing (SAC 2003). (2003) 747 – 752

9. Julstrom, B.A., Raidl, G.R.: Edge Sets: An Effective Evolutionary Coding of Spanning Trees. IEEE Trans. Evolutionary Computation **7** (2003) 225 – 239

10. Knowles, J.D., Corne, D.W.: A New Evolutionary Approach to the Degree-Constrained Minimum Spanning Tree Problem. IEEE Trans. Evolutionary Computation **4** (2000) 125 – 133

11. Knowles, J.D., Corne, D.W.: A Comparison of Encodings and Algorithms for Multiobjective Minimum Spanning Tree Problems. In: Proc. 2001 Congress on Evolutionary Computation (CEC-01). Volume 1. (2001) 544 – 551

12. Coello, C.A.C., Veldhuizen, D.A.V., Lamont, G.B.: Evolutionary Algorithms for Solving Multi-Objective Problems. Boston, MA: Kluwer (2002)

13. Deb, K.: Multiobjective Optimization Using Evolutionary Algorithms. Chichester, UK: Wiley (2001)

14. Purshouse, R.C., Fleming, P.J.: Elitism, Sharing and Ranking Choices in Evolutionary Multi-criterion Optimization. Research Report No. 815, Dept. Automatic Control & Systems Engineering, University of Sheffield (2002)

15. Kumar, R., Rockett, P.I.: Improved Sampling of the Pareto-front in Multiobjective Genetic Optimization by Steady-State Evolution: A Pareto Converging Genetic Algorithm. Evolutionary Computation **10** (2002) 283 – 314

16. Laumanns, M., Thiele, L., Deb, K., Zitzler, E.: Combining Convergence and Diversity in Evolutionary Multiobjective Optimization. Evolutionary Computation **10** (2002) 263 – 282

17. Zohu, G., Gen, M.: Genetic Algorithm Approach on Multi-Criteria Minimum Spanning Tree Problem. European J. Operations Research **114** (1999) 141–152

Simple Deadlock-Free Dynamic Network Reconfiguration

Olav Lysne[1], José Miguel Montañana[2], Timothy Mark Pinkston[3],
José Duato[2], Tor Skeie[1], and José Flich[2]

[1] Simula Research Laboratory Oslo, Norway
[2] Technical Univ. of Valencia Valencia, Spain
[3] University of Southern California Los Angeles, CA 90089-2562

Abstract. Dynamic reconfiguration of interconnection networks is defined as the process of changing from one routing function to another while the network remains up and running. The main challenge is in avoiding deadlock anomalies while keeping restrictions on packet injection and forwarding minimal. Current approaches fall in one of two categories. Either they require the existence of extra network resources like e.g. virtual channels, or their complexity is so high that their practical applicability is limited. In this paper we describe a simple and powerful method for dynamic networks reconfiguration. It guarantees a fast and deadlock-free transition from the old to the new routing function, it works for any topology and between any pair of old and new routing functions, and it guarantees in-order packet delivery when used between deterministic routing functions.

1   Introduction

System availability and reliability are becoming increasingly important as system size and demand increase. This is especially true for high-end servers (web, database, video-on-demand servers, data centers, etc.), which are currently based on clusters of PCs and/or highly parallel computers. In these systems, the network interconnecting the processing nodes among them and to I/O devices plays a very important role toward achieving high system availability.

Since the seminal work of Kermani and Kleinrock on *virtual cut-through* [8] and later Dally and Seitz on *wormhole switching* [4,5], we have seen an ever increasing body of research on these switching techniques. These techniques are in common use today in interprocessor communication (IPC) networks.

For a survey of interconnection networks we refer to [6].

In some situations the premises on which the routing algorithm and/or network topology are defined may break, which affects the network's dependability. This may happen when the topology of the network changes, either involuntarily due to failing/faulty components or voluntarily due to hot removal or addition of components. This normally requires the network routing algorithm (a.k.a., routing function) to be *reconfigured* in order to (re)establish full network connectivity among the attached nodes. In transitioning between the old

L. Bougé and V.K. Prasanna (Eds.): HiPC 2004, LNCS 3296, pp. 504–515, 2004.
© Springer-Verlag Berlin Heidelberg 2004

and new routing functions during network reconfiguration, additional dependencies among network resources may be introduced, causing what is referred to as *reconfiguration-induced deadlock*.

Current techniques typically handle this situation through *static reconfiguration*—meaning that application traffic is stopped and, usually, dropped from the network during the reconfiguration process (see, for example, [13, 14]). While this approach guarantees the prevention of reconfiguration-induced deadlock, it can lead to unacceptable packet latencies and dropping frequencies for many applications, particularly real-time and quality-of-service (QoS) applications [7].

With *dynamic reconfiguration*, the idea is to allow user traffic to continue uninterruptedly during the time that the network is reconfigured, thus reducing the number of packets that miss their real-time/QoS deadline. Recently, some key efforts have been put toward addressing the issue of deadlock-free dynamic reconfiguration within the context of link-level flow controlled interconnection networks. In [2], a *Partial Progressive Reconfiguration* (PPR) technique is proposed that allows arbitrary networks to migrate between two instantiations of up*/down* routing. The effect of load and network size on PPR performance is evaluated in [3]. Another approach is the NetRec scheme [11] which requires every switch to maintain information about switches some number of hops away. Yet another approach is the *Double Scheme* [12], where the idea is to use two required sets of virtual channels in the network which act as two disjoint virtual network layers during reconfiguration. The basic idea is first to drain one virtual network layer and reconfigure it while the other is fully up and running, then to drain and reconfigure the other virtual network layer while the first is up and running, thus allowing "always on" packet delivery during reconfiguration. An orthogonal approach which may be applicable on top of all of the above techniques is described in [9], where it is shown that for up*/down* routing, only parts of the network (i.e., the "skyline") need to be reconfigured on a network change. In [10] a methodology for developing dynamic network reconfiguration processes between any pair of routing functions is described.

All the approaches mentioned above suffer from different shortcomings. PPR [2] will only work between two routing functions that adhere to the up*/down* scheme. NetRec [11] is specially tailored for rerouting messages around a faulty node. It basically provides a protocol for generating a tree that connects all the nodes that are neighbors to a fault, and drops packets to avoid deadlocks in the reconfiguration phase. The Double Scheme is the most flexible of the above, in that it can handle any topology and make a transition between any pair of deadlock free routing functions. On the other hand it requires the presence of two sets of virtual channels.

In this paper we present a simple and powerful method for dynamic network reconfiguration. In contrast to the approaches mentioned above, our method is able to handle any topology and any pair of routing functions, regardless of the number of virtual channels available in the network. It is directly applicable when the new routing function is available, and it does not require a new reconfiguration method to be derived before it can be applied. Our technique guarantees in-order delivery of packets during reconfiguration, and can for that reason off-load much of the fault-handling burden from the higher-level protocols.

2   The Method

We assume familiarity with the standard notation and definitions of cut-through switching. In particular we assume that the basic notions of *deadlock freedom* in general, and *channel dependency graphs* in particular are known. The readers that are unfamiliar with these notions are referred to [6].

Our focus is on the transition from one routing function to another. We will denote these two routing functions R_{old} and R_{new} respectively, with the subscripts taking the obvious meaning. In what follows we simply assume that each of these are deadlock free and has a cycle free channel dependency graph, unless explicitly stated otherwise. Furthermore we assume that if R_{old} supplies any faulty channels, the packets destined for these channels are dropped rather than stalled, and that R_{new} supplies channels such that the faulty components are circumvented.

As we consider transitions from one routing function to another, channel dependency graphs are not a sufficient tool to detect absence from deadlocks. Even if the prevailing routing function at any given time supplies channels in a deadlock free manner during reconfiguration, there may be configurations of packets that are deadlocked. This is because a packet may have made previous routing decisions based on old routing choices that are no longer allowed in the current routing function, and by doing that it has ended up in a situation where it keeps a channel dependency from a previous routing function alive. Such dependencies are called *ghost dependencies* [10]. We therefore need a notion of deadlocks that encompasses more information than just channel dependencies. We use a simplified form of definition 10 in [15]:

Definition 1. *A set of packets is* deadlocked *if every packet in the set must wait for some other packet in the set to proceed before it can proceed itself.*

We shall use the above definition to show that our reconfiguration method will allow no deadlock to form.

In the following we describe the fundamentals of our simple reconfiguration algorithm. In the description we shall for simplicity assume that there will only be one single reconfiguration process active at a time, and that this reconfiguration process will complete before another one is started.

Our method is based on two pillars. The first is that we let every packet be routed either solely according to R_{old} or solely according to R_{new}. The packets that we route solely according to R_{old} will be called *old* packets, and the packets that are routed solely according to R_{new} are called *new* packets. It is very likely that a channel will be used by both old packets and new packets, so channel dependencies can form from the interaction between old and new packets.

The second pillar is the following lemma:

Lemma 1. *Assume a transition from R_{old} to R_{new} in which every packet is routed solely according to R_{old} or solely according to R_{new}. Any deadlocked set of packets in such a transition will have to contain old packets waiting behind new packets.*

Proof. The proof is by contradiction. Assume a deadlocked set of packets in which no old packets wait behind new packets.

Case 1: There are no old packets in the set. In that case the set must contain only new packets that should be able to reach their destination using R_{new}. This implies that R_{new} is not deadlock free, and we have contradiction.

Case 2: There are old packets in the set. Since we assume that no old packet wait behind new packets, the old packets must all be waiting behind each other. In that case there must exist a deadlocked set containing only old packets. This implies that R_{old} is not deadlock free, and we have contradiction.

A consequence of the above lemma is that if we make sure that packets routed according to R_{old} will never have to wait behind packets routed according to R_{new}, we will have achieved freedom from deadlock even if R_{new} packets wait behind R_{old} packets. This can be achieved by letting all channels transmit a token that indicates that all packets injected before this token shall be routed according to R_{old}, and all packets injected after this token shall be routed according to R_{new}. We let this token flood the network in the order of the channel-dependency graph of R_{old}, and for each channel it traverses, it means the same thing: all packets transmitted across the channel before this token shall be routed according to R_{old}, and all packets after this token shall be routed according to R_{new}. Every packet routed according to R_{new} will simply have to wait for the token to have passed before it enters a channel. That way no packet routed according to R_{old} will ever have to wait for a packet routed according to R_{new} to proceed, thus deadlock cannot form. A more formal description of one version of the process follows:

1. Let each injection link send a token onto all of its channels indicating that no packets that have been routed according to R_{old} will arrive on this channel.
2. Let each switch do the following:
   - For each input channel do the following:
     (a) Continue using R_{old} until a token has arrived at the head of the queue[1].
     (b) When the token has made it to the head of the queue, change into R_{new} for this input channel.
     (c) Thereafter forward packets only to those output channels that have transmitted the token.

---

[1] We make the usual assumption that a packet is routed only when it is at the head of the input queue.

– For each output channel do the following:
  (a) Wait until all input channels from which the output channel can expect to receive traffic according to R_{old} have processed the token.
  (b) Thereafter transmit a token on the output channel.

The following results can now be derived for this process:

Observation 1. *All input channels on all switches use R_{old} until they process the token and thereafter, use R_{new}.*

Lemma 2. *This process ensures that all packets are routed either solely according to R_{old} or solely according to R_{new}.*

Proof. Consider a packet that experiences routing according to both routing functions. On its path from source to destination there will be two consecutive switches, S_1 and S_2, where this packet is routed according to different routing functions. There are two cases.

Case 1: The packet was first routed according to R_{old} in S_1 and then routed according to R_{new} in S_2. According to observation 1 this packet must have arrived the switch S_2 an input channel that had already processed the token. Furthermore in S_1 it was routed according to R_{old}, so there it arrived on an input channel *before* the token was processed on that input channel. Therefore if S_2 received the token before the packet, S_1 must have sent the token out on the output channel going to S_2 before S_1 had processed the token on one input channel from which this output channel could expect to receive traffic according to R_{old}. According to bullet points 2a and 2b for output channels in the process description, this cannot happen.

Case 2: The packet was first routed according to R_{new} in S_1 and then routed according to R_{old} in S_2. According to observation 1 this packet must have arrived S_2 on an input link that had not yet processed the token. Furthermore, in S_1 it was routed according to R_{new}, so there it arrived after the token was processed. Therefore S_1 must have forwarded packets from an input link that had processed the token onto an output link where the token has not yet been transmitted. According to the procedure for input channels in the process description, this cannot happen.

Corollary 1. *Each channel will first transmit only old packets, then the token and then only new packets.*

Proof. Assume a channel for which the corollary does not hold. Since the method does not discriminate between channels that terminate in switches and channels that terminate in compute nodes, we may without loss of generality assume that this channel terminates in a switch[2].

---

[2] This means that if the corollary was not valid for a channel that terminates in a compute node, one could easily generate a topology where the same channel terminated in a switch instead.

This would either require a new packet to traverse the channel before the token, or an old packet to traverse the channel after the token. In this case the new packet would be routed according to R_{old} or the old packet would be routed according to R_{new} in the next switch. This contradicts Lemma 2.

Now we prove that the reconfiguration terminates. Termination requires that all channels will eventually have transmitted the token, thus all input channels in all switches will be using R_{new}.

Lemma 3. *The process terminates with all channels having transmitted the token if R_{old} has a cycle free dependency graph.*

Proof. We first prove the lemma for the case where there are no data-packets in the network. Observe that the tokens propagate through the network following the channel dependency graph of R_{old}. Let o be an arbitrary output channel attached to a switch. Whenever all input channels that have a dependency to o according to R_{old} have the token at their queue head, the token is transmitted on o. Since the dependency graph of R_{old} is acyclic, and there are no packets occupying queue space, the lemma follows.

The only case we need to worry about is when the propagation of the tokens are hindered by data packets. According to Corollary 1 the tokens can only be waiting behind old packets, and new packets can only wait behind tokens or other new packets. Since R_{old} is deadlock-free, all old packets will eventually reach their destination or be dropped due to component failure. Therefore no token can be indefinitely blocked, thus the lemma follows.

Lemma 4. *If both R_{old} and R_{new} are deterministic, in-order packet delivery is maintained during the reconfiguration process.*

Proof. This is a consequence of Corollary 1. Every channel entering a compute node will first transmit old packets that clearly arrive in order, followed by the token, and finally new packets, that will also arrive in order.

3   Simulation Experiments

In this section we will evaluate the proposed reconfiguration mechanism. First, we will present the evaluation model, describing all the simulation parameters and the networks we have used. Then, we will present the reconfiguration scenarios we have used to evaluate the basic performance of the mechanism. Finally, we will present the evaluation results.

3.1   Evaluation Model

In order to evaluate the mechanism, we have developed a detailed simulator that allows us to model the network at the register transfer level. The simulator models an IBA network, following the IBA specifications [1].

Packets are routed at each switch by accessing the forwarding table. This table contains the output port to be used at the switch for each possible destination.

The routing time at each switch will be set to 100 ns. This time includes the time to access the forwarding tables, the crossbar arbiter time, and the time to set up the crossbar connections.

Switches can support up to 16 virtual lanes (VLs). VLs can be used to form separate virtual networks. We will use a non-multiplexed crossbar on each switch. This crossbar supplies separate ports for each VL. Buffers will be used both at the input and the output side of the crossbar. Buffer size will be fixed in both cases to 1 KB.

Links in InfiniBand are serial. In the simulator, the link injection rate will be fixed to the 1X configuration [1]. 1X cables have a link speed of 2.5 Gbps. Therefore, a bit can be injected every 0.4 ns. With 8/10 coding [1] a new byte can be injected into the link every 4 ns. We also model the fly time (time required by a bit to reach the opposite link side). We will model 20 m copper cables with a propagation delay of 5 ns/m. Therefore, the fly time will be set to 100 ns.

The IBA specification defines a credit-based flow control scheme for each virtual lane with independent buffer resources. A packet will be transmitted over the link if there is enough buffer space (credits of 64 bytes) to store the entire packet. IBA allows the definition of different MTU (Maximum Transfer Unit) values for packets ranging from 256 to 4096 bytes. Additionally, the virtual cut-through switching technique is used.

For each simulation run, we assume that the packet generation rate is constant and the same for all the end-nodes. Except when specified, the number of simulated packets is 160,000 and results will be collected from the last 80,000 packets (that is, a transient state of 80,000 packets and a permanent state of 80,000 packets).

The uniform traffic pattern will be used. With this pattern, each source sends packets to all the destination with the same probability. Packet size will be fixed to 58 bytes. This includes the IBA packet header (20 bytes), the packet payload (32 bytes) and the IBA packet tail (6 bytes).

When using reconfiguration, two packet tokens will be used. The first token will be the *start-token* and will be transmitted from a random switch. This token will be sized in 1 byte and will be broadcasted to all the switches. The second token will be the *reconfiguration-token* and will be sized in 8 bytes. In all the cases, reconfiguration will be started in the middle of collecting results (in the middle of the permanent state).

In all the presented results, we will plot the average network latency[3] measured in nanoseconds versus the average accepted traffic[4] measured in bytes per nano-second per switch. Also, the evolution in time of the average packet latency and the average latency from generation time [5] will be plotted.

---

[3] Latency is the elapsed time between the injection of a packet until it is delivered at the destination end-node.

[4] Accepted traffic is the amount of information delivered by the network per time unit.

[5] Latency from generation time is the elapsed time between the generation of a packet at the source host until it is delivered at the destination end-node.

3.2   Evaluation Results

Different topologies and different combinations of routing algorithms (R_{old} and R_{new}) have been evaluated. Table 1 shows all the combinations we have analyzed. Due to space limitations we will only display results for case 1 (4×4 Torus). The results for all other cases were very similar.

Table 1. Reconfiguration scenarios. When we use terms like "upper left" and "center" to denote switches in a torus, this is relative to a visual view of the torus when it is laid out in two dimensions as a mesh with wraparound links

Case	Topology	R_{old} Routing	R_{old} Root	R_{new} Routing	R_{new} Root
1	4×4 Torus	UD	upper left	UD	center
2	4×4 Mesh	UD	upper left	UD	bottom right
3	8×8 Mesh	UD	upper left	UD	bottom right
4	4×4 Mesh	XY	-	UD	upper left
5	4×4 Mesh	XY	-	YX	-

When using the up*/down* routing, the root switch must be selected. In the case of the torus network, the root switch was the upper left switch of the network for the R_{old} routing, and the center switch of the network for the R_{new}.

Figure 1 shows the performance evaluation obtained in a 4×4 torus network when the reconfiguration is applied. R_{new} and R_{old} use up*/down* (with different root switches). For every simulated point, the reconfiguration mechanism is triggered and the average latency of packets is measured. The Figure also shows the performance when the reconfiguration is not triggered (R_{new} is used all the time). As can be noticed, the reconfiguration process slightly affects the performance of the network, and this is visible only at the saturation point.

However, these results should be put in context. Since the average latency of packets is measured from the latency of each simulated packet, as the number of simulated packets grows, the percentage of affected packets by reconfiguration will decrease. Therefore low differences in terms of average latency will appear. The results shown in Figure 1 were generated from simulation of 20,000 packets. The reconfiguration process was launched after 1,000 packets had been received. In order to measure stable states 20,000 packets were injected before collecting results.

In order to better view the impact of the reconfiguration, Figure 1.b shows the reconfiguration latency. That is, the time required by the mechanism to reconfigure the entire network (from the sending of the first *start-token* up to receiving the last *reconfiguration-token*). As can be noticed, the reconfiguration latency keeps low and constant for most of the traffic injection rates. For most of the traffic points, the reconfiguration latency is around 1,6 microseconds. The network in the simulations usually gets 666 microseconds to deliver the 20,000 useful packets (with traffic injection near the saturation knee). Therefore, the reconfiguration latency only affects to 0.25% of the simulated time.

However, near the saturation knee, the reconfiguration latency increases. At saturation knee, the reconfiguration latency is 49 microseconds, representing 7% of the total simulated time. The reason for this is that only when the network is partly congested, the delivery of reconfiguration tokens will suffer as some old packets will be waiting at some queues during a large amount of time due to congestion. In the normal situation, the queues will not be full, and therefore, the old packets will advance with no delay, so the reconfiguration packets will also quickly advance and propagate. Beyond saturation knee, the reconfiguration tokens also get congested and therefore the reconfiguration latency is excessive.

Figure 1.c shows the maximum packet latency for every simulated point. As we can observe, the mechanism practically does not introduce any penalty to packets as the latency for the packet with the maximum latency is practically the same. Indeed, only slight differences appear near the congested knee, where the maximum latency is slightly increased. This is mainly due to the congestion encountered by reconfiguration tokens. Notice that if the network is congested, the old packets will suffer long latencies regardless of the presence of new packets in the network.

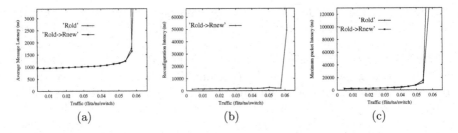

Fig. 1. Impact of the reconfiguration method in a 4×4 torus network. R_{old} is up*/down* (root switch is upper left) and R_{new} is up*/down* (root switch is at the center). (a) Network latency vs injected traffic, (b) reconfiguration latency, and (c) maximum packet latency

In order to obtain a closer view to the impact of the reconfiguration method on network performance, Figure 2 shows the evolution of packet latency at different traffic injection rates when the reconfiguration is triggered. The simulated network is a 4×4 torus network, R_{old} and R_{new} use up*/down* (with different root switches). In particular, the Figure shows the latency evolution in different traffic loads: low load, medium load and close to saturation (the previous point to saturation). The figure shows results for the average network latency and for the average packet latency from generation time. With vertical lines, the start and the finish of the reconfiguration process is shown.

As can be noticed, for low load (Figures 2.a and 2.d), the impact on network latency and packet latency from generation time is not significant. There is no variation in latency in the reconfiguration process. Even more, for medium traffic loads (Figures 2.b and 2.e) the impact on network latency and packet latency

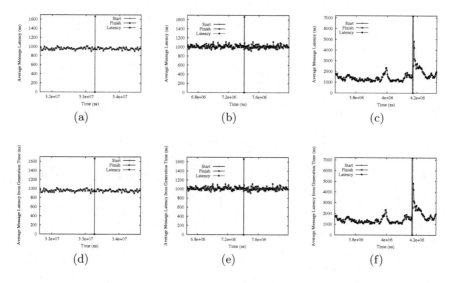

Fig. 2. Latency performance during simulation time for (a,d) low load, (b,e) medium load, and (c,f) near saturation. (a, b, c) Average network latency. (d, e, f) Average latency from generation time. 4×4 torus network. R_{old} is UD and R_{new} is UD

from generation time is also negligible. The reason for this is that there are few packets in the network, and therefore, the queues occupancy is also low, so, the reconfiguration tokens can advance quickly throughout the network and switches rapidly change from R_{old} to R_{new}. Indeed, the Figures show average results for every 100 packets delivered. In both cases, the reconfiguration process lasted 1.5 microseconds and 1.9 microseconds, respectively. In that time, less than 100 packets are delivered. Therefore, the impact is minimum for the traffic load.

However, for traffic near saturation of R_{old} (Figures 2.c and 2.f) a much clearer impact on latency is observed. In this case, the reconfiguration process lasted 5.5 microseconds. After the reconfiguration process, we can notice that average network latency sharply increases from 1.5 microseconds to 5 microseconds. Later, the average latency goes back and normalizes to 1.5 microseconds. This temporal increase in latency is due to some introduced blocking of new packets that must wait for *reconfiguration-tokens*. As the network is close to saturation, queues start to fill and therefore, *reconfiguration-tokens* start to encounter some degree of blocking. However, the negative effect of reconfiguration is low as the reconfiguration process is still fast.

Figure 3 shows the behavior in a congested scenario (beyond saturation) of R_{old}. In this situation, the reconfiguration latency has been increased to 929 microseconds. As can be observed, for the average network latency (Figure 3.a) there is a slight increase on latency in the start of the reconfiguration. However, by the end of the reconfiguration, there is an extremely sharp increase of the latency (from 5 microseconds up to 700 microseconds). The reconfiguration process finishes when the last *reconfiguration-token* arrives to the last switch. In this switch, there will be new packets blocked since the start of the reconfiguration (as all the switches have compute nodes attached to them and all send packets all the times). Therefore, once reconfiguration is finished, the most delayed new packets will be unblocked and therefore they will reach their destinations, thus extremely increasing latency. However, notice that this situation normalizes very fast, as the number of theses packets is quite low.

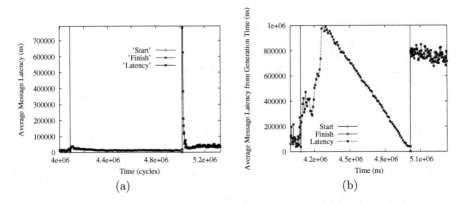

Fig. 3. Latency performance during simulation time for saturation. (a) Average network latency. (b) Average latency from generation time. 4×4 torus network. R_{old} is UD and R_{new} is UD

4   Conclusion

In this paper we have described a simple procedure for dynamic and deadlock free reconfiguration between two routing functions. Our method works for any topology and between any pair of routing functions. It guarantees in-order delivery of packets during reconfiguration, as long as the old and the new routing functions are deterministic. And it requires no additional virtual channels.

Preliminary evaluation results have shown that the mechanism works efficiently in different topologies and when using different routing algorithms (old and new). The reconfiguration latency is roughly constant for low and medium traffic loads and increases slightly for traffic loads near saturation. The peak latency experienced by packets when reconfiguring is only noticeable for traffic loads beyond saturation.

There are several interesting paths for further research that will be addressed in further work. One is the development of simulation results for fault tolerance. Another is connected to the fact that this method is based on routing tables being present in the switches. This means that adjustments need to be made before the method can be applied to source routing. What adjustments should be made, and the effect of such adjustments needs to be studied more closely.

References

1. InfiniBand Trade Association. *InfiniBand Architecture. Specification Volume 1. Release 1.0a.* Available at http://www.infinibandta.com, 2001.
2. R. Casado, A. Bermúdez, J. Duato, F. J. Quiles, and J. L. Sánchez. A protocol for deadlock-free dynamic reconfiguration in high-speed local area networks. *IEEE Transactions on Parallel and Distributed Systems*, 12(2):115–132, February 2001.
3. R. Casado, A. Bermúdez, F. J. Quiles, J. L. Sánchez, and J. Duato. Performance evaluation of dynamic reconfiguration in high-speed local area networks. In *Proceedings of the Sixth International Symposium on High-Performance Computer Architecture*, 2000.
4. W. J. Dally and C. L. Seitz. The torus routing chip. *Distributed Computing*, 1:187–196, 1986.
5. W. J. Dally and C. L. Seitz. Deadlock-free message routing in multiprocessor interconnection networks. *IEEE Transactions on Computers*, C-36(5):547–553, 1987.
6. José Duato, Sudhakar Yalamanchili, and Lionel Ni. *Interconnection Networks: An Engineering Approach*. Morgan Kaufmann Publishers, 2003.
7. J. Fernández, J. García, and J. Duato. A new approach to provide real-time services on high-speed local area networks. In *Proceedings of the 15th International Parallel and Distributed Processing Symposium (IPDPS-01)*, pages 124–124, Los Alamitos, CA, April 23–27 2001. IEEE Computer Society.
8. P. Kermani and L. Kleinrock. Virtual cut-through: A new computer communication switching technique. *Computer Networks*, 3:267–286, 1979.
9. O. Lysne and J. Duato. Fast dynamic reconfiguration in irregular networks. In *Proceedings of the 2000' International Conference of Parallel Processing, Toronto (Canada)*, pages 449–458. IEEE Computer Society, 2000.
10. O. Lysne, T. M. Pinkston, and J. Duato. A methodology for developing dynamic network reconfiguration processes. In *2003 International Conference on Parallel Processing (ICPP'03)*, pages 77–86. IEEE, 2003.
11. N. Natchev, D. Avresky, and V. Shurbanov. Dynamic reconfiguration in high-speed computer clusters. In *Proceedings of the International Conference on Cluster Computing*, pages 380–387, Los Alamitos, 2001. IEEE Computer Society.
12. T. Pinkston, R. Pang, and J. Duato. Deadlock-free dynamic reconfiguration schemes for increased network dependeability. *IEEE Transactions on Parallel and Distributed Systems*, 14(8):780–794, August 2003.
13. Thomas L. Rodeheffer and Michael D. Schroeder. Automatic reconfiguration in Autonet. In *Proceedings of 13th ACM Symposium on Operating Systems Principles*, pages 183–197. Association for Computing Machinery SIGOPS, October 1991.
14. Dan Teodosiu, Joel Baxter, Kinshuk Govil, John Chapin, Mendel Rosenblum, and Mark Horowitz. Hardware fault containment in scalable shared-memory multiprocessors. In *Proceedings of the 24th Annual International Symposium on Computer Architecture (ISCA-97)*, volume 25,2 of *Computer Architecture News*, pages 73–84, New York, 1997. ACM Press.
15. S. Warnakulasuriya and T. M. Pinkston. A formal model of message blocking and deadlock resolution in interconnection networks. *IEEE Transactions on Parallel and Distributed Systems*, 11(3):212–229, 2000.

Lock-Free Parallel Algorithms: An Experimental Study

Guojing Cong and David Bader*

Department of Electrical and Computer Engineering,
University of New Mexico, Albuquerque, NM 87131 USA
{cong, dbader}@ece.unm.edu

Abstract. Lock-free shared data structures in the setting of distributed computing have received a fair amount of attention. Major motivations of lock-free data structures include increasing fault tolerance of a (possibly heterogeneous) system and alleviating the problems associated with critical sections such as *priority inversion* and *deadlock*. For parallel computers with tightly-coupled processors and shared memory, these issues are no longer major concerns. While many of the results are applicable especially when the model used is shared memory multiprocessors, no prior studies have considered improving the performance of a parallel implementation by way of lock-free programming. As a matter of fact, often times in practice lock-free data structures in a distributed setting do not perform as well as those that use locks. As the data structures and algorithms for parallel computing are often drastically different from those in distributed computing, it is possible that lock-free programs perform better. In this paper we compare the similarity and difference of lock-free programming in both distributed and parallel computing environments and explore the possibility of adapting lock-free programming to parallel computing to improve performance. Lock-free programming also provides a new way of simulating PRAM and asynchronous PRAM algorithms on current parallel machines.

Keywords: Lock-free Data Structures, Parallel Algorithms, Shared Memory, High-Performance Algorithm Engineering.

1 Introduction

Mutual exclusion locks are widely used for interprocess synchronization due to their simple programming abstractions. However, they have an inherent weakness in a (possibly heterogeneous and faulty) distributed computing environment, that is, the crashing or delay of a process in a critical section can cause deadlock or serious performance degradation of the system [18, 28]. Lock-free data structures (sometimes called concurrent objects) were proposed to allow concurrent accesses of parallel processes (or threads) while avoiding the problems of locks.

* This work was supported in part by NSF Grants CAREER ACI-00-93039, ITR ACI-00-81404, DEB-99-10123, ITR EIA-01-21377, Biocomplexity DEB-01-20709, and ITR EF/BIO 03-31654; and DARPA Contract NBCH30390004.

L. Bougé and V.K. Prasanna (Eds.): HiPC 2004, LNCS 3296, pp. 516–527, 2004.
© Springer-Verlag Berlin Heidelberg 2004

1.1   Previous Results

Lock-free synchronization was first introduced by Lamport in [25] to solve the concurrent *readers and writers problem*. Early work on lock-free data structures focused on theoretical issues of the synchronization protocols, e.g., the power of various atomic primitives and impossibility results [3, 7, 11, 12, 13, 16], by considering the simple *consensus problem* where n processes with independent inputs communicate through a set of shared variables and eventually agree on a common value. Herlihy [20] unified much of the earlier theoretic results by introducing the notion of *consensus number* of an object and defining a hierarchy on the concurrent objects according to their consensus numbers. Consensus number measures the relative power of an object to reach distributed consensus, and is the maximum number of processes for which the object can solve the consensus problem. It is impossible to construct lock-free implementations of many simple and useful data types using any combination of atomic *read, write, test&set, fetch&add* and *memory-to-register swap* because these primitives have consensus numbers either one or two. On the other hand, *compare&swap* and *load-linked, store-conditional* have consensus numbers of infinity, and hence are *universal* meaning that they can be used to solve the consensus problem of any number of processes. Lock-free algorithms and protocols have been proposed for many of the commonly used data structures, e.g., linked lists [37], queues [21, 26, 35], set [27], union-find sets [2], heaps [6], and binary search trees [14, 36]. There are also efforts to improve the performance of lock-free protocols [1, 6].

While lock-free data structures and algorithms are highly resilient to failures, unfortunately, they seem to come at a cost of degraded performance. Herlihy *et al.* studied practical issues and architectural support of implementing lock-free data structures [19, 18], and their experiments with small priority queues show that lock-free implementations do not perform as well as lock-based implementations. LaMarca [24] developed an analytical model based on architectural observations to predict the performance of lock-free synchronization protocols. His analysis and experimental results show that the benefits of guaranteed progress come at the cost of decreased performance. Shavit and Touitou [32] studied lock-free data structures through *software transactional memory*, and their experimental results also show that on a simulated parallel machine lock-free implementations are inferior to standard lock-based implementations.

1.2   Asynchronous Parallel Computation

Cole and Zajichek [9] first introduced lock-free protocols into parallel computing when they proposed asynchronous PRAM (APRAM) as a more realistic parallel model than PRAM because APRAM acknowledges the cost of global synchronization. Their goal was to design APRAM algorithms with fault-resilience that perform better than straightforward simulations of PRAM algorithms on APRAM by inserting barriers. A parallel connected components algorithm without global synchronization was presented as an example. It turned out, however, according to the research of lock-free data structures in distributed computing, that it is impossible to implement many lock-free data structures on APRAM with only atomic register read/write [3, 20]. Attiya *et al.* [4] proved a lower bound of $\log n$ time complexity of any lock-free algorithm on a computational model that is essentially APRAM that achieves *approximate agreement* among n processes in contrast to constant time of non-lock-free algorithms. This suggests an $\Omega(\log n)$ gap between lock-free and non-lock-free computation models.

Currently lock-free data structures and protocols are still mainly used for fault-tolerance and seem to be inferior in performance to lock-based implementations. In this paper we consider adapting lock-free protocols to parallel computations where multiple processors are tightly-coupled with shared memory. We present novel applications of lock-free protocols where the performance of the lock-free algorithms beat not only lock-based implementations but also the best previous parallel implementations. The rest of the paper is organized as follows: section 2 discusses the potential advantages of using lock-free protocols for parallel computations; section 3 presents two lock-free algorithms as case studies; and section 4 is our conclusion.

2   Application of Lock-Free Protocols to Parallel Computations

Fault-tolerance typically is not a primary issue for parallel computing (as it is for distributed computing) especially when dedicated parallel computers with homogeneous processors are employed. Instead we are primarily concerned with performance when solving large instances. We propose novel applications of lock-free protocols to parallel algorithms that handle large inputs and show that lock-free implementations can have superior performance.

A parallel algorithm often divides into phases and in each phase certain operations are applied to the input with each processor working on portions of the data structure. For irregular problems there usually are overlaps among the portions of data structures partitioned onto different processors. Locks provide a mechanism for ensuring mutually exclusive access to critical sections by multiple working processors. Fine-grained locking on the data structure using system mutex locks can bring large memory overhead. What is worse is that many of the locks are never acquired by more than one processor. Most of the time each processor is working on distinct elements of the data structure due to the large problem size and relatively small number of processors. Yet still extra work of locking and unlocking is performed for each operation applied to the data structure, which results in a large execution overhead.

The access pattern of parallel processors to the shared data structures makes lock-free protocols via atomic machine operations an elegant solution to the problem. When there is work partition overlap among processors, usually it suffices that the overlap is taken care of by one processor. If other processors can detect that the overlap portion is already taken care of, they no longer need to apply the operations and can abort. Atomic operations can be used to implement this "test-and-work" operation. As the contention among processors is low, the overhead of using atomic operations is expected to be small. Note that this is very different from the access patterns to the shared data structures in distributed computing, for example, two producers attempting to put more work into the shared queues. Both producers must complete their operations, and when there is conflict they will retry until success.

In some recent experimental studies on symmetric multiprocessors (SMPs) [14, 36] the design and implementation of lock-free data structures involves mutual-exclusions and are not strictly lock-free in the sense that a crash inside the critical region prevents application progress. Mutual-exclusion is achieved using inline atomic operations and is transparent to users because it is hidden in the implementation of the data structures. Both Fraser [14] and Valois [36] show that for many search structures (binary tree, skip list, red-black tree) well-implemented algorithms using atomic primitives can match or surpass the performance of lock-based designs in many situations. However, their implementations comprise the guarantee of progress in case of failed processes. "Lock-free" here means free of full-fledged system mutex locks and are actually block-free using spinlocks. Spinlocks do make sense in the homogeneous parallel computing environment with dedicated parallel computers where no process is particularly slower than others and the program is computation intensive. It is a better choice to busy-wait than to block when waiting to enter the critical section. In this paper we also study the application of busy-wait spinlocks to parallel algorithms. We refer interested readers to [30, 35] for examples of block-free data structures.

3   Lock-Free Parallel Algorithms

In this section we present parallel algorithms that are either mutual-exclusion free or block-free to demonstrate the usage of lock-free protocols. Section 3.1 considers the problem of resolving work partition conflicts for irregular problems using a lock-free parallel spanning tree algorithm as an example. Section 3.2 presents an experimental study of a block-free minimum spanning tree (MST) algorithm. As both algorithms take graphs as input, before we present the algorithms, here we describe the the collection of graph generators and the parallel machine we used.

We ran our shared-memory implementations on the Sun Enterprise 4500, a uniform-memory-access shared memory parallel machine with 14 UltraSPARC II processors and 14 GB of memory. Each processor has 16 Kbytes of direct-mapped data (L1) cache and 4 Mbytes of external (L2) cache. The clock speed of each processor is 400 MHz.

Our graph generators include several employed in previous experimental studies of parallel graph algorithms for related problems. For instance, mesh topologies are used in the connected component studies of [15, 17, 22, 23], the random graphs are used by [8, 15, 17, 22], and the geometric graphs are used by [8, 15, 17, 22, 23].

- **Meshes.** Mesh-based graphs are commonly used in physics-based simulations and computer vision. The vertices of the graph are placed on a 2D or 3D mesh, with each vertex connected to its neighbors. **2DC** is a complete 2D mesh; **2D60** is a 2D mesh with the probability of 60% for each edge to be present; and **3D40** is a 3D mesh with the probability of 40% for each edge to be present.
- **Random Graph.** A random graph of n vertices and m edges is created by randomly adding m unique edges to the vertex set. Several software packages generate random graphs this way, including LEDA [29].
- **Geometric Graphs.** Each vertex has a fixed degree k. Geometric graphs are generated by randomly placing n vertices in a unit square and connecting each vertex with its nearest k neighbors. [31] use these in their empirical study of sequential MST algorithms. **AD3** ([23]) is a geometric graph with $k = 3$.

For MST, uniformly random weights are associated with the edges.

3.1   Lock-Free Protocols in Parallel Spanning Tree

We consider the application of lock-free protocols to the Shiloach-Vishkin parallel spanning tree algorithm [33, 34]. This algorithm is representative of several connectivity algorithms that adapt the graft-and-shortcut approach, and is implemented in prior experimental studies. For graph $G = (V,E)$ with $|V| = n$ and $|E| = m$, the algorithm achieves complexities of $O(\log n)$ time and $O((m+n)\log n)$ work under the arbitrary CRCW PRAM model.

The algorithm takes an edge list as input and starts with n isolated vertices and m processors. Each processor P_i ($1 \leq i \leq m$) inspects edge $e_i = (v_{i_1}, v_{i_2})$ and tries to graft vertex v_{i_1} to v_{i_2} under the constraint that $v_{i_1} < v_{i_2}$. Grafting creates $k \geq 1$ connected components in the graph, and each of the k components is then shortcutted to to a single supervertex. Grafting and shortcutting are iteratively applied to the reduced graphs $G' = (V',E')$ (where V' is the set of supervertices and E' is the set of edges among supervertices) until only one supervertex is left. For a certain vertex v with multiple adjacent edges, there can be multiple processors attempting to graft v to other smaller vertices. Yet only one grafting is allowed, and we label the corresponding edge that causes the grafting as a spanning tree edge. This is a partition conflict problem.

Two-phase election can be used to resolve the conflicts. The strategy is to run a race among processors, where each processor that attempts to work on a vertex v writes its id into a tag associated with v. After all the processors are done, each processor checks the tag to see whether it is the winning processor. If so, the processor continues with its operation, otherwise it aborts. Two-phase election works on platforms that provide write atomicity. A global barrier synchronization among processors is used instead of a possibly large number of fine-grained locks. The disadvantage is that two runs are involved.

A natural solution to the work partition problem is to use lock-free atomic instructions. When a processor attempts to graft vertex v, it invokes the atomic *compare&swap* operation to check whether v has been worked on. If not, the atomic nature of the operation also ensures that other processors will not work on v again. The detailed description of the algorithm and an inline assembly function for *compare&swap* can be found in [10].

We compare the performance of the lock-free Shiloach-Vishkin spanning tree implementation with four other implementations that differ only in how the conflicts are resolved. In Table 1 we describe the four implementations.

Table 1. Five implementations of Shiloach-Vishkin's parallel spanning tree algorithm

Implementation	Description
span-2phase	conflicts are resolved by two-phase election
span-lock	conflicts are resolved using system mutex locks
span-lockfree	no mutual exclusion, races are prevented by atomic updates
span-spinlock	mutual exclusion by spinlocks using atomic operations
span-race	no mutual exclusion, no attempt to prevent races

Among the four implementations, **span-race** is not a correct implementation and does not guarantee correct results. It is included as a baseline to show how much overhead is involved with using lock-free protocols and spinlocks.

Experimental Results. In Fig. 1 we see **span-lock** does not scale with the number of the processors, and is consistently the approach with the worst performance. **span-2phase**, **span-lockfree**, and **span-spinlock** scale well with the number of processors, and the execution time of **span-lockfree** and **span-spinlock** is roughly half of that of **span-2phase**. It is interesting to note that **span-lockfree**, **span-spinlock** and **span-race** are almost as fast as each other for various inputs, which suggests similar overhead for spinlocks and lock-free protocols, and the overhead is negligible.

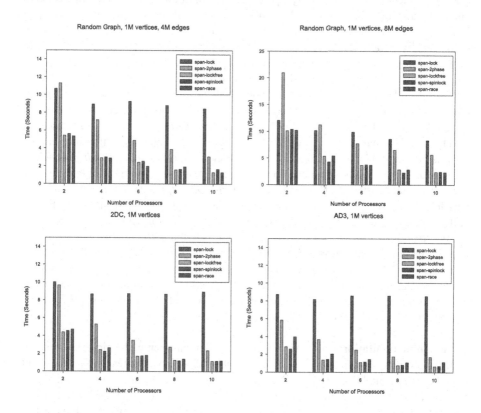

Fig. 1. The performance of the spanning tree implementations. The vertical bars from left to right are **span-lock**, **span-2phase**, **span-lockfree**, **span-spinlock**, and **span-race**, respectively

3.2   Block-Free Parallel Algorithms

For parallel programs that handle large inputs on current SMPs, spinlocks are a better choice than the blocking system mutex locks. Spinlocks are simpler, take less memory and do not involve the kernel. Due to the large inputs and relatively smaller number of processors available, most of the time each processor is working on distinct data elements, and the contention over a certain lock or data element is very low. Even in case of contention, as the expected time that a processor spends in the critical section is short, it is much cheaper to busy wait for a few cycles than to block. In the previous experimental study with the spanning tree algorithm, we have already seen that spinlock is a good candidate for mutual exclusion. As an example we next present a parallel minimum spanning tree (MST) algorithm using spinlocks for synchronization that achieves a more drastic performance improvement.

Parallel Borůvka's Algorithm and Previous Experimental Studies. Given an undirected connected graph G with n vertices and m edges, the minimum spanning tree (MST) problem finds a spanning tree with the minimum sum of edge weights. In our previous work [5], we studied the performance of different variations of parallel Borůvka's algorithm. Borůvka's algorithm is comprised of Borůvka iterations that are used in many parallel MST algorithms. A Borůvka iteration is characterized by three steps: *find-min*, *connected-components* and *compact-graph*. In *find-min*, for each vertex v the incident edge with the smallest weight is labeled to be in the MST; *connect-components* identifies connected components of the induced graph with the labeled MST edges; *compact-graph* compacts each connected component into a single supervertex, removes self-loops and multiple edges, and re-labels the vertices for consistency.

Here we summarize each of the Borůvka algorithms. The major difference among them is the input data structure and the implementation of *compact-graph*. **Bor-ALM** takes an adjacency list as input and compacts the graph using two parallel sample sorts plus sequential merge sort; **Bor-FAL** takes our *flexible adjacency list* as input and runs parallel sample sort on the vertices to compact the graph. For most inputs, **Bor-FAL** is the fastest implementation. In the *compact-graph* step, **Bor-FAL** merges each connected components into a single supervertex that gets the adjacency list of all the vertices in the component. **Bor-FAL** does not attempt to remove self-loops and multiple edges, and avoids runs of extensive sortings. Self-loops and multiple edges are filtered out in the *find-min* step instead. **Bor-FAL** greatly reduces the number of shared memory writes at the relatively small cost of an increased number of reads, and proves to be efficient as predicted on current SMPs.

A New Implementation. Now we present an implementation with spinlocks (denoted as **Bor-spinlock**) that further reduces the number of memory writes. In fact the input edge list is not modified at all in **Bor-spinlock**, and the *compact-graph* step is completely eliminated. The main idea is that instead of compacting connected components, for each vertex there is now an associated label *supervertex* showing to which supervertex it belongs. In each iteration all the vertices are partitioned among the processors. For each vertex v of its assigned partition, processor p finds the adjacent edge e with the smallest weight. If we compact connected components, e would belong to the supervertex v' of v in the new graph. Essentially processor p finds the adjacent edge with smallest weight for v'. As we do not compact graphs, the adjacent edges for v' are scattered among the adjacent edges of all vertices that share the same supervertex v', and different processors may work on these edges simultaneously. Now the problem is that these processors need to synchronize properly in order to find the edge with the mini-

mum weight. Again this is an example of the irregular work-partition problem. Fig. 2
illustrates the specific problem for the MST case.

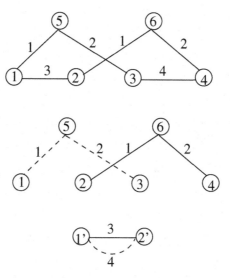

Fig. 2. Example of the race condition between two processors when Borůvka's algorithm is used
to solve the MST problem

On the top in Fig. 2 is an input graph with six vertices. Suppose we have two processors P_1 and P_2. Vertices 1, 2, and 3, are partitioned on to processor P_1 and vertices 4, 5,
and 6 are partitioned on to processor P_2. It takes two iterations for Borůvka's algorithm
to find the MST. In the first iteration, the *find-min* step of **Bor-spinlock** labels $< 1,5 >$,
$< 5,3 >$, $< 2,6 >$ and $< 6,4 >$ to be in the MST. *connected-components* finds vertices
1, 3, and 5, in one component, and vertices 2, 4, and 6, in another component. The MST
edges and components are shown in the middle of Fig. 2. Vertices connected by dashed
lines are in one component, and vertices connected by solid lines are in the other component. At this time, vertices 1, 3, and 5, belong to supervertex $1'$, and vertices 2, 4, and
6, belong to supervertex $2'$. In the second iteration, processor P_1 again inspects vertices
1, 2, and 3, and processor P_2 inspects vertices 4, 5, and 6. Previous MST edges $< 1,5 >$,
$< 5,3 >$, $< 2,6 >$ and $< 6,4 >$ are found to be edges inside supervertices and are ignored. On the bottom in Fig. 2 are the two supervertices with two edges between them.
Edges $< 1,2 >$ and $< 3,4 >$ are found by P_1 to be the edges between supervertices $1'$
and $2'$, edge $< 3,4 >$ is found by P_2 to be the edge between the two supervertices. For
supervertex $2'$, P_1 tries to label $< 1,2 >$ as the MST edge while P_2 tries to label $< 3,4 >$.
This is a race condition between the two processors, and locks are used in **Bor-spinlock**
to ensure correctness. The formal description of the algorithm is given in [10].

We compare the performance of **Bor-spinlock** with the best previous parallel implementations. The results are shown in Fig. 3. **Bor-FAL** is the fastest implementation
for sparse random graphs, **Bor-ALM** is the fastest implementation for meshes. From

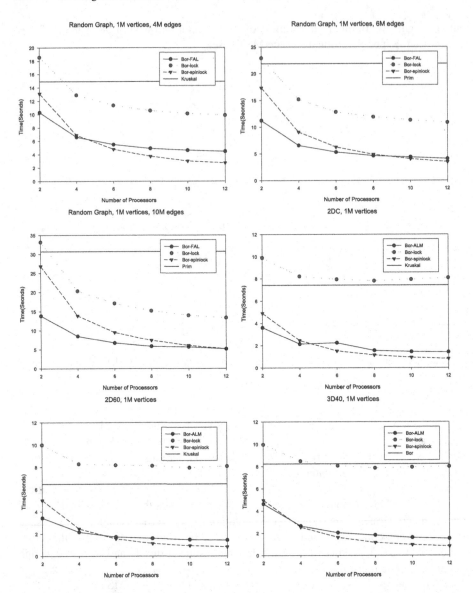

Fig. 3. Comparison of the performance of **Bor-spinlock** against the previous implementations. The horizontal line in each graph shows the execution time of the best sequential implementation

our results we see that with 12 processors **Bor-spinlock** beats both **Bor-FAL** and **Bor-ALM**, and performance of **Bor-spinlock** scales well with the number of processors. In Fig. 3, performance of **Bor-lock** is also plotted. **Bor-lock** is the same as **Bor-spinlock** except that system mutex locks are used. **Bor-lock** does not scale with the number of processors. The performance of the best sequential algorithms among the three candidates, Kruskal, Prim, and Borůvka, is plotted as a horizontal line for each input graph.

For all the input graphs shown in Fig. 3, **Bor-spinlock** tends to perform better than the previous best implementations when more processors are used. Note that a maximum speedup of 9.9 for 2D60 with 1M vertices is achieved with **Bor-spinlock** at 12 processors. Fig. 3 demonstrates the potential advantage of spinlock-based implementations for large and irregular problems. Aside from good performance, **Bor-spinlock** is also the simplest approach as it does not involve sorting required by the other approaches.

4   Conclusions

In this paper we present novel applications of lock-free and block-free protocols to parallel algorithms and show that these protocols can greatly improve the performance of parallel algorithms for large, irregular problems. As there is currently no direct support for invoking atomic instructions from most programming languages, our results suggest it necessary that there be orchestrated support for high performance algorithms from the hardware architecture, operating system, and programming languages. Two graph algorithms are discussed in this paper. In our future work, we will consider applying lock-free and block-free protocols to other types of algorithms, for example, parallel branch-and-bound.

References

1. J. Allemany and E.W. Felton. Performance issues in non-blocking synchronization on shared-memory multiprocessors. In *Proc. 11th ACM Symposium on Principles of Distributed Computing*, pages 125–134, Aug 1992.
2. R.J. Anderson and H. Woll. Wait-free parallel algorithms for the union-find problem. In *Proc. 23rd Annual ACM Symposium on Theory of Computing*, pages 370 – 380, May 1991.
3. J. Aspnes and M. P. Herlihy. Wait-free data structures in the asynchronous PRAM model. In *Proc. 2nd Ann. Symp. Parallel Algorithms and Architectures (SPAA-90)*, pages 340–349, Jul 1990.
4. H. Attiya, N. Lynch, and N. Shavit. Are wait-free algorithms fast? *J. ACM*, 41(4):725–763, 1994.
5. D. A. Bader and G. Cong. Fast shared-memory algorithms for computing the minimum spanning forest of sparse graphs. In *Proc. Int'l Parallel and Distributed Processing Symp. (IPDPS 2004)*, page to appear, Santa Fe, New Mexico, April 2004.
6. G. Barnes. Wait free algorithms for heaps. Technical Report TR-94-12-07, University of Washington, 1994.
7. B. Chor, A. Israeli, and M. Li. On processor coordination using asynchronous hardware. In *Proc. 6th ACM Symposium on Principles of Distributed Computing*, pages 86–97, Vancouver, British Columbia, Canada, 1987.
8. S. Chung and A. Condon. Parallel implementation of Borůvka's minimum spanning tree algorithm. In *Proc. 10th Int'l Parallel Processing Symp. (IPPS'96)*, pages 302–315, April 1996.
9. R. Cole and O. Zajicek. The APRAM : incorporating asynchrony into the PRAM model. In *Proc. 1st Ann. Symp. Parallel Algorithms and Architectures (SPAA-89)*, pages 169–178, Jun 1989.
10. G. Cong and D.A. Bader. Lock-free parallel algorithms: an experimental study. Technical report, Electrical and Computer Engineering Dept., University of Mexico, 2004.

11. C. Dwork, D. Dwork, and L. Stockmeyer. On the minimal synchronism needed for distributed consensus. *J. ACM*, 34(1):77–97, 1987.

12. C. Dwork, N. Lynch, and L. Stockmeyer. Consensus in the presence of partial synchrony. *J. ACM*, 35(2):288–323, 1988.

13. M. Fischer, N.A. Lynch, and M.S. Paterson. Impossibility of distributed commit with one faulty process. *J. ACM*, 32(2):374–382, 1985.

14. K.A. Fraser. *Practical lock-freedom*. PhD thesis, King's College, University of Cambridge, Sep 2003.

15. S. Goddard, S. Kumar, and J.F. Prins. Connected components algorithms for mesh-connected parallel computers. In S. N. Bhatt, editor, *Parallel Algorithms: 3rd DIMACS Implementation Challenge October 17-19, 1994*, volume 30 of *DIMACS Series in Discrete Mathematics and Theoretical Computer Science*, pages 43–58. American Mathematical Society, 1997.

16. A. Gottlieb, R. Grishman, C.P. Kruskal, K.P. McAuliffe, L. Rudolph, and M. Snir. The NYU ultracomputer — designing a MIMD, shared-memory parallel machine. *IEEE Trans. Computers*, C-32(2):175–189, 1984.

17. J. Greiner. A comparison of data-parallel algorithms for connected components. In *Proc. 6th Ann. Symp. Parallel Algorithms and Architectures (SPAA-94)*, pages 16–25, Cape May, NJ, June 1994.

18. M. Herlihy and J.E.B. Moss. Transactional memory: Architectural support for lock-free data structures. In *Proc. 20th Int'l Symposium in Computer Architecture*, pages 289–300, May 1993.

19. M.P. Herlihy. A methodology for implementing highly concurrent data objects. In *Proc. 2nd ACM SIGPLAN Symposium on Principles and Practice of Parallel Programming*, pages 197–206, Mar 1990.

20. M.P. Herlihy. Wait-free synchronization. *ACM Transactions on Programming Languages and Systems*, 13(1):124–149, 1991.

21. M.P. Herlihy and J.M. Wing. Axioms for concurrent objects. In *Proc. 14th ACM SIGACT-SIGPLAN Symposium on Principles of Programming Languages*, pages 13 – 26, Jan 1987.

22. T.-S. Hsu, V. Ramachandran, and N. Dean. Parallel implementation of algorithms for finding connected components in graphs. In S. N. Bhatt, editor, *Parallel Algorithms: 3rd DIMACS Implementation Challenge October 17-19, 1994*, volume 30 of *DIMACS Series in Discrete Mathematics and Theoretical Computer Science*, pages 23–41. American Mathematical Society, 1997.

23. A. Krishnamurthy, S. S. Lumetta, D. E. Culler, and K. Yelick. Connected components on distributed memory machines. In S. N. Bhatt, editor, *Parallel Algorithms: 3rd DIMACS Implementation Challenge October 17-19, 1994*, volume 30 of *DIMACS Series in Discrete Mathematics and Theoretical Computer Science*, pages 1–21. American Mathematical Society, 1997.

24. A. LaMarca. A performance evaluation of lock-free synchronization protocols. In *Proc. 13th Annual ACM Symposium on Principles of Distributed Computing*, pages 130 – 140, Aug 1994.

25. L. Lamport. Concurrent reading and writing. *Communications of the ACM*, 20(11):806–811, 1977.

26. L. Lamport. Specifying concurrent program modules. *ACM Transactions on Programming Languages and Systems*, 5(2):190–222, 1983.

27. V. Lanin and D. Shaha. Concurrent set manipulation without locking. In *Proc. 7th ACM SIGACT-SIGMOD-SIGART Symposium on Principles of Database Systems*, pages 211 – 220, Mar 1988.

28. H. Massalin and C. Pu. Threads and input/output in the synthesis kernel. In *Proc. 12th ACM Symposium on Operating Systems Principles*, pages 191–201, Dec 1989.

29. K. Mehlhorn and S. Näher. *The LEDA Platform of Combinatorial and Geometric Computing.* Cambridge University Press, 1999.
30. M.M. Michael and M.L. Scott. Nonblocking algorithms and preemption-safe locking on multiprogrammed shared memory multiprocessors. *Journal of Parallel and Distributed Computing,* 51(1):1–26, 1998.
31. B.M.E. Moret and H.D. Shapiro. An empirical assessment of algorithms for constructing a minimal spanning tree. In *DIMACS Monographs in Discrete Mathematics and Theoretical Computer Science: Computational Support for Discrete Mathematics 15,* pages 99–117. American Mathematical Society, 1994.
32. N. Shavit and D. Touitou. Software transactional memory. In *Proc. 14th annual ACM Symposium on Principles of Distributed Computing,* pages 204–213, Aug 1995.
33. Y. Shiloach and U. Vishkin. An $O(\log n)$ parallel connectivity algorithm. *J. Algs.,* 3(1):57–67, 1982.
34. R.E. Tarjan and J. Van Leeuwen. Worst-case analysis of set union algorithms. *J. ACM,* 31(2):245–281, 1984.
35. P. Tsigas and Y. Zhang. A simple, fast and scalable non-blocking concurrent FIFO queue for shared memory multiprocessor systems. In *Proc. 13th Ann. Symp. Parallel Algorithms and Architectures (SPAA-01),* pages 134–143, Sep 2001.
36. J. Valois. *Lock-free data structures.* PhD thesis, Rensselaer Polytechnic Institute, May 1995.
37. J.D. Valois. Lock-free linked lists using compare-and-swap. In *Proc. 14th Annual ACM Symposium on Principles of Distributed Computing,* pages 214 – 222, Aug 1995.

Author Index